T0320742

Advanced Concepts in Quantum Mechanics

Introducing a geometric view of fundamental physics, starting from quantum mechanics and its experimental foundations, this book is ideal for advanced undergraduate and graduate students in quantum mechanics and mathematical physics.

Focusing on structural issues and geometric ideas, this book guides readers from the concepts of classical mechanics to those of quantum mechanics. The book features an original presentation of classical mechanics, with the choice of topics motivated by the subsequent development of quantum mechanics, especially wave equations, Poisson brackets and harmonic oscillators. It also presents new treatments of waves and particles and the symmetries in quantum mechanics, as well as extensive coverage of the experimental foundations.

Giampiero Esposito is Primo Ricercatore at the Istituto Nazionale di Fisica Nucleare, Naples, Italy. His contributions have been devoted to quantum gravity and quantum field theory on manifolds with boundary.

Giuseppe Marmo is Professor of Theoretical Physics at the University of Naples Federico II, Italy. His research interests are in the geometry of classical and quantum dynamical systems, deformation quantization and constrained and integrable systems.

Gennaro Miele is Associate Professor of Theoretical Physics at the University of Naples Federico II, Italy. His main research interest is primordial nucleosynthesis and neutrino cosmology.

George Sudarshan is Professor of Physics in the Department of Physics, University of Texas at Austin, USA. His research has revolutionized the understanding of classical and quantum dynamics.

Advanced Concepts in Quantum Mechanics

GIAMPIERO ESPOSITO

GIUSEPPE MARMO

GENNARO MIELE

GEORGE SUDARSHAN

CAMBRIDGE
UNIVERSITY PRESS

Shaftesbury Road, Cambridge CB2 8EA, United Kingdom

One Liberty Plaza, 20th Floor, New York, NY 10006, USA

477 Williamstown Road, Port Melbourne, VIC 3207, Australia

314–321, 3rd Floor, Plot 3, Splendor Forum, Jasola District Centre, New Delhi – 110025, India

103 Penang Road, #05–06/07, Visioncrest Commercial, Singapore 238467

Cambridge University Press is part of Cambridge University Press & Assessment, a department of the University of Cambridge.

We share the University's mission to contribute to society through the pursuit of education, learning and research at the highest international levels of excellence.

www.cambridge.org
Information on this title: www.cambridge.org/9781107076044

First published 2015

A catalogue record for this publication is available from the British Library

Library of Congress Cataloging-in-Publication data
Esposito, Giampiero, author.
Advanced concepts in quantum mechanics / Giampiero Esposito,
Giuseppe Marmo, Gennaro Miele, George Sudarshan.
pages   cm.
Includes bibliographical references.
ISBN 978-1-107-07604-4 (Hardback)
1. Quantum theory.   I. Marmo, Giuseppe, author.   II. Miele, Gennaro, author.
III. Sudarshan, E. C. G., author.   IV. Title.
QC174.12.E94 2015
530.12–dc23   2014014735

ISBN   978-1-107-07604-4   Hardback

for Gennaro and Giuseppina; Patrizia; Arianna, Davide and Matteo; Bhamathi

Contents

Preface

In the course of teaching quantum mechanics at undergraduate and post-graduate level, we have come to the conclusion that there is another original book to be written on the subject. The abstract setting foreseen by Dirac and the geometric view pioneered by von Neumann are finding new realizations, leading to further progress both in physics and mathematics, while the applications to quantum computation are opening a new era in modern science. Our emphasis is mainly on structural issues and geometric ideas, moving the reader gradually from the concepts of classical mechanics to those of quantum mechanics, but we have also inserted many problems for students throughout the text, since the book is written, in the first place, for advanced undergraduate and graduate students, as well as for research workers.

The overall picture presented here is original, and also the parts in common with a previous monograph by some of us have been rewritten in most cases. The analysis of waves and particles (Chapter 3), the treatment of symmetries in quantum mechanics (in particular, the first half of Chapter 10), the assessment of modern pictures of quantum mechanics (Chapter 12) have never appeared before in any monograph, to the best of our knowledge. The material on experimental foundations is rather rich and it cannot easily be found to the same extent elsewhere. Our presentation of classical mechanics is original and the choice of topics is motivated by the subsequent development of quantum mechanics, expecially wave equations, Poisson brackets and harmonic oscillators. The examples in Chapters 6 and 7 are frequently discussed with a care not always used in many introductory presentations in the literature. We find it also useful to offer an unified view of approximation methods, as we do in Chapter 11, which is divided into three parts: perturbation theory, the JWKB method and scattering theory.

We hope that, having acquired familiarity with symbols of differential operators, geometric formulation and tomographic picture, the reader will find it easier to follow the latest developments in quantum theory, which embodies, in the broadest sense, all we know about guiding principles and fundamental interactions in physics.

Our friend Eugene Saletan, with whom some of us worked and corresponded on the subject of dynamical systems over many years, is deeply missed. Special thanks are due to our colleagues Fedele Lizzi, Francesco Nicodemi and Luigi Rosa for discussing various aspects of the manuscript, and to our students who, never being satisfied with our writing, helped us a lot in conceiving and completing the present monograph. Last, but not least, the Cambridge University Press staff, i.e. Nicholas Gibbons, Neeraj Saxena, Zoë Pruce, Lindsay Stewart, Jeethu Abraham, Sarah Payne and the copy-editor, Zoë Lewin, have provided invaluable help in the course of completing our task.

1 Introduction: the need for a quantum theory

1.1 Introducing quantum mechanics

Interference phenomena of material particles (say, electrons, neutrons, etc.) provide us with the most convincing evidence for the need to elaborate on a new mechanics that goes beyond and encompasses classical mechanics. At the same time, 'corpuscular' behaviour of radiation, i.e. light, as exhibited in phenomena like photoelectric and Compton effects (see Sections 2.2 and 2.3, respectively), shows that the description of radiation also has to undergo significant changes.

If we examine the relation between corpuscular-like and wave-like behaviour, we find that it is fully described by the following phenomenological equations:

$$E = h\nu = \hbar\omega, \ \vec{p} = \hbar\vec{k}, \tag{1.1.1}$$

which can be re-expressed in an invariant way with the help of 1-form notation (see Chapter 15) through the **Einstein–de Broglie** relation:

$$p_j \, \mathrm{d}x^j - E \, \mathrm{d}t = \hbar(k_j \, \mathrm{d}x^j - \omega \, \mathrm{d}t). \tag{1.1.2}$$

This relation between the 1-form $p_j \, \mathrm{d}x^j - E \, \mathrm{d}t$ on the phase space over space–time and the 1-form $\hbar\left(k_j \, \mathrm{d}x^j - \omega \, \mathrm{d}t\right)$ on the optical phase space establishes a relation between momentum and energy of the 'corpuscular' behaviour and the frequency of the 'wave' behaviour. The proportionality coefficient is the Planck constant. Such a relation likely summarizes one of the main new concepts encoded in quantum mechanics.

The way we use this relation is to predict under which experimental conditions light of a given wavelength and frequency will be detected as a corpuscle with a corresponding momentum and energy and vice-versa, i.e. when an electron will be detected as a wave in the appropriate experimental conditions. (To help dealing with orders of magnitude, we recall that the frequency associated with an electron of kinetic energy equal to 1 eV is $2.42 \cdot 10^{14}$ Hz, while the corresponding wavelength and wave number are $1.23 \cdot 10^{-9}$ m and $5.12 \cdot 10^{9}$ m^{-1}, respectively. Two standard length units are angstrom = Å= 10^{-10} m and fermi = Fm = 10^{-15} m.)

If we examine more closely an interference experiment, like the double-slit one, we find some peculiar aspects for which we do not have a simple interpretation in the classical setting.

If the experiment is performed in such a way that we make sure that, at each time, only one electron is present between the source and the screen, we find that the electron 'interferes with itself' and at the same time impinges on the screen at 'given points'.

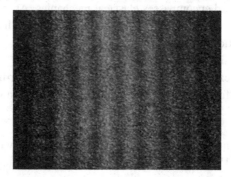

After a few hundred electrons have passed, we find a picture of random spots distributed on the screen (Figure 1.1). However, with several thousands electrons, a very clear typical interference pattern is obtained (Figure 1.2).

The same situation occurs again if we experiment with photons (light quanta), with an experimental setup that makes sure that only one photon is present at a time.

This experiment suggests that the new theory must include a wave character (to take into account the interference aspects) and, in addition, statistical–probabilistic, character along with an intrinsically discrete aspect, i.e. a corpuscular nature. All this is quite counterintuitive for particles, but it is even more unexpected for light. Within the classical setting we have to accept that it is not so simple to provide a single model capable of capturing these various aspects at the same time.

From the historical point of view, things developed differently because inconsistencies already arose in the derivation of the law for the spectral distribution of energy density of a black body. Planck conceived of the idea of emission and absorption of radiation by quanta in order to explain the finite energy density of black-body radiation (Section 2.1). The theory of classical electrodynamics gave an infinite density for this radiation. Indeed, the energy density per unit frequency was $8\pi\nu^2 KT/c^3$, as calculated on the basis of this theory, and the integral over the frequency ν is clearly divergent. Based in part on intuition, partly on experimental information and partly to agree with Wien's displacement law, Planck replaced the previous formula by

$$\frac{8\pi h\nu^3/c^3}{\left(e^{h\nu/KT} - 1\right)}.$$

To give an 'explanation' of it, he postulated that both emission and absorption of radiation occur instantaneously and in finite quanta.

Moreover, it was not possible to account for the stability of atoms and molecules along with the detected atomic spectra. To account for the experimental facts, Bohr postulated the quantum condition for electronic orbits. This hypothesis was highly successful in describing the spectrum of atomic hydrogen clearly and also in a qualitative way the periodic system, and hence some basic properties of all atoms. In spite of these partial successes, the absence of mathematically sound rules on the basis of which the electronic orbits, and therefore the energy levels, could be determined was greatly disturbing. It was also quite unclear how the electron jumps from one precisely defined orbit to another. The next chapter is devoted to a detailed description of some crucial experiments mentioned above, presented in their historical sequence, with the aim of providing the physical background from which the new theory of quanta emerged.

Eventually, the efforts of theoreticians gave rise to two alternative, but equivalent, formulations of quantum mechanics. They are usually called the Schrödinger picture and the Heisenberg picture. As will be seen in the coming chapters, the first one uses as a primary object the carrier space of states, while the latter uses as carrier space the space of observables. The former picture is built in analogy with wave propagation, the latter in analogy with Hamiltonian mechanics on phase space, i.e. the corpuscular behaviour.

The Schrödinger equation has the form

$$i\hbar\frac{\mathrm{d}}{\mathrm{d}t}\psi = \widehat{H}\psi. \tag{1.1.3}$$

The complex-valued function ψ is called the *wave function*, it is defined on the configuration space of the system we are considering, and it is interpreted as a probabilistic amplitude. This interpretation requires that ($\mathrm{d}\mu$ being the integration measure)

$$\int_D \psi^*\psi\,\mathrm{d}\mu = 1, \tag{1.1.4}$$

i.e. because of the probabilistic interpretation, $\psi^*\psi$ must be a probability density and therefore ψ is required to be square-integrable. Thus, wave functions must be elements of a Hilbert space of square-integrable functions. The operator $\widehat{H}$, acting on wave functions, is the infinitesimal generator of a 1-parameter group (see Chapter 15) of unitary transformations describing the evolution of the system under consideration. The unitarity requirement results from imposing that the evolution of an isolated system should be compatible with the probabilistic interpretation.

These are the basic ingredients appearing in the Schrödinger evolution equation. The presence of the new fundamental constant $\hbar$ within the new class of phenomena implies some fundamental aspects completely different from the previous classical ones. For instance, it is clear that any measurement process requires an exchange of energy (or information) between the object being measured and the measuring apparatus. The existence of $\hbar$ requires that these exchanges cannot be made arbitrarily small and therefore idealized to be negligible. Thus, the presence of $\hbar$ in the quantum theory means that in the measurement process we cannot conceive of a sharp separation between the 'object' and the 'apparatus' so that we may 'forget the apparatus' altogether.

In particular, it follows that even if the apparatus is described classically it should be considered as a quantum system with a quantum interaction with the object to be measured. Moreover, in the measurement process, there is an inherent ambiguity in the 'cut' between what we identify as the object and what we identify as apparatus.

The problem of measurement in quantum theory is a very profound one and goes beyond the scope of our manuscript. It is worth mentioning that, within the von Neumann formulation of quantum mechanics, the measurement problem gives rise to the so-called 'wave-function collapse'. The state vector of the system we are considering, when we measure some real dynamical variable A, i.e. a linear operator acting on the Hilbert space $\mathcal{H}$, is projected onto one of the eigenspaces of A, with some probability that can be computed. Since our aim is only to highlight the various structures occurring in the different formulations of quantum mechanics, we shall adhere to the von Neumann projection prescription.

Experimental foundations of quantum theory

The experimental foundations of quantum theory are presented in some detail in this chapter: on the one hand, the investigation of black-body radiation, which helps in developing an interdisciplinary view of physics, besides having historical interest; on the other hand, the energy and linear momentum of photons, atomic spectra, discrete energy levels, wave-like properties of electrons, interference phenomena and uncertainty relations.

2.1 Black-body radiation

Black-body radiation is not just a topic of historical interest. From a pedagogical point of view, it helps in developing an interdisciplinary view of physics, since it involves, among the other, branches of physics such as electrodynamics and thermodynamics, as well as a new constant of nature, the Planck constant, which is peculiar to quantum theory and quantum statistics. Moreover, looking at modern developments, the radiation that pervades the whole universe (Gamow 1946, Penzias and Wilson 1965, Smoot *et al.* 1992, Spergel *et al.* 2003) is a black-body radiation, and the expected emission of particles from black holes (space–time regions where gravity is so strong that no light ray can escape to infinity, and all nearby matter gets eaten up) is also (approximately) a black-body radiation (Hawking 1974, 1975).

In this section, relying in part on Born (1969), we are aiming to derive the law of heat radiation, following Planck's method. We think of a box for which the walls are heated to a definite temperature T. The walls of the box send out energy to each other in the form of heat radiation, so that within the box there exists a radiation field. This electromagnetic field may be characterized by specifying the average energy density u, which in the case of equilibrium is the same for every internal point; if we split the radiation into its spectral components, we denote by $u_\nu d\nu$ the energy density of all radiation components for which the frequency falls in the interval between ν and $\nu + d\nu$. (The spectral density is not the only specification; we need to know the state of the entire radiation field including the photon multiplicity.) Thus, the function u_ν extends over all frequencies from 0 to ∞, and represents a continuous spectrum. Note that, unlike individual atoms in rarefied gases, which emit line spectra, molecules, which consist of a limited number of atoms, emit narrow 'bands', which are often resolvable. A solid represents an infinite number of vibrating systems of all frequencies, and hence emits an effectively continuous spectrum. But inside a black cavity all bodies emit a continuous spectrum characteristic of the temperature.

The first important property in our investigation is a theorem by Kirchhoff (1860), which states that the ratio of the emissive and absorptive powers of a body depends only on the *temperature* of the body, and not on its nature (recall that the *emissive power* is, by definition, the radiant energy emitted by the body per unit time, whereas the *absorptive power* is the fraction of the radiant energy falling upon it that the body absorbs). A *black body* is meant to be a body with absorptive power equal to unity, i.e. a body that absorbs all of the radiant energy that falls upon it. The radiation emitted by such a body, called *black-body radiation*, is therefore a function of the temperature alone, and it is important to know the spectral distribution of the intensity of this radiation. Any object inside the black cavity emits the same amount of radiant energy. We are now aiming to determine the law of this intensity, but before doing so it is instructive to describe in detail some arguments in the original paper by Kirchhoff (cf. Stewart 1858).

2.1.1 Kirchhoff laws

The brightness $\mathcal{B}$ is the energy flux per unit frequency, per unit surface, for a given solid angle per unit time. Thus, if dE is the energy incident on a surface dS with solid angle $d\Omega$ in a time dt with frequency $d\nu$, we have (θ being the incidence angle)

$$dE = \mathcal{B} \, d\nu \, dS \, d\Omega \, \cos\theta \, dt. \qquad (2.1.1)$$

The brightness $\mathcal{B}$ is independent of position, direction and the nature of the material. This is proved as follows.

(i) $\mathcal{B}$ cannot depend on position, since otherwise two bodies absorbing energy at the same frequency and placed at different points P_1 and P_2 would absorb different amounts of energy, although they were initially at the same temperature T equal to the temperature of the cavity. We would then obtain the spontaneous creation of a difference of temperature, which would make it possible to realize a perpetual motion of the second kind, hence violating the second principle of thermodynamics, which is of course impossible.

(ii) $\mathcal{B}$ cannot depend on direction either. Let us insert into the cavity a mirror S of negligible thickness, and imagine we can move it along a direction parallel to its plane. In such a way no work is performed, and hence the equilibrium of radiation remains unaffected. Then let A and B be two bodies placed at different directions with respect to S and absorbing in the same frequency interval. If the amount of radiation incident upon B along the BS direction is smaller than that along the AS direction, bodies A and B attain spontaneously different temperatures, although they were initially in equilibrium at the same temperature! Thermodynamics forbids this phenomenon as well.

(iii) Once equilibrium is reached, $\mathcal{B}$ is also independent of the material the cavity is made of. Let the cavities C_1 and C_2 be made of different materials, and suppose they are at the same temperature and linked by a tube such that only radiation of frequency ν can pass through it. If $\mathcal{B}$ were different for C_1 and C_2 a non-vanishing energy flux through the tube would therefore be obtained. Thus, the two cavities would change their temperature spontaneously, against the second law of thermodynamics. Similar considerations prove $\mathcal{B}$ to be independent of the shape of the cavity as well.

By virtue of (i)–(iii) Eq. (2.1.1) reads, more precisely, as

$$dE = \mathcal{B}(\nu, T)d\nu \; dS \; d\Omega \; \cos\theta \; dt. \qquad (2.1.2)$$

Moreover, the energy absorbed by the surface element dS of the wall once equilibrium is reached is (x denoting all variables other than ν, T)

$$dE_{\text{abs}} = a_{\text{m}}(\nu, T, x)dE, \qquad (2.1.3)$$

while the emitted energy is

$$dE_{\text{em}} = e_{\text{m}}(\nu, T, x)d\nu \; dS \; d\Omega \; \cos\theta \; dt. \qquad (2.1.4)$$

Under equilibrium conditions, the amounts of energy dE_{em} and dE_{abs} are equal, and hence

$$\frac{e_{\text{m}}(\nu, T, x)}{a_{\text{m}}(\nu, T, x)} = \mathcal{B} = \mathcal{B}(\nu, T). \qquad (2.1.5)$$

Thus, the ratio of emissive and absorptive powers is equal to the brightness and hence can only depend on frequency and temperature, although both e_{m} and a_{m} can separately depend on the nature of materials.

As far as the production of black-body radiation is concerned, it has been proved by Kirchhoff that an enclosure (typically, an oven) at uniform temperature, in the wall of which there is a small opening, behaves as a black body. Indeed, all the radiation which falls on the opening from the outside passes through it into the enclosure, and is, after repeated reflection at the walls, completely absorbed by them. The radiation in the interior, and hence also the radiation which emerges again from the opening, should therefore possess exactly the spectral distribution of intensity, which is characteristic of the radiation of a black body.

2.1.2 Electromagnetic field in a hollow cavity

According to classical electrodynamics, a hollow cavity filled with electromagnetic radiation (possibly in thermodynamical equilibrium with the cavity surfaces) contains energy stored in the electromagnetic field as described by the expression[1]

$$\mathcal{E} = \frac{1}{8\pi} \int \left(|\vec{E}|^2 + |\vec{B}|^2 \right) dV, \qquad (2.1.6)$$

where the fields $\vec{E}$ and $\vec{B}$ satisfy the Maxwell equations

$$\vec{\nabla} \wedge \vec{E} = -\frac{1}{c}\frac{\partial}{\partial t}\vec{B}, \qquad\qquad \vec{\nabla} \cdot \vec{B} = 0, \qquad (2.1.7)$$

$$\vec{\nabla} \wedge \vec{B} = \frac{1}{c}\frac{\partial}{\partial t}\vec{E} + \frac{4\pi}{c}\vec{J}, \qquad\qquad \vec{\nabla} \cdot \vec{E} = 4\pi\rho, \qquad (2.1.8)$$

with ρ and $\vec{J}$ denoting the charge and current density, respectively. The most general solution of Eqs. (2.1.7) expresses the fields $\vec{E}$ and $\vec{B}$ in terms of scalar and vector potentials as

$$\vec{B} = \vec{\nabla} \wedge \vec{A}, \qquad \vec{E} = -\vec{\nabla}\phi - \frac{1}{c}\frac{\partial}{\partial t}\vec{A}. \qquad (2.1.9)$$

---

[1] Hereafter we will use Gaussian units, see for example Jackson (1975), for a detailed discussion.

Once the electromagnetic fields are given, Eqs. (2.1.9) do not fix ϕ and $\vec{A}$. In fact, if according to Eq. (2.1.9) ϕ and $\vec{A}$ yield $\vec{E}$ and $\vec{B}$, from the pair ϕ' and $\vec{A}'$ defined by

$$\vec{A}' \equiv \vec{A} - \vec{\nabla}\chi, \qquad \phi' \equiv \phi + \frac{1}{c}\frac{\partial}{\partial t}\chi, \qquad (2.1.10)$$

the same electromagnetic fields for every arbitrary χ function are obtained. Such a level of freedom in choosing the scalar and vector potentials associated with given electromagnetic fields, which makes the former physically unobservable, is commonly denoted as *gauge symmetry*. In the case of Maxwell equations in vacuum, a the gauge symmetry can be completely exploited by imposing simultaneously the conditions

$$\vec{\nabla} \cdot \vec{A} = \partial^i A_i = 0, \qquad \phi = 0, \qquad (2.1.11)$$

which is a particular case of *transverse gauge*. By substituting Eqs. (2.1.9) in (2.1.8) for the vacuum case and using the conditions (2.1.11) we get the wave equation for the transverse degrees of freedom of $\vec{A}$, i.e. (hereafter $\triangle \equiv \frac{\partial^2}{\partial x^2} + \frac{\partial^2}{\partial y^2} + \frac{\partial^2}{\partial z^2}$ if expressed in Cartesian coordinates)

$$\left(\triangle - \frac{1}{c^2}\frac{\partial^2}{\partial t^2}\right)\vec{A}_t \equiv \Box \vec{A}_t = 0, \qquad (2.1.12)$$

$$\vec{\nabla} \cdot \vec{A}_t = 0. \qquad (2.1.13)$$

As already proved in the previous subsection, the energy density of a hollow cavity filled of electromagnetic radiation in thermal equilibrium with the cavity surface *cannot depend on the nature and shape* of the cavity. For this reason, we can choose the particular case of a cubic cavity with periodic boundary conditions, which allows a simpler treatment of the electromagnetic problem.

Let us consider a cube with edge length L; the generic field $\vec{A}_t(\vec{r}, t)$ simultaneously periodic along the three coordinate directions can be expanded as

$$\vec{A}_t(\vec{r}, t) = \sum_{l,n,m\in\mathbb{Z}} \left\{ \vec{a}_{lnm}(t) \cos\left[\frac{2\pi}{L}(lx + my + nz)\right] \right.$$
$$\left. + \vec{b}_{lnm}(t) \sin\left[\frac{2\pi}{L}(lx + my + nz)\right] \right\}. \qquad (2.1.14)$$

By defining the propagation vector $\vec{k} \equiv (2\pi/L)(l, m, n)$, the condition $\vec{\nabla} \cdot \vec{A}_t = 0$ implies $\vec{k} \cdot \vec{a}_{lnm}(t) = \vec{k} \cdot \vec{b}_{lnm}(t) = 0$ (transverse condition).

Hence Eq. (2.1.14) can be rewritten

$$\vec{A}_t(\vec{r}, t) = \sum_{\vec{k},\mu} \left[\vec{a}_{\vec{k},\mu}(t) \cos\left(\vec{k} \cdot \vec{r}\right) + \vec{b}_{\vec{k},\mu}(t) \sin\left(\vec{k} \cdot \vec{r}\right) \right], \qquad (2.1.15)$$

where the index μ labels the two independent solutions of the transverse condition, and Eq. (2.1.12) gives

$$\left(\frac{d^2}{dt^2} + |\vec{k}|^2 c^2\right)\vec{a}_{\vec{k},\mu}(t) = 0,$$
$$\left(\frac{d^2}{dt^2} + |\vec{k}|^2 c^2\right)\vec{b}_{\vec{k},\mu}(t) = 0, \qquad (2.1.16)$$

which show that $\vec{a}_{\vec{k},\mu}$ and $\vec{b}_{\vec{k},\mu}$ behave as harmonic oscillators with angular frequency $\omega = |\vec{k}|\, c$, where $|\vec{k}| \equiv (2\pi/L)\sqrt{l^2 + m^2 + n^2}$.

By deriving from Eq. (2.1.15) the corresponding electromagnetic fields and substituting them in Eq. (2.1.6) we get

$$\mathcal{E} = \frac{L^3}{8\pi c^2} \sum_{\vec{k},\mu} \frac{1}{2} \left\{ \left[\left| \frac{\mathrm{d}}{\mathrm{d}t} \left(\vec{a}_{\vec{k},\mu} + \vec{a}_{-\vec{k},\mu} \right) \right|^2 + |\vec{k}|^2 c^2 \left| \left(\vec{a}_{\vec{k},\mu} + \vec{a}_{-\vec{k},\mu} \right) \right|^2 \right] \right.$$
$$\left. + \left[\left| \frac{\mathrm{d}}{\mathrm{d}t} \left(\vec{b}_{\vec{k},\mu} - \vec{b}_{-\vec{k},\mu} \right) \right|^2 + |\vec{k}|^2 c^2 \left| \left(\vec{b}_{\vec{k},\mu} - \vec{b}_{-\vec{k},\mu} \right) \right|^2 \right] \right\}. \qquad (2.1.17)$$

From Eq. (2.1.17) deduce that the electromagnetic energy in a hollow cavity receives contributions from the sum of countable and separate harmonic oscillator-type degrees of freedom with mass equal to $L^3/(8\pi c^2)$ and angular frequency ω. For each particular mode, i.e. for each $\vec{k}$, the two independent polarizations are labelled by μ. Note that the presence in Eq. (2.1.17) of terms proportional to $\vec{a}_{\vec{k},\mu} + \vec{a}_{-\vec{k},\mu}$ and $\vec{b}_{\vec{k},\mu} - \vec{b}_{-\vec{k},\mu}$, even though they have particular properties of symmetry with respect to $\vec{k} \to -\vec{k}$, ensures one independent degree of freedom for each value of $\vec{k}$ and μ.

By virtue of the isotropy expected for the radiation energy density in the hollow cavity describing the black body, the expression in square brackets on the right-hand side of Eq. (2.1.17) (total energy of the single harmonic oscillator) can depend on ω only, hence in the sum of Eq. (2.1.17) the directional degrees of freedom can be integrated out.

If we fix $\vec{k}$, the infinitesimal number of oscillators around this value is

$$\delta n = \mathrm{d}l\, \mathrm{d}m\, \mathrm{d}n = L^3/(2\pi)^3 \, \mathrm{d}k_x\, \mathrm{d}k_y\, \mathrm{d}k_z = L^3/(2\pi)^3 \, |\vec{k}|^2 \, \mathrm{d}|\vec{k}| \, \mathrm{d}\Omega. \qquad (2.1.18)$$

Once the angular integration is performed the total number of oscillators between the frequencies ν and $\nu + \mathrm{d}\nu$ is obtained, i.e.

$$\delta N = \frac{8\pi V}{c^3} \nu^2 \, \mathrm{d}\nu, \qquad (2.1.19)$$

where we have used the relation $\nu = |\vec{k}|\, c/(2\pi)$, added an extra factor 2 to Eq. (2.1.19) to take into account the different polarizations, and denoted with V the volume of the cavity $V = L^3$. By using Eqs. (2.1.17) and (2.1.19) we get for the cubic cavity

$$\frac{1}{V} \frac{\mathrm{d}\mathcal{E}}{\mathrm{d}\nu} = \frac{8\pi V}{c^3} \nu^2 \, e_{\mathrm{ho}}(\nu), \qquad (2.1.20)$$

where $e_{\mathrm{ho}}(\nu)$ denotes the energy contribution of the harmonic-oscillator-like degrees of freedom with frequency ν appearing on the right-hand side of Eq. (2.1.17).

The expression of $\mathcal{E}$ can then be obtained by determining the explicit expression of $e_{\mathrm{ho}}(\nu)$. In the following we will take a different approach, but we will revert to Eq. (2.1.20) to physically interpret our results.

2.1.3 Stefan and displacement laws

Remaining within the framework of thermodynamics and the electromagnetic theory of light, two laws can be deduced concerning the way in which black-body radiation depends on the temperature. First, the Stefan law states that the total emitted radiation is proportional to the fourth power of the temperature of the radiator. Thus, the hotter the body, the more it radiates. Second, Wien found the *displacement law* (1896), which states that the spectral distribution of the energy density is given by an equation of the form

$$u_\nu(\nu, T) = \nu^3 F(\nu/T), \tag{2.1.21}$$

where F is a function of the ratio of the frequency to the temperature, but cannot be determined more precisely with the help of thermodynamical methods. This formula can be proved by using the approach of previous subsection, i.e. describing the black body as a hollow cavity of volume V in the shape of a cube of edge length L. As shown before, the equilibrium radiation field will consist of standing waves and this leads to the following relation for the frequency:

$$\left(\frac{\nu L}{c}\right)^2 = l^2 + m^2 + n^2, \tag{2.1.22}$$

where l, m and n are integers. If an adiabatic change of volume is performed, the quantities l, m and n, being integers and hence unable to change infinitesimally, will remain invariant. Under an adiabatic transformation the product νL is therefore invariant, or, introducing the volume V instead of L,

$$\nu^3 V = \text{invariant}, \tag{2.1.23}$$

under adiabatic transformation. The result can be proved to be independent of the shape of the volume.

However, it is more convenient to have a relation between ν and T, and for this purpose the entropy of the radiation field must be considered. Classical electrodynamics tells us that the radiation pressure P is one-third of the total radiation energy density $u(T)$:

$$P = \frac{1}{3}u(T). \tag{2.1.24}$$

On combining Eq. (2.1.24) with the thermodynamic equation of state

$$\left(\frac{\partial U}{\partial V}\right)_T = T\left(\frac{\partial P}{\partial T}\right)_V - P, \tag{2.1.25}$$

and the relation $U = uV$, the differential equation,

$$u = \frac{1}{3}T\frac{du}{dT} - \frac{1}{3}u, \tag{2.1.26}$$

which is solved by using the Stefan law is obtained,

$$u(T) = aT^4. \tag{2.1.27}$$

Equations (2.1.24) and (2.1.27), when combined with the thermodynamic Maxwell relation

$$\left(\frac{\partial S}{\partial V}\right)_T = \left(\frac{\partial P}{\partial T}\right)_V, \tag{2.1.28}$$

yield

$$S = \frac{4}{3}aT^3 V. \tag{2.1.29}$$

From Eqs. (2.1.23) and (2.1.29) we find that, under an isentropic transformation, the ratio v/T must be invariant. Moreover, since the resolution of a spectrum into its components is a reversible process, the entropy s per unit volume can be written as the sum of contributions $s_v(T)$ corresponding to different frequencies. Each of these terms, being a function of v and with the entropy density corresponding to the specific frequency v, can depend on v and T only through the adiabatic invariant v/T, or (Ter Haar 1967)

$$s = \sum_v s(v/T). \tag{2.1.30}$$

Also, the total energy density can be expressed by a sum:

$$u(T) = \sum_v \mathcal{U}_v(T), \tag{2.1.31}$$

and Eqs. (2.1.27) and (2.1.29) show that

$$s = \frac{4}{3}\frac{u}{T}, \tag{2.1.32}$$

and hence

$$\mathcal{U}_v(T) = Tf_1(v/T) = vf_2(v/T), \tag{2.1.33}$$

so that

$$u(T) = \sum_v vf_2(v/T) = \int_0^\infty vZ(v)f_2(v/T)\mathrm{d}v. \tag{2.1.34}$$

Such an equality is simple but non-trivial: summation over v should be performed with the associated 'weight', and it should reduce to an integral over all values of v from 0 to ∞ to recover agreement with the formula $u(T) = \int_0^\infty u_v(v, T)\mathrm{d}v$, so that

$$\sum_v \cdot \to \int_0^\infty Z(v) \cdot \mathrm{d}v.$$

This implies the following equation for the spectral distribution of energy density:

$$u_v(v, T) = vZ(v)f_2(v/T), \tag{2.1.35}$$

where $Z(v)\mathrm{d}v$ is the number of frequencies in the radiation between v and $v + \mathrm{d}v$. By virtue of Eq. (2.1.22), this is proportional to the number of points with integral coordinates within the spherical shell between the spheres with radii vL/c and $(v+\mathrm{d}v)L/c$, from which it follows that

$$Z(v) = Cv^2, \tag{2.1.36}$$

for some parameter C independent of v. The laws expressed by Eqs. (2.1.35) and (2.1.36) therefore lead to the Wien law (2.1.21) (Ter Haar 1967).

At this stage, however, it is still unclear why such a formula is called the *displacement law*. The reason is as follows. It was found experimentally by Lummer and Pringsheim that the intensity of the radiation from an incandescent body, maintained at a definite temperature, was represented, as a function of the wavelength, by a curve (Figure 2.1)

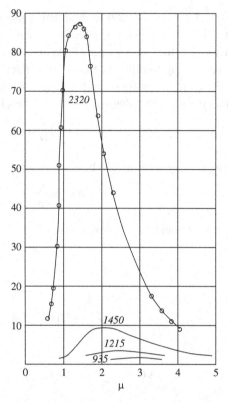

Fig. 2.1 Distribution of the intensity of thermal radiation as a function of wavelength according to the measurements of Lummer and Pringsheim. The y-axis corresponds to $u(\lambda, T) \times 10^{-11}$ in CGS units (Born 1969, by kind permission of Professor Blin-Stoyle).

such that the product of the temperature T and the wavelength $\lambda_{\max}$ for which the intensity attains its maximum, is constant:

$$\lambda_{\max} T = \text{constant.} \tag{2.1.37}$$

The Wien law makes it possible to understand why Eq. (2.1.37) holds. Indeed, so far we have referred to the energy distribution as a function of the frequency ν, with u_ν representing the radiation energy in the frequency interval $d\nu$. The displacement law, however, refers to a graph showing the intensity distribution as a function of λ, so that we now deal with u_λ, representing the energy in the wavelength interval $d\lambda$. Of course, it is required that

$$u_\nu d\nu = u_\lambda d\lambda. \tag{2.1.38}$$

Moreover, since $\lambda\nu = c$, for the relation between $d\nu$ and λ,

$$\frac{|d\nu|}{\nu} = \frac{|d\lambda|}{\lambda}. \tag{2.1.39}$$

Thus, for the spectral distribution of energy expressed as a function of the wavelength,

$$u_\lambda = \frac{c^4}{\lambda^5} F\left(\frac{c}{\lambda T}\right). \tag{2.1.40}$$

We are now in a position to immediately prove the displacement law, by evaluating the wavelength for which u_λ is a maximum. For this purpose, we set to zero the derivative of u_λ with respect to λ. This eventually yields

$$\frac{c}{\lambda T} F'\left(\frac{c}{\lambda T}\right) + 5F\left(\frac{c}{\lambda T}\right) = 0. \tag{2.1.41}$$

This is an equation in the variable $c/\lambda T$, the solution of which (we know it exists from the experimental data) has, of course, the form $\lambda T = \text{constant}$. Thus, the theorem concerning the displacement of the intensity maximum with temperature follows immediately from Wien's law. The value of the constant, however, cannot be determined until the special form of the function F is known.

2.1.4 Planck model

As far as the function F is concerned, thermodynamics is, by itself, unable to determine it. Still, it is clear that the form of the law given by the function F should be independent of the special mechanism. Thus, as the simplest *model* of a radiating body, Planck chose a collection of linear harmonic oscillators of proper frequency ν (Planck 1900). For each oscillator, on the one hand, it is possible to determine the energy radiated per second. This is being the radiation emitted by an oscillating dipole of charge q, given by

$$\delta\varepsilon = \frac{2q^2 \overline{(\ddot{r})^2}}{3c^3} = \frac{2q^2}{3mc^3} (2\pi\nu)^2 \overline{\varepsilon}, \tag{2.1.42}$$

where ε is the energy of the oscillator, and the bars denote mean values over times which, although large in comparison with the period of vibration, are still sufficiently small to allow the radiation emitted to be neglected. From the equation of motion $\ddot{r} = -(2\pi\nu)^2 r$ and

$$\overline{\varepsilon}_{\text{kin}} = \frac{1}{2} m \overline{\dot{r}^2} = \frac{1}{2} m \overline{(2\pi\nu r)^2} = \overline{\varepsilon}_{\text{pot}} = \frac{1}{2} \overline{\varepsilon}. \tag{2.1.43}$$

On the other hand, it is well known from classical electrodynamics that the work done on the oscillator per second by a radiation field with the spectral energy density u_ν is (see appendix)

$$\delta W = \frac{\pi q^2}{3m} u_\nu. \tag{2.1.44}$$

When energy balance is achieved, these two amounts of energy should be equal. Hence,

$$u_\nu = \frac{8\pi\nu^2}{c^3} \overline{\varepsilon}. \tag{2.1.45}$$

It is thus clear that, if we know the mean energy of an oscillator, we also know the spectral intensity distribution of the cavity radiation.

The value of $\bar{\varepsilon}$, as determined by the theorem of equipartition of energy of classical statistical mechanics, is $\bar{\varepsilon} = KT$. This happens because, according to a classical analysis, any term in the Hamiltonian that is proportional to the square of a coordinate or a momentum variable, contributes the amount $\frac{1}{2}KT$ to the mean energy. For each degree of freedom of the harmonic oscillator there are two such terms, and hence $\bar{\varepsilon}$ equals KT. Now if the classical mean value of the energy of the oscillator, as just determined, is inserted into the radiation formula (2.1.45), we obtain

$$u_\nu = \frac{8\pi \nu^2}{c^3} KT. \tag{2.1.46}$$

This is the Rayleigh–Jeans radiation formula proposed in Rayleigh (1900) and Jeans (1905) (actually Rayleigh forgot the polarization while Jeans corrected this). Comparing the previous expression with Eq. (2.1.20) we find that according to the classical treatment $e_{\mathrm{ho}}(\nu) = KT$. In other words, the harmonic oscillator degrees of freedom describing the radiation field in a cavity obey classical thermodynamics as well as being matter oscillators.

Some remarks are now in order.

(i) The Rayleigh–Jeans formula agrees with the Wien displacement law. This was expected to be the case, since the Wien law is deduced from thermodynamics, and hence should be of universal validity.

(ii) For long-wave components of the radiation, i.e. for small values of the frequency ν, the Rayleigh–Jeans equation reproduces the experimental intensity distribution very well.

(iii) For high frequencies, however, Eq. (2.1.46) is completely wrong. It is indeed known from experiments that the intensity function reaches a maximum at a definite frequency and then decreases again. In contrast, Eq. (2.1.46) fails entirely to show this maximum, and instead describes a spectral intensity distribution that becomes infinite as the frequency ν tends to infinity. The same is true of the total energy of radiation, obtained by integrating u_ν over all values of ν from 0 to ∞: the integral diverges. We are facing here what is called in the literature the *ultraviolet catastrophe*.

To overcome this serious inconsistency, Planck *assumed* the existence of *discrete, finite quanta of energy*, here denoted by ε_0. According to this scheme, the energy of the oscillators can only take values that are integer multiples of ε_0, including 0. We are now going to see how this hypothesis leads to the so-called Planck radiation law. The essential point is, of course, the determination of the mean energy $\bar{\varepsilon}$. The derivation differs from that resulting from classical statistical mechanics only in replacing integrals by sums. The individual energy values occur again with the 'weight' given by the Boltzmann factor, but one should bear in mind that only the energy values $n\varepsilon_0$ are admissible, n being an integer greater than or equal to 0. In other words, the Planck hypothesis leads to the following expression for the mean energy (the parameter β being equal to $1/KT$):

$$\bar{\varepsilon} = \frac{\sum_{n=0}^{\infty} n\varepsilon_0 e^{-\beta n \varepsilon_0}}{\sum_{n=0}^{\infty} e^{-\beta n \varepsilon_0}} = -\frac{\mathrm{d}}{\mathrm{d}\beta} \log \sum_{n=0}^{\infty} e^{-\beta n \varepsilon_0}. \tag{2.1.47}$$

At this stage, we recall an elementary property that follows from the analysis of the geometrical series:

$$\sum_{n=0}^{\infty} e^{-\beta n \varepsilon_0} = \lim_{N \to \infty} \sum_{n=0}^{N} (e^{-\beta \varepsilon_0})^n$$

$$= \lim_{N \to \infty} \frac{(1 - (e^{-\beta \varepsilon_0})^{N+1})}{(1 - e^{-\beta \varepsilon_0})} = \frac{1}{(1 - e^{-\beta \varepsilon_0})}. \tag{2.1.48}$$

The joint effect of Eqs. (2.1.47) and (2.1.48) is thus

$$\bar{\varepsilon} = \frac{\varepsilon_0 e^{-\beta \varepsilon_0}}{(1 - e^{-\beta \varepsilon_0})} = \frac{\varepsilon_0}{(e^{\beta \varepsilon_0} - 1)}. \tag{2.1.49}$$

Equations (2.1.45) and (2.1.49) therefore lead to the radiation formula

$$u_\nu = \frac{8\pi \nu^2}{c^3} \frac{\varepsilon_0}{(e^{\varepsilon_0/KT} - 1)}. \tag{2.1.50}$$

To avoid obtaining a formula that is inconsistent with the Wien displacement law, which, being derived from thermodynamics alone, is certainly valid, we have to assume that (the temperature being forced to appear only in the combination ν/T)

$$\varepsilon_0 = h\nu, \tag{2.1.51}$$

where h is the Planck constant. Hence, the fundamental Planck radiation law (Figure 2.2) is obtained:

$$u_\nu = \frac{8\pi h\nu^3}{c^3} \frac{1}{(e^{h\nu/KT} - 1)}. \tag{2.1.52}$$

Comparing the previous expression with Eq. (2.1.20) we get the quantum expression for $e_{ho}(\nu) = h\nu/(e^{h\nu/KT} - 1)$.

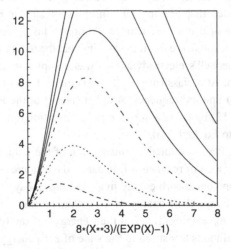

8*(X**3)/(EXP(X)−1)

Fig. 2.2 Spectral distribution of the intensity of thermal radiation according to Planck, for temperatures between 2000 and 4500 K. The abscissa corresponds to wavelength. The lowest curve is obtained for $T = 2000$ K, and the following curves correspond to values of $T = 2500, 3000, 3500, 4000$ and 4500 K, respectively. There is full agreement with the experimental results.

Bose gave an independent derivation of the Planck formula by considering photons as strictly identical particles (Bose 1924). This quantum derivation in 1924 before quantum theory was properly formulated (by Schrödinger, Heisenberg and Dirac) required a new method considering combinations of strictly interacting particles. This is the integer-spin analogue of the Pauli principle and is referred to as Bose statistics.

The radiation formula (2.1.52) is in very good agreement with all experimental results. In particular, for low values of the frequency, it reduces to the Rayleygh–Jeans formula (2.1.46), whereas, as ν tends to ∞, it takes the approximate form

$$u_\nu \sim \frac{8\pi h\nu^3}{c^3} e^{-h\nu/KT}, \qquad (2.1.53)$$

which agrees with a formula first derived, empirically, in Wien (1896), in the attempt to account for measurements in this region of the spectrum.

However, the derivation of the Planck radiation law given so far, following Planck's argument, is heuristic and unsatisfactory, since it does not provide in-depth reasons for the existence of discrete, finite quanta of energy. What was truly lacking, at the beginning of the twentieth century, was a consistent framework for *quantum statistical mechanics*, from which the result (2.1.52) should follow without *ad hoc* assumptions. History tells us that the Planck hypothesis met, at first, with violent opposition, and was viewed as a mathematical artifice, which might be interpreted without going outside the framework of classical physics. But the attempts to preserve the laws of classical physics failed. Instead, it became clear that one has to come to terms with a new constant of nature, h, and build a new formalism which, in particular, accounts for the Planck radiation law. More precisely, the following remarks are in order.

(i) The Planck assumption according to which the energy of the oscillator can only take values that are integer multiples of ε_0 contradicts completely what was known from classical electrodynamics. Although his argument clearly had a heuristic value, it nevertheless had the merit of showing that the theory of radiation–matter interactions based upon Maxwell's electrodynamics was completely unable to account for the law of heat radiation. At a classical level there is, in fact, no obvious link between the energy of the oscillator and its frequency. Another 'merit' of the Planck analysis was that of arriving at Eq. (2.1.52) by using a very simple assumption (among the many conceivable procedures leading to Eq. (2.1.52)).

When Planck studied the interaction of the radiation field with matter and represented matter by a set of resonators (i.e. damped harmonic oscillators) of frequency ν, he assumed that resonators absorb energy in a continuous way from the electromagnetic field and that they exchange their energy in a continuous way. The resonators would emit a radiation equal to $E_n = nh\nu$ only when their energy is exactly equal to E_n, in this way performing a discontinuous transition to the state of zero energy. Although, with hindsight, we know that such a model is incorrect, we should acknowledge that it contains ideas which had a profound influence. (a) The different orbits (with $H = E_n$) divide phase space into regions of area h. (b) The average energy of a quantum state turns out to be $\left(n+\frac{1}{2}\right)h\nu$, hence leading to the concept of zero-point energy for the first time (strictly, this will be discovered when the full apparatus of quantum mechanics is developed). (c) The emission of radiation is viewed, for the first time, as a probabilistic process (Parisi 2001).

(ii) Planck did not realize that at equilibrium, in classical mechanics, the properties of the resonators do not affect the black-body radiation, but he had the merit of isolating the physically relevant points, where progress could be made, not paying attention to the possible contradictions (Parisi 2001).

(iii) Since the electromagnetic field inside the box interacts with a very large number of oscillators, it was suggested for some time that the particular *collective* properties of matter, rather than energy exchanges with a single atom, can account for the Planck hypothesis without making it mandatory to give up the classical theory of electromagnetic phenomena. However, the work in Einstein (1905, 1917), see below, proved that the energy of the electromagnetic field (and the associated linear momentum) is localized, and hence radiation–matter interactions are localized, and cannot be understood by appealing to collective properties of material media.

(iv) It should be stressed that no thermal equilibrium can ever be reached within a box with reflecting walls. The single monochromatic components of the electromagnetic field do not interact with each other, and hence no process can lead to thermal equilibrium in such a case. Fortunately, there are *no* perfect reflectors.

(v) As far as the emission of radiation is concerned, all energy-balance arguments should take into account both induced emission and spontaneous emission. The former results from the interaction with an external field, whereas the latter may be due to energy acquired during previous collisions, or to previous interactions with an electromagnetic field.

2.1.5 Contributions of Einstein

In Einstein (1905), the author found that, in the region where the Wien law is valid, it can be said that, thermodynamically speaking, monochromatic radiation consists of independent energy quanta of magnitude $h\nu$. To prove this, he applied thermodynamical concepts to electromagnetic radiation, starting from the definition of temperature,

$$\frac{1}{T} \equiv \left(\frac{\partial S}{\partial U} \right)_V = \frac{\partial \sigma}{\partial u_\nu}, \tag{2.1.54}$$

where the entropy density σ refers to a constant volume, and the same holds for u_ν. If the Wien law holds, i.e. for $h\nu \gg KT$:

$$u_\nu = b\nu^3 e^{-h\nu/KT}, \tag{2.1.55}$$

Equation (2.1.54) leads to

$$\frac{\partial \sigma}{\partial u_\nu} = -\frac{K}{h\nu} \log \left(\frac{u_\nu}{b\nu^3} \right), \tag{2.1.56}$$

which is solved by

$$\sigma(\nu, T) = -\frac{K u_\nu}{h\nu} \left[\log \left(\frac{u_\nu}{b\nu^3} \right) - 1 \right]. \tag{2.1.57}$$

Thus, the entropy S in a volume V reads as

$$S = \sigma V = -\frac{K u_v V}{h v} \left[\log \left(\frac{u_v V}{b V v^3} \right) - 1 \right] = -\frac{KE}{h v} \left[\log \left(\frac{E}{b V v^3} \right) - 1 \right], \qquad (2.1.58)$$

where $E = u_v V$ is the total energy of monochromatic radiation in a volume V. If the energy is kept fixed while the volume is expanded from V_0 to V, the resulting variation of entropy is

$$S - S_0 = \frac{KE}{h v} \log \left(\frac{V}{V_0} \right). \qquad (2.1.59)$$

On the other hand, if the radiation is treated as an ideal gas undergoing an isothermal expansion, another formula for the variation of entropy can be written, i.e.

$$S - S_0 = \frac{1}{T} \int_{V_0}^{V} dQ = NK \int_{V_0}^{V} \frac{dy}{y} = NK \log \frac{V}{V_0}. \qquad (2.1.60)$$

Equations (2.1.59) and (2.1.60) express the same variation of entropy at fixed energy, and tell us that monochromatic radiation of frequency $v \gg \frac{KT}{h}$ behaves as a gas of N particles for which the total energy

$$E = Nhv. \qquad (2.1.61)$$

Thus, each particle can be thought of as a photon of energy hv.

In Einstein (1917), the author obtained a profound and elegant derivation of the Planck radiation formula by considering the canonical distribution of statistical mechanics for molecules which can be found only in a discrete set of states $Z_1, Z_2, \ldots, Z_n, \ldots$ with energies $E_1, E_2, \ldots, E_n, \ldots$:

$$W_n = p_n e^{-E_n/kT}, \qquad (2.1.62)$$

where W_n is the relative occurrence of the state Z_n, p_n is the statistical weight of Z_n and T is the temperature of the gas of molecules. On the one hand, a molecule might perform, without external stimulation, a transition from the state Z_m to the state Z_n (assuming $E_m > E_n$) while emitting the radiation energy $E_m - E_n$ of frequency v. The probability dW for this process of *spontaneous emission* in the time interval dt is

$$dW = A_{m \to n} dt, \qquad (2.1.63)$$

where $A_{m \to n}$ denotes a constant.

On the other hand, under the influence of a radiation density u_v, a molecule can make a transition from the state Z_n to the state Z_m by absorbing the radiative energy $E_m - E_n$, and the probability law for this process is

$$dW = B_{n \to m} u_v dt. \qquad (2.1.64)$$

Moreover, the radiation can also lead to a transition of the opposite kind, i.e. from state Z_m to state Z_n. The radiative energy $E_m - E_n$ is then freed according to the probability law

$$dW = B_{m \to n} u_v dt. \qquad (2.1.65)$$

In these equations, $B_{n \to m}$ and $B_{m \to n}$ are also constants.

Einstein then looked for that particular radiation density u_ν which guarantees that the exchange of energy between radiation and molecules preserves the canonical distribution (2.1.62) of the molecules. This is achieved if and only if, on average, as many transitions from Z_m to Z_n take place as of the opposite type, i.e.

$$\left(B_{m \to n} u_\nu + A_{m \to n}\right) p_m e^{-E_m/KT} = B_{n \to m} u_\nu p_n e^{-E_n/KT}. \tag{2.1.66}$$

Einstein also assumed that the B constants are related by

$$p_m B_{m \to n} = p_n B_{n \to m}, \tag{2.1.67}$$

to ensure that as the temperature tends to infinity u_ν also tends to infinity, and hence found

$$u_\nu = \frac{A_{m \to n}/B_{m \to n}}{(e^{(E_m - E_n)/KT} - 1)}, \tag{2.1.68}$$

which reduces to the Planck radiation formula (2.1.52), by virtue of the Wien displacement law (2.1.21), which implies

$$\frac{A_{m \to n}}{B_{m \to n}} = \alpha_1 \nu^3, \tag{2.1.69}$$

$$E_m - E_n = \alpha_2 \nu, \tag{2.1.70}$$

where α_1 and α_2 are constants which cannot be fixed at this stage (Einstein 1917).

2.1.6 Dynamic equilibrium of the radiation field

While spontaneous emission was known for a long time in atomic physics, it was Einstein who emphasized its role and derived the Planck distribution of spectral energy on a *dynamic* basis as we have just seen, in contrast with the original Planck derivation. Einstein considered a two-level atom and monochromatic radiation of frequency $\nu = \frac{(E_2 - E_1)}{h}$. But in actual fact there are many frequencies, many species of atoms and many energy levels (and populations of these levels). This generic problem was posed and solved by Bose. In the briefest outlook his derivation observes that, like in Maxwell's derivation of the velocity distribution in kinetic theory, the various populations enter through appropriate Lagrange multipliers. Dynamic balance of the entire complex demands that for *every* frequency we have the law (2.1.52). In both Einstein's and Bose's derivations the atomic population in each level was proportional to the Boltzmann factor $e^{-E_n/KT}$. Other important work on black-body radiation can be found in Mandel (1963) and Mandel *et al.* (1964).

2.2 Photoelectric effect

As we have said, in the analysis of black-body radiation Planck introduced, for the first time, the hypothesis of quanta: whenever matter emits or absorbs radiation, it does so in a sequence of elementary acts, in each of which an amount of energy ε is emitted or absorbed proportional to the frequency ν of the radiation: $\varepsilon = h\nu$, where h is the universal constant

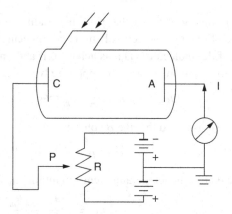

Fig. 2.3   The circuit used in the Millikan experiment. The energy with which the electron leaves the surface is measured by the product of its charge with the potential difference against which it is just able to drive itself before being brought to rest. Millikan was careful enough to use only light for which the illuminated electrode was photoelectrically sensitive, but for which the surrounding walls were not photosensitive.

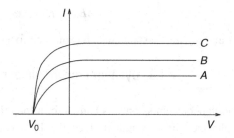

Fig. 2.4   Variation of the photoelectric current with voltage, for given values A, B, C of the intensity.

known as Planck's constant. We are now going to see how the ideas developed along similar lines make it possible to obtain a satisfactory understanding of the photoelectric effect. This provides part of the phenomenological foundations of the Einstein–de Broglie relation (1.1.2).

The photoelectric effect was discovered by Hertz and Hallwachs in 1887. The effect consists of the emission of electrons from the surface of a solid when electromagnetic radiation is incident upon it (Hughes and DuBridge 1932, DuBridge 1933, Holton 2000). The three empirical laws of such an effect are as follows (see Figures 2.3, 2.4; the Millikan experiment quoted therein should not be confused with the measurement of the electron charge $q_e = -|q_e|$, also due to Millikan).

(i) The electrons are emitted only if the frequency of the incoming radiation is greater than a certain value ν_0, which is a peculiar property of the metal used in the experiment, and is called *the photoelectric threshold*.

(ii) The velocities of the electrons emitted by the surface range from 0 to a maximum value v_{max}. The kinetic energy corresponding to v_{max} depends linearly on the frequency

v: $T_{\max} = k(v - v_0), k > 0$. $T_{\max}$ does not depend on the intensity of the incoming radiation.

(iii) For a given value of the frequency v of the incoming radiation, the *number* of electrons emitted per cm^2 per second is proportional to the intensity.

These properties cannot be understood if one assumes that classical electromagnetic theory rules the phenomenon. In particular, if one assumes that the energy is uniformly distributed over the metallic surface, it is unclear how the emission of electrons can occur when the intensity of the radiation is extremely low (which would require a long time before the electron would receive enough energy to escape from the metal). The experiments of Lawrence and Beans showed that the time lag between the incidence of radiation on a surface and the appearance of (photo)electrons is less than 10^{-9} s.

2.2.1 Classical model

Let us now discuss a model introduced by Sommerfeld and Debye (1913) to describe the photoelectric effect. Such a model is inspired by the Thomson atomic model of an electron elastically bound to the atom of size $R_A \approx 3 \cdot 10^{-10}$ m and subject to a viscous force of constant η. The value of η is determined as a function of the atomic relaxation time $\tau \equiv \frac{2m_e}{\eta} = 10^{-8}$ s. This is the time needed by the atom to release its energy through radiation or collisions. For simplicity, we consider a one-dimensional model where the electron displacement x obeys the differential equation

$$m_e \ddot{x} = f(x, \dot{x}, t) \approx \frac{\partial f}{\partial x} x + \frac{\partial f}{\partial \dot{x}} \dot{x} + f(0), \qquad (2.2.1)$$

where the right-hand side has been expanded to first order in x and $\dot{x}$, and the various terms depend on t. The damped oscillator equation is indeed of this kind:

$$m_e \ddot{x} = -kx - \eta\dot{x} - |q_e|E, \qquad (2.2.2)$$

where we may associate k with the atomic frequencies, and E is the magnitude of the applied electric field, possibly depending on time. We assume the presence of many atoms with different frequencies, continuously distributed about

$$\sqrt{\frac{k}{m}} = \omega_0 = 2\pi v_0 \approx 10^{15} \text{ s}^{-1}.$$

The applied electric field, corresponding to visible light, varies in time according to $E = E_0 \cos(\omega t)$. A generic solution of the equation of motion (2.2.2) reads as

$$x = x_0 \cos(\omega t + \phi) + A_1 e^{-\alpha_1 t} + A_2 e^{-\alpha_2 t}. \qquad (2.2.3)$$

The α values are obtained by solving the associated homogeneous equation, corresponding to $E = 0$ in Eq. (2.2.2), i.e.

$$m_e \alpha^2 - \eta\alpha + k = 0. \qquad (2.2.4)$$

The roots of Eq. (2.2.4) are

$$\alpha_{1,2} = \frac{\eta \pm \sqrt{\eta^2 - 4km_e}}{2m_e} = \frac{1}{\tau} \pm \sqrt{\frac{1}{\tau^2} - \omega_0^2}. \qquad (2.2.5)$$

They reduce to

$$\alpha_\pm \approx \frac{1}{\tau} \pm i\omega_0, \qquad (2.2.6)$$

under the assumption $\frac{1}{\tau} << \omega_0$.

The particular solution $x_0 \cos(\omega t + \phi)$ of the full equation (2.2.2) yields, upon substitution,

$$-m_e \omega^2 x_0 \cos(\omega t + \phi) = -kx_0 \cos(\omega t + \phi) + \eta \omega x_0 \sin(\omega t + \phi) - |q_e| E_0 \cos(\omega t). \qquad (2.2.7)$$

Recalling that $\cos(\omega t) = \cos(\omega t + \phi) \cos(\phi) + \sin(\omega t + \phi) \sin(\phi)$ Eq. (2.2.7) can be recast in the form

$$(-m_e \omega^2 x_0 + kx_0 + |q_e| E_0 \cos(\phi)) \cos(\omega t + \phi) = (\eta \omega x_0 - |q_e| E_0 \sin(\phi)) \sin(\omega t + \phi), \qquad (2.2.8)$$

which, to be satisfied for all t, implies

$$-m_e \omega^2 x_0 + kx_0 + |q_e| E_0 \cos(\phi) = 0, \qquad (2.2.9)$$

$$\eta \omega x_0 - |q_e| E_0 \sin(\phi) = 0, \qquad (2.2.10)$$

from which we get

$$\tan(\phi) = \frac{2\omega}{\tau(\omega^2 - \omega_0^2)}, \qquad (2.2.11)$$

and hence

$$\cos(\phi) = \frac{1}{\sqrt{1 + \tan^2(\phi)}} = \frac{(\omega^2 - \omega_0^2)}{\sqrt{(\omega^2 - \omega_0^2)^2 + \frac{4\omega^2}{\tau^2}}}, \qquad (2.2.12)$$

$$\sin(\phi) = \frac{\tan(\phi)}{\sqrt{1 + \tan^2(\phi)}} = \frac{\frac{2\omega}{\tau}}{\sqrt{(\omega^2 - \omega_0^2)^2 + \frac{4\omega^2}{\tau^2}}}. \qquad (2.2.13)$$

For the amplitude x_0 the resonant form is

$$x_0 = \frac{\frac{|q_e| E_0}{m_e}}{\sqrt{(\omega^2 - \omega_0^2)^2 + \frac{4\omega^2}{\tau^2}}}. \qquad (2.2.14)$$

The constants A_1 and A_2 in the solution (2.2.3) should be evaluated by using the initial conditions. By virtue of the reality of x, Eq. (2.2.3) can be re-expressed in the form

$$x = x_0 \cos(\omega t + \phi) + A e^{-\frac{t}{\tau}} \cos(\omega_0 t + \phi_0). \qquad (2.2.15)$$

If the electron is initially at rest, A and ϕ_0 can be determined by setting $x(t = 0) = 0 = \dot{x}(t = 0)$, i.e.

$$x_0 \cos(\phi) + A \cos(\phi_0) = 0, \qquad (2.2.16)$$

$$x_0 \omega \sin(\phi) = -A \left[\frac{1}{\tau} \cos(\phi_0) + \omega_0 \sin(\phi_0) \right]. \qquad (2.2.17)$$

In particular,

$$\tan(\phi_0) = \frac{\omega}{\omega_0} \tan(\phi) - \frac{1}{\omega_0 \tau}. \tag{2.2.18}$$

The electron will leave the atomic band if the electron displacement x is greater than the atomic radius. The expression for x is the sum of two parts, where the first is associated with stationary oscillations, while the second is a transient decaying with time constant τ. In principle, the maximum amplitude could take place during the transient or later: to decide which is the case we must compare the value of A with that of x_0. From Eq. (2.2.16) it is clear that the magnitude of A is of the same order as x_0 unless $\cos(\phi_0)$ is very different from $\cos(\phi)$.

We may consider separately the generic case and the resonant one. In the former, the displacement is of order $\frac{|q_e E_0|}{\omega^2 m_e}$. For the photoelectric effect to occur, it is necessary to fulfill the condition

$$\frac{|q_e E_0|}{\omega^2 m_e} \approx R_A. \tag{2.2.19}$$

The power density needed is therefore

$$P = c\varepsilon_0 E_0^2 \approx c\varepsilon_0 \left(\frac{R_A \omega^2 m_e}{q_e}\right)^2 \approx 10^{15} \frac{W}{m^2}. \tag{2.2.20}$$

This power density is too large, and any kind of electrode would be vaporized.

Thus, we have to consider instead the resonant case, and set $\omega = \omega_0$ therein. We find $\phi = \phi_0 = \frac{\pi}{2}, A = -x_0$, which implies that

$$x = -\frac{|q_e|E_0 \tau}{2m_e \omega_0} \left(1 - e^{-\frac{t}{\tau}}\right) \sin(\omega_0 t). \tag{2.2.21}$$

The oscillation amplitude should be greater than the atomic radius, i.e.

$$\frac{|q_e E_0|\tau}{2m_e \omega_0} \left(1 - e^{-\frac{t}{\tau}}\right) \geq R_A. \tag{2.2.22}$$

Thus, the threshold field is $\frac{2m_e \omega_0 R_A}{q_e \tau}$, with a corresponding power density

$$P = c\varepsilon_0 \left(\frac{2\omega_0 m_e R_A}{q_e \tau}\right)^2 \approx 100 \frac{W}{m^2}, \tag{2.2.23}$$

and the time to reach the runaway amplitude is of order τ.

Such a classical model requires a frequency tuned on the resonance frequency, and the photoelectric effect would no longer occur both below and above the resonance frequencies of the atoms in the electrode.

2.2.2 Quantum theory of the effect

The peculiar emission of electrons is however, naturally accounted for, if Planck's hypothesis is accepted. More precisely, it has to be assumed that the energy of radiation is quantized not only when emission or absorption occur, but that it can also travel in space in the form of elementary quanta of radiation with energy $h\nu$. Correspondingly, the photoelectric effect should be thought of as a collision process between the incoming quanta of radiation and the electrons belonging to the atoms of the metallic surface. According to this quantum scheme, the atom upon which the photon falls receives, all at once, the energy $h\nu$. As a result of this process, an electron can be emitted only if the energy $h\nu$ is greater than the work function W_0:

$$h\nu > W_0. \tag{2.2.24}$$

The first experimental law, (i), is therefore understood, provided the photoelectric threshold is identified with $\frac{W_0}{h}$:

$$\nu_0 = \frac{W_0}{h}. \tag{2.2.25}$$

If the inequality (2.2.24) is satisfied, the electron can leave the metallic plate with an energy which, at the very best, is $W = h\nu - W_0$, which implies

$$W_{\max} = h(\nu - \nu_0). \tag{2.2.26}$$

This agrees completely with the second law, (ii). Lastly, upon varying the intensity of the incoming radiation, the number of quanta falling upon the surface in a given time interval changes, but from the above formulae it is clear that the energy of the quanta, and hence of the electrons emitted, is not affected by the intensity.

In the experimental apparatus, ultraviolet or X-rays fall upon a clean metal cathode, and an electrode collects the electrons that are emitted with kinetic energy $T = h\nu - W_0$. If V_0 is the potential for which the current vanishes,

$$V_0 = \frac{h\nu}{|q_e|} - \frac{W_0}{|q_e|}. \tag{2.2.27}$$

The plot of $V_0(\nu)$ is a straight line (Figure 2.5) that intersects the ν-axis when $\nu = \nu_0$. The slope of the experimental curve makes it possible to measure Planck's constant (for this purpose, Millikan used monochromatic light). The value of $\frac{h}{|q_e|}$ is 4.14×10^{-15} V s, with $h = 6.6 \times 10^{-27}$ erg s.

Einstein made a highly non-trivial step, by postulating the existence of elementary quanta of radiation which travel in space. This was far more than what Planck had originally demanded in his attempt to understand black-body radiation. Note also that, strictly speaking, Einstein was not aiming to 'explain' the photoelectric effect. When he wrote his fundamental papers (Einstein 1905, 1917), the task of theoretical physicists was not quite that of having to understand a well-established phenomenology, since the Millikan measurements were made 10 years after the first Einstein paper. Rather, Einstein developed some far-reaching ideas which, in particular, can be applied to account for all known aspects of the photoelectric effect. Indeed, in Einstein (1905), the author writes as follows.

...The wave theory of light, which operates with continuous spatial functions, has worked well in the representation of purely optical phenomena and will probably never be replaced by another theory. It should be kept in mind, however, that the optical observations refer to time averages rather than instantaneous values. In spite of the complete experimental confirmation of the theory as applied to diffraction, reflection,

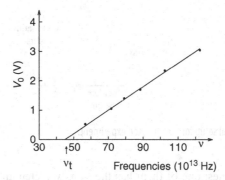

Fig. 2.5 Results of the Millikan experiment for the retarding potential V_0 expressed as a function of frequency. A linear relation is found between V_0 and ν, and the slope of the corresponding line is numerically equal to $\frac{h}{q_e}$. The intercept of such a line on the ν-axis is the lowest frequency at which the metal in question can be photoelectrically active. Reprinted with permission from R. A. Millikan, Phys. Rev. 7, 355–88 (1916). Copyright (1916) by the American Physical Society.

refraction, dispersion, etc., it is still conceivable that the theory of light which operates with continuous spatial functions may lead to contradictions with experience when it is applied to the phenomena of emission and transformation of light.

It seems to me that the observations associated with blackbody radiation, fluorescence, the production of cathode rays by ultraviolet light, and other related phenomena connected with the emission or transformation of light are more readily understood if one assumes that the energy of light is discontinuously distributed in space. In accordance with the assumption to be considered here, the energy of a light ray spreading out from a point source is not continuously distributed over an increasing space but consists of a finite number of energy quanta which are localized at points in space, which move without dividing, and which can only be produced and absorbed as complete units.

2.3 Compton effect

Classically, a monochromatic plane wave of electromagnetic nature carries momentum according to the relation $p = \frac{E}{c}$. Since E is quantized and we identify it as the fourth component of a 4-vector, one is naturally led to expect and ask whether the momentum is carried in the form of quanta with absolute value $\frac{h\nu}{c}$. The Compton effect (Compton 1923a,b) provides clear experimental evidence in favour of this conclusion, and supports the existence of photons. For this purpose, the scattering of monochromatic X- and γ-rays from gases, liquids and solids is studied in the laboratory. Under normal circumstances, the X-rays pass through a material of low atomic weight (e.g. coal). A spectrograph made out of crystal collects and analyses the rays scattered in a given direction (Figure 2.6). Jointly with the radiation is found, scattered by means of the process we are going to describe, yet another radiation which is scattered without any change of its wavelength. There exist two

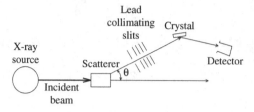

Fig. 2.6 Experimental setup for the Compton experiment.

nearby lines: one of them has the same wavelength λ as the incoming radiation, whereas the other line has a wavelength $\lambda' > \lambda$. The line for which the wavelength remains unaffected can be accounted for by thinking that the incoming photon also meets the 'deeper underlying' electrons of the scattering material. For such processes, the mass of the whole atom is involved, which reduces the value of the shift $\lambda' - \lambda$ significantly, so that it becomes virtually unobservable. We are now going to consider the scattering process involving *the external electron only*.

Let us assume that the incoming radiation consists of photons having frequency ν. Let m_e be the rest mass of the electron, $\vec{v}$ its velocity after collision with the photon and let ν' be the frequency of the scattered photon. The conservation laws that make it possible to obtain a theoretical description of the phenomenon are the conservation of energy and linear momentum, and the description has to be considered within a relativistic setting. By using the energy-momentum 4-vector,

$$P^\mu_{e\text{ in}} + P^\mu_{\text{Ph in}} = P^\mu_{e\text{ out}} + P^\mu_{\text{Ph out}}.$$

The notation is 'in' for incoming particles and 'out' for outgoing particles. Of course, the conservation law holds for each component using any axis. By selecting a split of space–time into space and time, we denote by $\hat{k}$ the unit vector along the direction of the incoming photon, and by $\hat{k}'$ the unit vector along the direction of emission of the scattered photon (see Figure 2.7, where, however, we refer to 3-vectors rather than 4-vectors).

The energy conservation reads, in our problem, as

$$m_e c^2 + h\nu = \frac{m_e c^2}{\sqrt{1 - \frac{v^2}{c^2}}} + h\nu'. \tag{2.3.1}$$

Moreover, taking into account that the linear momentum of the electron vanishes before the scattering takes place, the conservation of linear momentum leads to

$$\frac{h\nu}{c}\hat{k} = \frac{m_e \vec{v}}{\sqrt{1 - \frac{v^2}{c^2}}} + \frac{h\nu'}{c}\hat{k}'. \tag{2.3.2}$$

If Eq. (2.3.2) is projected onto the x- and y-axes it yields the equations

$$\frac{h\nu}{c} = \frac{h\nu'}{c}\cos\theta + p\cos\phi, \tag{2.3.3}$$

$$\frac{h\nu'}{c}\sin\theta = p\sin\phi, \tag{2.3.4}$$

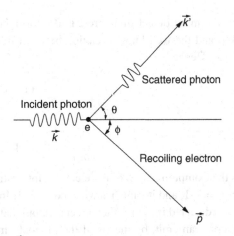

Fig. 2.7   A photon with linear momentum $\vec{k}$ collides with an electron at rest and is scattered with momentum $\vec{k'}$, while the electron recoils with momentum $\vec{p}$.

which, together with Eq. (2.3.1), are three equations from which we may evaluated ϕ, the frequency ν' of the scattered X-ray, and the linear momentum p of the electron as functions of the scattering angle θ. Here, attention is focused on the formula for wavelength shift. First, setting $\beta \equiv \frac{\nu}{c}$, from Eq. (2.3.2) it is found that

$$\frac{m_e^2 \beta^2 c^2}{(1-\beta^2)} = \frac{h^2 \nu^2}{c^2} + \frac{h^2 \nu'^2}{c^2} - 2\frac{h\nu}{c}\frac{h\nu'}{c}\cos\theta, \qquad (2.3.5)$$

where θ is the angle formed by the unit vectors $\hat{k}$ and $\hat{k}'$. Moreover, Eq. (2.3.1) leads to

$$\frac{m_e^2}{(1-\beta^2)} = \left(m_e + \frac{h\nu}{c^2} - \frac{h\nu'}{c^2}\right)^2. \qquad (2.3.6)$$

Thus, on using the identity

$$\frac{\beta^2 c^2}{(1-\beta^2)} = -c^2 + \frac{c^2}{m_e^2}\frac{m_e^2}{(1-\beta^2)}, \qquad (2.3.7)$$

the comparison of Eqs. (2.3.5) and (2.3.6) yields

$$-m_e^2 c^2 + c^2\left(m_e + \frac{h\nu}{c^2} - \frac{h\nu'}{c^2}\right)^2$$

$$= \frac{h^2 \nu^2}{c^2} + \frac{h^2 \nu'^2}{c^2} - 2\frac{h\nu}{c}\frac{h\nu'}{c}\cos\theta. \qquad (2.3.8)$$

A number of cancellations are now found to occur, which significantly simplifies the final result, i.e.

$$\nu - \nu' = \frac{h\nu\nu'}{m_e c^2}(1 - \cos\theta). \qquad (2.3.9)$$

However, the main object of interest is the formula for $\lambda' - \lambda$, which is obtained from Eq. (2.3.9) and the well-known relation between frequency and wavelength: $\nu/c = 1/\lambda$, $\nu'/c = 1/\lambda'$. Hence,

$$\lambda' - \lambda = \frac{h}{m_e c}(1 - \cos\theta), \tag{2.3.10}$$

where

$$\frac{h}{m_e c} = 0.0024 \text{ nm}, \tag{2.3.11}$$

which is the Compton length of the electron. Interestingly, the wavelength shift is maximal when $\cos\theta = -1$, and it vanishes when $\cos\theta = 1$. In the actual experiments, the scattered photons are detected if in turn they meet an atom that is able to absorb them (provided that such an atom can emit, by means of the photoelectric effect, an electron, the passage of which is visible on a photographic plate).

We can thus conclude that photons behave exactly as if they were particles with energy $h\nu$ and linear momentum $\frac{h\nu}{c}$. According to relativity theory, developed by Einstein and Poincaré, the equation $p = \frac{E}{c}$ is a peculiar property of massless particles. Thus, we can say that photons behave like relativistic massless particles. In particular, this means that the 'radiation' emitted by the electron does not propagate in all directions, as would happen for a wave, but goes in a specified direction exactly as would occur for particles.

The frequency shift is a peculiar property of a quantum theory which relies on the existence of photons, because in the classical electromagnetic theory no frequency shift occurs. To appreciate this, let us consider the classical description of the phenomenon. On denoting the position vector in $\mathbf{R}^3$ by $\vec{r}$, with Cartesian coordinates (x, y, z), and by the wave vector $\vec{k}$ with corresponding components (k_x, k_y, k_z), the electric field of the incoming plane wave of frequency $\nu = \frac{\omega}{2\pi}$ may be written in the form

$$\vec{E} = \vec{E}_0 \cos\left(\vec{k} \cdot \vec{r} - \omega t\right), \tag{2.3.12}$$

where the vector $\vec{E}_0$ has components $(E_{0_x}, E_{0_y}, E_{0_z})$ independent of position and time. Strictly speaking, one has then to build a wave packet from these elementary solutions of the Maxwell equations, but Eq. (2.3.12) is all we need to obtain the classical result. The electric field, which varies in space and time according to Eq. (2.3.12), generates a magnetic field that also varies in space and time in a similar way. This is clearly seen from the Faraday law

$$\vec{\nabla} \wedge \vec{E} + \frac{1}{c}\frac{\partial \vec{B}}{\partial t} = 0, \tag{2.3.13}$$

which can be integrated to find

$$\vec{B} = -c \int \vec{\nabla} \wedge \vec{E} \, dt = \vec{B}_0 \cos\left(\vec{k} \cdot \vec{r} - \omega t\right), \tag{2.3.14}$$

where we have defined

$$\vec{B}_0 \equiv \frac{c}{\omega}\vec{k} \wedge \vec{E}_0. \tag{2.3.15}$$

The force acting on the electron of charge q_e is, therefore,

$$\vec{F} = m_e \frac{\mathrm{d}}{\mathrm{d}t}\vec{v} = q_e \left(\vec{E} + \frac{1}{c}\vec{v} \wedge \vec{B} \right)$$

$$= q_e \left(\vec{E}_0 + \frac{1}{c}\vec{v} \wedge \vec{B}_0 \right) \cos\left(\vec{k} \cdot \vec{r} - \omega t \right), \qquad (2.3.16)$$

where, from Eq. (2.3.15),

$$\left(\vec{v} \wedge \vec{B}_0 \right)_x \equiv v_y B_{0_z} - v_z B_{0_y}$$

$$= \frac{c}{\omega}\left[v_y\left(k_x E_{0_y} - k_y E_{0_x} \right) - v_z\left(k_z E_{0_x} - k_x E_{0_z} \right) \right]$$

$$= \frac{c}{\omega}\left[k_x\left(\vec{v} \cdot \vec{E}_0 \right) - E_{0_x}\left(\vec{v} \cdot \vec{k} \right) \right]. \qquad (2.3.17)$$

Analogous equations hold for the other components of $\vec{v} \wedge \vec{B}_0$ so that, eventually,

$$\vec{v} \wedge \vec{B}_0 = \frac{c}{\omega}\left[\vec{k}\left(\vec{v} \cdot \vec{E}_0 \right) - \vec{E}_0\left(\vec{v} \cdot \vec{k} \right) \right], \qquad (2.3.18)$$

which implies (see Eqs. (2.3.12) and (2.3.16))

$$m_e \frac{\mathrm{d}}{\mathrm{d}t}\vec{v} = q_e \left\{ \vec{E} + \frac{1}{\omega}\left[\vec{k}\left(\vec{v} \cdot \vec{E} \right) - \vec{E}\left(\vec{v} \cdot \vec{k} \right) \right] \right\}, \qquad (2.3.19)$$

with

$$\frac{\mathrm{d}}{\mathrm{d}t}\vec{r} = \vec{v}. \qquad (2.3.20)$$

The magnetic forces are negligible compared with the electric forces, so that the acceleration of the electron reduces to $\frac{\mathrm{d}\vec{v}}{\mathrm{d}t} = \frac{q_e}{m_e}\vec{E}$. By virtue of its oscillatory motion, the electron starts radiating a field which, at a distance R, has components with magnitude (Jackson 1975)

$$|\vec{E}'| = |\vec{B}'| = \frac{|q_e \ddot{r}|}{c^2 R}\sin\phi, \qquad (2.3.21)$$

where c is the velocity of light and ϕ is the angle between the scattered beam and the line along which the electron oscillates. Substituting for the acceleration,

$$|\vec{E}'| = |\vec{B}'| = \frac{q_e^2 E \sin\phi}{m_e c^2 R}. \qquad (2.3.22)$$

Once again, the classical model cannot account for the experimental findings.

2.3.1 Thomson scattering

To sum up, in a classical model, the atomic electrons vibrate with the same frequency as the incident radiation. These oscillating electrons, in turn, radiate electromagnetic waves *of the same frequency*, leading to so-called Thomson scattering. This is a non-relativistic scattering process, which describes X-ray scattering from electrons and γ-ray scattering from protons. For a particle of charge q and mass m, the total Thomson scattering cross-section (recall that the cross-section describes basically the probability of the scattering process) reads as (Jackson 1975)

$$\sigma_T = \frac{8\pi}{3}\left(\frac{q^2}{mc^2}\right)^2. \tag{2.3.23}$$

For electrons, $\sigma_T = 0.665 \times 10^{-24}$ cm^2. The associated characteristic length is

$$\frac{q_e^2}{m_e c^2} = 2.82 \times 10^{-13} \text{ cm} \tag{2.3.24}$$

and is called the classical electron radius.

To sum up, we may say that both the photoelectric effect and the Compton effect call for a 'corpuscular' description of light and therefore the introduction of photons. This means reading the Einstein–de Broglie relation (1.1.2) from right to left. The interference phenomena for electrons provide the phenomenological support for reading the relation (1.1.2) from left to right.

2.4 Particle-like behaviour and the Heisenberg picture

2.4.1 Atomic spectra and the Bohr hypotheses

The frequencies that can be emitted by material bodies form their emission spectrum, whereas the frequencies that can be absorbed form their absorption spectrum. For simplicity, we consider gases and vapours.

A device to obtain the emission and absorption spectra works as follows. Some white light falls upon a balloon containing gas or a vapour; a spectrograph, e.g. a prism P_1, splits the light transmitted from the gas into monochromatic components, which are collected on a plate L_1 (Figure 2.8). On L_1 a continuous spectrum of light can be seen transmitted from

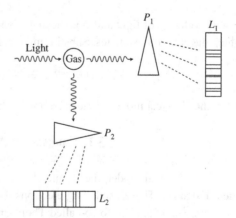

Fig. 2.8   Experimental setup used to obtain the emission and absorption spectra.

the gas, interrupted by dark lines corresponding to the absorption frequencies of the gas. These dark lines form the absorption spectrum. To instead obtain the emission spectrum, some energy has to be transmitted to the gas, which will eventually emit such energy in the form of electromagnetic radiation. This can be achieved in various ways: by heating the material, by an electrical discharge, or by sending light into the material as we outlined previously. By referring to this latter case for simplicity, if we want to analyse the emitted light, we shall perform our observations in a direction orthogonal to that of the incoming light (to avoid being disturbed by such light). A second prism P_2 is inserted to decompose the radiation emitted from the gas, and this is collected on plate L_2. On L_2, on a dark background, some bright lines can be seen corresponding to the frequencies emitted from the gas. These lines form the emission spectrum.

First, the observations show that the emission and absorption spectra are quite different: the emission spectrum contains far more lines than the absorption spectrum, and, within it, all lines of the absorption spectrum can be found. Moreover, if the incoming radiation has a spectrum of frequencies ν greater than a certain value ν_1, it is also possible to observe, in the emission spectrum, lines corresponding to frequencies smaller than ν_1. It was found that often a line may be associated with two others by the relation $\nu = \nu_1 + \nu_2$. More specifically, if lines are parametrized with two numbers, say $\nu_{i,j}$, then $\nu_{i,k} = \nu_{i,j} + \nu_{j,k}$. This rule was known as the Ritz combination principle. To account for the emission and absorption spectra, Bohr made some assumptions (Bohr 1913) that, as in the case of Einstein's hypothesis, disagree with classical physics, which was unable to account for the discreteness properties of the spectra. The basic idea was that privileged orbits for atoms exist that are stable. If the electrons in the atom lie on one of these orbits, they do not radiate. For these orbits, Maxwell's laws are suspended. Such orbits are discrete, and hence the corresponding energy levels are discrete as well. The full set of hypotheses are as follows (Bohr 1913, Herzberg 1944).

(i) An atom can only have a discrete set of energies starting from a minimal energy: $E_1 < E_2 < \cdots < E_n < \cdots$. These energy levels describe only the *bound states* of an atom, i.e. states that correspond to bounded classical orbits in phase space. The minimal energy state of an atomic system is called the *ground state*.

(ii) When an atomic system is in one of the above discrete energy levels it does not radiate. The emission (respectively, absorption) of radiation is associated with the transition of the electron from one orbit to another of lower (respectively, higher) energy.

(iii) The allowed orbits are those for which the integrals of $\oint p_i \mathrm{d}q^i$ along the orbit are integer multiples of the Planck constant. Such a generalization of Bohr's idea is better known as the Bohr–Sommerfeld quantization rule. This issue will be further discussed in Section 2.6.1.

We are now going to derive some consequences of the above assumptions.

(a) The spectra should be formed by lines for which the frequencies are given by

$$\nu_{n,m} = \frac{|E_n - E_m|}{h}, \tag{2.4.1}$$

with all possible values of E_n and E_m. Each material, however, also has to exhibit a continuous spectrum, which corresponds to transitions from a bound state to ionization states (also called *continuum states*, because Bohr's hypothesis of discrete energies does not hold for them).

(b) Bohr's assumptions are compatible with Einstein's hypothesis. Indeed, if an atom radiates energy in the form of discrete quanta, when the atom emits (or absorbs) a photon of frequency v, its energy changes by an amount hv.

(c) It is then clear why the emission spectra are richer than the absorption spectra. Indeed, at room temperature, the vast majority of atoms are in the ground state, and hence, in absorption, only the frequencies

$$v_{1,n} = \frac{E_n - E_1}{h} \qquad (2.4.2)$$

are observed, which correspond to transitions from the ground state E_1 to the generic level E_n. Over a very short time period (of the order of 10^{-8} or 10^{-9} s), radiation is re-emitted in one or more transitions to lower levels, until the ground state is reached. Thus, during the emission stage, the whole spectrum given by the previous formula may be observed.

(d) From what we have said it follows that all frequencies of the emission spectrum are obtained by taking differences of the frequencies of the absorption spectrum:

$$|v_n - v_m| = |(E_n - E_1)/h - (E_m - E_1)/h| = |(E_n - E_m)/h| = v_{n,m}. \qquad (2.4.3)$$

This property provides a simple interpretation of the Ritz combination principle. More precisely, for a complex atom the lines of the spectrum can be classified into series, each of them being of the form

$$\frac{1}{\lambda} = R\left(\frac{1}{m^2} - \frac{1}{n^2}\right), \qquad (2.4.4)$$

where n and m are integers, with m fixed and R being the Rydberg constant. From this experimental discovery it is found that, on the one hand, the frequency $v = \frac{c}{\lambda}$ is a 'more natural' parameter than the wavelength λ for indexing the lines of the spectrum and, on the other hand, *the spectrum is a set of differences of frequencies* (or spectral terms), i.e. there exists a set $\mathcal{F}$ of frequencies such that the spectrum is the set of differences

$$v_{i,j} \equiv v_i - v_j \qquad (2.4.5)$$

of arbitrary pairs of elements of $\mathcal{F}$. Therefore,

$$v_{i,k} = v_{i,j} + v_{j,k}, \qquad (2.4.6)$$

which provides a precise statement of the Ritz combination principle: the spectrum is endowed with a composition law, according to which the sum of the frequencies $v_{i,j}$ and $v_{l,k}$ is again a frequency of the spectrum only when $l = j$. Their combination is then expressed by Eq. (2.4.6).

(e) From the knowledge of the absorption spectrum the energies E_n can be derived, because, once the constant h is known, the absorption spectrum makes it possible to determine $E_2 - E_1, E_3 - E_1, \ldots, E_n - E_1$ and so on, up to $E_\infty - E_1$. Moreover, if the energy corresponding to the ionization threshold is set to zero, i.e. to the limit level E_∞, then $E_1 = -hv$, where v is the limit frequency of the spectrum (for frequencies greater than v is obtained a continuous spectrum is obtained and the atom is ionized).

(f) Spectroscopists had been able to group together the lines of a (emission) spectrum, in such a way that the frequencies, or, more precisely, the wave numbers $\frac{1}{\lambda} = \frac{\nu}{c}$ corresponding to the lines of a spectrum can be expressed as differences between 'spectroscopic terms' (Balmer 1885):

$$\frac{1}{\lambda} = \frac{\nu}{c} = T(n) - T(m), \tag{2.4.7}$$

where n and m are positive integers. Each series is picked out by a particular value of n, and by all values of m greater than n. Thus, for example, the first series (which is the absorption series) corresponds to the wave numbers

$$\frac{1}{\lambda} = T(1) - T(m), \quad m > 1. \tag{2.4.8}$$

Now according to Bohr the spectroscopic terms $T(n)$ are nothing but the energy levels divided by hc:

$$T(n) = -\frac{E_n}{hc}, \tag{2.4.9}$$

and the various series correspond to transitions that share the same final level.

Property (iii) makes it possible to determine the energy levels of a system. It is indeed possible to perform the analysis for the hydrogen atom and hydrogen-like atoms, i.e. those systems where only one electron is affected by the field of a nucleus of charge Zq_e, where Z is the atomic number. In this case the corresponding Hamiltonian can be written as

$$\mathcal{H} = \frac{|\vec{p}|^2}{2m_e} - \frac{Zq_e^2}{|\vec{r}|}. \tag{2.4.10}$$

Let us choose polar coordinates, $\vec{r} \equiv (r, \theta, \varphi)$. In this case $\mathcal{H}$ becomes

$$\mathcal{H} = \frac{1}{2m_e} \left(p_r^2 + \frac{p_\theta^2}{r^2} + \frac{p_\varphi^2}{r^2 \sin^2(\theta)} \right) - \frac{Zq_e^2}{r} = \frac{1}{2m_e} \left(p_r^2 + \frac{L^2}{r^2} \right) - \frac{Zq_e^2}{r}, \tag{2.4.11}$$

with L^2 denoting the squared modulus of angular momentum. The conjugate momenta to polar coordinates turn out to be $p_r = m_e \dot{r}$, $p_\varphi = m_e r^2 \dot{\varphi}$, $p_\theta = m_e r^2 \dot{\theta}$ and $p_\varphi = m_e r^2 \sin(\theta) \dot{\varphi}$. The Bohr–Sommerfeld condition on φ implies the quantization of the z-component of angular momentum $L_z = p_\varphi$, i.e.

$$\oint p_\varphi \, d\varphi = 2\pi L_z = mh \implies L_z = m\hbar, \tag{2.4.12}$$

with $\hbar \equiv h/(2\pi)$. Analogously, the condition on θ leads to the quantization of the modulus of the total angular momentum as $L = k\hbar$ where $k = 1, 2, 3 \ldots$ and corresponding m can assume the integer values $-k, -k+1, \ldots, k-1, k$. Lastly, $\oint p_r \, dr = n_r h$ implies that the allowed energy values of hydrogen-like atoms are

$$E_n = -\frac{m_e Z^2 q_e^4}{2(n_r + k)^2 \hbar^2}, \tag{2.4.13}$$

where $n_r + k$ is the *principal quantum number*.

2.5 Corpuscular character: the experiment of Franck and Hertz

The experiment of Franck and Hertz, performed for the first time in 1914 (Franck and Hertz 1914), was intended to test directly the first fundamental postulate of Bohr, according to which an atom can only have a discrete series of energy levels $E_0, E_1, E_2, \ldots$, corresponding to the frequencies $\nu_i = -\frac{E_i}{h}$, with $i = 0, 1, 2, \ldots$. The phenomenon under investigation is the collision of an electron with a monatomic substance. The atoms of such a substance are, to a large extent, in the ground state E_0. If an electron with a given kinetic energy collides with such an atom, which can be taken to be at rest both before the collision (by virtue of the small magnitude of its velocity, due to thermal agitation) and after this process (by virtue of its large mass), the collision is necessarily elastic if $T < E_1 - E_0$, where E_1 is the energy of the closest excited state. Thus, the atom remains in its ground state, and the conservation of energy implies that the electron is scattered in an arbitrary direction with the same kinetic energy, T. In contrast, if $T \geq E_1 - E_0$, inelastic collisions may occur that excite the atom to a level with energy E_1, while the electron is scattered with kinetic energy

$$T' = T - (E_1 - E_0). \tag{2.5.1}$$

The experiment is performed by using a glass tube filled with monatomic vapour. On one side of the tube there is a metal filament F, which is heated by an auxiliary electric current. Electrons are emitted from F via the thermionic effect. On the other side of the tube there is a grid G and a plate P. On taking the average potential of the filament as zero, one 'inserts' in between F and P an electromotive force $V - \varepsilon$ and a weak electromotive force ε in between G and P. A galvanometer, which is inserted in the circuit of P, makes it possible to measure the current at P and to study its variation as V is increased. Such a current is due to the electrons which, emitted from the filament, are attracted towards the grid, where they arrive with a kinetic energy $T = |q_e V|$, unless inelastic collisions occur. The electrons pass through the holes of the grid (overcoming the presence of the *counterfield*) and a large number of them reach the plate (despite the collisions occurring in between G and P). This occurs because the kinetic energy of the electrons is much larger than ε.

Since, for $|q_e V| < E_1 - E_0$, only elastic collisions may occur in between F and G, we have to expect that the higher the kinetic energy of the electrons, the larger the number of electrons reaching the plate will be. The experiment indeed shows that, for V increasing between 0 and the first excitation potential $\frac{E_1 - E_0}{|q_e|}$, the current detected at the plate increases continuously (see Figure 2.9). However, as soon as V takes on larger values, inelastic collisions may occur in the neighbourhood of the grid and, if the density of the vapour is sufficiently high, a large number of electrons lose almost all of their kinetic energy in such collisions. Hence, they are no longer able to reach the plate, because they do not have enough energy to overcome the *counterfield* between the grid and the plate. This leads to a substantial reduction of the current registered at P. The process is now repeated: a further increase of the potential enhances the current at P provided that V remains smaller than the second excitation potential, $\frac{E_2 - E_0}{|q_e|}$, and so on. It is thus clear that the excitation potentials

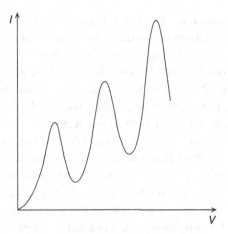

Fig. 2.9 Variation of the current I detected at the plate as the potential V is increased in the experiment of Franck and Hertz. For V increasing between 0 and the first excitation potential, the current I increases continuously. The subsequent decrease results from inelastic collisions occurring in the neighbourhood of the grid. All excitation potentials can be measured in this way.

are being measured, and the experimental data are in good agreement with the theoretical model.

It should be stressed that, when V increases so as to become larger than integer multiples of $\frac{E_1 - E_0}{|q_e|}$, the electron may undergo multiple collisions instead of a single inelastic collision. Then, the current at P starts decreasing for energies slightly larger than $(E_1 - E_0), 2(E_1 - E_0), 3(E_1 - E_0), \ldots$.

2.6 Wave-like behaviour and the Bragg experiment

Apart from the interference experiments with electrons that we have described at the beginning of Chapter 1, we are now going to outline the 'corpuscular' to 'wave' description implied by the Einstein–de Broglie relation when viewed from left to right.

2.6.1 Connection between the wave picture and the discrete-level picture

In the light of a number of experimental results, some key assumptions can be formulated:

(i) The existence of photons (Einstein 1905, 1917);

(ii) Bohr's assumption on the selection of classical orbits (Bohr 1913);

(iii) Bohr's formula for the frequency of the radiation emitted (see Eq. (2.4.1)) and for the difference of the allowed values of energy.

These ad hoc assumptions make it possible to account successfully for a large number of experimental results, and Arthur Haas was indeed able to compute the Rydberg constant from elementary atomic parameters (Haas 1910a,b, 1925). (The Balmer formula contained a new constant, the Rydberg constant, in terms of which the energy differences of atomic levels was expressed. To the extent that the atomic levels depend on the energy of the electron subject to the Coulomb-like potential of the nucleus and the kinetic energy of the electron, it should be possible to relate the Rydberg constant to the other fundamental constants. This relation was deduced by Arthur Haas.) The resulting theoretical picture, however, is not satisfactory, in that classical physics is amended just when we strictly need to do so, but remains otherwise unaffected.

In 1923, Louis de Broglie studied the following problem: the electromagnetic radiation, which has always been considered (quite legitimately) to be of a wave-like nature, also has a particle-like nature by virtue of the existence of photons. The link between the wave and particle aspects is obtained with the help of the Planck constant. The Planck constant, however, also plays a role, via the Bohr–Sommerfeld quantization conditions (see below, Eq. (2.6.1)), in problems where one deals with particles in the first place. Moreover, integer numbers occur in the quantization condition, and it is well known to physicists that their occurrence is associated with wave-like phenomena (e.g. interference, stationary waves). de Broglie was then led to study whether it was possible to follow the opposite path, i.e. *to look for a wave-like aspect in what we always considered from a purely particle-like point of view* (Holland 1993).

Let us consider, for example, an electron in a circular orbit in the hydrogen atom, for which the Bohr–Sommerfeld condition

$$\oint \vec{p} \cdot \mathrm{d}\vec{q} = nh \tag{2.6.1}$$

implies that (with $p = |\vec{p}|$)

$$pL = nh, \tag{2.6.2}$$

where L is the length of the orbit. (The concept of an orbit becomes inappropriate when the full formalism of wave mechanics is developed, but remains useful for our introductory remarks.)

Equation (2.6.1) is a particular case of a method which can be considered for all dynamical systems subject to forces that can be derived from a potential and are independent of time. Let the given dynamical system have N degrees of freedom, described by position coordinates $q^1, \ldots, q^N$, with corresponding momenta $p_1, \ldots, p_N$. On choosing the position coordinates so as to achieve separation of variables, in that $\int \sum_{j=1}^{N} p_j \mathrm{d}q^j$ is a sum of functions each of which depends on one variable only, the integral of each term $p_j \mathrm{d}q^j$ taken over a complete cycle of the variable q^j should be equal to an integer multiple of the Planck constant:

$$\oint p_j \mathrm{d}q^j = n_j h, \quad \forall j = 1, \ldots, N. \tag{2.6.3}$$

As already stated in a previous section these are the Bohr–Sommerfeld conditions. They can be stated in a form independent of the choice of variables, as was shown by Einstein in 1917. For this purpose; let us consider the Maupertuis action, for which $\mathrm{d}S = \sum_{j=1}^{N} p_j \mathrm{d}q^j$ (here there is some abuse of notation, since the right-hand side is not an exact differential on the full phase space, but becomes exact only on particular (dynamically invariant) sub-manifolds), and notice that such an expression is invariant under canonical transformations. For any closed curve C in the region R identified as the level set of a maximal family of pairwise commuting first integrals (the commutation property being valid with respect to Poisson brackets), the desired invariance principle can be stated in the form

$$\int_C \sum_{j=1}^N p_j \mathrm{d}q^j = nh. \tag{2.6.4}$$

- The level sets occurring here are either tori[2] or cylinders. For multiperiodic evolution they are tori. If the integration is performed in a system of coordinates where the separation of variables is obtained,

$$\sum_{j=1}^N K_j \oint p_j \mathrm{d}q^j = \sum_{j=1}^N K_j n_j h.$$

If the motions are bounded in space, the integral can be performed on any cycle of an invariant torus, and the quantum numbers correspond to the various independent cycles.

If we re-express Eq. (2.6.2) in the form $L = \frac{nh}{p}$, and bear in mind that, for a photon, $\frac{h}{p} = \lambda$, we obtain the following interpretation, known as the de Broglie hypothesis: *to every particle one can associate a wave. The relation between wavelength λ and momentum p is given, as in the case of photons, by*

$$\lambda = \frac{h}{p}. \tag{2.6.5}$$

The allowed orbits are those which contain an integer number of wavelengths.

The Bohr quantization condition is therefore studied under a completely different point of view. So far, it is not yet clear whether we have only introduced a new terminology, or whether a more substantial step has been made. In particular, we do not know what sort of wave we are dealing with, and we are unable to identify it with a meaningful physical concept. We can, however, consider some peculiar properties of wave-like phenomena. In particular, it is well known that waves interfere, and hence, if de Broglie waves exist, it should be possible to detect these wave-like aspects by means of interference experiments. Before thinking of an experiment that provides evidence of the wave-like behaviour of particles, it is appropriate to gain an idea of the wavelengths we are dealing with. For a free particle, the formulae (2.6.5) and $E = E_{\text{kin}} = \frac{p^2}{2m}$ yield

$$\lambda = \frac{h}{\sqrt{2mE}}, \tag{2.6.6}$$

which is the de Broglie wavelength. On expressing λ in nanometres and E in electronvolts, for an electron,

$$\lambda = \frac{1.24}{\sqrt{E}} \text{ nm}, \tag{2.6.7}$$

---

[2] A torus can be viewed as a closed surface defined as the product of two circles. It can also be seen as a surface of revolution generated by revolving a circle in three-dimensional space about an axis coplanar with the circle. If the axis of revolution does not touch the circle, the surface has a ring shape.

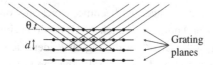

Fig. 2.10   A beam of monochromatic X-rays interacts with a grating.

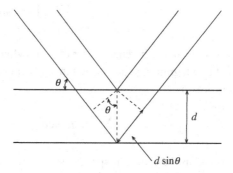

Fig. 2.11   Reflection of X-rays by grating planes.

and hence, for an electron with energy of 1 eV, $\lambda = 1.24$ nm. Note that, while for photons λ is inversely proportional to E, for massive particles λ is inversely proportional to the square root of E. Thus, for electrons with an energy of the order of 100 eV, λ is of the order of a few angstroms, i.e. the same order of magnitude as the wavelength of X-rays.

A way to provide evidence in favour of X-rays being (electromagnetic) waves consists in analysing their reflection from a crystal. This phenomenon is known as Bragg reflection. In a crystal, the regular distribution of atoms (or ions) determines some grating planes for which the mutual separation d is called the *grating step* (Figure 2.10). The grating step is usually of the order of an angstrom. If we let a beam of monochromatic X-rays fall upon a crystal, at an angle θ with respect to the surface of the crystal (see again Figure 2.10), and if we study the radiation emerging with the same inclination θ, we notice that the radiation is reflected only for particular values of θ, here denoted by $\theta_1, \theta_2, \ldots$. More precisely, we either find some sharp maxima in the intensity of the reflected radiation, corresponding to the angles $\theta_1, \theta_2, \ldots$, or some minima, the intensity of which is very small and, indeed, virtually zero. To account for this behaviour, one has to think of the X-rays as being reflected by the various grating planes (Figure 2.11). The waves reflected from two adjacent planes differ by a phase shift, since they follow optical paths which differ by an amount $2d \sin \theta$.

When $2d \sin \theta$ is an integer multiple of the wavelength λ constructive interference of the reflected waves occurs, i.e. a maximum of the intensity. If the crystal is slowly rotated, diffracted beams flash out momentarily every time the crystal satisfies the Bragg condition (see below). By contrast, if $2d \sin \theta$ is an odd multiple of $\frac{\lambda}{2}$, this leads to destructive interference of waves reflected from two adjacent planes, and hence no reflected radiation is observed. In the intermediate cases, the waves reflected from several grating planes interfere with each other, and this leads to an almost vanishing intensity. To sum up, the maxima in the intensity of the reflected radiation are obtained for θ such that

$$2d \sin \theta = n\lambda, \tag{2.6.8}$$

which is called the *Bragg condition*. The number of maxima observed is the maximum integer contained in $\frac{2d}{\lambda}$. It is hence clear why, to observe easily the phenomenon in the case of X-rays (for which $\lambda \cong 0.1$ nm), it is necessary to have d of the same order of magnitude as λ. Thus, crystals are preferred in this type of experiment. Moreover, if the angles θ occurring in the Bragg relation are known, λ can be determined if d is known, or vice versa.

The same experiment was performed by Davisson and Germer in 1927, but replacing X-rays by a beam of collimated electrons with the same energy (within the experimental limits). It was then found that the electrons were reflected only for particular values of θ, in agreement with the Bragg condition. This phenomenon provided strong experimental evidence in favour of the electrons having a wave-like behaviour. Moreover, the experiment of Davisson and Germer makes it possible to determine λ from the Bragg relation, so that Eq. (2.6.6) is verified. A description is given in the following section.

2.7 Experiment of Davisson and Germer

Shortly after the appearance of de Broglie's original papers on wave mechanics, Elsasser (1925) had predicted that evidence for the wave nature of particle mechanics would be found in the interaction between a beam of electrons and a single crystal. He believed that evidence of this sort was already provided by experimental curves found by Davisson and Kunsman, showing the angular distribution of electrons scattered by a target of polycrystalline platinum. However, this was not quite the case, because the maxima in the scattering curves for platinum are unrelated to crystal structure.

By virtue of the similarities between the scattering of *electrons* by the crystal and the scattering of *waves* by three- and two-dimensional gratings, a description of the occurrence and behaviour of the electron diffraction beams in terms of the scattering of an *equivalent wave radiation* by the atoms of the crystal, and its subsequent interference, was 'not only possible, but most simple and natural', as Davisson and Germer pointed out. A wavelength is then associated with the incident electron beam, which turns out to be in good agreement with the value $\frac{h}{mv}$ of wave mechanics.

To figure out how the electron wavelength is measured in the Davisson–Germer experiment (Figures 2.12, 2.13), let us consider, for simplicity, a one-dimensional model, where the incoming wave is diffracted from each atom (of the crystal), represented by a point on a line. Constructive or destructive interference may occur for the waves diffracted from the atoms. On denoting the separation between adjacent atoms by d, the angle formed by the directions of the incoming and reflected beams by θ and with n an integer ≥ 1, the condition for constructive interference is

$$d \sin \theta = n\lambda. \tag{2.7.1}$$

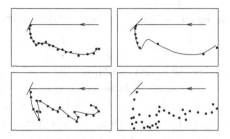

Fig. 2.12   The top two figures describe scattering of 75 V electrons from a block of nickel (many small crystals). The top two figures describe scattering of 75 V electrons from several large nickel crystals. Reprinted with permission from C. Davisson and L. H. Germer, Phys. Rev. 30, 705–40 (1927).

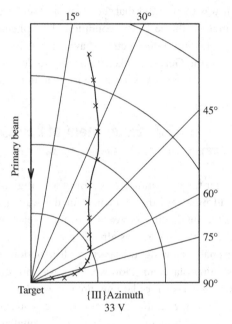

Fig. 2.13   Typical colatitude scattering curve for the single nickel crystal. Reprinted with permission from C. Davisson and L. H. Germer, Phys. Rev. 30, 705–40 (1927).

This relation says that the difference in path from two neighbouring atoms to the point of observation is an integer multiple of the wavelength. In a typical diffraction experiment, Davisson and Germer were dealing with $d = 2.15 \times 10^{-8}$ cm, a potential of 54 eV and a maximum was observed for $\theta = 50°$. Thus, when $n = 1$, these experimental data led to $\lambda = 0.165$ nm, which is in fairly good agreement with the theoretical value in the non-relativistic limit, i.e.

$$\lambda \cong \frac{h}{\sqrt{2m_e T}} = 0.167 \text{ nm}. \qquad (2.7.2)$$

Higher-order maxima, corresponding to greater values of the integer n, were also observed, and they were all in good agreement with the theoretical predictions. It is also clear, from

Eq. (2.7.2), why a beam of electrons was actually chosen: since they have a very small mass, the corresponding wavelength is expected to be sufficiently large.

The experiments we have described so far provided the phenomenological basis of the Einstein–de Broglie relation (1.1.2). We conclude this section with a historical remark, which relies on the Nobel Laureate speech delivered by Davisson in 1937. The experiment performed by him and Germer in 1925 was not, *at first*, a proof of the validity of wave mechanics. It was only in the summer of 1926 that the physics community came to appreciate the relevance of such an investigation for the wave-like picture of physical phenomena, after a number of discussions between Davisson, Richardson, Born, Franck and other distinguished scientists. This kind of experiment has been repeated with protons, neutrons, helium atoms and ions, and in all these cases the Bragg relation for material particles has been verified. The transition from the old paradigm to an accepted new one took the order of a quarter of a century.

2.8 Interference phenomena among material particles

The wave-like nature of light is proved by the interference phenomena it gives rise to. It is hence legitimate to ask the question: how can we accept the existence of interference phenomena, if we think of light as consisting of photons? There are, indeed, various devices that can produce interference fringes. For example, a source S of monochromatic light illuminates an opaque screen where two thin slits, close to each other, have been produced. In passing through the slits, light is diffracted. On a plate L located a distance from the slits, interference fringes are observed in the area where overlapping occurs of the diffraction patterns produced from the slits A and B, i.e. where light is simultaneously received from A and B (Figure 2.14).

Another device is the Fresnel biprism (Born and Wolf 1959): the monochromatic light emitted from S is incident on two coupled prisms P_1 and P_2; light rays are deviated from P_1 and P_2 as if they were emitted from two (virtual) coherent sources S' and S''. As in the

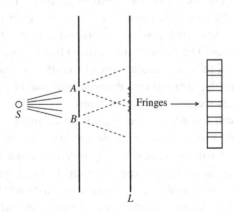

Fig. 2.14   Diffraction pattern from a double slit.

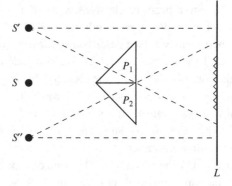

Fig. 2.15   Diffraction from a biprism.

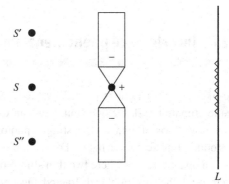

Fig. 2.16   Interference fringes with electrons, when they are deviated by an electric field.

previous device, interference fringes are observed where light emitted both from P_1 and P_2 is collected (Figure 2.15).

Interestingly, the Fresnel biprism makes it possible to produce interference fringes with electrons (Figure 2.16). The source S is replaced by an electron gun and the biprism is replaced by a metallic panel where a slit has been produced. At the centre of the slit, a wire of silver-plated quartz is maintained at a potential slightly greater than the potential of the screen. The electrons are deviated by the electric field of the slit, and they reach the screen as if they were coming from two different sources. For simplicity, we can consider the Fresnel biprism and talk about photons, but of course this discussion can be repeated in precisely the same way for electrons.

How can the interference experiment be interpreted in terms of photons? It is clear that bright fringes result from the arrival of several photons, whereas no photons arrive where dark fringes are observed. It therefore seems that the various photons interact with each other so as to give rise, on plate L, to an irregular distribution of photons, and hence bright as well as dark fringes are observed. If this is the case, what is going to happen if we reduce the intensity of the light emitted by S until only one photon at a time travels from the source S to the plate L? The answer is that we have then to increase the exposure time of the plate L, but eventually we will find the same interference fringes as discussed previously. Thus, the interpretation based upon the interaction among photons is incorrect: photons do not interfere with each other, but the only possible conclusion is that the interference involves the single photon, just as in the case of the superposition for polarization. However, according to a particle picture, a photon (or an electron) starting from S and arriving at L, *either* passes through A *or* passes through B. We shall say that, if it passes through A, it is in the state ψ_A (strictly speaking, the concept of state will be spelled out only in Chapter 4), whereas if it passes through B it is in the state ψ_B. But if this were the correct description of the possible options, we would be unable to account for the interference fringes. Indeed, if the photon is in the state ψ_A this means, according to what we said above, that slit B can be neglected (it is as if it had been closed down). Under such conditions, it should be possible for the photon to arrive at all points on plate L of the diffraction pattern produced from A, and hence also at those points where dark fringes occur. The same holds, with A replaced by B, if we say that the photon is in the state ψ_B. This means that a third option should be possible, inconceivable from the classical viewpoint, and different from ψ_A and ψ_B. We shall then say that photons are in a state ψ_C, different from both ψ_A and ψ_B, but ψ_C should 'be related', somehow, with both ψ_A and ψ_B. In other words, it is incorrect to say that photons pass through A or through B, but it is as if each of them were passing, at the same time, through both A and B. This conclusion is suggested by the wave-like interpretation of the interference phenomenon: if only slit A is open, there exists a wave $A(x, y, z, t)$ in between the screen and L, whereas, if only slit B is open, there exists a wave $B(x, y, z, t)$ in between the screen and L. If now both slits are opened up, the wave involved is neither $A(\vec{r}, t)$ nor $B(\vec{r}, t)$, but $C(\vec{r}, t) = A(\vec{r}, t) + B(\vec{r}, t)$.

However, if the photon is passing 'partly through A and partly through B', what should we expect if we place two photomultipliers F_1 and F_2 in front of A and B, respectively (Figure 2.17), and a photon is emitted from S? Should we expect that F_1 and F_2 register, at the same time, the passage of the photon? If this were the case, we would have achieved, in the laboratory, the 'division' of a photon! What happens, however, is that only one of the two photomultipliers registers the passage of the photon, and upon repeating the experiment several times we find that, *on average*, half of the events can be ascribed to F_1 and half of the events can be ascribed to F_2. Does this mean that the existence of the state ψ_C is an incorrect assumption? Note, however, that the presence of photomultipliers has made it impossible to observe the interference fringes, since the photons are completely absorbed by such devices.

At this stage, we might think that, with the help of a more sophisticated experiment, we could still detect which path has been followed by photons, while maintaining the ability to observe interference fringes. For this purpose, we might think of placing a mirror S_1 behind slit A, and another mirror S_2 behind slit B (Figure 2.18). Such mirrors can be freely moved by hypothesis, so that, by observing their recoil, we could (in principle) understand whether the photon passed through A or, instead, through slit B. Still, once again, the result of the experiment is negative: if we manage to observe the recoil of a mirror, no interference fringes are detected. The wave-like interpretation of the failure is as follows: the recoil of the mirror affects the optical path of one of the rays, to the extent that interference fringes are destroyed. In summary, we can make some key statements.

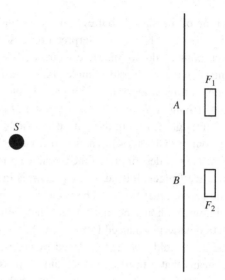

Fig. 2.17   Double-slit experiment with a photon.

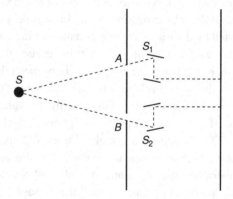

Fig. 2.18   Double-slit experiment supplemented by mirrors.

(i) Interference fringes are also observed by sending only one photon at a time. Thus, the single photon is found *to interfere with itself*.

(ii) It is incorrect to say that the single photon passes through slit A or through slit B. Instead, there is a third option, represented by a state ψ_C, and deeply intertwined with both ψ_A and ψ_B.

(iii) A measurement which shows whether the photon passed through A or through B perturbs the state of the photon to such an extent that no interference fringes are detected. Thus, *either we know which slit the photon passed through, or we observe interference fringes*. We cannot achieve both goals: the two possibilities are incompatible.

Similar conclusions hold for electrons. In particular, an experimental setup aimed at detecting which path the electrons follow (Figures 2.19, 2.20) on their way from the source to the screen uses a conducting plate and the mirror image of the charge to find out which

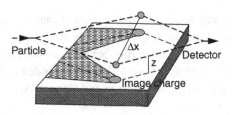

Fig. 2.19 The electron beam is split into two parts. Both beams travel over a plate made of a highly resistive material at the same, small height. The electrons induce charge inside the plate. Along with the electron beam, induced charges move through the plate, giving rise to currents inside the plate. Such currents encounter ohmic resistance, and hence dissipation occurs. The state of the electron gas inside the plate is disturbed, being different for the 'left' and 'right' path of the electron, since heating occurs at different locations. This provides a *which-path information*, a peculiar property of particle-like behaviour. The electron beam and the plate are then said to be entangled. The interference obtained is irreversible, since a record of the electron's path remains. When the beams are eventually recombined, the interference contrast is reduced.

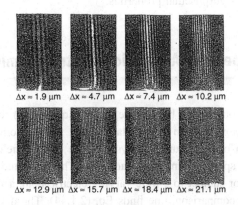

Fig. 2.20 Decoherence as a function of height z of the beams over the plate for increasing lateral separations Δx of the beams. Reprinted with permission from F. Hasselbach and P. Sonnentag, Phys. Rev. Lett. 98, 200402 (2007).

region will be heated by the passage of the electron. In the setup described in Figure 2.19, for $z \approx 0$, the interference pattern disappears.

To sum up, the main findings of the experiments we are interested in are as follows.

(i) Electrons arrive at the screen in identical 'lumps', i.e. there is no detection of a 'half-electron'; the whole electron is detected, with the full electronic charge and electronic mass.

(ii) There is a wave associated with the electrons to describe the interference pattern.

(iii) There is a probability for the electron to impinge on the screen within a pre-assigned region.

(iv) Under some appropriate experimental conditions we recover a classical-like behaviour of the electrons, i.e. the interference pattern disappears. More specifically, the more visible the interference pattern is, the less distinguishable the paths are. This behaviour is usually referred to as the *complementarity principle* formulated by Bohr in 1928. In particular, this requires that our formulation of the new mechanics should contain a way to recover the classical behaviour under appropriate limiting conditions. This requirement is often referred to as the *correspondence principle*. We should stress that electrons have wave properties but they are not classical waves: the electron wave cannot be split like a classical wave. History unfolded differently, the wave nature of electrons was investigated in the Davisson and Germer experiment that was tailored after the Bragg experiment.

What we have described is quite counter-intuitive for particles, but it is even more surprising for light. We have to accept that it is not so simple to elaborate a model capable of accounting for these results within the classical framework. Indeed, light provides the most relevant example of the dual nature of quantum objects: its oscillatory properties were suitable to verify the electromagnetic theory of Maxwell, while its 'lumpiness' (the photons) signalled the birth of modern quantum theory. The dualism between the two pictures may appear to be in contradiction, however, it constitutes the first example of what is known as complementarity in quantum theory, i.e. the possibility displaying both wave and corpuscular properties.

Appendix 2.A   Classical electrodynamics and the Planck formula

In Section 2.1, we have made use of Eq. (2.1.44) for the work δW done per second by a radiation field on an oscillator. Now we are going to prove it. For this purpose, we first perform a Fourier analysis of the electric field, from which we derive a relation for the spectral density of radiation. Then, by studying the equation of motion of the oscillator subject to a time-varying electric field, a formula can be derived for the work δW. In comparison, one finds Eq. (2.1.44). The details are as follows. Usually, Fourier transformations can be carried out only for absolutely integrable functions. In the case of a stationary situation (such as black-body radiation) the 'signal' $E_x(t)$ is not absolutely integrable since it is integrated over an infinite time. To deal with such cases, Wiener (1930) introduced generalized harmonic analysis. The main idea is to evaluate the autocorrelation function

$$C(t - t') \equiv \langle E(t)E(t') \rangle$$

and take its Fourier transform. This spectrum is called the *power spectrum* of the stationary process.

The radiation field is defined by specifying how the electric field depends on time. To ensure the convergence of the integrals that we are going to use, we assume that the radiation field is a function with compact support, and hence is non-vanishing only in the closed time interval $[0, T]$. The limit $T \to \infty$ may be taken eventually. The Fourier analysis of the x-component of the electric field is

$$E_x(t) = \int_{-\infty}^{\infty} f(v)e^{2\pi i v t} \, dv, \qquad (2.A.1)$$

where the amplitudes $f(v)$ are determined by (the field having compact support)

$$f(v) \equiv \int_0^T E_x(t)e^{-2\pi i vt}\, dt. \tag{2.A.2}$$

By virtue of the reality of E_x, $f^*(v) = f(-v)$, and an analogous analysis may be performed for the y- and z-components. Thus, the total energy of the radiation field is given by

$$u = \frac{1}{8\pi}\overline{\left(E^2 + H^2\right)} = \frac{1}{4\pi}\overline{E^2} = \frac{3}{4\pi}\overline{E_x^2}, \tag{2.A.3}$$

because the time averages (see the comments after Eq. (2.A.7)) have the property

$$\overline{E_x^2} = \overline{E_y^2} = \overline{E_z^2}.$$

Now we can compute these time averages, starting from the identity

$$\overline{E_x^2} = \frac{1}{T}\int_0^T E_x^2\, dt = \frac{1}{T}\int_0^T E_x dt \int_{-\infty}^\infty f(v)e^{2\pi i vt}\, dv \tag{2.A.4}$$

and changing the order of integration. This yields

$$\overline{E_x^2} = \frac{1}{T}\int_{-\infty}^\infty f(v)\, dv \int_0^T E_x e^{2\pi i vt}\, dt = \frac{1}{T}\int_{-\infty}^\infty f(v)f^*(v)\, dv$$

$$= \frac{2}{T}\int_0^\infty |f(v)|^2\, dv. \tag{2.A.5}$$

Equations (2.A.3) and (2.A.5) imply that the total density of radiation reads as

$$u = \int_0^\infty u_v\, dv = \frac{3}{2\pi T}\int_0^\infty |f(v)|^2\, dv, \tag{2.A.6}$$

so that the spectral density is

$$u_v = \frac{3}{2\pi T}|f(v)|^2. \tag{2.A.7}$$

So far, we have presented the simplified argument given by Born in his book on atomic physics. However, the averaging process is a crucial point, and a number of comments are in order before we can continue our investigation. The calculation of the averages $\langle E^k(t)\rangle$, $\langle H^k(t)\rangle$, where the integer $k = 1, 2, \ldots$, makes it necessary to use the theory of electromagnetic signals. The physical problem consists of several atoms, which emit electromagnetic radiation over the whole range of frequencies $v \in \;]-\infty, \infty[$. The results that we need are as follows.

(i) Denoting by N the average number of events per unit time, and by $[0, T]$ the time interval during which the observations are performed,

$$\langle E(t)\rangle = N\int_0^T E(\tau)\, d\tau = NT\int_0^T \frac{E(\tau)}{T}\, d\tau = 0. \tag{2.A.8}$$

The mean value of the electric field is then said to be equal to the product of the average number of events per unit time, N, with the time integral of E, or, equivalently, to the product of the average number of events, NT, with the temporal average of E. This mean value vanishes, because E is a rapidly varying function, represented, hereafter, by the infinite sum

$$\sum_{i=-\infty}^{\infty} \varepsilon(t - t_i),$$

where each ε is non-vanishing only in a finite interval.

(ii) In general, in the interval $[0, T]$ a number M of distinct signals are detected. Thus, for each component of the electric field, the mean value of $E^2(t)$ turns out to be (the index for the component is omitted, for simplicity of notation)

$$\langle E^2(t) \rangle = \sum_{i=1}^{M} N_i \int_{-\infty}^{\infty} \varepsilon^2(\tau) \, d\tau = \sum_{i=1}^{M} N_i \int_{-\infty}^{\infty} |\widetilde{\varepsilon}(\nu)|^2 \, d\nu, \tag{2.A.9}$$

where the Parseval formula has been used to obtain the second equality. The integrals in (2.A.9) do not exist for a steady beam, but following Norbert Wiener the *power spectrum* can be obtained from the autocorrelation function which is square-integrable. Now although the atoms emit at different frequencies, for all of them, $\varepsilon(\nu)$ may be approximated by a curve having the shape of a bell. More precisely, the behaviour of $\varepsilon(\nu)$ is well approximated by a curve which changes rapidly in the neighbourhood of some value ν_i, where it attains its maximum: $\varepsilon(\nu) = \varepsilon(|\nu - \nu_i|)$. Bearing in mind that the atoms may emit over the whole range of frequencies, so that, in (2.A.9), the sum

$$\sum_{i=1}^{M} N_i$$

should be replaced by the integral $\int_{-\infty}^{\infty} N(\nu') \, d\nu'$, we find that

$$\langle E^2(t) \rangle = \int_{-\infty}^{\infty} \overline{N}(\nu) e_0(\nu) \, d\nu = \int_{-\infty}^{\infty} e(\nu) \, d\nu = 2 \int_0^{\infty} e(\nu) \, d\nu, \tag{2.A.10}$$

where $\overline{N}(\nu)$ is the average number of events per unit time:

$$\overline{N}(\nu) \equiv \frac{\int_{-\infty}^{\infty} N(\nu') |\widetilde{\varepsilon}(|\nu - \nu'|)|^2 \, d\nu'}{\int_{-\infty}^{\infty} |\widetilde{\varepsilon}(|\nu - \nu'|)|^2 \, d\nu'}, \tag{2.A.11}$$

and $e_0(\nu)$ is the total energy of the signal corresponding to the frequency ν:

$$e_0(\nu) \equiv \int_{-\infty}^{\infty} |\widetilde{\varepsilon}(|\nu - \nu'|)|^2 d\nu'. \tag{2.A.12}$$

It is thus crucial to appreciate, hereafter, that all mean values we refer to should be obtained, strictly speaking, by combining the operation of a statistical average with the temporal average of the function under investigation. It is only upon considering the joint effect of these two operations that mean values are obtained that are independent of the instant of time when the measurements begin.

To complete our analysis, we have now to consider the vibrations of the linear harmonic oscillator of mass m and charge q under the action of the electric field. We can assume, for simplicity, that the oscillator vibrates only in the x-direction (Born 1969):

$$m\ddot{x} + ax = qE_x(t), \tag{2.A.13}$$

where $a \equiv 4\pi^2 \nu_0^2 m$. The general solution of Eq. (2.A.13) consists of the general solution of the homogeneous equation plus a particular solution of the full equation. The former reads as

$$x_1(t) = x_0 \sin(2\pi \nu_0 t + \phi), \tag{2.A.14}$$

with x_0 and ϕ being arbitrary constants, while the latter is

$$x_2(t) = \frac{q}{2\pi \nu_0 m} \int_0^t E_x(\tau) \sin[2\pi \nu_0(t - \tau)]d\tau, \tag{2.A.15}$$

if the initial conditions are $x(0) = \dot{x}(0) = 0$. Of course, to derive Eq. (2.A.15) the method of *variation of parameters* has to be applied. This requires that looking for the solution of Eq. (2.A.13) in the form (here $\omega \equiv 2\pi \nu$)

$$x_2(t) = A_1(t) \cos(\omega t) + A_2(t) \sin(\omega t), \tag{2.A.16}$$

where A_1 and A_2 are two functions of the time variable subject to the conditions

$$\dot{A}_1 \cos(\omega t) + \dot{A}_2 \sin(\omega t) = 0, \tag{2.A.17}$$

$$-\omega \dot{A}_1 \sin(\omega t) + \omega \dot{A}_2 \cos(\omega t) = \frac{q}{m} E_x, \tag{2.A.18}$$

the solution of which is, by elementary integration,

$$A_1(t) = -\frac{q}{m\omega} \int_0^t E_x(\tau) \sin(\omega \tau) \, d\tau + A_1^0, \tag{2.A.19}$$

$$A_2(t) = \frac{q}{m\omega} \int_0^t E_x(\tau) \cos(\omega \tau) \, d\tau + A_2^0. \tag{2.A.20}$$

The initial conditions $x_2(0) = 0, \dot{x}_2(0) = 0$, imply that both A_1^0 and A_2^0 should vanish, and hence Eqs. (2.A.16), (2.A.19) and (2.A.20) lead to the result in the form (2.A.15).

We are now in a position to evaluate the work done by the field on the oscillator. More precisely, we are interested in the work per unit time (here $x(t) = x_1(t) + x_2(t)$)

$$\delta W = \frac{q}{T} \int_0^T \dot{x}(t) E_x(t) \, dt. \tag{2.A.21}$$

By construction, the contribution of x_1 to the integral (2.A.21) vanishes, and

$$\delta W = \frac{q}{T} \frac{q}{m} \int_0^T E_x(t) \, dt \int_0^t E_x(\tau) \cos[2\pi \nu_0(t - \tau)] \, d\tau. \tag{2.A.22}$$

The integrand in (2.A.22) is a symmetric function of t and τ, and hence,

$$\delta W = \frac{q^2}{mT} \int_0^T E_x(\tau) \, d\tau \int_\tau^T E_x(t) \cos[2\pi \nu_0(t - \tau)] \, dt, \tag{2.A.23}$$

which is re-expressed, after interchanging t and τ, as

$$\delta W = \frac{q^2}{mT} \int_0^T E_x(t) \, dt \int_t^T E_x(\tau) \cos[2\pi \nu_0(t - \tau)] \, d\tau. \tag{2.A.24}$$

The comparison of Eqs. (2.A.22) and (2.A.24) leads to the formula

$$\delta W = \frac{q^2}{2mT} \int_0^T E_x(t) \, dt \left(\int_0^t + \int_t^T \right) E_x(\tau) \cos[2\pi \nu_0(t-\tau)] \, d\tau. \tag{2.A.25}$$

Note that the cosine function is conveniently re-expressed in the form

$$\frac{1}{2} \left[e^{2\pi i \nu_0(t-\tau)} + e^{-2\pi i \nu_0(t-\tau)} \right],$$

and hence the work per unit time turns out to be

$$\begin{aligned}
\delta W &= \frac{q^2}{4mT} \int_0^T E_x(t) e^{2\pi i \nu_0 t} \, dt \int_0^T E_x(\tau) e^{-2\pi i \nu_0 \tau} \, d\tau \\
&\quad + \frac{q^2}{4mT} \int_0^T E_x(t) e^{-2\pi i \nu_0 t} \, dt \int_0^T E_x(\tau) e^{2\pi i \nu_0 \tau} \, d\tau \\
&= \frac{q^2}{4mT} \left[f^*(\nu_0) f(\nu_0) + f(\nu_0) f^*(\nu_0) \right] \\
&= \frac{q^2}{2mT} |f(\nu_0)|^2.
\end{aligned} \tag{2.A.26}$$

From Eqs. (2.A.7) and (2.A.26) the desired result is obtained:

$$\delta W = \frac{q^2}{2mT} \frac{2\pi T}{3} u_\nu = \frac{\pi q^2}{3m} u_\nu. \tag{2.A.27}$$

Waves and particles

The various phenomenological aspects we have considered, for instance interference phenomena along with the 'lumpiness' registered at the detectors, require that we have to elaborate a theory that can account for wave and corpuscular behaviour.

Among theories we are already familiar with, wave optics seems to be able to deal with interference on one hand and rays (beams) on the other hand. We may therefore look for inspiration in such a theory to envisage a description of the variety of phenomena we have seen in the previous chapter. Having this is mind, it is appropriate to consider the relevant theoretical aspects entering the description of light propagation.

3.1 Waves: d'Alembert equation

The propagation of electromagnetic fields is described by Maxwell's equations, which we already encountered in the course of studying the electromagnetic field in a hollow cavity in Chapter 2. The fields occurring in this description are $\vec{E}$ and $\vec{B}$, the electric field and magnetic induction, respectively, which, under appropriate regularity conditions, obey the equations (here, unlike all other chapters, we denote with words the differential operators curl, divergence and gradient, since an intrinsic notation is preferable to go beyond vector calculus[1] on $\mathbf{R}^3$)

$$\mathrm{curl}\vec{E} + \frac{1}{c}\dot{\vec{B}} = 0, \ \mathrm{div}\vec{B} = 0, \tag{3.1.1}$$

where the dot denotes differentiation with respect to time. Material fields are $\vec{D}$ and $\vec{H}$, electric displacement and magnetic vector respectively, along with the electric current

---

[1] Let us consider $\mathbf{R}^3$ endowed with the Euclidean metric

$$g_E = \delta_{ij}\mathrm{d}x^i \otimes \mathrm{d}x^j,$$

and let Cartesian coordinates be $x^1 = x, x^2 = y, x^3 = z$. using g_E, the scalar product of vectors is defined as

$$\vec{A} \cdot \vec{B} \equiv A_i B^i = \delta_{ij} A^i B^j = g_E(\vec{A}, \vec{B}).$$

density $\vec{j}$ and charge density ρ. They all satisfy the following differential equations when appropriate regularity conditions are satisfied:

$$\text{curl}\vec{H} - \frac{1}{c}\dot{\vec{D}} = \frac{4\pi}{c}\vec{j}, \ \text{div}\vec{D} = 4\pi\rho. \tag{3.1.2}$$

It follows that $\text{div}\vec{j} = -\frac{1}{4\pi}\text{div}\dot{\vec{D}}$, or

$$\frac{\partial\rho}{\partial t} + \text{div}\vec{j} = 0, \tag{3.1.3}$$

called the continuity equation.

To take into account the properties of the sub-systems we are dealing with, along with these fields, we have to supplement these equations by the so-called constitutive equations, i.e. equations relating *vacuum fields* and *material particles*. These constitutive equations have a phenomenological nature, and in general may be non-local and non-linear. In limiting cases they acquire the simple form of linear relations

$$\vec{B} = \mu\vec{H}, \ \vec{D} = \varepsilon\vec{E}, \ \vec{j} = \sigma\vec{E}, \tag{3.1.4}$$

where σ is the specific conductivity, and the equation involving it represents the differential form of Ohm's law. The magnetic permeability is denoted by μ and the electric permittivity is denoted by ε.

The differential of a smooth function f reads as

$$df = \frac{\partial f}{\partial x^k}dx^k,$$

and the vector operator $\vec{\nabla}$ such that

$$(\vec{\nabla})_k = \frac{\partial}{\partial x^k}$$

is the gradient operator used in our book. The gradient can be used in two ways to act upon vectors. First, we can define the divergence of a vector $\vec{A}$, i.e.

$$\vec{\nabla} \cdot \vec{A} \equiv \frac{\partial}{\partial x^k}A^k = \delta_i^j\frac{\partial}{\partial x^j}A^i = g_E(\vec{\nabla}, \vec{A}),$$

which means that

$$\text{div} \equiv g_E(\vec{\nabla}, \cdot).$$

Second, bearing in mind the fully anti-symmetric Levi–Civita symbol used in the volume 3-form $V_3 \equiv \varepsilon_{ijk}dx^i \wedge dx^j \wedge dx^k$, we can instead take the 'vector product of $\vec{\nabla}$ with a vector', which means building the action of the curl operator on $\vec{A}$ according to

$$(\vec{\nabla} \wedge \vec{A})^i \equiv \varepsilon^{ij}{}_k\frac{\partial}{\partial x^j}A^k \implies \text{curl} \equiv \vec{\nabla} \wedge \cdot .$$

Key ingredients of vector calculus on $\mathbf{R}^3$ are therefore the Euclidean metric g_E, the gradient $\vec{\nabla}$ and the volume 3-form V_3. One then defines the divergence and curl according to the previous prescriptions, and eventually the Laplacian $P \equiv -\triangle$, where

$$\triangle \equiv \partial^i\partial_i = \delta^{ij}\partial_j\partial_i = g_E(\vec{\nabla}, \vec{\nabla}) = \text{div} \ \vec{\nabla}.$$

The gradient, divergence and Laplacian can be defined for arbitrary manifolds of whatever finite dimension, but more advanced geometric tools are then necessary.

Electromagnetic theory considers the light intensity as the energy flux of the field. By introducing the electric energy density W_e, and the magnetic energy density W_m, in the simplified situation we are considering we have

$$W_e = \frac{1}{8\pi}\vec{E} \cdot \vec{D}, \; W_m = \frac{1}{8\pi}\vec{H} \cdot \vec{B}, \tag{3.1.5}$$

and the balance equation is

$$\frac{d}{dt}\int \left(W_e + W_m\right)dV + \int \vec{j} \cdot \vec{E}dV + \frac{1}{4\pi}\int \left(\vec{E} \wedge \vec{H}\right) \cdot \vec{n}\, dS = 0. \tag{3.1.6}$$

By introducing the Poynting vector $\vec{S} = \frac{1}{4\pi}\vec{E} \wedge \vec{H}$ (the density of energy flux) and by assuming that there are no charges and that we are dealing with a non-conducting medium, we find

$$\frac{\partial}{\partial t}\left(W_e + W_m\right) + \mathrm{div}\vec{S} = 0. \tag{3.1.7}$$

We recall that the field $(\vec{E}, \vec{B})$ acts on particles of electric charge q, moving with velocity $\vec{v}$, with a force (Lorentz force)

$$\vec{F} = q\left(\vec{E} + \frac{1}{c}\vec{v} \wedge \vec{B}\right). \tag{3.1.8}$$

Thus, by assuming that we have no charges, we find that there is no mechanical work done by the electromagnetic field. If we confine our attention to the particular situation of absence of charges and currents ($\vec{j} = 0, \; \rho = 0$), we derive the following equations for the electric field $\vec{E}$ and the magnetic vector $\vec{H}$ in the presence of a homogeneous medium:

$$\left(\Delta - \frac{\varepsilon\mu}{c^2}\frac{\partial^2}{\partial t^2}\right)\vec{E} = 0, \; \left(\Delta - \frac{\varepsilon\mu}{c^2}\frac{\partial^2}{\partial t^2}\right)\vec{H} = 0. \tag{3.1.9}$$

By introducing $v = \frac{c}{\sqrt{\varepsilon\mu}} = \frac{c}{n}$ (n being the absolute refractive index), we find the equation for each Cartesian component of the field in the form of a scalar wave equation, the d'Alembert equation

$$\left(\Delta - \frac{1}{v^2}\frac{\partial^2}{\partial t^2}\right)u = 0. \tag{3.1.10}$$

The operator $\Delta - \frac{1}{v^2}\frac{\partial^2}{\partial t^2} \equiv \Box$ is called the d'Alembert operator.

To analyze simple solutions of this equation we consider a Cartesian coordinate system (x_0, x, y, z) and write the previous equation in the form

$$\Box u = \left(\frac{\partial^2}{\partial x^2} + \frac{\partial^2}{\partial y^2} + \frac{\partial^2}{\partial z^2} - n^2\frac{\partial^2}{\partial x_0^2}\right)u = 0. \tag{3.1.11}$$

The simplest solutions of this equation have the form

$$u = f(\vec{k} \cdot \vec{r} - \omega t) + g(\vec{k} \cdot \vec{r} + \omega t), \tag{3.1.12}$$

with the condition $\vec{k} \cdot \vec{k} - \frac{\omega^2}{v^2} = 0$ usually called the dispersion relation. At any instant of time, $\vec{k} \cdot \vec{r} = $ constant represents a plane surface and u is constant on it, i.e. u represents a plane wave. It is possible to find a different family of solutions by considering the d'Alembert equation in radial coordinates:

$$\Box u = \frac{1}{r^2}\frac{\partial}{\partial r}\left(r^2\frac{\partial u}{\partial r}\right) + \frac{1}{r^2\sin\theta}\frac{\partial}{\partial\theta}\left(\sin\theta\frac{\partial u}{\partial\theta}\right) + \frac{1}{r^2\sin^2\theta}\frac{\partial^2 u}{\partial\varphi^2}$$
$$- \frac{1}{v^2}\frac{\partial^2 u}{\partial t^2}. \tag{3.1.13}$$

From the function ru,

$$\frac{1}{r^2}\frac{\partial}{\partial r}\left(r^2\frac{\partial}{\partial r}ru\right) = \frac{1}{r}\frac{\partial^2}{\partial r^2}(ru)$$

and

$$\frac{1}{v^2}\frac{\partial^2}{\partial t^2}(ru) = \frac{\partial^2}{\partial r^2}(ru).$$

Solutions representing spherical waves are easily written in the form

$$u(r,t) = \frac{f(r+vt)}{r} + \frac{g(r-vt)}{r}, \tag{3.1.14}$$

with $r = \sqrt{x^2 + y^2 + z^2}$. The first term represents a spherical wave converging towards the origin, the second term represents a spherical wave diverging from the origin. We may further analyze plane waves and introduce familiar concepts occurring in the description of light. At a fixed point, say $P \equiv (\vec{r}_0)$, the wave disturbance is a function of time only, i.e. $u(\vec{r}_0, t) = F(t)$. When $\vec{u}$ is periodic in time we may consider $F(t) = A\cos(\omega t + \varphi_0)$. We recall that $A > 0$ is the amplitude, and $\omega t + \varphi_0$ is the phase. The quantity $v = \frac{\omega}{2\pi} = \frac{1}{T}$ is the frequency and represents the number of vibrations per second, ω is the angular frequency, T is the period. These particular solutions of the wave equations are said to be harmonic in time. A monochromatic plane wave has the special form

$$u(\vec{r}, t) = A\cos\left[(\omega t - \vec{k}\cdot\vec{r}) + \varphi_0\right]. \tag{3.1.15}$$

Remark A monochromatic plane wave should be considered as an abstraction, since it fills the whole of space and lasts throughout all time, carrying an infinite energy. Realistic waves always occupy a limited region of space at each particular instant of time, and at any particular point they have a beginning and an end. A realistic wave is called a wave packet. Such a packet is not merely a single plane wave, but rather an assembly of such waves. It may be represented as a discrete sum of plane waves or as a continuous superposition of plane waves, the coefficients being appropriate weights. Each individual monochromatic plane wave satisfies the wave equation, and, on account of the linearity of this equation, the superposition is also a solution.

On reverting now to the monochromatic plane wave we find that, when $\vec{r}\cdot\vec{k}$ is increased by the amount $\lambda = v\frac{2\pi}{\omega} = vT$, the function is unchanged. The length λ is called the wavelength; $\vec{k}$ is usually associated with the wave vector and is normal to the wave front, it is the direction of propagation. Thus, the wavelength is associated with the minimum increase of $\vec{r}\cdot\vec{k}$ which does not change the phase. A general time-harmonic, monochromatic wave, has the form

$$u(\vec{r}, t) = A(\vec{r})\cos(\omega t - k_0 S(\vec{r})). \tag{3.1.16}$$

The surfaces $S(\vec{r}) = $ constant replace now our plane surfaces or spherical surfaces and are called wave surfaces or geometrical wave fronts. Note, however, that now the surfaces of constant amplitude of the wave do not coincide with the surfaces of constant phase. These waves are usually called inhomogeneous waves.

Instead of cosine or sine functions, it is convenient to use complex-valued functions, solutions of the wave equations, say

$$u(\vec{r}, t) = A(\vec{r})e^{i(\omega t - k_0 S(\vec{r}))}. \tag{3.1.17}$$

Unlike a plane harmonic wave, the more general form of the wave function is not periodic with respect to space. The characteristic theorem of geometrical optics says that every sufficiently small portion of a wave surface can be regarded as a plane wave which propagates itself independently in the direction of its normal, with the phase velocity it has at the point in question. Thus, the phase $\omega t - k_0 S(\vec{r})$ suggests we are dealing with the infinitesimal form, i.e. we introduce the 1-form

$$\omega dt - k_0 \mathrm{grad} S \cdot d\vec{r},$$

called the phase 1-form, and we may set $\omega dt - k_0 \mathrm{grad} S \cdot d\vec{r} = 0$ to express the fact that the phase $\omega t - k_0 S(\vec{r})$ is the same at $(\vec{r}, t)$ and $(\vec{r} + d\vec{r}, t + dt)$. By using a unit vector $\widehat{P}$ in the direction of $d\vec{r}$, say $d\vec{r} = \widehat{P} df$, we find

$$\frac{df}{dt} = \frac{\omega}{k_0(\mathrm{grad} S \cdot \widehat{P})}, \tag{3.1.18}$$

obtained from $\omega dt - k_0 \widehat{P} \cdot \mathrm{grad} S \, df = 0$. When

$$\widehat{P} = \frac{\mathrm{grad} S}{|\mathrm{grad} S|}, \tag{3.1.19}$$

then

$$\frac{df}{dt} = \frac{\omega \, | \, \mathrm{grad} S \, |}{k_0(\mathrm{grad} S \cdot \mathrm{grad} S)}. \tag{3.1.20}$$

This is called the *phase velocity* and represents the speed at which each of the wave-front surfaces proceeds. By using the Einstein–de Broglie identification

$$(E - \hbar\omega)dt - (p_i - \hbar k_i)dr^i = 0, \tag{3.1.21}$$

it is clear that the phase velocity does not behave as a vector. It is, rather,

$$\frac{dt}{d\vec{r}} = \frac{k_0 \mathrm{grad} S}{\omega}, \tag{3.1.22}$$

which has a vector behaviour. The quantity (3.1.22) is sometimes called the phase-vector slowness. We notice that the phase velocity in certain cases may be greater than the speed of light c. For plane waves this will be so when $n = \sqrt{\varepsilon\mu}$ is smaller than 1, as in the case of dispersive media in regions of anomalous dispersion. Our experience from phase space or the optical phase space suggests that instead of the 1-form $\omega \, dt - \vec{k} \cdot d\vec{r}$ we may consider $t \, d\omega - \vec{r} \cdot d\vec{k}$; they give rise to the same symplectic form. The transformation

$$(t, \omega; \vec{r}, \vec{k}) \rightarrow (-\omega, t; -\vec{k}, \vec{r})$$

is a canonical transformation and is provided by the space–time Fourier transform which exchanges the role of independent variables. By using the conjugate 1-form and setting $t\,\mathrm{d}\omega - \vec{r}\cdot\mathrm{d}\vec{k} = 0$, we find

$$\vec{r} = t\frac{\mathrm{d}\omega}{\mathrm{d}\vec{k}}, \qquad (3.1.23)$$

which represents a motion with velocity $\frac{\mathrm{d}\omega}{\mathrm{d}\vec{k}}$. This vector is much more in line with the phase-vector slowness and represents the *group velocity* of the wave packet. It may be considered as the velocity of the propagation of a group of waves as a whole; it also represents the velocity at which energy is propagated. (One has to be aware, however, of regions of anomalous dispersion where the group velocity may exceed the speed of light in vacuum or become negative.)

A highly simplified model, where some properties of the medium where propagation takes place may be given, is ruled by the equation

$$\left(\frac{\partial^2}{\partial t^2} - c_x^2\frac{\partial^2}{\partial x^2} - c_y^2\frac{\partial^2}{\partial y^2} - c_z^2\frac{\partial^2}{\partial z^2}\right)u = 0. \qquad (3.1.24)$$

This medium could be a triassic crystal having a different speed of light in the three different directions.

To deal with the transition from wave optics to ray optics we consider the wave equation (3.1.10) and we write a generic u in a polar form

$$u = A\,\mathrm{e}^{\mathrm{i}F}, \qquad (3.1.25)$$

and consider the splitting of the partial differential equation into real and imaginary parts. For this purpose we need a closer look at the d'Alembert operator. Indeed, when written in terms of second derivatives in Cartesian coordinates, it reads as $\eta^{\mu\nu}\frac{\partial}{\partial x^\mu}\frac{\partial}{\partial x^\nu}$, where $\eta^{\mu\nu}$ are the contravariant components of the Minkowski metric. This is formally analogous to the Laplacian, with the understanding that $\eta^{\mu\nu}$ should be replaced by the Euclidean metric $\mathrm{diag}(1, 1, 1, 1)$ on $\mathbf{R}^4$. However, the geometric definition of the Laplacian P involves the div and grad operators, i.e. $P = -\mathrm{div}\,\mathrm{grad}$ (that the undergraduate student normally encounters only for vector calculus on $\mathbf{R}^3$). Thus, to be able also to express the geometric form of the d'Alembert operator in analogous form, we are led to introduce the following notation for the d'Alembert operator acting on functions:

$$\Box\,u = \widetilde{\mathrm{div}}\,\widetilde{\mathrm{grad}}u, \qquad (3.1.26)$$

where (the operator below being defined using the Minkowski metric)

$$\widetilde{\mathrm{grad}}u = \frac{\partial u}{\partial x}\frac{\partial}{\partial x} + \frac{\partial u}{\partial y}\frac{\partial}{\partial y} + \frac{\partial u}{\partial z}\frac{\partial}{\partial z} - \frac{1}{v^2}\frac{\partial u}{\partial t}\frac{\partial}{\partial t}, \qquad (3.1.27)$$

which maps functions into vector-valued differential operators and satisfies the Leibniz rule, while

$$\widetilde{\mathrm{div}}\left(A_\mu\frac{\partial}{\partial x_\mu}\right) = \sum_\mu \frac{\partial A_\mu}{\partial x_\mu}, \qquad (3.1.28)$$

which maps vector-valued homogeneous differential operators into functions and does not satisfy the Leibniz rule. Let us therefore consider, using (3.1.25) and (3.1.26),

$$\widetilde{\text{grad}}(A\,e^{iF}) = \left(\widetilde{\text{grad}}A\right)e^{iF} + i\,A\widetilde{\text{grad}}F\,e^{iF}, \tag{3.1.29}$$

$$\widetilde{\text{div}}\,\widetilde{\text{grad}}(A\,e^{iF}) = \widetilde{\text{div}}\,\widetilde{\text{grad}}A\,e^{iF} + \widetilde{\text{grad}}A\,\widetilde{\text{grad}}e^{iF}$$

$$+ i[\widetilde{\text{grad}}A\,\widetilde{\text{grad}}F + A\widetilde{\text{div}}\,\widetilde{\text{grad}}F]e^{iF}$$

$$+ i^2\,A\widetilde{\text{grad}}F\,\widetilde{\text{grad}}F\,e^{iF} = 0. \tag{3.1.30}$$

On separating real and imaginary parts we obtain, from

$$\frac{\widetilde{\text{div}}\,\widetilde{\text{grad}}A}{A}u - \left(\widetilde{\text{grad}}F\right)^2 u$$

$$+ i[2\widetilde{\text{grad}}A\,\widetilde{\text{grad}}F + A\widetilde{\text{div}}\,\widetilde{\text{grad}}F]e^{iF} = 0, \tag{3.1.31}$$

the following equations:

$$\frac{\widetilde{\text{div}}\,\widetilde{\text{grad}}A}{A} - (\widetilde{\text{grad}}F)^2 = 0, \tag{3.1.32}$$

$$2A\widetilde{\text{grad}}A\,\widetilde{\text{grad}}F + A^2\widetilde{\text{div}}\,\widetilde{\text{grad}}F = 0. \tag{3.1.33}$$

The original linear wave equation is split into two non-linear real equations of the form

$$(\widetilde{\text{grad}}F)^2 = \frac{\Box A}{A}, \quad \widetilde{\text{div}}(A^2\widetilde{\text{grad}}F) = 0. \tag{3.1.34}$$

The second equation is a continuity equation on space–time. The first equation reduces to an equation only for F when $\frac{\Box A}{A}$ can be neglected.

To better grasp this hypothesis, let us assume that $A = A(x, y, z)$, i.e. it depends only on spatial coordinates in a given splitting of space–time. Under such circumstances, being negligible means

$$\frac{|\triangle A|}{|A|} << \frac{1}{\lambda^2}, \tag{3.1.35}$$

and in this splitting of space–time

$$(\widetilde{\text{grad}}F)^2 = \left(\frac{\partial F}{\partial x}\right)^2 + \left(\frac{\partial F}{\partial y}\right)^2 + \left(\frac{\partial F}{\partial z}\right)^2 - \frac{1}{v^2}\left(\frac{\partial F}{\partial t}\right)^2 = 0. \tag{3.1.36}$$

We re-express this equation as the Hamilton–Jacobi equation on phase space associated with the relation

$$\vec{p} \cdot \vec{p} - p_0^2 = 0. \tag{3.1.37}$$

We recall that, with any energy-momentum 4-vector, in a given splitting of space–time we identify the energy as $E = p_0 c$. The relation we are considering is the particular case of zero-mass of the *mass-shell relation*

$$p_0^2 - \vec{p} \cdot \vec{p} = m_0^2 c^2. \tag{3.1.38}$$

Therefore, if we consider the Hamiltonian function $\mathcal{H} \equiv p_0^2 - \vec{p} \cdot \vec{p}$, we find that it describes a family of free particles of arbitrary mass. Interestingly, to describe a particle of a given mass we are dealing with Hamilton equations associated with a mass-shell relation, rather than with a function. It is only on the level set ruled by Eq. (3.1.38), the mass shell, that the Hamilton equations of matter describe a particle of mass m_0. The solutions of this equation, restricted to the surface of phase space described by

$$p_\mu - \frac{\partial F}{\partial x^\mu} = 0,$$

reproduce the propagation of rays when projected on the spatial part of the space and time splitting of space–time.

The three-dimensional surface of space–time identified by $F = $ constant, when intersected with the hyperplane $x_0 = ct = $ constant, gives a two-dimensional surface for each value of t. When all these two-dimensional surfaces are projected on space they appear as moving surfaces, i.e. wave fronts, ruled by the rays which on space appear to be orthogonal to these wave fronts.

To sum up, starting from a differential operator on space–time, we have constructed an equation for the phase of the wave function, which in some appropriate limit appears like the Hamilton–Jacobi equation associated with a function on phase space. If we consider this function as the Hamiltonian for some Hamilton equations, we get a ray equation on space. This procedure describes the transition from the wave propagation to the ray propagation.

In some sense, the construction of wave mechanics uses this analogy and tries to invert the procedure. From the equations of motion in Hamiltonian form one tries to construct a partial differential equation whose 'appropriate' limit (i.e. for $h \to 0$) reproduces the Hamilton equations. One should be aware of the fact that the partial differential equation we construct contains *much more information* than the Hamilton equations we start with. For this reason, the transition from the Hamiltonian function to the second-order partial differential equation has only a heuristic value and its adequacy relies on its experimental predictions. The association of differential operators with functions on phase space relies on the notion of the symbol of a differential operator. This will be considered in the next chapter.

In the next section we present a brief summary of the Hamiltonian formalism for particles, specifically adapted to the arguments we have given to go from waves to rays.

3.2 Particles: Hamiltonian equations

We consider the phase space (or space of states or carrier space) $\mathbf{R}^4 \times (\mathbf{R}^4)^*$ $((\mathbf{R}^n)^*$ denoting the dual of $\mathbf{R}^n$ for all n), parametrized with coordinates (q^a, p_a). Any function on this space, say $H \in \mathcal{F}(\mathbf{R}^4 \times (\mathbf{R}^4)^*)$, defines equations of motion (the Hamilton equations of motion) by setting

$$\frac{dq^a}{ds} = \frac{\partial H}{\partial p_a}, \quad \frac{dp_a}{ds} = -\frac{\partial H}{\partial q^a}, \quad a = 1, \ldots, 4. \tag{3.2.1}$$

Here the parameter s is not identified with the time coordinate of $\mathbf{R}^4$. These equations are a first-order differential equation in normal form. For H satisfying some regularity conditions, for any initial condition (Cauchy data) $q(0) = q_0$, $p(0) = p_0$ there exists a unique solution $(q(s), p(s))$.

If, for each initial condition, solutions exist for $s \in]-\infty, +\infty[$, we say that the equations are a complete system and define a 1-parameter group of transformations

$$\Phi : \mathbf{R} \times \mathbf{R}^4 \times (\mathbf{R}^4)^* \to \mathbf{R}^4 \times (\mathbf{R}^4)^*,$$

$$(s, q_0, p_0) \to (q(s), p(s)).$$

Remark The Hamiltonian function is often identified with the energy of the physical system which is being described. However, the procedure can be applied to any function, *independently of the meaning* (physical interpretation) associated with it. Therefore, appropriate regular functions will generate one-parameter groups of transformations of phase space. We often say that real-valued functions on phase space represent not only measurable physical quantities, they also act as generating functions of transformations.

Examples

Example 1: On $\mathbf{R}^3 \times (\mathbf{R}^3)^*$ we consider the function

$$L = q^1 p_2 - q^2 p_1. \tag{3.2.2}$$

The Hamilton equations associated with L are given by

$$\frac{dq^1}{ds} = -q^2, \ \frac{dq^2}{ds} = q^1, \ \frac{dq^3}{ds} = 0, \tag{3.2.3}$$

$$\frac{dp_1}{ds} = -p_2, \ \frac{dp_2}{ds} = p_1, \ \frac{dp_3}{ds} = 0. \tag{3.2.4}$$

On fixing initial conditions $q(0) = q_0, p(0) = p_0$ we find a solution reading as

$$\begin{pmatrix} q^1(s) \\ q^2(s) \end{pmatrix} = \begin{pmatrix} \cos(s) & -\sin(s) \\ \sin(s) & \cos(s) \end{pmatrix} \begin{pmatrix} q^1(0) \\ q^2(0) \end{pmatrix}, \ q^3(s) = q^3(0), \tag{3.2.5}$$

$$\begin{pmatrix} p_1(s) \\ p_2(s) \end{pmatrix} = \begin{pmatrix} \cos(s) & -\sin(s) \\ \sin(s) & \cos(s) \end{pmatrix} \begin{pmatrix} p_1(0) \\ p_2(0) \end{pmatrix}, \ p_3(s) = p_3(0). \tag{3.2.6}$$

Example 2: The linear function $f = b^a p_a$ gives rise to the equations of motion

$$\frac{dq^a}{ds} = b^a, \ \frac{dp_a}{ds} = 0. \tag{3.2.7}$$

Once again, we find a general solution with Cauchy data

$$(q^a(0), p_a(0))$$

such that

$$q^a(s) = q^a(0) + b^a s, \ p_a(s) = p_a(0). \tag{3.2.8}$$

Example 3: Lorentz transformations. The function $f \equiv x_1 p_2 + x_2 p_1$ gives rise to the equations

$$\dot{x}_1 = x_2, \ \dot{x}_2 = x_1, \ \dot{p}_1 = p_2, \ \dot{p}_2 = p_1. \tag{3.2.9}$$

By fixing initial conditions $q(0) = q_0, p(0) = p_0$, we find

$$\begin{pmatrix} q^1(s) \\ q^2(s) \end{pmatrix} = \begin{pmatrix} \cosh s & \sinh s \\ \sinh s & \cosh s \end{pmatrix} \begin{pmatrix} q^1(0) \\ q^2(0) \end{pmatrix}, \tag{3.2.10}$$

$$\begin{pmatrix} p_1(s) \\ p_2(s) \end{pmatrix} = \begin{pmatrix} \cosh s & \sinh s \\ \sinh s & \cosh s \end{pmatrix} \begin{pmatrix} p_1(0) \\ p_2(0) \end{pmatrix}. \tag{3.2.11}$$

In all cases we have obtained a 1-parameter group of transformations. The main property derived from the group property is the associativity condition

$$\phi(s_1, \phi(s_2, q_0, p_0)) = \phi(s_1 + s_2, q_0, p_0). \tag{3.2.12}$$

It is also easy to see that $\phi(-s)$ is the inverse transformation of $\phi(s)$, and the identity transformation is obtained when $s = 0$.

The Hamilton equation can be very conveniently written by introducing the skew-symmetric matrix

$$\Lambda = \left(\Lambda_{\alpha\beta} \right) = \begin{pmatrix} 0 & I \\ -I & 0 \end{pmatrix}, \tag{3.2.13}$$

where $\alpha, \beta = 1, \ldots, 2n$ and 0 and I represent the null $n \times n$ matrix and the identity $n \times n$ matrix, respectively.

By using collective coordinates, i.e. we forget about the distinction between vectors and covectors (a necessary step if we want to perform generic coordinate transformations on $\mathbf{R}^4 \times (\mathbf{R}^4)^*$) and introduce $(\xi) = (q, p)$, we can define column vectors

$$\dot{\xi} \equiv \begin{pmatrix} \dot{\xi}_1 \\ \cdot \\ \cdot \\ \cdot \\ \dot{\xi}_{2n} \end{pmatrix}, \ \vec{\nabla}H \equiv \begin{pmatrix} \frac{\partial H}{\partial \xi_1} \\ \cdot \\ \cdot \\ \cdot \\ \frac{\partial H}{\partial \xi_{2n}} \end{pmatrix}, \tag{3.2.14}$$

and write the equations of motion in the concise form

$$\dot{\xi}_\alpha = \Lambda_{\alpha\beta} \frac{\partial H}{\partial \xi_\beta}, \tag{3.2.15}$$

or, with an even more concise notation,

$$\dot{\xi} = \vec{\Lambda} \cdot \vec{\nabla}H. \tag{3.2.16}$$

The matrix $\left(\Lambda_{\alpha\beta} \right)$ is called the *symplectic matrix*. If we replace ξ_α with a generic function f and H with a generic function g, we find

$$\frac{df}{ds} = \frac{\partial f}{\partial \xi_\alpha} \frac{d\xi_\alpha}{ds} = \frac{\partial f}{\partial \xi_\alpha} \Lambda_{\alpha\beta} \frac{\partial g}{\partial \xi_\beta}. \tag{3.2.17}$$

The right-hand side of this relation makes it possible for us to introduce a binary bilinear product, defined on differentiable functions f and g by setting

$$\{f,g\} \equiv \frac{\partial f}{\partial \xi_\alpha} \Lambda_{\alpha\beta} \frac{\partial g}{\partial \xi_\beta}. \tag{3.2.18}$$

This bracket is called the Poisson bracket associated with Λ. From the skew-symmetry of Λ it follows that

$$\{f,g\} = -\{g,f\}. \tag{3.2.19}$$

Moreover it is possible to obtain, by direct computation, the Jacobi identity

$$\{f,\{g,h\}\} = \{\{f,g\},h\} + \{g,\{f,h\}\}. \tag{3.2.20}$$

From the definition we also have the very important property

$$\{f,gh\} = \{f,g\}\,h + g\,\{f,h\}. \tag{3.2.21}$$

This *derivation* property is what makes it possible to write

$$\{f,g\} = \frac{\partial f}{\partial \xi_\alpha} \{\xi_\alpha,\xi_\beta\} \frac{\partial g}{\partial \xi_\beta}, \tag{3.2.22}$$

i.e. it is sufficient to give the Poisson bracket on coordinate functions. Thus, with any function f we associate the first-order differential operator X_f given by

$$X_f = \frac{\partial f}{\partial \xi_\alpha} \{\xi_\alpha,\xi_\beta\} \frac{\partial}{\partial \xi_\beta}. \tag{3.2.23}$$

Remark   It is possible to define a Poisson bracket abstractly, by using the previous properties as defining relations. If we denote by M any carrier space and by $\mathcal{F}(M)$ the set of smooth functions on M, we say that the map

$$\{\,,\,\} : \mathcal{F}(M) \times \mathcal{F}(M) \to \mathcal{F}(M)$$

defines a Poisson bracket if the following properties hold:

$$\{f_1,f_2\} = -\{f_2,f_1\}, \tag{3.2.24}$$

$$\{f_1,\lambda f_2 + \mu f_3\} = \lambda\,\{f_1,f_2\} + \mu\,\{f_1,f_3\}, \quad \lambda,\mu \in \mathbf{R}, \tag{3.2.25}$$

$$\{f_1,\{f_2,f_3\}\} = \{\{f_1,f_2\},f_3\} + \{f_2,\{f_1,f_3\}\}, \tag{3.2.26}$$

$$\{f_1,f_2 f_3\} = \{f_1,f_2\}f_3 + f_2\,\{f_1,f_3\}. \tag{3.2.27}$$

These equations express anti-symmetry, bilinearity, Jacobi identity and the Leibniz rule (or derivation property with respect to the pointwise product), respectively. If

$$X_f \cdot g \equiv \{f,g\},$$

a theorem by Willmore (1960) ensures that X_f is a first-order homogeneous differential operator. Therefore, the tensor field representing the bracket gives a bidifferential operator.

It should be stressed that no requirement on the dimension of M has been made. For instance, it is easy to show (we leave it as an exercise) that on $\mathbf{R}^3$, with coordinates s_1, s_2, s_3, a Poisson bracket is obtained by setting

$$\{s_1, s_2\} = s_3, \quad \{s_2, s_3\} = s_1, \quad \{s_3, s_1\} = s_2. \tag{3.2.28}$$

It is interesting to derive the differential operators associated with the linear functions s_1, s_2, s_3. They read as

$$s_1 \to s_3 \frac{\partial}{\partial s_2} - s_2 \frac{\partial}{\partial s_3}, \quad s_2 \to s_1 \frac{\partial}{\partial s_3} - s_3 \frac{\partial}{\partial s_1}, \quad s_3 \to s_2 \frac{\partial}{\partial s_1} - s_1 \frac{\partial}{\partial s_2}. \tag{3.2.29}$$

From here, it is easy to find that the derived first-order differential operators satisfy the commutation relations

$$\left[s_3 \frac{\partial}{\partial s_2} - s_2 \frac{\partial}{\partial s_3}, s_1 \frac{\partial}{\partial s_3} - s_3 \frac{\partial}{\partial s_1} \right] = s_2 \frac{\partial}{\partial s_1} - s_1 \frac{\partial}{\partial s_2}, \tag{3.2.30}$$

and similarly for the other pairs. We have therefore found an isomorphism between the commutation relations (3.2.28) and those just derived.

We notice that $s_1^2 + s_2^2 + s_3^2$ has a vanishing Poisson bracket with any other function, i.e. it commutes with any function $f(s_1, s_2, s_3)$. It is called a central element or a Casimir function. With the given brackets, the Hamiltonian $I_1 s_1^2 + I_2 s_2^2 + I_3 s_3^2$ would lead to the equations of motion

$$\dot{s}_1 = 2 s_2 s_3 (I_3 - I_2), \tag{3.2.31}$$

$$\dot{s}_2 = 2 s_3 s_1 (I_1 - I_3), \tag{3.2.32}$$

$$\dot{s}_3 = 2 s_1 s_2 (I_2 - I_1). \tag{3.2.33}$$

3.2.1 Poisson brackets among velocity components for a charged particle

In classical electrodynamics, a particle of charge q and mass m is described by the Lagrangian

$$\mathcal{L} = \frac{m}{2} \delta_{ij} v^i v^j - q \left[\phi(\vec{x}, t) - \frac{1}{c} v^j A_j(\vec{x}, t) \right], \tag{3.2.34}$$

where ϕ is the scalar potential and A_j are the components of the vector potential. The Lorentz force is indeed easily obtained from such a Lagrangian because (recall that $\frac{\partial v^j}{\partial v^i} = \delta_i^j$)

$$\frac{\partial \mathcal{L}}{\partial v^i} = m v_i + \frac{q}{c} A_i, \tag{3.2.35}$$

$$\frac{\mathrm{d}}{\mathrm{d}t} \frac{\partial \mathcal{L}}{\partial v^i} = \frac{\mathrm{d}}{\mathrm{d}t}(m v_i) + \frac{q}{c} \left(\frac{\partial A_i}{\partial t} + \frac{\partial A_i}{\partial x^j} \frac{\mathrm{d}x^j}{\mathrm{d}t} \right)$$

$$= \frac{\mathrm{d}}{\mathrm{d}t}(m v_i) + \frac{q}{c} \left(\frac{\partial A_i}{\partial t} + v^j \frac{\partial A_i}{\partial x^j} \right), \tag{3.2.36}$$

$$\frac{\partial \mathcal{L}}{\partial x^i} = -q\left(\frac{\partial \phi}{\partial x^i} - \frac{1}{c} v^j \frac{\partial A_j}{\partial x^i}\right), \tag{3.2.37}$$

and hence the Euler–Lagrange equations read as

$$\frac{d}{dt}(mv_i) = q\left(-\frac{\partial \phi}{\partial x^i} - \frac{1}{c}\frac{\partial A_i}{\partial t}\right) + \frac{q}{c}v^j\left(\frac{\partial A_j}{\partial x^i} - \frac{\partial A_i}{\partial x^j}\right)$$

$$= q\left[E_i + \frac{1}{c}\left(\vec{v} \wedge \vec{B}\right)_i\right], \tag{3.2.38}$$

which are the three components of the Lorentz force, but expressed in a form involving covariant vectors (or covectors). Note that, to obtain the last line of Eq. (3.2.38), we have used the simple but non-trivial identity

$$\left(\vec{v} \wedge \vec{B}\right)_i = \varepsilon_i^{\ jk} v_j B_k = \varepsilon_{ij}^{\ \ k} v^j \varepsilon_k^{\ lp} \frac{\partial}{\partial x^l} A_p$$

$$= \varepsilon_{ij}^{\ \ k} \varepsilon_k^{\ lp} v^j \frac{\partial}{\partial x^l} A_p = \left(\delta_i^{\ l}\delta_j^{\ p} - \delta_i^{\ p}\delta_j^{\ l}\right) v^j \frac{\partial A_p}{\partial x^l}$$

$$= v^j\left(\frac{\partial A_j}{\partial x^i} - \frac{\partial A_i}{\partial x^j}\right). \tag{3.2.39}$$

Note now that, by imposing the standard Poisson brackets

$$\{q^i, q^j\} = 0, \tag{3.2.40}$$

$$\{p_i, p_j\} = 0, \tag{3.2.41}$$

$$\{q^j, p_k\} = \delta_k^j, \tag{3.2.42}$$

we can derive the Poisson brackets $\{v_i, q^j\}$ and $\{v_i, v_j\}$ adapted to the particular Lagrangian we are considering, bearing in mind that $p_i = mv_i + \frac{q}{c}A_i$, where q is the charge of the particle.

Indeed, the basic Poisson bracket among canonical momenta implies that

$$\{v_i, v_j\} = \frac{q}{mc}\left[\{A_j, v_i\} + \{v_j, A_i\}\right]. \tag{3.2.43}$$

Now for any pair of functions F and G of the (q, v) coordinates

$$\{F, G\} = \frac{\partial F}{\partial q^k}\left\{q^k, q^l\right\}\frac{\partial G}{\partial q^l} + \frac{\partial F}{\partial v^k}\left\{v^k, v^l\right\}\frac{\partial G}{\partial v^l}$$

$$+ \frac{\partial F}{\partial q^k}\left\{q^k, v^l\right\}\frac{\partial G}{\partial v^l} + \frac{\partial F}{\partial v^k}\left\{v^k, q^l\right\}\frac{\partial G}{\partial q^l}, \tag{3.2.44}$$

and hence

$$\{p_i, A_j\} = -\frac{\partial A_j}{\partial q^i}, \tag{3.2.45}$$

because

$$\{v_i, q^j\} = \frac{1}{m}\{p_i, q^j\} = -\frac{1}{m}\delta_i^j. \tag{3.2.46}$$

Eventually, these equations lead to

$$\{v_i, v_j\} = \frac{q}{mc}\left[-\frac{1}{m}\frac{\partial A_i}{\partial q^j} + \frac{1}{m}\frac{\partial A_j}{\partial q^i}\right] = \frac{q}{m^2 c}\varepsilon_{ij}{}^k B_k, \qquad (3.2.47)$$

which have the interesting feature of being gauge-invariant, unlike the more familiar Poisson brackets among canonical momenta. We also find that the Hamiltonian function, in these coordinates, is

$$H = E_L = v_j\frac{\partial L}{\partial v_j} - L = \frac{m}{2}\delta_{ij}v^i v^j + q\phi(\vec{x}, t). \qquad (3.2.48)$$

3.3 Homogeneous linear differential operators and equations of motion

If f is any function on $\mathbf{R}^n \times (\mathbf{R}^n)^*$, we may consider the derivation of f along the integral curves of the Hamiltonian system associated with H, i.e.

$$\frac{df}{dt} = \frac{\partial f}{\partial \xi_\alpha}\frac{d\xi_\alpha}{dt} = \frac{\partial f}{\partial \xi_\alpha}\{\xi_\alpha, H\} = \frac{\partial f}{\partial \xi_\alpha}\Lambda_{\alpha\beta}\frac{\partial H}{\partial \xi_\beta}. \qquad (3.3.1)$$

Since in these formulae f plays only the role of function acted upon by the operator $\frac{d}{dt}$, we may remove it from the formulae and get

$$\frac{d}{dt} = \frac{\partial H}{\partial \xi_\beta}\Lambda_{\alpha\beta}\frac{\partial}{\partial \xi_\alpha}, \qquad (3.3.2)$$

i.e. with any set of Hamilton equations of motion we may associate a first-order homogeneous differential operator, as already stressed with Eq. (3.2.23).

On reverting to the use of canonical (symplectic) coordinates, it is possible to write the Poisson bracket in the form

$$\{f, g\} = \left(\frac{\partial f}{\partial q}, \frac{\partial f}{\partial p}\right)\begin{pmatrix} 0 & I \\ -I & 0 \end{pmatrix}\begin{pmatrix} \frac{\partial g}{\partial q} \\ \frac{\partial g}{\partial p} \end{pmatrix} = \sum_{a=1}^{n}\left(\frac{\partial f}{\partial q^a}\frac{\partial g}{\partial p_a} - \frac{\partial f}{\partial p_a}\frac{\partial g}{\partial q^a}\right). \qquad (3.3.3)$$

Remark The method used in writing Eq. (3.3.2) suggests removing, at this stage, both f and g. We may then say that the Poisson bracket is associated with the bidifferential operator

$$\{\cdot, \cdot\} = \sum_{a=1}^{n}\left(\frac{\overleftarrow{\partial}}{\partial q^a}\frac{\overrightarrow{\partial}}{\partial p_a} - \frac{\overleftarrow{\partial}}{\partial p_a}\frac{\overrightarrow{\partial}}{\partial q^a}\right). \qquad (3.3.4)$$

This is a bidifferential operator because it acts on a pair of functions (f, g). It is often called a *Poisson tensor*. With our notation, $f\{\cdot, \cdot\}g$ will be meant to be the Poisson bracket $\{f, g\}$.

3.4 Symmetries and conservation laws

We can exploit many of our previously derived formulae to show how symmetries of the Hamiltonian function are connected with conservation laws or constants of motion.

We have seen that with any function F on phase space, by means of Eq. (3.3.2), it is possible to associate a first-order differential equation. On generic functions h,

$$\frac{dh}{dt} = \frac{\partial F}{\partial \xi_b} \Lambda_{ab} \frac{\partial h}{\partial \xi_a} = \{h, F\}. \tag{3.4.1}$$

In particular, if h is the Hamiltonian function H,

$$\frac{dH}{dt} = \{H, F\}. \tag{3.4.2}$$

If H is constant along the transformations defined by F, $\frac{dH}{dt} = 0$. Of course, since the Poisson bracket is anti-symmetric, it is also the case that

$$\frac{dF}{dt} = \{F, F\} = 0, \tag{3.4.3}$$

along the integral curves associated with H. Thus, a conserved quantity like F defines a transformation which preserves H. Conversely, if F generates a transform that preserves H then it is a constant of motion. What we have stated is the Poisson version of the Noether theorem connecting symmetries of the Lagrangian function $\mathcal{L}$ and conservation laws defined using $\mathcal{L}$. Let us consider now a few examples.

On the phase space associated with a particle with three degrees of freedom, say described by $H = \frac{p^2}{2m}$, we notice that H is invariant under translations in $\mathbf{R}^3$. If we consider the translation on phase space defined by

$$\vec{x} \to \vec{x} + \vec{a}s, \ \vec{p} \to \vec{p}, \tag{3.4.4}$$

for any function f on phase space we have the transformed

$$f_s'(\vec{x}, \vec{p}) = f(\vec{x} + \vec{a}s, \vec{p}), \tag{3.4.5}$$

and we get

$$\frac{d}{ds} f_s'(s = 0) = \frac{\partial f}{\partial \vec{x}} \vec{a}, \tag{3.4.6}$$

i.e. $\frac{d}{ds} = \vec{a} \frac{\partial}{\partial \vec{x}}$. By means of Eq. (3.3.2) we derive therefore

$$\frac{d}{ds} = \frac{\partial}{\partial \xi_a} (\vec{a} \cdot \vec{P}) \Lambda_{ab} \frac{\partial}{\partial \xi_b} = \vec{a} \frac{\partial}{\partial \vec{x}}. \tag{3.4.7}$$

We have found that $\vec{a} \cdot \vec{p}$ is the generating function of infinitesimal generators of translations in the direction $\vec{a}$. We also derive easily that $\vec{a} \cdot \vec{p}$ is a constant of motion for $H = \frac{\vec{p} \cdot \vec{p}}{2m}$.

Space rotations. If Λ is the bidifferential operator associated with canonical Poisson brackets on phase space, say

$$\Lambda = \frac{\partial}{\partial \vec{x}} \wedge \frac{\partial}{\partial \vec{p}}, \tag{3.4.8}$$

we consider the infinitesimal generator associated with the orbital angular momentum $\vec{L} \equiv \vec{x} \wedge \vec{p}$. We find the infinitesimal generators for each component given by

$$R_x = y\frac{\partial}{\partial z} - z\frac{\partial}{\partial y} + p_y\frac{\partial}{\partial p_z} - p_z\frac{\partial}{\partial p_y}, \tag{3.4.9}$$

$$R_y = z\frac{\partial}{\partial x} - x\frac{\partial}{\partial z} + p_z\frac{\partial}{\partial p_x} - p_x\frac{\partial}{\partial p_z}, \tag{3.4.10}$$

$$R_z = x\frac{\partial}{\partial y} - y\frac{\partial}{\partial x} + p_x\frac{\partial}{\partial p_y} - p_y\frac{\partial}{\partial p_x}. \tag{3.4.11}$$

We now show that rotations in the (y, z) plane are associated with R_x. Then, by the same procedure, we consider the corresponding differential equation on phase space. We find rotations both in the coordinates and momenta associated with R_x.

By letting these operators act on linear polynomials in x, y, z, or separately on linear polynomials of p_x, p_y, p_z, their action is provided by the following matrices:

$$R_x = \begin{pmatrix} 0 & 0 & 0 \\ 0 & 0 & -1 \\ 0 & 1 & 0 \end{pmatrix}, \tag{3.4.12}$$

$$R_y = \begin{pmatrix} 0 & 0 & 1 \\ 0 & 0 & 0 \\ -1 & 0 & 0 \end{pmatrix}, \tag{3.4.13}$$

$$R_z = \begin{pmatrix} 0 & -1 & 0 \\ 1 & 0 & 0 \\ 0 & 0 & 0 \end{pmatrix}. \tag{3.4.14}$$

The associated 1-parameter group of transformations is obtained simply by exponentiation

$$D_x(\alpha) = e^{\alpha R_x} = I + \alpha R_x + \frac{\alpha^2}{2!}R_x^2 + \frac{\alpha^3}{3!}R_x^3 + \cdots \tag{3.4.15}$$

By collecting separately odd and even powers we find again

$$D_x(\alpha) = \begin{pmatrix} 1 & 0 & 0 \\ 0 & \cos\alpha & -\sin\alpha \\ 0 & \sin\alpha & \cos\alpha \end{pmatrix}, \tag{3.4.16}$$

and similarly for the other rotations. We obtain similar 1-parameter groups of rotations $D_y(\beta)$ and $D_z(\gamma)$. Products of these matrices, in any order, define the rotation group, which is a relevant example of a Lie group (see Chapter 15). It is also possible to perform coordinate transformations from Cartesian to spherical polar coordinates, so as to obtain

$$R_x = -\sin\varphi\frac{\partial}{\partial\theta} - \cos\varphi\cot\theta\frac{\partial}{\partial\varphi}, \tag{3.4.17}$$

$$R_y = \cos\varphi \frac{\partial}{\partial\theta} - \sin\varphi \cot\theta \frac{\partial}{\partial\varphi}, \tag{3.4.18}$$

$$R_z = \frac{\partial}{\partial\varphi}, \tag{3.4.19}$$

where $x = r\sin\theta\cos\varphi, y = r\sin\theta\sin\varphi, z = r\cos\theta$. Usually, it is possible to extend coordinate transformations to coordinates and momenta transformations by requiring the invariance of the phase 1-form, i.e.

$$p_x dx + p_y dy + p_z dz = p_r dr + p_\theta d\theta + p_\varphi d\varphi. \tag{3.4.20}$$

Hence,

$$\begin{pmatrix} p_r \\ p_\theta \\ p_\varphi \end{pmatrix} = \begin{pmatrix} \sin\theta\cos\varphi & \sin\theta\sin\varphi & \cos\theta \\ r\cos\theta\cos\varphi & r\cos\theta\sin\varphi & -r\sin\theta \\ -r\sin\theta\sin\varphi & r\sin\theta\cos\varphi & 0 \end{pmatrix} \begin{pmatrix} p_x \\ p_y \\ p_z \end{pmatrix}, \tag{3.4.21}$$

obtained from the transformation of the differentials, which reads as

$$\begin{pmatrix} dx \\ dy \\ dz \end{pmatrix} = \begin{pmatrix} \sin\theta\cos\varphi & r\cos\theta\cos\varphi & -r\sin\theta\sin\varphi \\ \sin\theta\sin\varphi & r\cos\theta\sin\varphi & r\sin\theta\cos\varphi \\ \cos\theta & -r\sin\theta & 0 \end{pmatrix} \begin{pmatrix} dr \\ d\theta \\ d\varphi \end{pmatrix}. \tag{3.4.22}$$

It is now a simple exercise to show that the Hamiltonian of any particle in some external potential depending only on r, say $H = \frac{p^2}{2m} + U(r)$, is invariant under rotation and therefore the corresponding infinitesimal generator associated with the vector function $\vec{L} = \vec{r} \wedge \vec{p}$ is a constant of motion. Conversely, the constant of motion $\vec{L} = \vec{r} \wedge \vec{p}$ has generated transformations which preserve the Hamiltonian function.

3.4.1 Homomorphism between $SU(2)$ and $SO(3)$

Let us elaborate further on the rotation group in $\mathbf{R}^3$. As a manifold, the set of invertible 3×3 matrices defines a nine-dimensional space, and the particular structure of the rotation matrices gives rise to a sub-manifold of $\mathbf{R}^9$. To investigate its topological properties, it is more convenient to consider the group $SU(2)$ of 2×2 unitary matrices with unit determinant. The present subsection is therefore devoted to the relation between these two groups. The unitary group, as we shall see, plays a relevant role in the study of the harmonic oscillator and will play an important role in quantum mechanical systems.

We shall see that the two groups are related by a homomorphism. In general, for any two groups G_1 and G_2, a *homomorphism* is a map $\phi : G_1 \to G_2$ such that

$$\phi(g_1 \cdot g_2) = \phi(g_1)\phi(g_2),$$

$$\phi(g^{-1}) = (\phi(g))^{-1}.$$

We follow closely the presentation in Wigner (1959), which relies in turn on a method suggested by H. Weyl. To begin, let us recall some elementary results in the theory of matrices, which turn out to be very useful for our purposes.

(i) A matrix which transforms every real vector into a real vector is itself real, i.e. all its elements are real. If this matrix is applied to the kth unit vector (which has kth component $=1$, with all others vanishing), the result is the vector which forms the kth column of the matrix. Thus, this column must be real. But this argument can be applied to all k, and hence all columns of the matrix must be real.

(ii) A matrix $\mathcal{O}$ is said to be orthogonal if it preserves the scalar product of two arbitrary vectors, i.e. if

$$\vec{a} \cdot \vec{b} = \mathcal{O}\vec{a} \cdot \mathcal{O}\vec{b}. \tag{3.4.23}$$

An equivalent condition can be stated in terms of an arbitrary vector: a matrix $\mathcal{O}$ is orthogonal if the length of every single arbitrary vector $\vec{v}$ is left unchanged under transformation by $\mathcal{O}$. Consider now two arbitrary vectors $\vec{a}$ and $\vec{b}$, and write $\vec{v} = \vec{a} + \vec{b}$. Then our condition for the orthogonality of $\mathcal{O}$ is

$$\vec{v} \cdot \vec{v} = \mathcal{O}\vec{v} \cdot \mathcal{O}\vec{v}. \tag{3.4.24}$$

By virtue of the symmetry of the scalar product: $\vec{a} \cdot \vec{b} = \vec{b} \cdot \vec{a}$, this yields

$$(\vec{a} + \vec{b}) \cdot (\vec{a} + \vec{b}) = \vec{a} \cdot \vec{a} + \vec{b} \cdot \vec{b} + 2\vec{a} \cdot \vec{b}$$
$$= \mathcal{O}\vec{a} \cdot \mathcal{O}\vec{a} + \mathcal{O}\vec{b} \cdot \mathcal{O}\vec{b} + 2\mathcal{O}\vec{a} \cdot \mathcal{O}\vec{b}. \tag{3.4.25}$$

However, orthogonality also implies that $\vec{a} \cdot \vec{a}$ equals $\mathcal{O}\vec{a} \cdot \mathcal{O}\vec{a}$, and the same for $\vec{a}$ replaced by $\vec{b}$. It then follows from Eq. (3.4.25) that

$$\vec{a} \cdot \vec{b} = \mathcal{O}\vec{a} \cdot \mathcal{O}\vec{b}, \tag{3.4.26}$$

which implies that $\mathcal{O}$ is orthogonal. It can be shown, in a similar way, that $\mathcal{U}$ is unitary only if $\vec{v} \cdot \vec{v} = \mathcal{U}\vec{v} \cdot \mathcal{U}\vec{v}$ holds for every vector (here the scalar product is the Hermitian product in the corresponding complex vector space).

By definition, a matrix that leaves each real vector real, and leaves the length of every vector unchanged and preserves the origin, is a *rotation*. Indeed, when all lengths are equal in the original and transformed figures, the angles must also be equal; hence the transformation is merely a rotation.

(iii) We now want to determine the general form of a two-dimensional unitary matrix

$$\mathbf{u} = \begin{pmatrix} a & b \\ c & d \end{pmatrix} \tag{3.4.27}$$

of determinant $+1$ by considering the elements of the product

$$\mathbf{u}\mathbf{u}^{\dagger} = \mathbb{I}. \tag{3.4.28}$$

This time, since the action is on the complex vector space $\mathbf{C}^2$ with Hermitian product

$$\bar{z}_1 z_1 + \bar{z}_2 z_2,$$

similarly to the case of rotations, the column vectors of the matrix $\mathbf{u}$ should be orthonormal vectors, and the same for the row vectors. These properties are encoded in the condition (3.4.28).

Recall that the † operation means taking the complex conjugate and then the transpose of the original matrix. Thus, the condition (3.4.28) implies that

$$a^*c + b^*d = 0, \tag{3.4.29}$$

which leads to $c = -b^*d/a^*$. The insertion of this result into the condition of unit determinant:

$$ad - bc = 1, \tag{3.4.30}$$

yields $(aa^* + bb^*)d/a^* = 1$. Moreover, since $aa^* + bb^* = 1$ from the product (3.4.28), it follows that $d = a^*$ and $c = -b^*$. The general two-dimensional unitary matrix with unit determinant is hence

$$\mathbf{u} = \begin{pmatrix} a & b \\ -b^* & a^* \end{pmatrix}, \tag{3.4.31}$$

where, of course, we still have to require that $aa^* + bb^* = 1$. Note that, if $a = y_0 + iy_3$ and $b = y_1 + iy_2$, then

$$\det \mathbf{u} = y_0^2 + y_1^2 + y_2^2 + y_3^2 = 1.$$

This is the equation of a unit 3-sphere centred at the origin, which means that $SU(2)$ has 3-sphere topology and is hence *simply connected* (see Chapter 15 for the definition and examples; the n-sphere is simply connected for all $n > 1$). More precisely, $SU(2)$ is homeomorphic to $S^3 \subset \mathbf{R}^4$.

Consider now the Pauli matrices

$$\sigma_x \equiv \begin{pmatrix} 0 & 1 \\ 1 & 0 \end{pmatrix}, \tag{3.4.32}$$

$$\sigma_y \equiv \begin{pmatrix} 0 & -i \\ i & 0 \end{pmatrix}, \tag{3.4.33}$$

$$\sigma_z \equiv \begin{pmatrix} 1 & 0 \\ 0 & -1 \end{pmatrix}, \tag{3.4.34}$$

which are at the same time Hermitian and unitary. Every traceless two-dimensional matrix, here denoted by $\mathbf{h}$, can be expressed as a linear combination of these matrices:

$$\mathbf{h} = x\sigma_x + y\sigma_y + z\sigma_z = r^i\sigma_i. \tag{3.4.35}$$

Explicitly,

$$\mathbf{h} = \begin{pmatrix} z & x - iy \\ x + iy & -z \end{pmatrix}. \tag{3.4.36}$$

In particular, if x, y, and z are real, then $\mathbf{h}$ is Hermitian.

If $\mathbf{h}$ is transformed by an arbitrary unitary matrix $\mathbf{u}$ with unit determinant, a matrix with zero trace is again obtained, $\bar{\mathbf{h}} = \mathbf{uhu}^\dagger$. Thus, $\bar{\mathbf{h}}$ can also be written as a linear combination of $\sigma_x, \sigma_y, \sigma_z$:

$$\bar{\mathbf{h}} = \mathbf{uhu}^\dagger = \mathbf{u}r^i\sigma_i\mathbf{u}^\dagger = x'\sigma_x + y'\sigma_y + z'\sigma_z = r'^i\sigma_i, \tag{3.4.37}$$

and in matrix form we write

$$\begin{pmatrix} a & b \\ -b^* & a^* \end{pmatrix} \begin{pmatrix} z & x - iy \\ x + iy & -z \end{pmatrix} \begin{pmatrix} a^* & -b \\ b^* & a \end{pmatrix}$$

$$= \begin{pmatrix} z' & x' - iy' \\ x' + iy' & -z' \end{pmatrix}. \tag{3.4.38}$$

Equation (3.4.38) determines x', y', z' as linear functions of x, y, z. The transformation R_u, which carries $r = (x, y, z)$ into $R_u r = r' = (x', y', z')$, can be found from Eq. (3.4.38):

$$x' = \frac{1}{2}\left(a^2 + a^{*2} - b^2 - b^{*2}\right)x + \frac{i}{2}\left(a^{*2} - a^2 + b^{*2} - b^2\right)y$$

$$- \left(a^*b^* + ab\right)z, \tag{3.4.39}$$

$$y' = \frac{i}{2}\left(a^2 - a^{*2} + b^{*2} - b^2\right)x + \frac{1}{2}\left(a^2 + a^{*2} + b^2 + b^{*2}\right)y$$

$$+ i\left(a^*b^* - ab\right)z, \tag{3.4.40}$$

$$z' = (a^*b + ab^*)x + i(a^*b - ab^*)y + (aa^* - bb^*)z. \tag{3.4.41}$$

The particular form of the matrix R_u does not matter; it is important only that

$$x'^2 + y'^2 + z'^2 = x^2 + y^2 + z^2, \tag{3.4.42}$$

since

$$\det \overline{\mathbf{h}} = \det \mathbf{h},$$

$\mathbf{u}$ being an element of $SU(2)$. According to the analysis in (ii), this implies that the transformation R_u must be orthogonal. Such a property can also be seen directly from Eqs. (3.4.39)–(3.4.41).

Moreover, $\overline{\mathbf{h}}$ is Hermitian if $\mathbf{h}$ is. In other words, $r' = (x', y', z')$ is real if $r = (x, y, z)$ is real. This implies, by virtue of (i), that R_u is pure real, as can also be seen directly from Eqs. (3.4.39)–(3.4.41). Thus, R_u is a rotation: every two-dimensional unitary matrix $\mathbf{u}$ of unit determinant corresponds to a three-dimensional rotation R_u; the correspondence is given by Eqs. (3.4.37) or (3.4.38).

It should be stressed that the determinant of R_u is $+1$, since as $\mathbf{u}$ is changed continuously into a unit matrix, R_u goes continuously into the three-dimensional unit matrix. If its determinant were -1 at the beginning of this process, it would have to make the jump to $+1$. This is impossible, since the function 'det' is continuous, and hence the matrices with negative determinant cannot be connected to the identity of the group. As a corollary of these properties, R_u is a *pure rotation* for all $\mathbf{u}$.

The above correspondence is such that the product $\mathbf{qu}$ of two unitary matrices $\mathbf{q}$ and $\mathbf{u}$ corresponds to the product $R_{qu} = R_q \cdot R_u$ of the corresponding rotations. According to Eq. (3.4.37), applied to $\mathbf{q}$ instead of $\mathbf{u}$,

$$\mathbf{q}r^i\sigma_i\mathbf{q}^\dagger = (R_q r)^i\sigma_i, \tag{3.4.43}$$

and upon transformation with $\mathbf{u}$ this yields

$$\mathbf{u}\mathbf{q}r^i\sigma_i\mathbf{q}^\dagger\mathbf{u}^\dagger = \mathbf{u}(R_q r)^i\sigma_i\mathbf{u}^\dagger = (R_u R_q r)^i\sigma_i = (R_{uq}r)^i\sigma_i, \tag{3.4.44}$$

using Eq. (3.4.37) again, with $R_q r$ replacing r and $\mathbf{uq}$ replacing $\mathbf{u}$. Thus, a homomorphism exists between the group of two-dimensional unitary matrices of determinant $+1$ (the group $SU(2)$) and three-dimensional rotations; the correspondence is given by Eq. (3.4.37) or Eqs. (3.4.39)–(3.4.41). However, we note that so far we have not shown that the homomorphism exists between the $SU(2)$ group and the *whole* connected component of the rotation group (called *pure* rotation group by Wigner). This would imply that R_u ranges over all rotations as $\mathbf{u}$ covers the whole unitary group. This will be proved shortly. It should also be noticed that the homomorphism *is not an isomorphism*, since more than one unitary matrix corresponds to the same rotation (see below).

We first assume that $\mathbf{u}$ is a diagonal matrix, here denoted by $\mathbf{u}_1(\alpha)$ (i.e. we set $b = 0$, and, for convenience, we write $a = e^{-\frac{i}{2}\alpha}$). Then $|a^2| = 1$ and α is real:

$$\mathbf{u}_1(\alpha) = \begin{pmatrix} e^{-\frac{i}{2}\alpha} & 0 \\ 0 & e^{\frac{i}{2}\alpha} \end{pmatrix}. \tag{3.4.45}$$

From Eqs. (3.4.39)–(3.4.41) we can see that the corresponding rotation:

$$R_{u_1} = \begin{pmatrix} \cos\alpha & \sin\alpha & 0 \\ -\sin\alpha & \cos\alpha & 0 \\ 0 & 0 & 1 \end{pmatrix} \tag{3.4.46}$$

is a rotation about Z through an angle α. Next, we assume that $\mathbf{u}$ is real:

$$\mathbf{u}_2(\beta) = \begin{pmatrix} \cos\frac{\beta}{2} & -\sin\frac{\beta}{2} \\ \sin\frac{\beta}{2} & \cos\frac{\beta}{2} \end{pmatrix}. \tag{3.4.47}$$

From Eqs. (3.4.39)–(3.4.41) the corresponding rotation is found to be

$$R_{u_2} = \begin{pmatrix} \cos\beta & 0 & -\sin\beta \\ 0 & 1 & 0 \\ \sin\beta & 0 & \cos\beta \end{pmatrix}, \tag{3.4.48}$$

i.e. a rotation about Y through an angle β. The product of the three unitary matrices $\mathbf{u}_1(\alpha)\mathbf{u}_2(\beta)\mathbf{u}_1(\gamma)$, with $\alpha, \beta, \gamma \in S^1$, corresponds to the product of a rotation about Z through an angle γ, about Y through β, and about Z through α, in other words, to a rotation with Euler angles α, β, γ. It follows from this that the correspondence defined in Eq. (3.4.37) not only specified a three-dimensional rotation for every two-dimensional unitary matrix, but also *at least one unitary matrix* for every pure rotation. Specifically, the matrix

$$\begin{pmatrix} e^{-\frac{i}{2}\alpha} & 0 \\ 0 & e^{\frac{i}{2}\alpha} \end{pmatrix} \begin{pmatrix} \cos\frac{\beta}{2} & -\sin\frac{\beta}{2} \\ \sin\frac{\beta}{2} & \cos\frac{\beta}{2} \end{pmatrix} \begin{pmatrix} e^{-\frac{i}{2}\gamma} & 0 \\ 0 & e^{\frac{i}{2}\gamma} \end{pmatrix}$$

$$= \begin{pmatrix} e^{-\frac{i}{2}\alpha}\cos\frac{\beta}{2}e^{-\frac{i}{2}\gamma} & -e^{-\frac{i}{2}\alpha}\sin\frac{\beta}{2}e^{\frac{i}{2}\gamma} \\ e^{\frac{i}{2}\alpha}\sin\frac{\beta}{2}e^{-\frac{i}{2}\gamma} & e^{\frac{i}{2}\alpha}\cos\frac{\beta}{2}e^{\frac{i}{2}\gamma} \end{pmatrix} \tag{3.4.49}$$

corresponds to the rotation $\{\alpha\beta\gamma\}$. Thus, the homomorphism is, in fact, a homomorphism of the group $SU(2)$ onto the *whole* three-dimensional connected component of the rotation group.

The question remains of the *multiplicity* of the homomorphism, i.e. how many unitary matrices $\mathbf{u}$ correspond to the same rotation. For this purpose, it is sufficient to check how many unitary matrices $\mathbf{u}_0$ correspond to the identity of the rotation group, i.e. to the transformation $x' = x, y' = y, z' = z$. For these particular $\mathbf{u}_0$s, the identity $\mathbf{u}_0\mathbf{h}\mathbf{u}_0^{\dagger} = \mathbf{h}$ should hold for all $\mathbf{h}$; this can only be the case when $\mathbf{u}_0$ is a constant matrix: $b = 0$ and $a = a^*$, $\mathbf{u}_0 = \pm\mathbb{I}$ (since $|a|^2 + |b|^2 = 1$). Thus, the two unitary matrices $+\mathbb{I}$ and $-\mathbb{I}$, and *only these*, correspond to the identity of the rotation group. These two elements form an invariant subgroup of the group $SU(2)$, and the elements $\mathbf{u}$ and $-\mathbf{u}$, and only those, correspond to the same rotation. Indeed, on defining

$$\chi \equiv \begin{pmatrix} -1 & 0 \\ 0 & -1 \end{pmatrix},$$

we can express the 2×2 identity matrix as

$$\begin{pmatrix} 1 & 0 \\ 0 & 1 \end{pmatrix} = \chi^2,$$

and hence from the identity

$$\mathbf{u}r^i\sigma_i\mathbf{u}^{\dagger} = \mathbf{u}\chi r^i\sigma_i\chi\mathbf{u}^{\dagger}$$

it follows that, if $\mathbf{u} \to R_u$, then also $\mathbf{u}\chi \to R_u$, so that our homomorphism is 2-to-1.

Alternatively, one can simply note that only the half-Euler angles occur in Eq. (3.4.49). The Euler angles are determined by a rotation only up to a multiple of 2π; the half-angles, only up to a multiple of π. This implies that the trigonometric functions in Eq. (3.4.49) are determined only up to a sign.

A very important result has been thus obtained: there exists a 2-to-1 homomorphism of the group of two-dimensional unitary matrices with determinant 1 *onto* the three-dimensional connected component of the rotation group: there exists a 1-to-1 correspondence between *pairs* of unitary matrices $\mathbf{u}$ and $-\mathbf{u}$ and rotations R_u in such a way that, from $\mathbf{u}\mathbf{q} = \mathbf{t}$, it also follows that $R_uR_q = R_t$; conversely, from $R_uR_q = R_t$, $\mathbf{u}\mathbf{q} = \pm\mathbf{t}$. If the unitary matrix $\mathbf{u}$ is known, the corresponding rotation is best obtained from Eqs. (3.4.39)–(3.4.41). Conversely, the unitary matrix for a rotation $\{\alpha\beta\gamma\}$ is best found from Eq. (3.4.49), showing that $SU(2)$ covers the whole connected component of the rotation group.

3.5 Motivations for studying harmonic oscillators

In classical mechanics, linear oscillations are introduced by means of a mass attached to a spring with some elastic constant k. In more general terms, a linear system with bounded motions behaves like a family of oscillators. The ubiquitous role of harmonic oscillators has to do with the fact that a system described by kinetic energy plus potential energy, say $V(x)$, assumed to have a minimum at a point x_0, may be approximated by an oscillator in the following manner. We consider a Taylor expansion of $V(x)$ around the equilibrium point x_0; on taking only the first two non-vanishing terms,

$$V(x) = V(x_0) + \frac{1}{2}(x - x_0)^2 V''(x_0). \tag{3.5.1}$$

By using $\omega = \sqrt{\frac{V''(x_0)}{m}}$, m being the mass of the particle, and removing the constant value $V(x_0)$,

$$V(x) = \frac{1}{2}m\omega^2 x^2. \tag{3.5.2}$$

The total Hamiltonian function has the form

$$H = \frac{p^2}{2m} + \frac{1}{2}m\omega^2 x^2, \tag{3.5.3}$$

and for several oscillators it reads as

$$H = \sum_k \left(\frac{p_k^2}{2m_k} + \frac{1}{2}m_k\omega_k^2 x_k^2 \right). \tag{3.5.4}$$

We have seen, in dealing with the electromagnetic field in a cavity, that we may consider it as described by an infinite number of oscillators. It is this circumstance that places the harmonic oscillator in a pole position in the quantum theory we are going to develop in the following chapters.

The level sets of the Hamiltonian for single oscillator are ellipses, and it is possible to change variables so that level sets are represented by circles. We set

$$P = \frac{p}{\sqrt{m\omega}}, \quad Q = \sqrt{m\omega}x \tag{3.5.5}$$

to obtain

$$H = \frac{\omega}{2}(P^2 + Q^2). \tag{3.5.6}$$

Note that both P^2 and Q^2 have the dimension of action, i.e. time $\times$ energy. The associated equations of motion, in Hamiltonian form, are

$$\frac{d}{dt}\begin{pmatrix} Q \\ P \end{pmatrix} = \omega \begin{pmatrix} 0 & 1 \\ -1 & 0 \end{pmatrix} \begin{pmatrix} Q \\ P \end{pmatrix}, \tag{3.5.7}$$

the solution of which are

$$Q(t) = Q_0 \cos(\omega t) + P_0 \sin(\omega t), \tag{3.5.8}$$

$$P(t) = P_0 \cos(\omega t) - Q_0 \sin(\omega t). \tag{3.5.9}$$

3.6 Complex coordinates for harmonic oscillators

It is also possible to introduce complex coordinates, say

$$Z = \frac{1}{\sqrt{2}}(Q + iP), \ \overline{Z} = \frac{1}{\sqrt{2}}(Q - iP), \tag{3.6.1}$$

so that

$$H = \omega \overline{Z} Z. \tag{3.6.2}$$

This point of view replaces the even-dimensional real vector space representing phase space with a complex vector space. The Euclidean scalar product in the real vector space is replaced by the Hermitian scalar product. In complex coordinates we find the equations of motion

$$\frac{d}{dt}Z = -i\omega Z = \{Z, H\}, \tag{3.6.3}$$

$$\frac{d}{dt}\overline{Z} = i\omega \overline{Z} = \{\overline{Z}, H\}, \tag{3.6.4}$$

bearing in mind that

$$\{Z, \overline{Z}\} = i. \tag{3.6.5}$$

Note that, by doing so, we are considering an extension of Poisson brackets to complex coefficients. The resulting solutions are

$$Z(t) = e^{-i\omega t} Z(0), \ \overline{Z}(t) = e^{i\omega t} \overline{Z}(0). \tag{3.6.6}$$

It is now possible to extend these new coordinates also to higher dimensions or many oscillators, finding

$$H = \sum_k \omega_k \overline{Z}_k Z_k, \tag{3.6.7}$$

$$\frac{d}{dt}Z_k = -i\omega_k Z_k = \{Z_k, H\}, \tag{3.6.8}$$

$$\frac{d}{dt}\overline{Z}_k = i\omega_k \overline{Z}_k = \{\overline{Z}_k, H\}. \tag{3.6.9}$$

We notice that, by moving the imaginary unit to the left-hand side,

$$i\frac{d}{dt}Z_k = \omega_k Z_k, \tag{3.6.10}$$

$$-i\frac{d}{dt}\overline{Z}_k = \omega_k \overline{Z}_k. \tag{3.6.11}$$

What we learn from previous considerations is that the study of the classical harmonic oscillator uses the same mathematical equations we shall encounter in dealing with finite-dimensional Hilbert spaces in quantum mechanics. Of course, this should not be confused with the mathematical aspects of the quantum harmonic oscillator, which requires in any case an infinite-dimensional Hilbert space.

Let us now make some further considerations on classical harmonic oscillators with many degrees of freedom. We notice that $\overline{Z}_k Z_k$ are constants of motion for any $k \in \{1, \ldots, n\}$. If some of the frequencies are pairwise commensurable, i.e.

$$\omega_k = \omega n_k, \ \omega_j = \omega n_j, \tag{3.6.12}$$

we find that

$$\frac{d}{dt} \left(\overline{Z}_k \right)^{n_j} \left(Z_j \right)^{n_k} = i \sum_k \left(n_j \omega n_k - n_k \omega n_j \right) \overline{Z}_k^{n_j} Z_j^{n_k} = 0. \tag{3.6.13}$$

In particular, if $\omega = \omega_k$ for any $k \in \{1, \ldots, n\}$, the oscillator is called isotropic and all monomials $\overline{Z}_k Z_j$ are constants of motion. It follows that, for any Hermitian matrix A, the real-valued quadratic expression $\overline{Z}_k A^{jk} Z_j$ is a constant of motion. It is easily seen that the equations of motion for the isotropic harmonic oscillator are invariant under the full general linear group $GL(n, C)$, while the constants of motion are associated with Hermitian matrices. In particular, this means that there are many more symmetries than constants of motion. In other words, while constants of motion always generate symmetries, *there are symmetries of the equations of motion which do not give rise to constants of motion*.

3.7 Canonical transformations

Any coordinate transformation

$$\Phi : \mathbf{R}^n \times (\mathbf{R}^n)^* \to \mathbf{R}^n \times (\mathbf{R}^n)^*$$

which preserves the Poisson bracket is called a *canonical transformation*. If we consider the action of Φ on functions, i.e. $(\Phi^* f)(\xi) = f(\Phi^{-1}(\xi))$, we say that Φ is canonical with respect to the Poisson bracket if

$$\Phi^*(\{f, g\}) = \{\Phi^* f, \Phi^* g\}. \tag{3.7.1}$$

There are many equivalent ways to formulate the theory of canonical transformations; we note that, by virtue of our previous remarks on the derivation properties of the Poisson bracket, it is enough to check the canonicity of the transformation on symplectic coordinates (q^a, p_a), i.e. if

$$\Phi(q^a, p_a) = (Q^b, P_b), \tag{3.7.2}$$

Φ is canonical provided that

$$\left\{ Q^b, Q^c \right\} = 0, \ \{P_b, P_c\} = 0, \ \left\{ Q^b, P_c \right\} = \delta^b_{\ c}, \tag{3.7.3}$$

whenever the starting coordinates satisfy the relations

$$\left\{ q^b, q^c \right\} = 0, \ \{p_b, p_c\} = 0, \ \left\{ q^b, p_c \right\} = \delta^b_{\ c}. \tag{3.7.4}$$

We have seen that functions on phase space may generate 1-parameter groups of transformations. We may wonder if transformations which are not part of a 1-parameter group may also admit a generating function. This is indeed the case, and a more convenient characterization of the canonical transformation is given in terms of the 'symplectic potential', i.e. the 1-form $\theta \equiv p_a dq^a$. We say that a transformation $\Phi : (q,p) \to (Q,P)$ is canonical if there is a differentiable function S_Φ such that

$$p_a dq^a - P_b dQ^b = dS_\Phi(q, Q). \tag{3.7.5}$$

The function S_Φ is called the *generating function* of the canonical transformation Φ. From this relation we find

$$\left(p_a - \frac{\partial S_\Phi}{\partial q^a}\right) dq^a - \left(P_b + \frac{\partial S_\Phi}{\partial Q^b}\right) dQ^b = 0. \tag{3.7.6}$$

Thus, by virtue of the arbitrariness of dq^a, dQ^b,

$$p_a = \frac{\partial S_\Phi}{\partial q^a}, \ P_b = -\frac{\partial S_\Phi}{\partial Q^b}. \tag{3.7.7}$$

We have to solve the first equation in terms of Q^b expressed as a function of (q^a, p_a), and insert it into the second equation in (3.7.7). Of course, for us to be able to solve for Q^b, the implicit function theorem yields a non-singularity condition, i.e.

$$\det\left(\frac{\partial^2 S_\Phi}{\partial q^a \partial Q^b}\right) \neq 0, \tag{3.7.8}$$

which has to be satisfied on the region of phase space where the canonical transformation may be described by a generating function.

3.8 Time-dependent Hamiltonian formalism

In a time-dependent formalism, the configuration space $\mathbf{R}^n$ is replaced by an extended configuration space $\mathbf{R}^n \times \mathbf{R}$, where an additional time variable is considered. The associated extended phase space is $\left(\mathbf{R}^n \times \mathbf{R}\right) \times \left(\mathbf{R}^n \times \mathbf{R}\right)^*$.

Often, by time-dependent formalism one simply means an odd-dimensional carrier space given by

$$\mathbf{R} \times \left[\mathbf{R}^n \times \mathbf{R}^n\right]^*,$$

i.e. there is no conjugate variable for the time coordinate. In such an approach we consider $H = H(q, p; t)$ and define the corresponding equations of motion as being given by

$$\frac{dt}{dt} = 1, \ \frac{dq^a}{dt} = \{q^a, H\}, \ \frac{dp_a}{dt} = \{p_a, H\}. \tag{3.8.1}$$

The Poisson bracket is evaluated, as usual, with the time variable considered as a parameter; very often these brackets are called *equal-time* brackets.

Canonical transformations are a 'kinematical' way to go from one canonical coordinate system to another. It is possible to investigate if such a change of canonical coordinates may lead us from one Hamiltonian dynamical system to another, showing that two apparently different systems are described by the same canonical equations written in different coordinate systems. In this way our transformation takes also into account dynamical information. To achieve this result, the symplectic potential $p_a dq^a$ to $p_a dq^a - H dt$ is extended; it is still possible to define generating functions of canonical transformations by setting

$$(p_a dq^a - H dt) - (P_b dQ^b - K dt) = dS(q^a, Q^b; t). \tag{3.8.2}$$

Once again, by an argument similar to the previous one used in Eq. (3.7.6), we find that

$$p_a = \frac{\partial S}{\partial q^a}, \ P_b = \frac{\partial S}{\partial Q^b}, \ K - H = \frac{\partial S}{\partial t}. \tag{3.8.3}$$

These equations, as previously, must be understood as expressed either in terms of $(q^a, p_a; t)$ or in terms of $(Q^b, P_b; t)$. In either case it is clear that the time variable is not affected by the transformation. Such a set of transformations make it possible to preserve the notion of simultaneity, as it occurs for instance in Galilean relativity. As a result, the two systems, described by the Hamiltonians K and H, are diffeomorphic (see Chapter 15) in the region where an S exists satisfying Eq. (3.8.2).

In those situations, e.g. special relativity, where we would like to have the possibility of mixing position and time variables, we have to consider an enlarged notion of extended phase space. For this purpose, we introduce the carrier space

$$\left(\mathbf{R} \times \mathbf{R}^n \right) \times \left(\mathbf{R} \times \mathbf{R}^n \right)^*,$$

and we also consider a conjugate variable for the time coordinate. The previous Hamiltonian function $H(q, p; t)$ is extended to $\mathcal{H} = H(q, p; t) + p_0$, where p_0 is the 'momentum' variable conjugate to $t = q^0$, and we extend earlier Poisson brackets by adding $\{q^0, p_0\} = 1$. The equations of motion associated with a generic Hamiltonian function $\mathcal{H}$ read as

$$\frac{dq^a}{ds} = \frac{\partial \mathcal{H}}{\partial p_a}, \ \frac{dq^0}{ds} = \frac{\partial \mathcal{H}}{\partial p_0}, \tag{3.8.4}$$

$$\frac{dp_a}{ds} = -\frac{\partial \mathcal{H}}{\partial q^a}, \ \frac{dp_0}{ds} = -\frac{\partial \mathcal{H}}{\partial q^0}, \tag{3.8.5}$$

where the evolution parameter s is no longer identified with the coordinate time function. Note that, while in the odd-dimensional time-dependent situation $\frac{d}{dt}H = \frac{\partial H}{\partial t}$, in this extended framework $\frac{d}{ds}\mathcal{H} = 0$. If we choose the arbitrary additive constant in $\mathcal{H}$ in such a way that $p_0 = -H(q, p; t)$, we find that these extended Hamilton equations are equivalent to the earlier set provided that we restrict the Cauchy data of the extended equations to be elements of the sub-manifold

$$\Sigma_0 \equiv \{(q, p; t, p_0) : p_0 + H(q, p; t) = 0\}. \tag{3.8.6}$$

To be more specific, the solutions of the equations of motion in our previous odd-dimensional time-dependent formalism depend on $2n + 1$ initial data. Thus, by restricting the analysis to Σ_0, we still have equations of motion depending on $2n + 1$ parameters.

The symplectic potential $p_a dq^a + p_0 dq^0$, when restricted to Σ_0, becomes $p_a dq^a - H \, dt$. It should be clear now that, as long as we perform only transformations which preserve

Σ_0, we are also allowed to mix the time coordinate with other phase-space coordinates. In particular, we may perform Lorentz or Poincaré transformations without changing the Hamiltonian description we have offered. An instance of this extended formalism is given by the phase space $\mathbf{R}^4 \times \left(\mathbf{R}^4\right)^*$ with Hamiltonian function

$$\mathcal{H} = -p_0^2 + p_1^2 + p_2^2 + p_3^2 + m_0^2 c^2.$$

The equations of motion depend on the differential of $\mathcal{H}$, therefore the mass term does not play any role. To take into account the information on the mass we restrict to initial conditions which satisfy $\mathcal{H}(q(0), p(0)) = 0$, i.e. they are taken on the hypersurface $\Sigma_0 \equiv \{(q, p) : \mathcal{H}(q, p) = 0\}$ that identifies the mass-shell relation of a point particle. This means that the equations of motion we derive from $\mathcal{H}$ are appropriate for any free particle of any mass. We restrict our considerations to a particular particle with a fixed mass by selecting Cauchy data belonging to the corresponding mass-surface. Σ_0 is clearly preserved by all canonical transformations associated with the inhomogeneous Lorentz group.

3.9 Hamilton–Jacobi equation

Given a Hamiltonian function $H = H(q, p)$ on the phase space $\mathbf{R}^n \times \left(\mathbf{R}^n\right)^*$, we have constructed a first-order differential equation by means of the standard Poisson bracket, i.e.

$$\frac{\mathrm{d}q^a}{\mathrm{d}t} = \frac{\partial H}{\partial p_a}, \quad \frac{\mathrm{d}p_a}{\mathrm{d}t} = -\frac{\partial H}{\partial q^a}. \tag{3.9.1}$$

It is possible to associate with H another differential equation, a first-order partial differential equation, called the Hamilton–Jacobi equation associated with H, defined by replacing the variable p with $\frac{\partial S}{\partial q}$, i.e.

$$H\left(q^a, \frac{\partial S}{\partial q^a}\right) = E. \tag{3.9.2}$$

In this way, we get a first-order partial differential equation for $S : \mathbf{R}^n \to \mathbf{R}$.

When the Hamiltonian function depends also on the time variable t, Eq. (3.9.2) is replaced by

$$H\left(q^a, \frac{\partial S}{\partial q^a}; t\right) = -\frac{\partial S}{\partial t}. \tag{3.9.3}$$

We notice that this equation is a particular instance of Eq. (3.8.3) when $K = 0$; in this particular case $\dot{Q} = 0, \dot{P} = 0$, hence the Q variables in K appear as parameters. On dealing with time-dependent Hamiltonians, the unknown function S is often denoted by $W : \mathbf{R} \times \mathbf{R}^n \to \mathbf{R}$, which is called the Hamilton *principal function*, while the name of Hamilton's *characteristic function* is reserved for S.

A solution of Eq. (3.9.2) is said to be a *complete solution* if S depends on n additional parameters, which include E, such that, when these parameters vary, the correspondence $q \to \frac{\partial S}{\partial q} = p$ is a diffeomorphism between the configuration space and the momentum space. In this case, $S : \mathbf{R}^n \times \mathbf{R}^n \to \mathbf{R}$ and the essential dependence on additional parameters λ means

$$\det\left(\frac{\partial^2 S}{\partial q^a \partial \lambda^b}\right) \neq 0,$$

i.e. the same conditions we encountered when dealing with generating functions of canonical transformations, with λ^b playing the role of Q^b.

By means of solutions of Eq. (3.9.2), it is possible to associate with the Hamilton equation defined on phase space a first-order differential equation on configuration space, so that

$$\frac{\mathrm{d}q^a}{\mathrm{d}t} = \left.\frac{\partial H}{\partial p_a}\right|_{p_a = \frac{\partial S}{\partial q^a}}. \tag{3.9.4}$$

Integral curves of this equation define a congruence of lines on configuration space. If we take any solution of this equation, e.g. $q^a(t) = q^a(q^0; t)$, we may define a *curve in phase space* by setting

$$q^a(t) = q^a(q^0; t), \; p_a(t) = \frac{\partial S}{\partial q^a}(q^a(t)). \tag{3.9.5}$$

These curves, so constructed, will be solutions of the Hamilton equation. When we let q^0 vary over the whole configuration space the family of curves we generate on phase space will span an n-dimensional surface, which may be identified with the graph of the map

$$q^a \in \mathbf{R}^n \to \left(q^a, \frac{\partial S}{\partial q^a}\right) \in \mathbf{R}^n \times \left(\mathbf{R}^n\right)^*.$$

Occasionally, we may denote this surface by Σ_S. This surface is invariant under the evolution defined by Eq. (3.9.1), and the restriction to it of the symplectic potential is equal to dS. Therefore, looking for invariant surfaces on which the symplectic potential becomes an exact 1-form gives another way to find solutions of the Hamilton–Jacobi equation; they are often called 'generalized solutions'.

By using our previous considerations, we can understand why, in the Hamilton–Jacobi theory, it is possible to show that a solution may be found by integrating the Lagrangian function, which describes the dynamical system we are considering, along solutions of the Euler–Lagrange equations of motion, i.e.

$$W(q_f, t_f; q_i, t_i) = \int_{q_i, t_i}^{q_f, t_f} \mathcal{L}(q, \dot{q}, t')\mathrm{d}t'. \tag{3.9.6}$$

As a matter of fact, if we note that

$$\mathcal{L}\mathrm{d}t = (p\dot{q} - H)\mathrm{d}t = p\mathrm{d}q - H\mathrm{d}t,$$

it is clear that, along integral curves of the equations of motion,

$$\mathrm{d}W(q_f, t_f; q_i, t_i) = (p_f \mathrm{d}q_f - H_f \mathrm{d}t) - (p_i \mathrm{d}q_i - H_i \mathrm{d}t), \tag{3.9.7}$$

where $H_f = H_i$ if the Hamiltonian is a constant of motion (i.e. H is time-independent). Thus, W becomes the generating function of the canonical transformation associated with time evolution, i.e. the unfolding in time of the initial conditions.

We now provide some simple examples of solutions for some Hamilton–Jacobi equations.

Example 1: Free particle on a line

For a free particle on a line, the Hamiltonian is $H = \frac{p^2}{2m}$ and we consider two complete integrals of the Hamilton–Jacobi equation. The integral curves read as

$$q(t) = \frac{p}{m}t + q_i, \; p(t) = p_i, \; v = \frac{p}{m}, \tag{3.9.8}$$

and hence

$$W_1 = \int_{q_i,t_i}^{q_f,t_f} \frac{mv^2}{2} \mathrm{d}t = \int_{q_i,t_i}^{q_f,t_f} \frac{m}{2}\left(\frac{q(t)-q_i}{t_f-t_i}\right)^2 \mathrm{d}t = \frac{m}{2}\frac{(q_f-q_i)^2}{(t_f-t_i)}. \tag{3.9.9}$$

This solution is a complete integral, with 'parameters' (q_i, t_i).

For the second integral we write $\frac{mv^2}{2} = \frac{pv}{2}$, then

$$\int_{q_i,t_i}^{q_f,t_f} \frac{mv^2}{2} \mathrm{d}t = \frac{p_f v_f}{2}(t_f-t_i) = \frac{k}{2}(q_f-q_i) = W_2, \tag{3.9.10}$$

with $p_f = p_i = k$, the parameters being (k, q_i).

Example 2: Carrier space $\mathbf{R} \times \mathbf{R}^*$

We consider a Hamiltonian function given by

$$H = \frac{p^2}{2m} - mtkq. \tag{3.9.11}$$

The solutions of the equations of motion are therefore

$$q(t) = q_0 + \frac{p_0}{m}t + \frac{1}{6}kt^3, \tag{3.9.12}$$

$$p(t) = m\dot{q}(t) = p_0 + \frac{1}{2}mkt^2. \tag{3.9.13}$$

The Lagrangian $\mathcal{L}$ corresponding to H reads as

$$\mathcal{L}(t) = \frac{p_0^2}{2m} + mkq_0t + \frac{3}{2}kp_0t^2 + \frac{7}{24}mk^2t^4. \tag{3.9.14}$$

Integration of $\mathcal{L}$ along the solutions yields

$$W(q_f, t_f; q_i, 0) = \frac{p_i^2}{2m}t + \frac{1}{2}mkq_it^2 + \frac{1}{2}kp_it^3 + \frac{7}{120}mk^2t^5. \tag{3.9.15}$$

A practical way to find solutions of the Hamilton–Jacobi equation for the Hamilton equations of motion is to consider constants of motion. If $f_1, f_2, \ldots, f_n$ are constants of motion functionally independent along the momenta, i.e.

$$\{f_j, H\} = 0, \; \forall j = 1, \ldots, n, \tag{3.9.16}$$

and moreover

$$\det\left(\frac{\partial f_j}{\partial p_k}\right) \neq 0, \tag{3.9.17}$$

we may fix values for the constants of motion, e.g. $f_j = \lambda_j$, and solve for the p, i.e.

$$p_j = p_j(q, \lambda_1, \ldots, \lambda_n), \ \forall j = 1, \ldots, n. \tag{3.9.18}$$

On inserting these functions into $p_a \mathrm{d}q^a$ we find

$$p_a(q, \lambda_1, \ldots, \lambda_n)\mathrm{d}q^a = \mathrm{d}S(q, \lambda_1, \ldots, \lambda_n) \tag{3.9.19}$$

if and only if the functions f_j satisfy

$$\{f_j, f_k\} = 0, \ \forall j, k \in \{1, \ldots, n\}. \tag{3.9.20}$$

This is a practical way of finding the invariant surface Σ_S that we have introduced after Eq. (3.9.5). Therefore, if we have started with pairwise commuting functions, by solving with respect to momenta we may construct a solution of the Hamilton–Jacobi equation from Eq. (3.9.19).

3.10 Motion of surfaces

To be able to visualize our considerations, we consider a two-dimensional configuration space, i.e. $\mathbf{R}^2$. On the plane we consider a function $W(x, y; t)$ and fix a value of it, e.g. $W(x, y; t) = c$. For each value of t we get a line, and assume it is closed. As t changes and takes the values $t_1 < t_2 < \cdots < t_n$, we obtain a family of concentric lines. For each line, we consider

$$\mathrm{grad}\, W_t = \left(\frac{\partial W_t}{\partial x}, \frac{\partial W_t}{\partial y} \right),$$

which is orthogonal to the line $W(x, y; t) = c$. The first-order differential equation

$$\frac{\mathrm{d}x}{\mathrm{d}t} = \frac{\partial W_t}{\partial x}, \frac{\mathrm{d}y}{\mathrm{d}t} = \frac{\partial W_t}{\partial y} \tag{3.10.1}$$

has flux lines orthogonal to the surfaces. Thus, any solution of the Hamilton–Jacobi equation determines wave fronts, $W(x, y; t) \equiv c$, transversal to $\dot{\xi} = \mathrm{grad}\, W$.

The case of the Hamilton–Jacobi associated with the free particle in $\mathbf{R}^4$, with solutions

$$W = \frac{m}{2t}\left[(x - x_0)^2 + (y - y_0)^2 + (z - z_0)^2 \right], \tag{3.10.2}$$

provides, by setting $W = c$ and cutting this hypersurface of $\mathbf{R}^4$ with the hyperplane $t = k$, a family of concentric spheres with centre (x_0, y_0, z_0). The gradient gives the radial vector field. The equations of motion on $\mathbf{R}^4$ are obtained by starting with

$$\frac{\mathrm{d}\vec{x}}{\mathrm{d}t} = \frac{\vec{p}}{m}, \frac{\mathrm{d}\vec{p}}{\mathrm{d}t} = 0, \frac{\mathrm{d}x_0}{\mathrm{d}t} = 1. \tag{3.10.3}$$

Thus,

$$\frac{\mathrm{d}\vec{x}}{\mathrm{d}t} = \frac{1}{t}(\vec{x} - \vec{x}_0), \frac{\mathrm{d}x_0}{\mathrm{d}t} = 1 \tag{3.10.4}$$

is the equation of motion on $\mathbf{R}^4$ obtained by replacing $(\vec{p}, p_0)$ with

$$\left(\operatorname{grad} W, \frac{\partial W}{\partial t} \right).$$

we find (summation over repeated indices is understood)

$$\frac{dW}{dt} = \frac{\partial W}{\partial \vec{x}} \frac{d\vec{x}}{dt} + \frac{\partial W}{\partial t} = p_i \dot{x}^i - H = \mathcal{L}. \tag{3.10.5}$$

By considering now an Hamiltonian function of the classical type, the 'wave front' moves with velocity

$$|\operatorname{grad} W| \, n_i \frac{dx^i}{dt} = -\frac{\partial W}{\partial t}, \tag{3.10.6}$$

i.e.

$$u = \frac{E}{|\operatorname{grad} W|}, \tag{3.10.7}$$

where

$$|\operatorname{grad} W| = |\vec{p}| = \sqrt{2m(E - V(x))}, \tag{3.10.8}$$

which implies that

$$u = \frac{E}{\sqrt{2m(E - V(x))}}. \tag{3.10.9}$$

To sum up, the motion of a codimension-1 surface[2] on configuration space may be described by the differential equation defined by its gradient. On the other hand, given a differential equation on $\mathbf{R}^4$, if this is the gradient of a function, we may associate with it a family of codimension-1 surfaces. Thus, for any Hamiltonian system described by the Hamiltonian function of mechanical type, i.e. $H = \frac{\vec{p} \cdot \vec{p}}{2m} + V(\vec{x})$, we find half of the Hamilton equations $\frac{d\vec{q}}{dt} = \frac{\vec{p}}{m}$ to be of the form

$$\frac{d\vec{q}}{dt} = \frac{\operatorname{grad} S}{m}, \tag{3.10.10}$$

whenever we replace p by means of the gradient of a solution of the Hamilton–Jacobi equation.

Once more, in conclusion, from a solution S of the Hamilton–Jacobi equation we construct the first-order differential equation (3.10.10) on the configuration space. Having solved this equation to find $q(t) = q(t; q_0)$, we construct

$$p(t) = \frac{\partial S}{\partial q}(q(t; q_0)). \tag{3.10.11}$$

The curve $(q(t), p(t))$ in phase space obtained in such a way provides a solution of the Hamilton equations. Thus, *at least locally, there is a one-to-one correspondence between solutions of Hamilton–Jacobi and solutions of the Hamilton equations*. On reverting to solutions $u = A e^{iF}$ of the wave equation, we find that the phase of a wave solution is related to the action, which we obtain by integrating the Lagrangian function along solutions of the equations of motion.

---

[2] A p-dimensional surface embedded in n-dimensional space is said to have codimension $(n - p)$.

Appendix 3.A　Space–time picture

The unification of space and time into a four-dimensional continuum has been one of the greatest achievements of the last century, mostly due to the work of Einstein boosted by the work of Minkowski. Thus, space–time is a four-dimensional manifold M^4 whose points are *events*, every event being associated with an instant of time and a location in space. The description of the external world by any observer requires identifications of instants of time and location. Therefore, a reference frame $\mathcal{F}$ on space–time will require first an association of an instant of time with any event, i.e. a projection

$$t : M^4 \to \mathbf{R}, \; t_r = t(m) \in \mathbf{R}. \tag{3.A.1}$$

The inverse image $t^{-1}(a), a \in \mathbf{R}$, is the space of simultaneous events in M^4. By letting $a \in \mathbf{R}$ vary over the whole of $\mathbf{R}$, we define a family of leaves (see Chapter 15 for the definition), each leaf $t^{-1}(a)$ being diffeomorphic (see Chapter 15) with $\mathbf{R}^3$. Even though we require that all leaves are diffeomorphic among them, there is no natural isomorphism (by natural one means a map whose existence is guaranteed by the mere existence of the spaces related by it) between the leaves. To compare different locations at different instants of time any reference frame postulates the existence of a first-order differential equation on M^4 admitting solutions for any time, from $-\infty$ to $+\infty$, and for any initial location. In this way, by selecting some initial simultaneity surface $t^{-1}(a_0)$ as a family of initial conditions, we require that the evolution of this leaf after a time τ should coincide with the simultaneity leaf $t^{-1}(a_0 + \tau)$.

Thus, every frame defines a splitting of space–time M^4 into time $\mathbf{R}$ and space $\mathbf{R}^3$ so that $M^4 \equiv \mathbf{R} \times \mathbf{R}^3$, the splitting depending on the particular reference frame, i.e. the projection t and the chosen differential equation. Since a family of integral curves with the stated requirements is equivalent to a vector field, the field of velocities, represented by a homogeneous first-order differential operator, say $X = v_\mu \frac{\partial}{\partial x_\mu}$, we could say that a reference frame is represented by a pair (dt, X), with the requirement

$$dt(X) = 1. \tag{3.A.2}$$

We have replaced the projection $t : M^4 \to \mathbf{R}$ with its differential so that the origin of time plays no privileged role. It is possible to define alternative frames by selecting different pairs (dt', X').

Remark　When relativistic concepts are extended to include gravitational phenomena and hence all laws of classical physics, the resulting framework requires that some of the assumptions on X and the time 1-form $\alpha = dt$ should be released. For details we refer to the work in de Ritis *et al.* (1999), Marmo and Preziosi (2006).

Compatible reference frames identify alternative ways of describing the external world in such a way that experimental observations in one frame may be compared with those of the other frame. A minimal requirement, stated for any simplified assumption, is that

$$dt(X') \neq 0, \; dt'(X) \neq 0. \tag{3.A.3}$$

In general, the attention is restricted to pairwise compatible frames which may be obtained one from the other by means of the action of a group of transformations on M^4. The selected group is usually called a relativity group for the situation we are examining. In this book we shall restrict our attention to the inhomogeneous Lorentz group, or Poincaré group; and the Galilei group; the reference frames used to arrive at these groups are called inertial frames. Let us discuss very briefly a dynamical definition of inertial frames.

3.A.1 Inertial frames and comparison dynamics

To avoid many circular definitions of inertial frames encountered in the literature we present here a dynamical definition, closer in spirit to the conceptual framework of general relativity, which is the extension of special relativity to include all gravitational phenomena. On M^4 we consider a second-order differential equation to be viewed as a *comparison dynamics*. In some arbitrary coordinate system,

$$\frac{d^2 x^\mu}{ds^2} = f^\mu \left(x, \frac{dx}{ds} \right). \tag{3.A.4}$$

If we set $v^\mu = \frac{dx^\mu}{ds}$, we associate with our equation the following tensor field (we do not prove here that it is a tensor field on the space of positions and velocities, for a proof we may refer to the work in Morandi *et al.* (1990)):

$$R_{\mu\nu}{}^\rho = \frac{1}{2} \left[\frac{\partial^2 f^\rho}{\partial x^\mu \partial v^\nu} - \frac{\partial^2 f^\rho}{\partial v^\mu \partial x^\nu} + \frac{1}{2} \left(\frac{\partial^2 f^\rho}{\partial u^\nu \partial u^\sigma} \frac{\partial f^\sigma}{\partial u^\mu} - \frac{\partial^2 f^\rho}{\partial u^\mu \partial u^\sigma} \frac{\partial f^\sigma}{\partial u^\nu} \right) \right]. \tag{3.A.5}$$

Any second-order differential equation on M^4 such that $R_{\mu\nu}^\rho = 0$ may be put in the form

$$\frac{d^2 \xi^\mu}{ds^2} = 0 \tag{3.A.6}$$

for an appropriate choice of coordinates (ξ, μ) on M^4. The second-order differential equation is equivalent to the family of first-order differential equations

$$\frac{d\xi^\mu}{ds} = v^\mu, \tag{3.A.7}$$

one equation for each choice of (v^1, v^2, v^3, v^4). Clearly, solutions of this equation are

$$\xi^\mu = \xi^\mu(0) + v^\mu s. \tag{3.A.8}$$

Solutions are defined from $-\infty$ to $+\infty$ in the parameter s and for any initial condition $(\xi^1, \xi^2, \xi^3, \xi^4)$. Therefore, letting v^μ vary we obtain all solutions of the starting second-order differential equation. These vector fields are candidates to represent reference frames once the 1-form dt will be chosen. According to our previous considerations, let us first look for putative relativity groups. First, we notice that our comparison dynamics, written in the form $\frac{d^2 \xi^\mu}{ds^2} = 0$, is invariant under the full inhomogeneous general linear group in four dimensions, denoted by $IGL(4, \mathbf{R})$. The family of all constants of motion for our equation is generated by v^μ and $x^\mu v^\nu - v^\mu x^\nu$. Let us look now for a Lagrangian description of our comparison dynamics.

3.A.2 Lagrangian descriptions of second-order differential equations

We say that a second-order differential equation

$$\frac{d^2 x^\mu}{ds^2} = f^\mu(x, v), \quad \frac{dx^\mu}{ds} = v^\mu \tag{3.A.9}$$

admits a Lagrangian description if the Euler–Lagrange equations, viewed as a partial differential equation in the following way:

$$\frac{\partial^2 \mathcal{L}}{\partial v^\mu \partial v^\nu} f^\nu + \frac{\partial^2 \mathcal{L}}{\partial v^\mu \partial x^\nu} v^\nu - \frac{\partial \mathcal{L}}{\partial x^\mu} = 0 \tag{3.A.10}$$

have solutions satisfying

$$\det \left\| \frac{\partial^2 \mathcal{L}}{\partial v^\mu \partial v^\nu} \right\| \neq 0. \tag{3.A.11}$$

Indeed, when this condition is satisfied, the Euler–Lagrange equation

$$\frac{d}{ds} \frac{\partial \mathcal{L}}{\partial v^\mu} - \frac{\partial \mathcal{L}}{\partial x^\mu} = 0, \quad \frac{dx^\mu}{ds} - v^\mu = 0, \tag{3.A.12}$$

which defines an implicit differential equation, may be put into explicit (or normal) form

$$\frac{dv^\mu}{ds} - f^\mu = 0, \quad \frac{dx^\mu}{ds} - v^\mu = 0. \tag{3.A.13}$$

On reverting to our present setting, where $f^\mu = 0$, it is clear that any function which depends only on the velocities will solve the associated partial differential equation for $\mathcal{L}$. Particular solutions are provided by $\mathcal{L} = g_{\mu\nu} v^\mu v^\nu$, with $g_{\mu\nu} \in \mathbf{R}$ and $\det \| g_{\mu\nu} \| \neq 0$. For instance

$$\mathcal{L} = (v^0)^2 - (v^1)^2 - (v^2)^2 - (v^3)^2.$$

If we drop the requirement that the Hessian matrix in Eq. (3.A.11) should be non-singular, then another solution is provided by

$$\mathcal{L} = \sqrt{(v^0)^2 - (v^1)^2 - (v^2)^2 - (v^3)^2}.$$

The solutions we are discussing pertain to those which are associated with metric tensors $g_{\mu\nu} dx^\mu \otimes dx^\nu$, or, in the particular case,

$$\eta_{\mu\nu} dx^\mu \otimes dx^\nu = dx^0 \otimes dx^0 - dx^1 \otimes dx^1 - dx^2 \otimes dx^2 - dx^3 \otimes dx^3.$$

Now it is possible to select a subgroup of the inhomogeneous general linear group by requiring that it also preserves the Lagrangian function we may select to describe our comparison dynamics. In this way, with our last choice, we end up with the inhomogeneous Lorentz group, i.e. the Poincaré group.

A family of pairwise compatible reference frames is obtained by selecting

$$X = v^j \frac{\partial}{\partial x^j} + \frac{\partial}{\partial x^0}, \quad dt = dx^0, \tag{3.A.14}$$

and transforming this pair with the Poincaré group. If we introduce the speed of light c and write $x^0 = ct$, we find that the metric tensor associated with the Lagrangian is

$$\eta = c^2 \mathrm{d}t \otimes \mathrm{d}t - \mathrm{d}\vec{x} \otimes \mathrm{d}\vec{x}. \tag{3.A.15}$$

When we write this tensor in contravariant form we get

$$G = \frac{1}{c^2} \frac{\partial}{\partial t} \otimes \frac{\partial}{\partial t} - \frac{\partial}{\partial \vec{x}} \otimes \frac{\partial}{\partial \vec{x}}. \tag{3.A.16}$$

This contravariant form is particularly interesting because it may be thought of as a symmetric bidifferential operator, yielding

$$G(\mathrm{d}f_1, \mathrm{d}f_2) = \frac{1}{c^2} \frac{\partial f_1}{\partial t} \frac{\partial f_2}{\partial t} - \frac{\partial f_1}{\partial \vec{x}} \cdot \frac{\partial f_2}{\partial \vec{x}}. \tag{3.A.17}$$

Note that, for $f_1 = f_2 = S$, by equating the result to zero we get the Hamilton–Jacobi equation. We may also regard the above equation as defining a second-order differential operator, possibly giving

$$\frac{1}{c^2} \frac{\partial}{\partial t} \left(\frac{\partial U}{\partial t} \right) - \frac{\partial}{\partial \vec{x}} \left(\frac{\partial U}{\partial \vec{x}} \right) = 0. \tag{3.A.18}$$

We may say that the Poincaré group emerges either as a symmetry group of the Lagrangian function describing the comparison dynamics, or as a symmetry group of the d'Alembert equation for the scalar function U, or as a symmetry group of the associated Hamilton–Jacobi equation. We have here an example of the profound connection between ordinary differential equations (equations of motion for point particles) and partial differential equations (equations for waves).

Remark   The choice $\mathcal{L} = \sqrt{v_0^2 - \vec{v} \cdot \vec{v}}$, instead of $\mathcal{L} = v_0^2 - \vec{v} \cdot \vec{v}$, while both are solutions of the equation associated with the free comparison dynamics, has other nontrivial different consequences. For instance, the first requires that $v_0^2 \geq \vec{v} \cdot \vec{v}$. The associated action functional for any path γ in M^4 reads as

$$A_{\mathcal{L}}(\gamma) = \int_\gamma \sqrt{\left(\frac{\mathrm{d}x_0}{\mathrm{d}s} \right)^2 - \left(\frac{\mathrm{d}\vec{x}}{\mathrm{d}s} \right)^2} \, \mathrm{d}s, \tag{3.A.19}$$

and is invariant under reparametrization of the evolution parameter s. This invariance is an instance of a new concept, i.e. gauge invariance of the action. If we elaborate further on the covariant and contravariant expressions of the Minkowski metric we find that, while the two expressions contain exactly the same information when c has any finite value, for $c \to \infty$ the covariant expression reduces to $\mathrm{d}t \otimes \mathrm{d}t$, while the contravariant one reduces to $-\frac{\partial}{\partial \vec{x}} \otimes \frac{\partial}{\partial \vec{x}}$, and hence they are no longer equivalent in the $c \to \infty$ limit. The subgroup of the general linear group which preserves both tensors, i.e. $\mathrm{d}t \otimes \mathrm{d}t$ and $\frac{\partial}{\partial \vec{x}} \otimes \frac{\partial}{\partial \vec{x}}$, defines the Galilei group and is considered to be a 'contraction' of the Poincaré group when $c \to \infty$. Again, this group may be used to define a family of pairwise compatible reference frames starting with a given one, say

$$\left(\mathrm{d}t, \frac{\partial}{\partial t} + v_j \frac{\partial}{\partial x_j} \right).$$

It should be noticed that this time, because of the required invariance of $\mathrm{d}t \otimes \mathrm{d}t$, the simultaneity will be preserved when going from one reference frame to another, unlike the case for the Poincaré group.

In this case, if we use the Lagrangian $\mathcal{L} = \frac{m}{2}\vec{v} \cdot \vec{v}$ and the requirement $\frac{\mathrm{d}x_0}{\mathrm{d}t} = c$ to describe our comparison equations of motion, we find that the Lagrangian will not be invariant under Galilei transformations. Let us now consider a specific realization of Galilei transformations. We look only at pure boost transformations, i.e.

$$\vec{x}' = \vec{x} - \vec{x}_0 - \vec{v}t, \; t' = t, \tag{3.A.20}$$

which imply

$$\dot{\vec{x}}' = \dot{\vec{x}} - \vec{v}. \tag{3.A.21}$$

The Lagrangian undergoes the transformation

$$\mathcal{L} = \frac{m}{2}(\dot{\vec{x}}')^2 = \frac{m}{2}\dot{\vec{x}}^2 + \frac{m}{2}\vec{v}^2 - m\dot{\vec{x}} \cdot \vec{v} = \mathcal{L} - \frac{\mathrm{d}}{\mathrm{d}t}\left(m\vec{x} \cdot \vec{v} - \frac{1}{2}m\vec{v}^2 t\right), \tag{3.A.22}$$

while

$$\frac{\partial}{\partial t'} = \frac{\partial t}{\partial t'}\frac{\partial}{\partial t} + \frac{\partial \vec{x}}{\partial t'}\frac{\partial}{\partial \vec{x}} = \frac{\partial}{\partial t} + \vec{v}\frac{\partial}{\partial \vec{x}}, \tag{3.A.23}$$

$$\frac{\partial}{\partial \vec{x}} = \frac{\partial t'}{\partial \vec{x}}\frac{\partial}{\partial t'} + \frac{\partial \vec{x}'}{\partial \vec{x}}\frac{\partial}{\partial \vec{x}'} = \frac{\partial}{\partial \vec{x}'}, \tag{3.A.24}$$

and we verify that $\mathrm{d}t = \mathrm{d}t'$ along with

$$\frac{\partial}{\partial \vec{x}} \otimes \frac{\partial}{\partial \vec{x}} = \frac{\partial}{\partial \vec{x}'} \otimes \frac{\partial}{\partial \vec{x}'}.$$

Remark The equations of motion define solutions, paths in M^4, which render the action stationary. Thus, under Galilei transformations, the action

$$A_{\mathcal{L}}(\gamma) = \int_\gamma \frac{m}{2}\vec{v} \cdot \vec{v}\mathrm{d}t \tag{3.A.25}$$

changes only by boundary terms, i.e. $mxv - \frac{m}{2}v^2 t$ evaluated at the boundary of γ, therefore stationary curves do not change.

We notice also that the phase of a plane wave $\frac{\mathrm{i}}{\hbar}(\vec{p} \cdot \vec{x} - Et)$ will change under Galilei transformations according to

$$\vec{p}' \cdot \vec{x}' - Et' = (\vec{p} \cdot \vec{x} - Et) + \frac{m}{2}\vec{v}^2 t - m\vec{v} \cdot \vec{x}. \tag{3.A.26}$$

From the relation

$$p_i \mathrm{d}x^i - H\mathrm{d}t = \mathcal{L}\mathrm{d}t$$

it follows that the phase changes exactly by the quantity occurring in the variation of the Lagrangian function.

In conclusion, compatible inertial frames associated with the free comparison dynamics detect the same solutions both in the Poincaré and Galilei relativity; however, those associated with the Lorentz group describe the equations of motion in terms of an invariant Lagrangian, while those associated with the Galilei group will use a Lagrangian that changes by a total time derivative. We are going to analyze how this different behaviour affects the association between constants of motion and symmetries.

3.A.3 Symmetries and constants of motion

Let us consider the association of symmetries with constants of motion in the Lagrangian and Hamiltonian formalism. We examine this first in general terms, and then in the particular situation we are interested in. We recall that, given any differential equation (either an ordinary differential equation or a partial differential equation), a transformation on the carrier space is a symmetry if it maps solutions into solutions. For an ordinary differential equation, a function is a constant of motion if, when evaluated along a solution, it acquires a constant value depending only on the chosen solution. For partial differential equations we usually talk about conservation laws rather than constants of motion. A correspondence between symmetries and constants of motion may be established only after we have chosen a Lagrangian function or a Poisson bracket in the Hamiltonian formalism. Let us assess first the Lagrangian formalism. We start from the Euler–Lagrange equations

$$\frac{\mathrm{d}}{\mathrm{d}t}\frac{\partial \mathcal{L}}{\partial v^j} - \frac{\partial \mathcal{L}}{\partial x^j} = 0, \, j = 1, 2, \ldots, n. \tag{3.A.27}$$

We consider a generic vector-valued function $(a^1, a^2, \ldots, a^n)$ and evaluate

$$\begin{aligned} 0 &= a^j \left(\frac{\mathrm{d}}{\mathrm{d}t}\frac{\partial \mathcal{L}}{\partial v^j} - \frac{\partial \mathcal{L}}{\partial x^j} \right) \\ &= \frac{\mathrm{d}}{\mathrm{d}t}\left(a^j \frac{\partial \mathcal{L}}{\partial v^j} \right) - a^j \frac{\partial \mathcal{L}}{\partial x^j} - \frac{\mathrm{d}a^j}{\mathrm{d}t}\frac{\partial \mathcal{L}}{\partial v^j}. \end{aligned} \tag{3.A.28}$$

Whenever the vector-valued function $(a^1, a^2, \ldots, a^n)$ is such that

$$a^j \frac{\partial \mathcal{L}}{\partial x^j} + \frac{\mathrm{d}a^j}{\mathrm{d}t}\frac{\partial \mathcal{L}}{\partial v^j} = \frac{\mathrm{d}}{\mathrm{d}t}F, \tag{3.A.29}$$

we find that $\frac{\mathrm{d}}{\mathrm{d}t}\left(a^j \frac{\partial \mathcal{L}}{\partial v^j} - F \right) = 0$, i.e. we obtain a constant of motion. We have obtained here a generalized version of Noether's theorem, where variations $\delta x^j = \varepsilon a^j$ are not required to be independent of velocities. Of course, when the a^j depend only on the positions, we recover the usual formulation of Noether's theorem. If, on the contrary, a^j also depends on velocities,

$$\frac{\mathrm{d}a^j}{\mathrm{d}t} = \frac{\partial a^j}{\partial x^k}\frac{\mathrm{d}x^k}{\mathrm{d}t} + \frac{\partial a^j}{\partial v^k}\frac{\mathrm{d}v^k}{\mathrm{d}t}; \tag{3.A.30}$$

thus, our expression would also depend on the acceleration. We take care of this drawback by replacing $\frac{\mathrm{d}v^k}{\mathrm{d}t}$ with the forces f^k. In this way we obtain a genuine generalization of the known theorem (Marmo and Mukunda 1986). As a simple example, consider $a^j = v^j$. We then find

$$v^j \frac{\partial \mathcal{L}}{\partial x^j} + f^j \frac{\partial \mathcal{L}}{\partial v^j} = \frac{\mathrm{d}}{\mathrm{d}t}\mathcal{L} \tag{3.A.31}$$

whenever $\frac{\partial \mathcal{L}}{\partial t} = 0$, i.e. $v^j \frac{\partial \mathcal{L}}{\partial v^j} - \mathcal{L}$ is a constant of motion.

On reverting to our 'free comparison dynamics', when we select

$$\mathcal{L} = \frac{m}{2}(v_0^2 - \vec{v} \cdot \vec{v})$$

the Poincaré group preserves the Lagrangian function and we find that transformations associated with constants of motion v_0, v_1, v_2, v_3 are translations (they correspond to what in the old literature are known as cyclic variables). Transformations associated with $x^\mu v^\nu - v^\mu x^\nu$ are those of the Lorentz group. When these transformations are written in terms of infinitesimal generators, we have

$$\frac{\partial}{\partial x_0}, \frac{\partial}{\partial \vec{x}}$$

associated with linear momenta $mv_0, m\vec{v}$, while

$$\vec{x}\frac{\partial}{\partial x_0} + x_0\frac{\partial}{\partial \vec{x}} + \vec{v}\frac{\partial}{\partial v_0} + v_0\frac{\partial}{\partial \vec{v}}$$

are associated with $x_0\vec{v} - \vec{x}v_0$ and are called boosts $\vec{k}$. Finally,

$$x_j\frac{\partial}{\partial x_k} - x_k\frac{\partial}{\partial x_j} + v_j\frac{\partial}{\partial v_k} - v_k\frac{\partial}{\partial v_j}$$

are associated with $m(x_jv_k - x_kv_j)$ and are the infinitesimal rotations associated with angular momentum.

Very often, in choosing the Lagrangian $\mathcal{L} = mc^2\sqrt{1 - \frac{v^2}{c^2}}$ a non-degenerate Lagrangian is obtained out of the degenerate one, $mc\sqrt{v_0^2 - \vec{v} \cdot \vec{v}}$. The particular choice is fixing the parametrization by selecting $x_0 = ct$, and therefore the Lagrangian obtained is no longer invariant under reparametrization of s, it is defined on the phase space of $\mathbf{R}^3$, not on the one associated with $\mathbf{R}^4$. This time, it is clear that, with this Lagrangian, boosts cannot be associated with point transformations. We have the following infinitesimal generators:

$$X_{P_j} = -\frac{\partial}{\partial q_j}, \tag{3.A.32}$$

$$X_{J_j} = -\varepsilon_{jkl}\left(q_k\frac{\partial}{\partial q_l} + v_k\frac{\partial}{\partial v_l}\right), \tag{3.A.33}$$

$$X_{K_j} = q_jv_k\frac{\partial}{\partial q_k} + (v_jv_k - \delta_{jk})\frac{\partial}{\partial v_k}. \tag{3.A.34}$$

Similarly, if we write the free Lagrangian $\mathcal{L} = \frac{m}{2}\vec{v} \cdot \vec{v}$, it is possible to show that the Galilei group will be generated by the infinitesimal generators associated with the following functions:

$$m\vec{v} = \vec{p}, \text{ translations} \tag{3.A.35}$$

$$\frac{m}{2}\vec{v} \cdot \vec{v} = \frac{\vec{p} \cdot \vec{p}}{2m} = H, \text{ time translations} \tag{3.A.36}$$

$$m(\vec{r} \wedge \vec{v}) = \vec{r} \wedge \vec{p} = \vec{L}, \text{ rotations} \tag{3.A.37}$$

$$\vec{K} = m\vec{x}, \text{ boosts} \tag{3.A.38}$$

so as to obtain what is known as the canonical realization of the Lie algebra of the Galilei group, i.e.

$$\{L_j, L_k\} = \varepsilon_{jkl}L_l, \ \{L_j, P_k\} = \varepsilon_{jkl}P_l, \ \{L_j, K_l\} = \varepsilon_{jkm}K_m, \tag{3.A.39}$$

$$\{H, K_j\} = -P_j, \tag{3.A.40}$$

while all other Poisson brackets vanish, where we use the canonical Poisson brackets

$$\{q_j, p_k\} = \delta_{jk}. \tag{3.A.41}$$

The transition to the Hamiltonian formalism is obtained via the Legendre transform

$$\left(q^\mu, v^\mu\right) \rightarrow \left(q^\mu, \frac{\partial \mathcal{L}}{\partial v^\mu} = p_\mu\right). \tag{3.A.42}$$

This transform takes us from TM (see Chapter 15) to T^*M; the image of the Legendre map is the whole T^*M if the Lagrangian is regular. When the Hessian matrix is singular, i.e.

$$\det \left\| \frac{\partial^2 \mathcal{L}}{\partial v^\mu \partial v^\nu} \right\| = 0, \tag{3.A.43}$$

the image will be a proper subset of T^*M and the description on T^*M requires the Dirac–Bergmann procedure (Bergmann 1949, Bergmann and Brunings 1949, Dirac 1964), which we are not going to consider.

It should be remarked that, when the equations of motion we want to describe carry constraints (see below) in addition to explicit equations of motion, the Lagrangian description, if it exists, necessarily requires a degenerate Lagrangian. To see how constraints arise, we consider the example of Maxwell equations in the absence of sources; here the equations of motion are

$$\frac{d}{dt}\vec{B} = -\text{curl}\,\vec{E}, \ \text{div}\,\vec{B} = 0, \tag{3.A.44}$$

$$\frac{d}{dt}\vec{E} = \text{curl}\,\vec{B}, \ \text{div}\,\vec{E} = 0. \tag{3.A.45}$$

We see that two of the equations do not contain time derivatives, they are simply constraints on the possible initial conditions. Starting with Euler–Lagrange equations

$$\frac{\partial^2 \mathcal{L}}{\partial v^\mu \partial v^\nu} \frac{dv^\nu}{ds} + \frac{\partial^2 \mathcal{L}}{\partial v^\mu \partial x^\nu} v^\nu - \frac{\partial \mathcal{L}}{\partial x^\mu} = 0, \tag{3.A.46}$$

if the Hessian matrix is non-singular there will be non-trivial solutions of the equation

$$b^\mu \frac{\partial^2 \mathcal{L}}{\partial v^\mu \partial v^\nu} = 0. \tag{3.A.47}$$

If we choose any such solution and multiply on the left the expanded version of the Euler–Lagrange equations we find

$$b^\mu \frac{\partial^2 \mathcal{L}}{\partial v^\mu \partial v^\nu} v^\nu - b^\mu \frac{\partial \mathcal{L}}{\partial x^\mu} = 0, \tag{3.A.48}$$

which does not contain the acceleration and therefore gives rise to constraints on the allowed initial conditions. This will be the case for the free relativistic particle if we choose the Lagrangian description provided by $\mathcal{L} = mc\sqrt{v_0^2 - \vec{v} \cdot \vec{v}}$. Having made these considerations, it may be convenient to *consider the Hamiltonian formalism by itself, without considering first the Lagrangian description* and then the transition to the phase space via the Legendre transform.

3.A.4 Symmetries and constants of motion in the Hamiltonian formalism

On phase space, say $T^*\mathbf{R}^4$, with coordinates (x^μ, p_ν), we consider the Poisson bracket defined by

$$\left\{x^\mu, p_\nu\right\} = \delta_\nu^\mu, \quad \left\{x^\mu, x^\nu\right\} = 0, \quad \left\{p_\mu, p_\nu\right\} = 0. \tag{3.A.49}$$

Given the equations of motion on phase space

$$\frac{\mathrm{d}}{\mathrm{d}t}x^\mu = g^\mu(x,p), \quad \frac{\mathrm{d}}{\mathrm{d}t}p_\mu = f_\mu(x,p), \tag{3.A.50}$$

we say that they may be given a Hamiltonian description if we have

$$g^\mu = \left\{x^\mu, H\right\}, \quad f_\mu = \left\{p_\mu, H\right\}. \tag{3.A.51}$$

When this is the case, H is called the Hamiltonian function or the generating function of the infinitesimal generator of the 1-parameter group of transformations describing the dynamical evolution. A function F is a constant of motion for the Hamiltonian equations of motion if

$$\begin{aligned} 0 &= \frac{\mathrm{d}F}{\mathrm{d}s} = \frac{\partial F}{\partial x^\mu}\frac{\mathrm{d}x^\mu}{\mathrm{d}s} + \frac{\partial F}{\partial p_\mu}\frac{\mathrm{d}p_\mu}{\mathrm{d}s} \\ &= \frac{\partial F}{\partial x^\mu}\left\{x^\mu, H\right\} + \frac{\partial F}{\partial p_\mu}\left\{p_\mu, H\right\}, \end{aligned} \tag{3.A.52}$$

which, by using the derivation property of the Poisson brackets, may be written

$$0 = \frac{\mathrm{d}F}{\mathrm{d}s} = \{F, H\}, \tag{3.A.53}$$

i.e. F is a constant of motion if it has a vanishing Poisson bracket with the Hamiltonian.

Given an infinitesimal canonical transformation,

$$\delta x^\mu = \varepsilon \left\{x^\mu, G\right\}, \quad \delta p_\mu = \varepsilon \left\{p_\mu, G\right\}, \tag{3.A.54}$$

this maps solutions into solutions of

$$\frac{\mathrm{d}x^\mu}{\mathrm{d}s} = \left\{x^\mu, H\right\}, \quad \frac{\mathrm{d}}{\mathrm{d}s}p_\mu = \left\{p_\mu, H\right\} \tag{3.A.55}$$

if

$$\frac{\mathrm{d}}{\mathrm{d}s}\delta x^\mu = \delta \frac{\mathrm{d}}{\mathrm{d}s}x^\mu, \tag{3.A.56}$$

i.e. the variation of a solution may be added to a solution to provide a new solution. By expanding we find

$$\{\delta x^\mu, H\} = \varepsilon \left\{ \frac{\mathrm{d}}{\mathrm{d}s} x^\mu, G \right\} = \varepsilon \{\{x^\mu, H\}, G\}, \qquad (3.\mathrm{A}.57)$$

and an analogous equation for δp_μ. The relation implies

$$\{x^\mu, \{G, H\}\} = 0, \quad \{p_\mu, \{G, H\}\} = 0, \qquad (3.\mathrm{A}.58)$$

i.e. $\{G, H\} = $ constant. Thus, a necessary and sufficient condition for a canonical transformation to map solutions into solutions is that

$$\mathrm{d}\{G, H\} = 0. \qquad (3.\mathrm{A}.59)$$

Clearly, it is sufficient that $\{G, H\} = 0$, i.e. constants of motion always generate symmetries.

For canonical transformations to map solutions into solutions, i.e. to be symmetries, it is sufficient that Eq. (3.A.59) is fulfilled. Very often it is required that the canonical transformation is a symmetry for H, not for the equation of motion. Let us give an example. On $T^*\mathbf{R}^2$, with coordinates $(x, y; p_x, p_y)$ we consider the Hamiltonian equations defined by $H = \frac{p^2}{2m} + xE$. We find that the equations are

$$\frac{\mathrm{d}x}{\mathrm{d}t} = \frac{p_x}{m}, \frac{\mathrm{d}p_x}{\mathrm{d}t} = -E, \qquad (3.\mathrm{A}.60)$$

$$\frac{\mathrm{d}y}{\mathrm{d}t} = \frac{p_y}{m}, \frac{\mathrm{d}p_y}{\mathrm{d}t} = 0. \qquad (3.\mathrm{A}.61)$$

Clearly, the equations of motion are invariant under translations, with infinitesimal generators $\frac{\partial}{\partial x}, \frac{\partial}{\partial y}$; however, the Hamiltonian changes by a constant because $\frac{\partial H}{\partial x} = E$, therefore p_x is not a constant of motion but

$$\mathrm{d}\{p_x, H\} = 0, \quad \{p_x, H\} \neq 0. \qquad (3.\mathrm{A}.62)$$

The corresponding constant of motion is time-dependent, i.e. $p_x + Et$. We are now ready to revert to our reference frames and define a notion of equivalence.

3.A.5 Equivalent reference frames

We would like to discuss the relations existing between the description provided by different reference frames for the same physical system. The theory of a physical system is fully specified only if we also provide the connection existing between the descriptions of all possible states of the system when observed by different frames. In other terms, we need the transformation theory of what is being observed to be able to compare the experimental findings in different frames.

Within the mathematical context, these observations were made into a paradigm by F. Klein, in the so-called Erlangen Programme, which states that every geometry is specified by its group of transformations and, conversely, every group of transformations specifies a geometry. Following this programme, we say that the Lagrangian formalism is connected with transformations that map second-order differential equations into second-order ones. The Hamiltonian formalism is characterized by the group of transformations which map Hamiltonian systems into Hamiltonian ones, which very often is understood in a restricted sense by requiring that transformations should preserve the Poisson brackets, possibly up to a conformal factor which would be constant on each connected component.

In evolutionary partial differential equations we can consider transformations which map hyperbolic equations into hyperbolic ones. Let us come to a precise formulation of the problem. We consider two frames $\mathcal{F}$ and $\mathcal{F}'$, which differ in the way they describe the physical properties of a given physical system $\mathcal{S}$. They can be displaced and/or rotated relative to each other in space–time, or they can move with respect to each other with a certain relative velocity, or they can use different sticks and clocks. Following our previous construction, they can be two inertial frames defined by a free comparison dynamical system. This point of view is usually called the *passive point of view*. The transformation that leads from one frame to another may also be applied to the observed system and therefore we may consider a dual point of view where the same reference frame is describing two different systems; they are each a transformation of the other in space–time. This is called the *active point of view*. In the Hamiltonian formalism we require that the transformation is a canonical transformation when considered on the space of states or on the space of physical observables. We can discuss the transformation properties of the equations of motion; if in one frame we have the Hamiltonian equations

$$\frac{d}{dt}f = \{f, H\}, \tag{3.A.63}$$

we may require that in the other frame we have

$$\frac{d}{dt'}f' = \{f', H'\}. \tag{3.A.64}$$

Since all various inertial frames are connected by transformations of the Poincaré group or of the Galilei group, we expect there to be no physical system able to discriminate one inertial frame from another, therefore all measured quantities in one frame are in one-to-one correspondence with those of the other frame, and the 'transformation' defined by such correspondence is the proper way to implement the transformation connecting one frame with the other.

Of course, by means of electromagnetic experiments, since the Maxwell equations are not invariant under Galilei transformations, we are able to find an experiment which discriminates one frame from another (e.g. think of the ether controversy). Einstein was the one who elevated the impossibility of revealing the uniform state of motion of a frame by means of physical experiments to a principle of nature. Here, the constancy of the speed of light for all observers in uniform motion will favour the d'Alembert wave equation with respect to other tensorial objects on space–time.

4   Schrödinger picture, Heisenberg picture and probabilistic aspects

This chapter begins by exploiting the Einstein–de Broglie relation and the notion of symbol. The following steps are the local and global conservation laws associated with the Schrödinger equation, probabilistic interpretation of the wave function and probability distribution, the spreading of wave packets and transformation properties of wave functions. In the last part, the Heisenberg picture is outlined at an introductory level.

4.1 From classical to wave mechanics

In the previous chapter we have seen how to go from wave equations on space–time to Hamilton equations on phase space by means of an appropriate approximation requirement on wave solutions. In more formal language, we could say we have considered a procedure to associate a Hamiltonian ordinary differential equation on phase space starting with a partial differential equation on configuration space. This association produces a correspondence between differential operators and polynomials on phase space, which relies on the notion of symbol (see below). Now we are aiming to pass from a function on phase space, i.e. the classical Hamiltonian, to an operator which, by analogy, is called the Hamiltonian operator. Such a transition is physically non-trivial because it amounts to building a wave equation from the equation on rays. From the formal point of view it appears to contain some ambiguities. Such a transition is always possible if polynomials in p have constant coefficients. When this is not the case, i.e. if the coefficients are functions on the configuration space, there is an obvious ambiguity in the association of an operator with xp or px, for instance. A first way out would be to consider ordered polynomials, i.e. all p should occur to the right before a differential operator is associated with it. In this way we would obtain an isomorphism of vector spaces between ordered polynomials of a given degree and differential operators of a given order. However, the correspondence, even though one to one, will not be an algebra isomorphism. The pointwise product of polynomials is commutative while the operator product is not commutative (indeed this property is responsible for the uncertainty relations as treated in Section 4.2).

If we use the Einstein–de Broglie identification in the form

$$\vec{p} \cdot \mathrm{d}\vec{x} - E\mathrm{d}t = \hbar(\vec{k} \cdot \mathrm{d}\vec{x} - \omega \mathrm{d}t),$$

the symbol of an operator reads, in general, as

$$\sigma(D) = \mathrm{e}^{-\frac{\mathrm{i}}{\hbar}(\vec{p}\cdot\vec{x}-Et)} \, D \, \mathrm{e}^{\frac{\mathrm{i}}{\hbar}(\vec{p}\cdot\vec{x}-Et)}, \tag{4.1.1}$$

which yields

$$\sigma\left(\frac{\partial}{\partial t}\right) = -\frac{i}{\hbar}E \tag{4.1.2}$$

and

$$\sigma\left(\frac{\partial}{\partial \vec{x}}\right) = \frac{i}{\hbar}\vec{p}. \tag{4.1.3}$$

From the property

$$e^{-\frac{i}{\hbar}(\vec{p}\cdot\vec{x}-Et)}D_1 \cdot D_2 e^{\frac{i}{\hbar}(\vec{p}\cdot\vec{x}-Et)}$$

$$= e^{-\frac{i}{\hbar}(\vec{p}\cdot\vec{x}-Et)}D_1 e^{\frac{i}{\hbar}(\vec{p}\cdot\vec{x}-Et)} \cdot e^{-\frac{i}{\hbar}(\vec{p}\cdot\vec{x}-Et)}D_2 e^{\frac{i}{\hbar}(\vec{p}\cdot\vec{x}-Et)}$$

it would follow that

$$\sigma(D_1 \cdot D_2) = \sigma(D_1)\sigma(D_2) = \sigma(D_2)\sigma(D_1) = \sigma(D_2 \cdot D_1).$$

This result would imply

$$\sigma(D_1 \cdot D_2 - D_2 \cdot D_1) = 0,$$

which cannot be accepted. Therefore a necessary condition for the multiplicative property of σ, i.e.

$$\sigma(D_1 \cdot D_2) = \sigma(D_1)\sigma(D_2),$$

is that D_1 and D_2 commute. If we restrict the use of the multiplicative property to decompositions in terms of pairwise commuting differential operators, we may use this condition. In this restriction there should be the prescription that *quantization* should be performed in Cartesian coordinates. In this manner, operators expressed by partial derivatives with constant coefficients will satisfy the requirement.

Therefore, once we know the association for first-order operators, we can extend it to higher-order differential operators. The associations

$$E \to i\hbar\frac{\partial}{\partial t}, \quad \vec{p} \to -i\hbar\frac{\partial}{\partial \vec{x}},$$

make it possible to construct a differential operator associated with a polynomial (relation) on phase space. Moreover, the symbol of a multiplication operator is the operator itself, viewed as a function. We therefore write, using a hat to denote operators,

$$V(\vec{x}) \to \hat{V}(\vec{x}), \quad x^i \to \hat{x}^i,$$

and we associate with the classical Hamiltonian

$$H = \frac{p^2}{2m} + V(\vec{x})$$

the differential operator

$$\frac{1}{2m}\left(-i\hbar\frac{\partial}{\partial \vec{x}}\right)^2 + \hat{V}(\vec{x}).$$

Thus, since classically $\frac{p^2}{2m} + V(x) = E$, we eventually obtain the partial differential equation

$$i\hbar \frac{\partial \psi}{\partial t} = \left[-\frac{\hbar^2}{2m} \triangle + \hat{V}(\vec{x}) \right] \psi, \qquad (4.1.4)$$

bearing in mind that operator equations like ours can be turned into partial differential equations, once the operators therein are viewed as acting on suitable 'functions'. This partial differential equation is called the Schrödinger equation, and gives rise to the *dispersion relation* $E = \frac{p^2}{2m} + V$. The passage from the classical Hamiltonian to the Hamiltonian differential operator is called quantization, and is the counterpart of the quantization rules of the old quantum theory. The symbol map, however, is severely affected by the coordinate system. For example, the symbol map is a product-preserving map (homomorphism) if the coefficients of the differential operator D are constant but, on changing coordinates, in general a variable-coefficient operator is obtained, so that the map does not preserve the product any longer. We are therefore *forced to quantize in Cartesian coordinates*, although nothing prevents us from changing coordinates once we are dealing with differential operators. The behaviour of the map with respect to multiplication is required, for instance, if we want the operator associated with p^2 to be the square of the one associated with $\vec{p}$. The symbol map is therefore a procedure to construct a differential operator out of a polynomial function in the momenta. The physical relevance of the constructed operator comes only from the experimental aspect, after the physical identification has been made. We notice that the commutator $\hat{x}\hat{p} - \hat{p}\hat{x}$ is

$$[\hat{x}, \hat{p}] = i\hbar I.$$

It is usually called the canonical commutation relation.

The ground is now ready for the investigation of all properties of a quantum theory based on a wave equation, and this is the object of the next subsections. Here we are going to show that, in wave mechanics, the wave function of Eq. (4.1.4) obeys a local and a corresponding global conservation law. The evolution equation (4.1.4) is then in the Hilbert space of square-integrable functions on $\mathbf{R}^3$. This result leads to the physical interpretation of the wave function. Some key properties of wave packets are then studied.

4.1.1　Properties of the Schrödinger equation

Continuity equation

In the Hilbert space of square-integrable functions $\mathcal{L}^2(\mathbf{R}^3)$, we denote the scalar product or Hermitian product by

$$\langle \psi | \varphi \rangle \equiv \int_{\mathbf{R}^3} \psi^*(x)\varphi(x)\mathrm{d}^3x.$$

We have $\langle \psi | \varphi \rangle = \left(\langle \varphi | \psi \rangle \right)^*$, along with complex linearity in the second element and complex anti-linearity in the first element. Now we consider the Schrödinger equation (4.1.4) and its complex conjugate equation, which is automatically satisfied:

$$-i\hbar\frac{\partial\psi^*}{\partial t} = \left(-\frac{\hbar^2}{2m}\triangle +V\right)\psi^*(\vec{x},t), \tag{4.1.5}$$

and we multiply Eq. (4.1.4) by $\psi^*(\vec{x},t)$, and Eq. (4.1.5) by $\psi(\vec{x},t)$. On subtracting one equation from the other we then find

$$-\frac{\hbar^2}{2m}\left(\psi^*\triangle\psi - \psi\triangle\psi^*\right) = i\hbar\left(\psi^*\frac{\partial\psi}{\partial t} + \psi\frac{\partial\psi^*}{\partial t}\right). \tag{4.1.6}$$

Interestingly, on defining

$$\vec{j} \equiv \frac{\hbar}{2mi}\left[\psi^*\vec{\nabla}(\psi) - \psi\vec{\nabla}(\psi^*)\right], \tag{4.1.7}$$

and $\rho \equiv \psi^*\psi$, Eq. (4.1.6) takes the form of the *continuity equation* for a current:

$$\frac{\partial\rho}{\partial t} + \partial^k j_k = 0. \tag{4.1.8}$$

This is a *local conservation law*. If the continuity equation is integrated on a volume $\mathcal{V}$, the divergence theorem yields

$$\int_{\mathcal{V}}\frac{\partial}{\partial t}\psi^*\psi\,d^3x = \frac{\partial}{\partial t}\int_{\mathcal{V}}\psi^*\psi\,d^3x$$

$$= \frac{d}{dt}\int_{\mathcal{V}}\psi^*\psi\,d^3x$$

$$= -\frac{\hbar}{2mi}\int_{\Sigma}\left(\psi^*\frac{\partial\psi}{\partial n} - \psi\frac{\partial\psi^*}{\partial n}\right)d\sigma, \tag{4.1.9}$$

where Σ is the boundary surface of $\mathcal{V}$ and $\frac{\partial}{\partial n}$ denotes differentiation along the direction normal to Σ. Thus, if ψ vanishes in a sufficiently rapid way as $|\vec{x}| \to \infty$, and if its first derivatives remain bounded in that limit, the integral on the right-hand side of Eq. (4.1.9) vanishes when the surface Σ is pushed off to infinity. The volume $\mathcal{V}$ extends then to the whole of $\mathbf{R}^3$, giving the *global conservation property*

$$\frac{d}{dt}\int_{\mathbf{R}^3}\psi^*\psi\,d^3x = 0,$$

which shows that the integral of $|\psi|^2$ over the whole space is independent of t. By virtue of the linearity of the Schrödinger equation, if ψ is square-integrable, the wave function can then be rescaled so that the integral is set to 1:

$$\int_{\mathbf{R}^3}\psi^*\psi\,d^3x = 1.$$

It is therefore clear that we are considering a theory for which the wave functions belong to the space of square-integrable functions on $\mathbf{R}^3$. This is a Hilbert space. The probabilistic interpretation of ψ, presented in Subsection 4.1.2, is then tenable.

Remark   If we write the wave function in polar form, i.e. $\psi = Ae^{i(S-Et)}$, Eq. (4.1.7) becomes $\vec{j} = \frac{\hbar}{m}A^2\frac{\partial S}{\partial x}$. On defining

$$\rho \equiv A^2, \quad \vec{v} = \frac{\hbar}{m}\vec{\nabla}S,$$

we find

$$\frac{\partial \rho}{\partial t} + \partial^k (\rho v_k) = 0.$$

We can solve the free equation

$$i\hbar \frac{\partial \psi}{\partial t} = -\frac{\hbar^2}{2m} \triangle \psi \tag{4.1.10}$$

by means of $\psi = N e^{-i\frac{Et}{\hbar}} e^{\pm i\frac{\vec{p}\cdot\vec{x}}{\hbar}}$. Indeed, this requires the condition $E = \frac{p^2}{2m}$. We see that, for each value of $E > 0$, there are two corresponding eigenfunctions of the Hamiltonian operator. For $E < 0$ there are no bounded solutions on the whole real axis.

We notice that the solutions do not belong to the set of square-integrable functions on the real line and, therefore, we cannot associate with them a probability distribution (but the wave packet remains normalizable). Note also that the canonical commutation relation does not depend on the representation we choose. For example, one might equally well use the spatial Fourier transform $\varphi(\vec{p}, t)$ of the wave function. We would then find the evolution equation for a free particle

$$i\hbar \frac{\partial}{\partial t} \varphi(\vec{p}, t) = \frac{p^2}{2m} \varphi(\vec{p}, t), \tag{4.1.11}$$

giving

$$\varphi(\vec{p}, t) = A(\vec{p}) e^{-i\frac{p^2 t}{2m\hbar}}, \tag{4.1.12}$$

with $\varphi(\vec{p}, 0) = A(\vec{p})$. By using the Fourier transform, a wave packet can be constructed by writing

$$\psi(\vec{x}, t) = (2\pi\hbar)^{-\frac{3}{2}} \int_{\mathbf{R}^3} A(\vec{p}) e^{i\frac{(\vec{p}\cdot\vec{x} - Et)}{\hbar}} d^3p, \tag{4.1.13}$$

and its complex conjugate is

$$\psi^*(\vec{x}, t) = (2\pi\hbar)^{-\frac{3}{2}} \int_{\mathbf{R}^3} A^*(\vec{p}) e^{-i\frac{(\vec{p}\cdot\vec{x} - Et)}{\hbar}} d^3p, \tag{4.1.14}$$

where $E = E(p)$. Now by setting

$$\varphi(\vec{p}, t) \equiv A(\vec{p}) e^{-i\frac{Et}{\hbar}} \tag{4.1.15}$$

the above formulae can be re-expressed as

$$\psi(\vec{x}, t) = (2\pi\hbar)^{-\frac{3}{2}} \int_{\mathbf{R}^3} \varphi(\vec{p}, t) e^{i\frac{\vec{p}\cdot\vec{x}}{\hbar}} d^3p, \tag{4.1.16}$$

$$\psi^*(\vec{x}, t) = (2\pi\hbar)^{-\frac{3}{2}} \int_{\mathbf{R}^3} \varphi^*(\vec{p}, t) e^{-i\frac{\vec{p}\cdot\vec{x}}{\hbar}} d^3p. \tag{4.1.17}$$

These equations imply that the spatial Fourier transform of the wave function can be expressed as

$$\varphi(\vec{p}, t) = (2\pi\hbar)^{-\frac{3}{2}} \int_{\mathbf{R}^3} \psi(\vec{x}, t) e^{-i\frac{\vec{p}\cdot\vec{x}}{\hbar}} d^3x, \tag{4.1.18}$$

Classical dynamical variable	Coordinate repr.	Momentum repr.
E	$i\hbar \frac{\partial}{\partial t}$	E
x	x	$i\hbar \frac{\partial}{\partial p}$
p	$-i\hbar \frac{\partial}{\partial x}$	p
x^n	x^n	$(i\hbar)^n \frac{\partial^n}{\partial p^n}$
p^n	$(-i\hbar)^n \frac{\partial^n}{\partial x^n}$	p^n
$\frac{p^2}{2m}$	$-\frac{\hbar^2}{2m}\triangle$	$\frac{p^2}{2m}$

Fig. 4.1   Coordinate vs. momentum representation.

and hence, from Eq. (4.1.15),

$$A(\vec{p}) = (2\pi\hbar)^{-\frac{3}{2}} \int_{\mathbf{R}^3} \psi(\vec{x}, t) e^{-i\frac{(\vec{p}\cdot\vec{x} - Et)}{\hbar}} \, d^3x. \tag{4.1.19}$$

Moreover, the following relation, known as the Parseval lemma (Hörmander 1983), holds:

$$\int_{\mathbf{R}^3} \psi^*\psi \, d^3x = \int_{\mathbf{R}^3} \varphi^*\varphi \, d^3p = \int_{\mathbf{R}^3} A^*A \, d^3p. \tag{4.1.20}$$

It should be emphasized that, on passing from the x- to the p-integration via a Fourier transform, the position operator becomes a first-order differential operator:

$$\widehat{x} : \varphi \to i\hbar \frac{\partial \varphi}{\partial p},$$

whereas the momentum operator acts in a multiplicative way:

$$p : \varphi \to p\,\varphi.$$

These equations define the *momentum representation* (see Figure 4.1), which is the wave-mechanical counterpart of the well-known canonical transformation in classical mechanics:

$$(p, q) \to (P = -q, Q = p),$$

i.e. it is canonical because of the identity

$$p\,dq = -q\,dp + d(pq).$$

The momentum representation amounts to considering the momenta p instead of the positions q as independent variables, in order to express our operators. This alternative option looks rather natural if one bears in mind that symbols can be polynomials in the q or p variables depending on which basis is chosen for the cotangent bundle and hence for square-integrable functions on the dual of $\mathbf{R}^4$ or on $\mathbf{R}^4$ itself. From these considerations it is now clear that, if the operator D acts on functions of the momenta by means of differentiation with respect to p, the symbol map will define a polynomial in (x, t).

4.1.2 Physical interpretation of the wave function

Let us now consider the physical meaning which can be attributed to $\psi(\vec{x},t)$. Born's proposal was to regard

$$\rho(\vec{x},t)\mathrm{d}^3x = \psi^*(\vec{x},t)\psi(\vec{x},t)\mathrm{d}^3x \qquad (4.1.21)$$

as a probability measure. In other words, this quantity represents the probability of observing a particle at time t within the volume element d^3x around the point for which the position vector is $\vec{x}$. Similarly, the quantity

$$\rho(\vec{p},t)\mathrm{d}^3p = \varphi^*(\vec{p},t)\varphi(\vec{p},t)\mathrm{d}^3p \qquad (4.1.22)$$

represents the probability of observing a particle at time t within the 'volume element' d^3p around the point with momentum $\vec{p}$. This interpretation is possible if ψ is square-integrable and normalized. Born was led to his postulate by comparing how the scattering of a particle by a potential is described in classical and quantum mechanics. For example, if an electron is allowed to interact with a short-range potential, and if a screen is placed at a distance from the interaction region of a few metres, the electron is always detected at some point on the screen. On repeating the experiment, the electron is detected at a different point, and so on. In other words, only one detector reveals the passage of the electron, but the detector changes while the experiment is performed. After several experiments, the fraction of the number of times the electron is detected at $\vec{x}$ at time t is proportional to $|\psi(\vec{x},t)|^2$, which is therefore the probability density of such an event. Remarkably, the electron exhibits both a corpuscular and a wave-like nature by virtue of the probabilistic interpretation of its wave function. This means that the *quantum electron* does not coincide with the notion of a *classical electron* we may have developed in a classical framework. The Schrödinger picture that we are describing unifies both these aspects.

We will entertain the idea that, in quantum mechanics, *no elementary phenomenon is a phenomenon unless it is registered* (Wheeler and Zurek 1983). For example, it would be meaningless to say that a particle passed through a hole in a wall unless one has a measuring device which provides evidence that it did so. Ultimately, we have to give up the attempt to describe the particle motion as if we could use what we are familiar with from everyday experience. We have instead to limit ourselves to an abstract mathematical description, which makes it possible to extract information *from the experiments that we are able to perform*. This crucial point (which is still receiving careful consideration in the current literature) is well emphasized by the interference from a double slit. Now, in the light of the previous paragraphs on the properties and meaning of the wave function, we are in a position to perform a more careful investigation, which should be compared with Chapter 2.

Suppose that a beam of electrons emitted from the source S, and having a sufficiently well-known energy, hits a screen Σ_1 with two slits F_1 and F_2. A second screen Σ_2 is placed thereafter (see Figure 4.2). To each electron of the beam a wave can be associated, which is partially reflected from Σ_1 and partially diffracted through the two slits. Now let ψ_1 and ψ_2 represent the two diffracted waves. The full wave in the neighbourhood of Σ_2 is given by

$$\psi(\vec{x},t) = \psi_1(\vec{x},t) + \psi_2(\vec{x},t). \qquad (4.1.23)$$

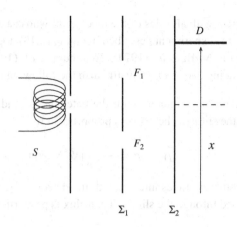

Fig. 4.2   Double-slit experiment with two screens.

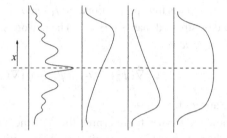

Fig. 4.3   Probability distribution in a Young experiment in the four cases: (i) no detector is placed; (ii) a detector in front of slit 1; (iii) a detector in front of slit 2; (iv) detectors in front of both slits.

Thus, bearing in mind what is known from the analysis of the continuity equation, we can say that the probability that a detector D placed on Σ_2 detects, in a given time interval, the particle, is given by

$$\vec{j} \cdot \vec{n} \mathrm{d}\sigma = -\frac{i\hbar}{2m}\left[\psi^*\vec{\nabla}(\psi) - \psi\vec{\nabla}(\psi^*)\right] \cdot \vec{n}\mathrm{d}\sigma$$
$$= -\frac{i\hbar}{2m}\Big[(\psi_1^* + \psi_2^*)\vec{\nabla}(\psi_1 + \psi_2)$$
$$- (\psi_1 + \psi_2)\vec{\nabla}(\psi_1^* + \psi_2^*)\Big] \cdot \vec{n}\mathrm{d}\sigma, \qquad (4.1.24)$$

where, according to a standard notation, $\vec{n}$ is the normal to Σ_2 and $\mathrm{d}\sigma$ denotes the section of the detector, which is viewed as being 'infinitesimal'. Since all electrons are (essentially) under the same conditions, the temporal average of Eq. (4.1.24) represents the effective flux of electrons through D per unit time. As the position of the detector on the screen varies, the observed flux is expected to exhibit the sequence of maxima and minima, which is typical of an interference process.

However, one may want to supplement the experimental apparatus, by inserting yet more detectors, C_1 and C_2, in front of the slits F_1 and F_2, respectively. From now on, we are describing what is, strictly, a *gedanken experiment*, i.e. a conceptual construction which is consistent with all rules of the theory, but whose actual implementation is difficult (see, however, the experiment described in Figure (2.19), together with the encouraging progress described in Merli *et al.* (1974), Tonomura *et al.* (1989), Scully *et al.* (1991)). With this understanding, we can now distinguish the following cases (see Figure 4.2).

(i) Some particles interact with the detector C_1, and hence we can say that they passed through the slit F_1. Their flux is proportional to

$$\vec{j}_1 \cdot \vec{n} d\sigma = -\frac{i\hbar}{2m} \left[\psi_1^* \vec{\nabla}(\psi_1) - \psi_1 \vec{\nabla}(\psi_1^*) \right] \cdot \vec{n} d\sigma. \tag{4.1.25}$$

(ii) The other particles interact with the detector C_2, and hence it is legitimate to say that they passed through the slit F_2. Their flux is proportional to

$$\vec{j}_2 \cdot \vec{n} d\sigma = -\frac{i\hbar}{2m} \left[\psi_2^* \vec{\nabla}(\psi_2) - \psi_2 \vec{\nabla}(\psi_2^*) \right] \cdot \vec{n} d\sigma. \tag{4.1.26}$$

(iii) The total flux is thus proportional to $\vec{j}_1 \cdot \vec{n} d\sigma + \vec{j}_2 \cdot \vec{n} d\sigma$.

It should be stressed that $\vec{j}_1 + \vec{j}_2 \neq \vec{j}$. The reason is that $\vec{j}$ also contains, from Eq. (4.1.24), the *interference terms*

$$\psi_1^* \vec{\nabla}(\psi_2), \quad \psi_2^* \vec{\nabla}(\psi_1), \quad \psi_1 \vec{\nabla}(\psi_2^*), \quad \psi_2 \vec{\nabla}(\psi_1^*),$$

which are absent from $\vec{j}_1 + \vec{j}_2$.

It is now necessary to interpret the scheme just outlined. If the electron were a particle without wave-like properties, the 'conventional wisdom' resulting from everyday experience would suggest that, to reach the detector D on Σ_2, it should always pass through F_1 or F_2. This would imply, in turn, that the total flux is always proportional to $(\vec{j}_1 \cdot \vec{n} + \vec{j}_2 \cdot \vec{n}) d\sigma$, *whether or not* the detectors C_1 and C_2 are placed in front of the two slits. However, this conclusion contradicts all that is known from diffraction and interference experiments on electrons. The crucial point is, as we stated before, that one can only say that the electron passed through F_1 or F_2 upon placing the detectors C_1 and C_2. Thus, the particle picture, without any reference to wave-like properties, is inconsistent (for a related discussion, see Sudarshan and Rothman (1991)).

On the other hand, the purely wave-like description is also inappropriate, because it disagrees with the property of the detectors of being able to register the passage of only one electron at a time. In other words, when the detectors C_1 and C_2 are placed in front of F_1 and F_2, *only one of them* (*either C_1 or C_2*) provides evidence of the passage of the electron. Moreover, if the single detector D on the screen Σ_2 is replaced by a set of detectors distributed all over the surface of Σ_2, only one detector at a time is able to indicate the passage of electrons, and this phenomenon occurs in a random way. The statistical distribution of countings, however, agrees with the predictions obtained from the Schrödinger equation. Furthermore, the current of electrons can be reduced so that no more than one electron is present in the apparatus at any time, thus showing that an electron 'interferes with itself'.

For theoretical developments on interference experiments, we refer the reader to the work in Bimonte and Musto (2003a,b) and references therein.

4.1.3 Mean values

The probabilistic interpretation of the wave function ψ makes it possible to define mean values, and hence the formalism developed so far leads to a number of simple properties, which can be tested against observation. For example, the mean values of position and momentum operators can be defined:

$$\langle x_l \rangle_\psi \equiv \int_{\mathbf{R}^3} \psi^* x_l \psi \, d^3 x, \tag{4.1.27}$$

$$\langle p_l \rangle_\varphi \equiv \int_{\mathbf{R}^3} \varphi^* p_l \varphi \, d^3 p. \tag{4.1.28}$$

The corresponding standard deviations can also be defined, i.e.

$$\langle (\Delta x_l)^2 \rangle_\psi \equiv \int_{\mathbf{R}^3} \psi^* (x_l - \langle x_l \rangle)^2 \psi \, d^3 x, \tag{4.1.29}$$

$$\langle (\Delta p_l)^2 \rangle_\varphi \equiv \int_{\mathbf{R}^3} \varphi^* (p_l - \langle p_l \rangle)^2 \varphi \, d^3 p. \tag{4.1.30}$$

The formulae described so far can be generalized to *entire rational functions* of x_l and p_l:

$$\langle F(x_l) \rangle_\psi = \int_{\mathbf{R}^3} \psi^* F(x_l) \psi \, d^3 x = \int_{\mathbf{R}^3} \varphi^* F\left(i\hbar \frac{\partial}{\partial p_l} \right) \varphi \, d^3 p$$

$$= \langle F(x_l) \rangle_\varphi, \tag{4.1.31}$$

$$\langle F(p_l) \rangle_\psi = \int_{\mathbf{R}^3} \psi^* F\left(\frac{\hbar}{i} \frac{\partial}{\partial x_l} \right) \psi \, d^3 x = \int_{\mathbf{R}^3} \varphi^* F(p_l) \varphi \, d^3 p$$

$$= \langle F(p_l) \rangle_\varphi. \tag{4.1.32}$$

These steps rely on the comment we have made after Eq. (4.1.20). Indeed, weaker conditions on F can also be considered, e.g. C^2 functions with a rapid fall off. Moreover, elementary differentiation, e.g.

$$\psi^* \frac{\partial^2 \psi}{\partial x_l^2} = \frac{\partial}{\partial x_l} \left(\psi^* \frac{\partial \psi}{\partial x_l} \right) - \frac{\partial \psi^*}{\partial x_l} \frac{\partial \psi}{\partial x_l},$$

can be used to compute the following mean values:

$$\langle p_l^2 \rangle_\varphi = \int_{\mathbf{R}^3} \varphi^* p_l^2 \varphi \, d^3 p = \int_{\mathbf{R}^3} \psi^* \left(-\hbar^2 \frac{\partial^2}{\partial x_l^2} \right) \psi \, d^3 x$$

$$= \hbar^2 \int_{\mathbf{R}^3} \frac{\partial \psi^*}{\partial x_l} \frac{\partial \psi}{\partial x_l} d^3 x = \langle p_l^2 \rangle_\psi, \tag{4.1.33}$$

$$\langle x_l^2 \rangle_\psi = \int_{\mathbf{R}^3} \psi^* x_l^2 \psi \, d^3 x = \int_{\mathbf{R}^3} \varphi^* \left(-\hbar^2 \frac{\partial^2}{\partial p_l^2} \right) \varphi \, d^3 p$$

$$= \hbar^2 \int_{\mathbf{R}^3} \frac{\partial \varphi^*}{\partial p_l} \frac{\partial \varphi}{\partial p_l} d^3 p = \langle x_l^2 \rangle_\varphi. \tag{4.1.34}$$

In the course of deriving the results (4.1.33) and (4.1.34) we have imposed, once more, suitable fall-off conditions at infinity on the wave function, so that all total derivatives in the integrand give a vanishing contribution to the integral.

To understand how the mean value $\langle x_l \rangle$ evolves in time with a dynamics ruled by $H = \frac{\vec{p}^2}{2m} + V(\vec{x})$, we now perform the following steps. By taking the Schrödinger equation and its complex conjugate, we consider the rate of change of the mean value of position, and we find

$$
\frac{\mathrm{d}}{\mathrm{d}t} \langle x_l \rangle_\psi = (2\pi\hbar)^{-\frac{3}{2}} \int_{\mathbf{R}^3} \left(\frac{\partial \psi^*}{\partial t} x_l \psi + \psi^* x_l \frac{\partial \psi}{\partial t} \right) \mathrm{d}^3 x
$$

$$
= (2\pi\hbar)^{-\frac{3}{2}} \int_{\mathbf{R}^3} \left[\frac{\mathrm{i}}{\hbar}(H\psi)^* x_l \psi + \psi^* x_l \left(-\frac{\mathrm{i}}{\hbar} H\psi \right) \right] \mathrm{d}^3 x
$$

$$
= (2\pi\hbar)^{-\frac{3}{2}} \int_{\mathbf{R}^3} \frac{\hbar}{2m\mathrm{i}} \left(\psi^* \frac{\partial \psi}{\partial x_l} - \psi \frac{\partial \psi^*}{\partial x_l} \right) \mathrm{d}^3 x
$$

$$
= \frac{1}{m} \langle p_l \rangle_\psi = \left\langle \frac{\partial H}{\partial p_l} \right\rangle_\psi , \tag{4.1.35}
$$

where use has been made of the Hamiltonian operator in the position representation:

$$
H = -\frac{\hbar^2}{2m} \triangle + V(\vec{x}).
$$

For the rate of change of the mean value of momentum,

$$
\frac{\mathrm{d}}{\mathrm{d}t} \langle p_l \rangle_\psi = \frac{1}{2} \int_{\mathbf{R}^3} \left[(H\psi)^* \frac{\partial \psi}{\partial x_l} - \psi^* \frac{\partial}{\partial x_l}(H\psi) \right.
$$

$$
\left. + (H\psi) \frac{\partial \psi^*}{\partial x_l} - \psi \frac{\partial}{\partial x_l}(H\psi)^* \right] \mathrm{d}^3 x
$$

$$
= \frac{\hbar^2}{4m} \int_{\mathbf{R}^3} \left[-(\triangle\psi^*) \frac{\partial \psi}{\partial x_l} + \psi^* \frac{\partial}{\partial x_l}(\triangle\psi) \right.
$$

$$
\left. - (\triangle\psi) \frac{\partial \psi^*}{\partial x_l} + \psi \frac{\partial}{\partial x_l}(\triangle\psi)^* \right] \mathrm{d}^3 x
$$

$$
+ \frac{1}{2} \int_{\mathbf{R}^3} \left[V\psi^* \frac{\partial \psi}{\partial x_l} - \psi^* \frac{\partial}{\partial x_l}(V\psi) \right.
$$

$$
\left. + V\psi \frac{\partial \psi^*}{\partial x_l} - \psi \frac{\partial}{\partial x_l}(V\psi^*) \right] \mathrm{d}^3 x. \tag{4.1.36}
$$

By using the identity

$$
\chi \triangle \eta - \eta \triangle \chi = \partial^k (\chi \vec{\nabla}\eta - \eta \vec{\nabla}\chi)_k, \tag{4.1.37}
$$

and exploiting the behaviour at infinity, i.e. the fall-off conditions therein,

$$
\frac{\mathrm{d}}{\mathrm{d}t} \langle p_l \rangle_\psi = m \frac{\mathrm{d}^2}{\mathrm{d}t^2} \langle x_l \rangle_\psi = -\left\langle \frac{\partial V}{\partial x_l} \right\rangle_\psi = -\left\langle \frac{\partial H}{\partial x_l} \right\rangle_\psi . \tag{4.1.38}
$$

In the above derivations, we may rely equally well on the momentum representation, for example

$$\frac{\mathrm{d}}{\mathrm{d}t}\langle x_l\rangle_\psi = \frac{\mathrm{d}}{\mathrm{d}t}\langle x_l\rangle_\varphi = \int_{\mathbf{R}^3}\left[\frac{\partial\varphi^*}{\partial t}\mathrm{i}\hbar\frac{\partial\varphi}{\partial p_l} + \varphi^*\mathrm{i}\hbar\frac{\partial}{\partial p_l}\frac{\partial\varphi}{\partial t}\right]\mathrm{d}^3p$$

$$= \int_{\mathbf{R}^3}\left[\frac{\mathrm{i}}{\hbar}\frac{p^2}{2m}\psi^*\mathrm{i}\hbar\frac{\partial\varphi}{\partial p_l} - \frac{\mathrm{i}}{\hbar}\varphi^*\mathrm{i}\hbar\frac{\partial}{\partial p_l}\left(\frac{p^2}{2m}\varphi\right)\right]\mathrm{d}^3p$$

$$= \int_{\mathbf{R}^3}\varphi^*\frac{p_l}{m}\varphi\mathrm{d}^3p = \frac{1}{m}\langle p_l\rangle_\psi. \tag{4.1.39}$$

Similarly, for a system of N particles, for which the Hamiltonian reads

$$\hat{H} = \sum_{j=1}^{N}\frac{\vec{p}_j^2}{2m_j} + V(\vec{x}_1,\ldots,\vec{x}_N), \tag{4.1.40}$$

we find

$$\frac{\mathrm{d}}{\mathrm{d}t}\langle\vec{x}_j\rangle_\varphi = \frac{1}{m_j}\langle\vec{p}_j\rangle_\varphi, \tag{4.1.41}$$

jointly with

$$\frac{\mathrm{d}}{\mathrm{d}t}\langle\vec{p}_j\rangle_\varphi = -\left\langle\frac{\partial V}{\partial\vec{x}_j}\right\rangle_\varphi. \tag{4.1.42}$$

The result expressed by Eqs. (4.1.41) and (4.1.42) is known as the Ehrenfest theorem (Ehrenfest 1927). It is crucial to compute first the partial derivatives with respect to x_j and then to take the mean value in Eq. (4.1.42), because

$$\left\langle\frac{\partial V}{\partial\vec{x}_j}\right\rangle_\varphi \neq \frac{\partial}{\partial\vec{x}_j}\langle V\rangle_\varphi.$$

If the operations of mean value and partial derivative are performed in the opposite order, the time derivative of the mean value of p_j is no longer obtained if V is an arbitrary function of $\vec{x}$ and $\vec{p}$. The theorem can be extended to functions of $\vec{x}$ and $\vec{p}$ which are at most quadratic in $\vec{p}$, provided the Hamiltonians are of the form $\frac{\vec{p}^2}{2m} + V(\vec{x})$. The modifications necessary for generic dynamical variables were discovered in Moyal (1949). To appreciate the points raised here, consider a one-dimensional model where the first derivative of the potential can be expanded in a power series:

$$V'(x) = V'(\langle x\rangle_t) + (x - \langle x\rangle_t)V''(\langle x\rangle_t)$$

$$+ \frac{1}{2!}(x - \langle x\rangle_t)(x - \langle x\rangle_t)V'''(\langle x\rangle_t) + \cdots. \tag{4.1.43}$$

The resulting mean value of $V'(x)$ is then

$$\langle V'(x)\rangle_t = V'(\langle x\rangle_t) + \frac{(\Delta x)_t^2}{2}V'''(\langle x\rangle_t) + \cdots. \tag{4.1.44}$$

The first correction to the formula expressing complete analogy between equations of motion for mean values and classical equations of motion for dynamical variables is therefore found to involve the third derivative of the potential, evaluated at a point equal to the mean value of position at time t.

Of course, if the mean values of x and p vanish, the previous analysis provides the time rate of change of the 'localization':

$$\langle (\triangle x_l)^2 \rangle \quad \text{and} \quad \langle (\triangle p_l)^2 \rangle.$$

When a wave packet is considered that, at $t = 0$, is localized in a certain interval, what is going to happen during the time evolution? It is clear that, if the wave packet were to remain localized in an interval of the same dimensions, we could then think of such a packet as describing a 'particle'.

4.1.4 Eigenstates and eigenvalues

In our introductory presentation of wave mechanics we can now call *eigenstates* of any measurable quantity A depending on the q and p variables those particular wave functions for which the standard deviation of A vanishes. The mean value of A in such a state can therefore be viewed as its *eigenvalue*. This would imply that the probability distribution is concentrated on a specific value. Moreover, if the mean value of A is constant in time on all states ψ for which the Hamiltonian has vanishing dispersion, A is said to be a constant of motion.

4.2 Probability distributions associated with vectors in Hilbert spaces

In the previous section, we have considered wave functions as elements of a Hilbert space. We have associated with them probability distributions either on the *configuration space*, the space of eigenvalues of the position operators, or on the *momentum space*, the space of eigenvalues of *momentum operators*. This association is quite general and can be considered on any Hilbert space, connected with the spectrum of any operator.

We can construct probability distributions associated with states. To begin, we introduce here a notation due to Dirac, which splits the bracket $\langle \psi | \varphi \rangle$ into 'bra' $\langle \psi |$ and 'ket' $|\varphi\rangle$. They represent vectors of the Hilbert space $\mathcal{H}$, or its dual $\mathcal{H}^*$, called ket and bra, respectively. The identification of the space $\mathcal{H}$ with its dual $\mathcal{H}^*$ is done by means of the scalar product, Hermitian product $\langle \psi | \varphi \rangle$. With any state $|\psi\rangle$ corresponding to the wave function ψ (see Section 4.3), and any diagonalizable operator A, it is possible to associate a probability distribution. Assume first that A acts on a finite-dimensional Hilbert space $\mathcal{H}$. If the associated matrix A is diagonalizable and all eigenvalues are simple (i.e. without degeneracy), the space $\mathcal{H}$ has a basis of right eigenvectors $\{v_j\} : Av_j = \varepsilon_j v_j$, and the dual space has a basis of left eigenvectors v^j satisfying $v^j A = \varepsilon_j v^j$. Note that left and right eigenvectors belonging to different eigenvalues are orthogonal, i.e.

$$\langle v^j | v_k \rangle = \delta_k^j.$$

With any normalized state $|\psi\rangle$ and any orthonormal basis of $\mathcal{H}$ we associate a probability distribution by setting

$$p_\psi(k) = \langle\psi|v_k\rangle\langle v^k|\psi\rangle \geq 0, \tag{4.2.1}$$

and

$$\sum_k p_k = 1. \tag{4.2.2}$$

We notice that this discrete probability distribution can be defined also for an infinite-dimensional Hilbert space, and relies on the property

$$\sum_j v^j \otimes v_j = I, \tag{4.2.3}$$

called the partition, or resolution, or decomposition of the identity. When the Hermitian operator A has a continuous spectrum, if we use again the notation $A|a\rangle = a|a\rangle$ and $\langle a|A = a\langle a|$, we have a continuous partition of the identity written in the form

$$\int |a\rangle\,\mathrm{d}a\langle a| = I. \tag{4.2.4}$$

Now with any normalized vector it is possible to associate a continuous probability distribution by setting

$$\int \langle\psi|a\rangle\,\mathrm{d}a\langle a|\psi\rangle = \int p_\psi(a)\,\mathrm{d}a, \tag{4.2.5}$$

where, again, $p_\psi(a) \geq 0$ and

$$\int p_\psi(a)\,\mathrm{d}a = 1. \tag{4.2.6}$$

Having a probability distribution it is possible to define mean values, or expectation values

$$\langle A\rangle_\psi = \sum_j a_j p_j \tag{4.2.7}$$

in the discrete situation, and

$$\langle A\rangle_\psi = \int a p_\psi(a)\,\mathrm{d}a \tag{4.2.8}$$

in the continuous case. The same is true for the square of A, A^2, and more generally for any function f of A:

$$\langle f(A)\rangle_\psi = \sum_j f(a_j) p_j, \tag{4.2.9}$$

or

$$\langle f(A)\rangle_\psi = \int f(a) p_\psi(a)\,\mathrm{d}a, \tag{4.2.10}$$

provided the right-hand side of these expressions can be defined. If the spectrum of A contains both a discrete and a continuous part, both contributions have to be used, its form relying on the decomposition of the identity.

In the previous derivation of mean values we have used the classical definition given by

$$\langle x \rangle_\psi = \int \psi^*(x) x \psi(x) \mathrm{d}x,$$

and similarly, for quantities depending on the momentum p, we have used the probability distribution associated with the eigenbasis provided by the momentum eigenbasis. We have also shown by means of the Fourier transform that a unitary transformation, i.e. a change of orthonormal basis for the Hilbert space, does not affect the expectation values. Therefore, it makes sense to denote by

$$e_A(\psi) \equiv \frac{\langle \psi | A | \psi \rangle}{\langle \psi | \psi \rangle} \qquad (4.2.11)$$

the expectation value, independently of any choice of basis to compute it.

What happens if we want to describe the mean value of an observable quantity with a probability distribution associated with a basis of vectors that is not made of eigenvectors of the observable? We consider the expectation value of an observable B in the orthonormal basis associated with A:

$$
\begin{aligned}
\langle B \rangle_\psi &= \langle \psi | B | \psi \rangle = \sum_{j,k} \langle \psi | a_j \rangle \langle a_j | B | a_k \rangle \langle a_k | \psi \rangle \\
&= \sum_{j,k} B_{j,k} \langle \psi | a_j \rangle \langle a_k | \psi \rangle.
\end{aligned}
\qquad (4.2.12)
$$

Thus, when the operator B is not diagonal in the basis associated with A, to express the expectation value of B we are forced to use the probability amplitudes $\langle \psi | a_j \rangle$ and $\langle a_k | \psi \rangle$, and it is not enough to know $p_\psi^A(j) \equiv \langle \psi | a_j \rangle \langle a_j | \psi \rangle$. By using the abstract Dirac notation of bra and ket, the probability distribution $p_\psi(k) = \langle \psi | v_k \rangle \langle v_k | \psi \rangle$ may be written as

$$\mathrm{Tr}\left(\frac{|\psi\rangle\langle\psi|}{\langle\psi|\psi\rangle} \frac{|v_k\rangle\langle v_k|}{\langle v_k|v_k\rangle} \right) = p_\psi(v_k), \qquad (4.2.13)$$

where we replace the vector $|\psi\rangle$ with its associated rank-1 projector.

In the example we have considered earlier we went from the basis associated with positions to the basis associated with momenta by means of the Fourier transform, a unitary transformation. By using the two partitions of unity

$$\int |x\rangle \mathrm{d}x \langle x| = I = \int |p\rangle \mathrm{d}p \langle p|, \qquad (4.2.14)$$

it is easy to derive that

$$\langle \psi | x \rangle = \int \langle \psi | p \rangle \mathrm{d}p \langle p | x \rangle, \qquad (4.2.15)$$

$$\langle p | \psi \rangle = \int \langle p | x \rangle \mathrm{d}x \langle x | \psi \rangle, \qquad (4.2.16)$$

so that $\langle p|x \rangle$ plays the role[1] of 'plane waves' in the Fourier transform. Indeed

$$\int \langle x|x' \rangle \mathrm{d}x' \langle x'|\psi \rangle = \int \langle x|p \rangle \mathrm{d}p \langle p|\psi \rangle, \qquad (4.2.17)$$

which implies that

$$\langle x|\psi \rangle = \int \langle x|p \rangle \mathrm{d}p \psi(p), \qquad (4.2.18)$$

having used $\langle x|x' \rangle = \delta(x - x')$.

In conclusion, the probabilistic interpretation of quantum mechanics means that, with any state vector and any orthonormal basis of the Hilbert space (proper or improper), it is possible to associate a probability distribution. We are not restricted to use only position or momentum representation.

4.3 Uncertainty relations for position and momentum

We have seen that wave functions in $\mathbf{R}^3$ can be written in the form (4.1.13). Moreover, on using the factorization (4.1.15) for the spatial Fourier transform of the wave function, such a transform can also be expressed as in Eq. (4.1.18). We have also learned that the quantities $\psi^*\psi$ and $\varphi^*\varphi$ are probability densities that make it possible to define mean values of functions of x and p.

The inequalities we call *uncertainty relations* result directly from Fourier analysis and hence are not an exclusive property of quantum mechanics. A possible formulation of such relations is as follows: *a non-vanishing function and its Fourier transform cannot both be localized with precision*. Indeed, in the framework of classical physics, if $f(t)$ represents the amplitude of a signal (e.g. an acoustic wave or an electromagnetic wave) at time t, its Fourier transform $\tilde{f}$ shows how f is constructed from sine waves of various frequencies. The uncertainty relation expresses a restriction with respect to the measurement in which the signal can be bounded in time and in the frequency band.

The previous statement can be made more explicit by considering the mean values (4.1.27) and (4.1.28) with quadratic deviations (4.1.29) and (4.1.30), respectively, as is done for example in Esposito *et al.* (2004). Here we focus instead on a general, abstract point of view. In this framework, we consider a generic Hilbert space $\mathcal{H}$ instead of $\mathcal{L}^2$, the space of square-integrable functions on some vector space. In doing this, and following Dirac, ket-vectors are elements of the Hilbert space $\mathcal{H}$, while bra-vectors are elements of the dual space $\mathcal{H}^*$. With this notation, if we denote by $\hat{x}$ and $\hat{p}$ the position and momentum operator, respectively, we can write in terms of improper eigenfunctions

$$\hat{x}|x \rangle = x|x \rangle, \; \hat{p}|p \rangle = p|p \rangle. \qquad (4.3.1)$$

---

[1] For the 'component' of the vector $|x\rangle$ on the momentum basis $|p\rangle$, see the considerations leading to Eqs. (4.3.1) and (4.3.2).

Thus, with this notation, wave functions appear as the 'component' of $|\psi\rangle$ in the basis of vectors $|x\rangle$ or $|p\rangle$, say

$$\langle x|\psi\rangle = \psi(x), \ \langle p|\psi\rangle = \psi(p). \tag{4.3.2}$$

We might replace $(\hat{x} \cdot \psi)(x) = x\psi(x)$ with a symmetric operator S, for which therefore

$$\langle S\psi|\psi\rangle = \langle \psi|S\psi\rangle, \tag{4.3.3}$$

and $\frac{i}{\hbar}\hat{p}\psi(x) = \frac{d}{dx}\psi(x)$ with $A|\psi\rangle$, A being anti-symmetric, i.e.

$$\langle A\psi|\psi\rangle = -\langle \psi|A\psi\rangle. \tag{4.3.4}$$

Then we obtain

$$\langle (S+A)\psi|(S+A)\psi\rangle \geq 0. \tag{4.3.5}$$

By expanding this expression we find

$$\langle S\psi|S\psi\rangle + \langle A\psi|A\psi\rangle - \langle \psi|(AS-SA)|\psi\rangle \geq 0. \tag{4.3.6}$$

The inequality is saturated for

$$(S+A)|\psi\rangle = 0. \tag{4.3.7}$$

For such a $|\psi\rangle$

$$(S-A)(S+A)|\psi\rangle = 0 \implies (S^2 - A^2)|\psi\rangle = (AS-SA)|\psi\rangle, \tag{4.3.8}$$

which represents an eigenvalue equation for $(S^2 - A^2)$ when $AS - SA$ is a multiple of the identity. In our explicit case, with

$$S = \hat{x}, \ A = \frac{d}{dx},$$

we find the scheme for the harmonic oscillator Hamiltonian, and ψ in the form of a Gaussian state becomes a solution for the harmonic oscillator eigenvalue problem. If we consider the transformation

$$A \to \pm \mu A, \ S \to \frac{1}{\mu}S, \tag{4.3.9}$$

we get

$$\left\langle \left(\frac{S}{\mu} \pm \mu A \right)\psi \middle| \left(\frac{S}{\mu} \pm \mu A \right)|\psi\right\rangle \geq 0. \tag{4.3.10}$$

By expanding the expression we get

$$|\langle (AS-SA)\psi|\psi\rangle|^2 \leq 4\langle S\psi|S\psi\rangle\langle A\psi|A\psi\rangle. \tag{4.3.11}$$

We notice that the inequality also remains valid if we perform the transformation, which we call a translation by a multiple of the identity,

$$A \to A - \langle \psi|A\psi\rangle I, \ S \to S - \langle \psi|S\psi\rangle I, \tag{4.3.12}$$

which does not change the commutation relations. With this transformation we get the replacements

$$\langle \psi | A^2 | \psi \rangle \rightarrow \langle \psi | A^2 | \psi \rangle - \langle \psi | A \psi \rangle^2, \tag{4.3.13}$$

$$\langle \psi | S^2 | \psi \rangle \rightarrow \langle \psi | S^2 | \psi \rangle - \langle \psi | S \psi \rangle^2, \tag{4.3.14}$$

and we recover the uncertainty relations.

4.4 Transformation properties of wave functions

To understand how the standard deviation evolves in time one can use the mean values of $\hat{x}^2$ and $\hat{p}^2$ provided that the mean values of $\hat{x}$ and $\hat{p}$ vanish. We shall perform this calculation by transforming our description from one reference frame to another by using the Galilei transformations. To achieve this it is necessary first to discuss the transformation properties of the Schrödinger equation, as we do hereafter.

When the transformation properties of the Schrödinger equation and of the wave function are analysed, it can be useful to assess the role played by the action $S \equiv \int \mathcal{L} dt = \int (p dq - H dt)$. Our analysis begins with classical considerations. For this purpose, let us consider the map (f being a smooth function depending on x and t)

$$(x, p) \rightarrow \left(x, p + \frac{\partial f}{\partial x} \right), \tag{4.4.1}$$

which implies, from the definition $\mathcal{L} dt \equiv p dx - H dt$, the transformation property

$$\mathcal{L}' = \mathcal{L} + \frac{df}{dt}, \tag{4.4.2}$$

provided that the velocity v is identified with $\frac{dx}{dt}$, because then

$$p' = \frac{\partial \mathcal{L}'}{\partial v} = p + \frac{\partial}{\partial v} \left(\frac{dx}{dt} \frac{\partial f}{\partial x} \right) = p + \frac{\partial f}{\partial x}.$$

Thus, the momentum variable, p, is *gauge-dependent* (see Eq. (4.4.1)), whereas the resulting Lagrangians lead to the same equations of motion. A physically significant transformation of this type is generated by the map $\vec{p} \rightarrow \vec{p} + \vec{a}$, $\vec{x} \rightarrow \vec{x}$. This amounts to using $(\vec{p} + \vec{a}) \cdot d\vec{x}$ as the phase 1-form on $T^* \mathbf{R}^3$.

In wave mechanics, if one evaluates the mean values of the quantum-mechanical operators, which correspond to the left- and right-hand sides of Eq. (4.4.1),

$$\langle \vec{p} \rangle = \int_{\mathbf{R}^3} d^3 x \, \psi^* \frac{\hbar}{i} \frac{\partial}{\partial \vec{x}} \psi, \tag{4.4.3}$$

$$\langle \vec{p}' \rangle = \int_{\mathbf{R}^3} d^3 x \, \psi^* \left(\frac{\hbar}{i} \frac{\partial}{\partial \vec{x}} \psi + \frac{\partial f}{\partial \vec{x}} \psi \right), \tag{4.4.4}$$

and hence the mean values of p and p' do not coincide. This is a clear indication that the wave function cannot remain unaffected, but has to change if we want to make sure that the mean value of $\langle p \rangle$ remains the same. The desired transformation law is written here in the form

$$\psi' = e^{-\frac{if}{\hbar}} \psi. \tag{4.4.5}$$

Note that it is possible to arrive at the same conclusion, i.e. the form of Eq. (4.4.5), using Eq. (4.4.2) and the definition of the action $S' \equiv \int \mathcal{L}' dt$. By explicit calculation

$$\langle \vec{p}' \rangle = \int_{\mathbf{R}^3} d^3 x (\psi')^* \left(\frac{\hbar}{i} \frac{\partial}{\partial \vec{x}} \psi' \right) + \int_{\mathbf{R}^3} d^3 x \, \psi^* \frac{\partial f}{\partial \vec{x}} \psi$$

$$= \int_{\mathbf{R}^3} d^3 x \psi^* \frac{\hbar}{i} \frac{\partial \psi}{\partial \vec{x}} = \langle \vec{p} \rangle. \tag{4.4.6}$$

Remark In higher-dimensional configuration spaces, the operators associated with $p_k + \frac{\partial f}{\partial x_k}$, i.e. $\frac{\hbar}{i} \frac{\partial}{\partial x_k} + \frac{\partial f}{\partial x_k}$, constitute a commuting set of differential operators and therefore can be used in setting the correspondence between polynomials and differential operators. This amounts to defining the symbol with the exponential of $\alpha k_\mu x^\mu + f(x)$, then

$$e^{-\alpha k_\mu x^\mu - f(x)} \frac{\partial}{\partial x^\nu} e^{\alpha k_\mu x^\mu + f(x)} = \alpha k_\nu + \frac{\partial f}{\partial x^\nu}.$$

The occurrence of a 'gauge transformation' in the way it is usually understood is clearly seen and also that it depends on the way we define the symbol. Our argument means that we attribute 'physical meaning' with $\frac{\hbar}{i} \frac{\partial}{\partial x}$ rather than with its symbol, the canonical momentum p, which is gauge variant.

Let us now study point transformations that change the Lagrangian by a total time derivative, as in Eq. (4.4.2). In particular, bearing in mind that we are interested in transformations such that the mean value of p vanishes, we consider the Galilei transformations. For this purpose, let us consider two frames Σ and Σ' with coordinates $(\vec{x}, t)$ and $(\vec{x}', t')$, respectively. We assume that the coordinate transformation is that resulting from the Galilei group:

$$\vec{x}' = \vec{x} - \vec{x}_0 - \vec{v}t, \quad t' = t, \tag{4.4.7}$$

which implies that $\frac{d}{dt}\vec{x}' = \frac{d}{dt}\vec{x} - \vec{v}$. Thus, the Lagrangian $\mathcal{L} = \frac{1}{2}m\frac{d}{dt}\vec{x} \cdot \frac{d}{dt}\vec{x}$ transforms as follows:

$$\mathcal{L}' = \frac{1}{2}m\frac{d}{dt}\vec{x}' \cdot \frac{d}{dt}\vec{x}' = \frac{1}{2}m\frac{d}{dt}\vec{x} \cdot \frac{d}{dt}\vec{x} + \frac{1}{2}m\vec{v}^2 - m\frac{d}{dt}\vec{x} \cdot \vec{v}$$

$$= \mathcal{L} - \frac{d}{dt}\left(m\vec{x} \cdot \vec{v} - \frac{1}{2}m\vec{v}^2 t \right). \tag{4.4.8}$$

The comparison with Eq. (4.4.5) and the consideration of physical dimensions therefore shows that the wave function undergoes the transformation

$$\psi' = e^{\frac{i}{\hbar}(m\vec{x} \cdot \vec{v} - \frac{1}{2}m\vec{v}^2 t)} \psi. \tag{4.4.9}$$

In particular, under transformations like (4.4.7), the wave function does not behave as a 'scalar function' but also picks up a phase.

4.4.1 Direct approach to the transformation properties of the Schrödinger equation

Let us now consider the effect of the transformations (4.4.7) on the differential operator occurring in the Schrödinger equation. Indeed, the elementary rules of differentiation of composite functions yield

$$\frac{\partial}{\partial t'} = \frac{\partial t}{\partial t'}\frac{\partial}{\partial t} + \frac{\partial \vec{x}}{\partial t'}\frac{\partial}{\partial \vec{x}} = \frac{\partial}{\partial t} + \vec{v}\frac{\partial}{\partial \vec{x}}, \tag{4.4.10}$$

$$\frac{\partial}{\partial \vec{x}} = \frac{\partial t'}{\partial \vec{x}}\frac{\partial}{\partial t'} + \frac{\partial \vec{x}'}{\partial \vec{x}}\frac{\partial}{\partial \vec{x}'} = \frac{\partial}{\partial \vec{x}'}, \tag{4.4.11}$$

which implies

$$i\hbar\frac{\partial}{\partial t'} + \frac{\hbar^2}{2m}\Delta' = i\hbar\left(\frac{\partial}{\partial t} + \vec{v}\frac{\partial}{\partial \vec{x}}\right) + \frac{\hbar^2}{2m}\Delta. \tag{4.4.12}$$

From Eqs. (4.4.5) and (4.4.12),

$$\psi'(\vec{x}',t') = e^{\frac{i}{\hbar}\left(\frac{1}{2}mv^2t - m\vec{v}\cdot\vec{x}\right)}\,\psi(\vec{x},t). \tag{4.4.13}$$

Therefore, in the frame Σ', $\psi'(x',t')$ solves the equation

$$\left(i\hbar\frac{\partial}{\partial t'} + \frac{\hbar^2}{2m}\Delta'\right)\psi' = 0, \tag{4.4.14}$$

if ψ solves the equation

$$\left(i\hbar\frac{\partial}{\partial t} + \frac{\hbar^2}{2m}\Delta\right)\psi = 0 \tag{4.4.15}$$

in the frame Σ. This happens because the left-hand sides of Eqs. (4.4.14) and (4.4.15) differ by the function

$$e^{\frac{if}{\hbar}}\left\{-\frac{\partial f}{\partial t}\psi - \vec{v}\frac{\partial f}{\partial \vec{x}}\psi + i\hbar\vec{v}\frac{\partial \psi}{\partial \vec{x}}\right.$$
$$\left. + \frac{\hbar^2}{2m}\sum_{l=1}^{3}\left[\frac{2i}{\hbar}\frac{\partial f}{\partial x_l}\frac{\partial \psi}{\partial x_l} + \frac{i}{\hbar}\frac{\partial^2 f}{\partial x_l^2}\psi - \frac{1}{\hbar^2}\left(\frac{\partial f}{\partial x_l}\right)^2\psi\right]\right\}, \tag{4.4.16}$$

which is found to vanish if $f = \frac{1}{2}mv^2t - m\vec{v}\cdot\vec{x}$.

Remark   Note that the phase of a plane wave: $\frac{i}{\hbar}\left(\vec{p}\cdot\vec{x} - Et\right)$ is not invariant under Galilei transformations, because

$$\vec{p}'\cdot\vec{x}' - E't' = (\vec{p}\cdot\vec{x} - Et) + \frac{1}{2}mv^2t - m\vec{v}\cdot\vec{x}. \tag{4.4.17}$$

From the relation $\vec{p}\cdot d\vec{x} - H dt \equiv \mathcal{L}dt$, it follows that the phase changes exactly by the quantity occurring in the variation of the Lagrangian. The transformation (4.4.5) is called a *gauge transformation*, and the physical quantities that remain invariant under such transformations are called *gauge-invariant*. The analysis performed in this section shows therefore that the wave function of quantum mechanics does not change as a (scalar) function under transformations of reference frames. Its deeper meaning can only become clear after a thorough investigation of the geometrical formulation of modern physical theories, but this task goes beyond the limits of an introductory course.

4.4.2 Width of the wave packet

As an application of the transformation properties that we have just derived, let us now consider how the mean quadratic deviation of position and momentum evolves in time *for a free particle*. This calculation is quite important to understanding whether a wave packet 'localized' in a certain interval $[x - \delta x, x + \delta x]$ remains localized and hence may be identified with a sort of particle. Of course, if the mean values of p_l and x_l vanish, the mean quadratic deviations coincide with $\langle p_l^2 \rangle$ and $\langle x_l^2 \rangle$, respectively. We can thus think of choosing a reference frame where $\langle p_l \rangle = 0$, and then perform a translation of coordinates so that $\langle x_l \rangle = 0$ as well. For this purpose, as we anticipated after Eq. (4.4.6), we define a suitable change of coordinates consisting of the Galilei transformations (4.4.7). We then consider

$$x'_l \equiv x_l - \langle x_l \rangle_\psi$$

$$= x_l - \int_{\mathbf{R}^3} A^* i\hbar \frac{\partial A}{\partial p_l} \mathrm{d}^3 p - t \int_{\mathbf{R}^3} \frac{\partial E}{\partial p_l} A^* A \mathrm{d}^3 p, \tag{4.4.18}$$

where A coincides with $A(\vec{p})$ in Eq. (4.1.12), and we perform the identifications

$$(x_0)_l \equiv \int_{\mathbf{R}^3} A^* i\hbar \frac{\partial A}{\partial p_l} \mathrm{d}^3 p, \tag{4.4.19}$$

$$v_l \equiv \int_{\mathbf{R}^3} \frac{\partial E}{\partial p_l} A^* A \mathrm{d}^3 p. \tag{4.4.20}$$

We now take into account that we are dealing with a free particle:

$$E = \frac{1}{2m} p^2 \implies \frac{\partial E}{\partial \vec{p}} = \frac{\vec{p}}{m} = \vec{v}.$$

The Galilei transformation induces the following transformation on the momenta:

$$p'_l = p_l - m v_l = p_l - m \int_{\mathbf{R}^3} \frac{\partial E}{\partial p_l} A^* A \mathrm{d}^3 p$$

$$= p_l - \int_{\mathbf{R}^3} p_l A^* A \mathrm{d}^3 p = p_l - \langle p_l \rangle_\psi. \tag{4.4.21}$$

In the frame Σ' we hence obtain

$$\langle x'_l \rangle_\psi = 0, \quad \langle p'_l \rangle_\psi = 0. \tag{4.4.22}$$

We now evaluate the mean values of p'^2_l and x'^2_l, omitting hereafter, for simplicity of notation, the prime. For the former

$$\langle p_l^2 \rangle_\psi = \int_{\mathbf{R}^3} p_l^2 \varphi^* \varphi \mathrm{d}^3 p = \int_{\mathbf{R}^3} 2mE \varphi^* \varphi \mathrm{d}^3 p$$

$$= \int_{\mathbf{R}^3} 2mE A^* A \mathrm{d}^3 p, \tag{4.4.23}$$

which is a constant in time. Moreover, from Eq. (4.1.15),

$$\langle x_l^2 \rangle_\psi = \hbar^2 \int_{\mathbf{R}^3} \frac{\partial \varphi^*}{\partial p_l} \frac{\partial \varphi}{\partial p_l} \mathrm{d}^3 p = \hbar^2 \int_{\mathbf{R}^3} \frac{\partial A^*}{\partial p_l} \frac{\partial A}{\partial p_l} \mathrm{d}^3 p$$

$$+ \mathrm{i}\hbar t \int_{\mathbf{R}^3} \frac{\partial E}{\partial p_l} \left(A^* \frac{\partial A}{\partial p_l} - A \frac{\partial A^*}{\partial p_l} \right) \mathrm{d}^3 p$$

$$+ t^2 \int_{\mathbf{R}^3} \left(\frac{\partial E}{\partial p_l} \right)^2 A^* A \mathrm{d}^3 p. \tag{4.4.24}$$

On using the well-known property $\dot{\vec{x}} = \frac{\partial E}{\partial \vec{p}} = \frac{\vec{p}}{m}$, the result (4.4.24) can be re-expressed in the form

$$\langle (\vec{x})^2 \rangle_\psi = \hbar^2 \int_{\mathbf{R}^3} \frac{\partial A^*}{\partial \vec{p}} \cdot \frac{\partial A}{\partial \vec{p}} \mathrm{d}^3 p + \mathrm{i}\frac{\hbar t}{m} \int_{\mathbf{R}^3} \vec{p} \cdot \left(A^* \frac{\partial A}{\partial \vec{p}} - A \frac{\partial A^*}{\partial \vec{p}} \right) \mathrm{d}^3 p$$

$$+ \frac{t^2}{m^2} \langle (\vec{p})^2 \rangle_\psi. \tag{4.4.25}$$

Thus, the wave packet has a 'width' that grows rapidly after the passage through a minimum; it also grows rapidly for earlier times.

The result (4.4.25) can also be expressed in terms of the wave function $\psi(\vec{x}, t)$. Let us denote by ψ_0 the initial value of the wave function: $\psi_0 \equiv \psi(\vec{x}, 0)$. The mean value of x^2 at the time $t = 0$ is given by

$$\langle x^2(0) \rangle_\psi = \int_{\mathbf{R}^3} \psi_0^* x^2 \psi_0 \mathrm{d}^3 x, \tag{4.4.26}$$

while

$$\vec{j}(0) = \frac{\hbar}{2mi} \left(\psi_0^* \vec{\nabla} \psi_0 - \psi_0 \vec{\nabla} \psi_0^* \right) \tag{4.4.27}$$

is the value of the current $\vec{j}(t)$ at $t = 0$. Thus,

$$\langle x^2(t) \rangle_\psi = \langle x^2(0) \rangle_\psi + 2t \int_{\mathbf{R}^3} \vec{x} \cdot \vec{j}(0) \mathrm{d}^3 x + \frac{t^2}{m^2} \langle p^2 \rangle_\psi. \tag{4.4.28}$$

It is also possible to re-express the result in the original frame Σ:

$$\langle (\Delta x)^2 \rangle_\psi = \langle (\Delta x(0))^2 \rangle_\psi + 2t \int_{\mathbf{R}^3} \left(\vec{x} - \langle \vec{x} \rangle \right) \cdot \left[\vec{j}(0) - \frac{\rho(0)}{m} \vec{p} \right] \mathrm{d}^3 x$$

$$+ \frac{t^2}{m^2} \langle [\Delta p(0)]^2 \rangle_\psi. \tag{4.4.29}$$

This relation holds for both positive and negative times.

4.5 Heisenberg picture

The evolution equation (4.1.4) is linear, therefore it allows for superposition of solutions, i.e. if ψ_1 and ψ_2 are solutions of the Schrödinger equation, the linear combination

$\lambda_1\psi_1 + \lambda_2\psi_2$ is also a solution with λ_1, λ_2 complex numbers. We have seen that the following quadratic expression:

$$\left(\lambda_1\psi_1 + \lambda_2\psi_2\right)^* \left(\lambda_1\psi_1 + \lambda_2\psi_2\right)$$

accounts for the probabilistic interpretation and it contains the interference term

$$\lambda_2\overline{\lambda}_1\psi_1^*\psi_2 + \overline{\lambda}_2\lambda_1\psi_2^*\psi_1.$$

We stress here that this interference term appears because we associate a physical meaning to the probability density rather than to the amplitude, and it is quadratic in the wave function. To account for the probabilistic interpretation we have considered $\psi^*\psi$ as a probability density so that the condition

$$\int_D \psi^*\psi \, \mathrm{d}\mu = 1$$

says that the probability to find the system somewhere in the total domain is 1. The Hermitian scalar product on these complex-valued wave functions is given by

$$\langle\psi|\varphi\rangle \equiv \int_D \psi^*\varphi \, \mathrm{d}\mu. \tag{4.5.1}$$

We notice that, from

$$\frac{\mathrm{d}}{\mathrm{d}t}\psi = \frac{\hat{H}}{\mathrm{i}\hbar}\psi,$$

it follows, under mild assumptions, that

$$\begin{aligned}
\frac{\mathrm{d}}{\mathrm{d}t}\left(\langle\psi|\varphi\rangle\right) &= \left\langle\frac{\mathrm{d}}{\mathrm{d}t}\psi\Big|\varphi\right\rangle + \left\langle\psi\Big|\frac{\mathrm{d}}{\mathrm{d}t}\varphi\right\rangle \\
&= \int_D \left(\frac{\partial}{\partial t}\psi^*\varphi + \psi^*\frac{\partial}{\partial t}\varphi\right)\mathrm{d}\mu \\
&= \int_D \left[-\left(\psi^*\frac{\hat{H}}{\mathrm{i}\hbar}\right)\varphi + \psi^*\left(\frac{\hat{H}}{\mathrm{i}\hbar}\varphi\right)\right]\mathrm{d}\mu,
\end{aligned} \tag{4.5.2}$$

where we have used the complex conjugate of the Schrödinger equation, i.e.

$$-\mathrm{i}\hbar\frac{\mathrm{d}}{\mathrm{d}t}\psi^* = \psi^*\hat{H}, \tag{4.5.3}$$

which follows from the requirement

$$(\hat{H}\psi)^* = \psi^*\hat{H}. \tag{4.5.4}$$

This means that $\hat{H}$ is a *real* operator, as it should be appropriate for a physical observable. If the requirement is fulfilled,

$$\frac{\mathrm{d}}{\mathrm{d}t}\langle\psi|\varphi\rangle = 0 \tag{4.5.5}$$

whenever $\psi(t)$ and $\varphi(t)$ are solutions of Eq. (13.4.2). In particular, if we start with some wave function satisfying the initial condition $\langle \psi(0)|\psi(0)\rangle = 1$, we find that the solution $\psi(t)$ of the equation of motion satisfies

$$\langle \psi(t)|\psi(t)\rangle = 1. \tag{4.5.6}$$

Assuming that we already have Schrödinger's equations on vectors of the Hilbert space, we are now aiming to comment on how to obtain equations of motion in terms of quantities which are more directly related to the probabilistic interpretation, i.e. rank-1 projectors (when applied twice, they remain equal to themselves)

$$\rho_\psi = \frac{|\psi\rangle\langle\psi|}{\langle\psi|\psi\rangle}. \tag{4.5.7}$$

By using the Schrödinger equation

$$\frac{\mathrm{d}}{\mathrm{d}t}\psi = \frac{\hat{H}}{\mathrm{i}\hbar}\psi$$

and its complex conjugate

$$\frac{\mathrm{d}}{\mathrm{d}t}\psi^* = \psi^*\frac{\hat{H}}{-\mathrm{i}\hbar}$$

we obtain the equations of motion

$$\mathrm{i}\hbar\frac{\mathrm{d}}{\mathrm{d}t}\rho_\psi = [\hat{H}, \rho]. \tag{4.5.8}$$

It is possible to generalize this construction from a given vector to a family of vectors of which we consider convex combinations. If

$$p_j \geq 0, \quad \sum_j p_j = 1, \tag{4.5.9}$$

we may consider

$$\rho = \sum_j p_j \rho_{\psi_j} \tag{4.5.10}$$

and derive the equation of motion

$$\mathrm{i}\hbar\frac{\mathrm{d}}{\mathrm{d}t}\rho = [\hat{H}, \rho]. \tag{4.5.11}$$

From here, we can further generalize to generic operators instead of restricting to convex combinations of rank-1 projectors.

With any self-adjoint operator A, and assumed to have a discrete spectrum

$$(a_1, a_2, \ldots, a_n, \ldots),$$

we can associate a basis of orthonormal eigenvectors denoted by $\{|a_j\rangle\}$. In this basis the operator has the form

$$A = \sum_j a_j |a_j\rangle\langle a_j|, \tag{4.5.12}$$

and it is possible to repeat all steps that we have considered for the convex combination ρ, called a mixed state, i.e. we get the equations of motion

$$i\hbar\frac{d}{dt}A = \widehat{H}A - A\widehat{H}. \tag{4.5.13}$$

Clearly, this expression remains valid even if the operator is not self-adjoint, or the real and imaginary part do not commute, and therefore do not have a common basis of eigenvectors. The equation of motion we have derived in the space of operators represents the equation of motion in the Heisenberg picture.

Historically, the Heisenberg approach to quantum mechanics can be seen to have its roots in the Ritz combination principle for the emission and absorption frequencies, according to which $v_{j,k} = v_{j,l} + v_{l,k}$, whereas otherwise the sum is not defined (e.g. $v_{j,l} + v_{m,k}$ with $l \neq m$). We already know, from the Einstein–de Broglie identification, that frequencies are proportional to the energies. Moreover, the evolution of a quantum system is described from the exponential of the Hamiltonian operator; it is therefore reasonable to consider the expression $e^{iv_{j,k}}$. By taking the combinations

$$A = \sum_{j,k} a_{jk}e^{iv_{j,k}}, \; B = \sum_{r,s} b_{rs}e^{iv_{r,s}} \tag{4.5.14}$$

and using a product rule according to which

$$e^{iv_{j,k}}e^{iv_{r,s}} = \delta_{kr}e^{iv_{j,s}}, \tag{4.5.15}$$

or, with the understanding that the left-hand side is not defined if $k \neq r$, it is found that the resulting product $A \cdot B$ is equivalent to the row-by-column product

$$(A \cdot B)_{js} = \sum_k a_{jk}b_{ks}. \tag{4.5.16}$$

Thus, on relying upon the Ritz combination principle, we obtain an operation that reproduces the product of matrices. The Heisenberg formulation consists therefore in considering matrices associated with measurable quantities and describing the evolution of a quantum system through equations of motion that involve directly observable quantities. If A and H represent a dynamical observable and the Hamiltonian operator, respectively,

$$i\hbar\frac{dA_{jk}}{dt} = H_{jl}A_{lk} - A_{jl}H_{lk}. \tag{4.5.17}$$

These equations on the space of matrices, which is possibly infinite-dimensional, read eventually, in operator form,

$$i\hbar\frac{d}{dt}A = [H, A]. \tag{4.5.18}$$

When A and H are viewed as operators on a Hilbert space with an orthonormal basis $\{|e_j\rangle\}$, A_{jk} can be expressed as the matrix element $\langle e_j|A|e_k\rangle$. Moreover, the expectation values of A, the mean values that we have introduced in the Schrödinger formulation, should be real-valued if A represents a measurable quantity. It therefore follows that operators associated to observables should be Hermitian, hence the Heisenberg equation of motion should be viewed as a differential equation on the space of Hermitian operators.

Digression *Hermitian operators and the group of unitary transformations.* In the Schrödinger picture, the evolution of a quantum system on the space of states is provided, at least formally, from the relation

$$|\psi(t)\rangle = e^{-\frac{iHt}{\hbar}} |\psi(0)\rangle. \tag{4.5.19}$$

The operator $e^{-\frac{iHt}{\hbar}}$ is a unitary operator if H is a Hermitian operator. A particular Hermitian operator, a rank-1 projector in the case of normalized vectors, is given by $|\psi(t)\rangle\langle\psi(t)|$, i.e.

$$e^{-\frac{iHt}{\hbar}} |\psi(0)\rangle\langle\psi(0)| e^{-\frac{iHt}{\hbar}} = |\psi(t)\rangle\langle\psi(t)|. \tag{4.5.20}$$

From this relation we are led to assume that the operator A evolves in time according to

$$A(t) = e^{-\frac{iHt}{\hbar}} A(0) e^{-\frac{iHt}{\hbar}}. \tag{4.5.21}$$

On taking the time derivative of $A(t)$,

$$i\hbar\frac{d}{dt}A = HA - AH = [H, A]. \tag{4.5.22}$$

By exploiting the uniqueness theorem of differential equations where it holds, which is certainly the case for finite-dimensional matrices, we can conclude that the solution of Eq. (4.5.22) is given, at least formally, by the formula (4.5.21).

4.6 States in the Heisenberg picture

In this picture, the way it was derived by Heisenberg from the Ritz combination principle, the observables, given by Hermitian operators, are a primary concept and states are a derived concept. They are defined as elements of the dual vector space (over the reals) of the vector space of observables, i.e. linear functionals with the additional requirements of positivity and normalization, respectively:

$$\omega(A^\dagger A) \geq 0, \tag{4.6.1}$$

$$\omega(\mathbb{I}) = 1. \tag{4.6.2}$$

A theorem by Gleason ensures that any such a state may be represented by means of an Hermitian operator and with the help of the trace, i.e.

$$\omega(A) = \mathrm{Tr}\,(\rho_\omega A). \tag{4.6.3}$$

We notice here that the vector space of operators may be endowed with a scalar product given by

$$\langle A|B\rangle = \mathrm{Tr}\,(A^\dagger B). \tag{4.6.4}$$

This relation may be thought of as defining the scalar product if we already have a meaning for the trace, or it can be viewed as defining the trace if we are given a scalar product. Hence, we conclude that the association $\omega \to \rho_\omega$ depends on the specific scalar product we use. From the equation for $\omega(A)$ it follows that ρ_ω is Hermitian with respect to the considered scalar product and positive along with the normalization condition $\text{Tr}\,\rho_\omega = 1$. From the equation of motion (4.5.22), it is possible to derive Eq. (4.5.11) by using the duality described by means of the trace operation. It is important to note that from Eq. (4.6.3), because of the adjoint of A appearing in Eq. (4.6.4), the equations of motion on the operators, considered on the dual space of states, differ from Eq. (4.5.22) for the overall sign on the right-hand side. To arrive at the equations of motion on the Hilbert space, one has to rely on the so-called Gel'fand–Naimark–Segal construction, or on the Wigner theorem that we shall mention in Chapter 10.

4.7 'Conclusions': relevant mathematical structures

To put the Schrödinger and the Heisenberg pictures better into perspective, we are going to make a few general considerations on the minimal mathematical structure required for the description of a physical system.

From a general point of view, the description of any physical system, either classical or quantum, requires three basic ingredients:

(i) a space of states, that we denote as $\mathcal{S}$,
(ii) a space of observables, that we denote as $\mathcal{O}$ and
(iii) a real-valued pairing $\mathcal{O} \times \mathcal{S} \to \mathbf{R}$. This pairing, which produces a real number out of an observable and a state, represents the measuring operation.

In quantum mechanics, we have two main pictures.

(I) In the Schrödinger picture, $\mathcal{S}$ is associated with a Hilbert space $\mathcal{H}$ and the set of dynamical variables (the observables) is a derived concept. Observables are identified with self-adjoint operators on $\mathcal{H}$. The pairing provides real numbers with a probability distribution, i.e. a computable probability to observe the given real numbers in the measurement of a selected observable (physical quantity).

(II) In the Heisenberg picture the situation is complementary: the set of dynamical variables (the observables) is the primary concept. They are assumed to be (the real part of) an algebra $\mathcal{A}$ (endowed with some structure); an instance of such an algebra is provided by all matrices with complex entries. The states, on the other hand, are a derived concept defined as a proper subset of the set of linear functionals on $\mathcal{A}$. The pairing with states provides again a real number and a probability for its occurrence.

It should be stressed, however, that a physical system requires, in addition to either one of the two primary carrier spaces, *a concrete realization of it to allow us to identify the physical variables*. This last requirement is often overlooked in the literature. We can clarify an aspect of this last point with a specific example taken from classical mechanics but that applies equally well within quantum mechanics.

Example 1: Let us consider the carrier space for a classical system to be a phase space $(\mathbf{R}^3 - \{\vec{0}\}) \times \mathbf{R}^3$ endowed with a Poisson bracket. On considering coordinates $(\vec{\xi}, \vec{\eta})$ we define the Poisson structure in the form

$$\{\xi_j, \xi_k\} = 0\,, \quad \{\eta_j, \eta_k\} = \lambda\epsilon_{jkl}\frac{\xi_l}{\|\xi\|^3}\,, \quad \{\xi_j, \eta_k\} = \delta_{jk}\,. \tag{4.7.1}$$

This carrier space is appropriate to describe an electron–monopole system or a massless particle with helicity. Indeed, if we set

$$\xi_j = x_j(\text{position}), \quad \eta_j = p_j(\text{momentum}), \tag{4.7.2}$$

the resulting Poisson brackets take the form required in the electron–monopole system. The brackets of the momenta are thus proportional to the magnetic field of the monopole. If we set $\xi_j = p_j$ and $\eta_j = x_j$, on the other hand, we endow the carrier space with the Poisson structure required to model the dynamical behaviour of a massless spinning particle. The cubic term in the denominator of the bracket of two position coordinates accounts then for the fact that a zero-rest-mass particle cannot be reduced to rest. And the non-vanishing of these brackets is taking into account that massless particles cannot be localized in space. A very similar situation prevails in the corresponding quantum case (Balachandran *et al.* 1983).

In conclusion, the description of a physical system requires not only an abstract mathematical model (a Poisson manifold, a Hilbert space, a C^*-algebra,...) but also a specific realization with an identification of the physical variables.

We have mentioned the mathematical structures used for the description of quantum physical systems to stress that the interplay of particle, wave and probability concepts will require that we learn how to master methods and techniques from different fields of mathematics. Moreover, the dimensions and durations of processes involved are so tiny that we can hope to develop an understanding only with the help of mathematics on one hand, while relying on the 'classical limit' on the other. For these reasons, our previous knowledge of propagation of waves, and Hamiltonian mechanics will turn out to be very useful and, to some extent, indispensable.

The quantum-classical interplay is illustrated by the wave-particle interplay. According to this interpretation, we may consider a situation which is already familiar in classical physics, i.e. the transition from wave optics to ray optics. A similar relation turns out also to be very useful to understand how to construct a wave mechanics for particles which would reproduce classical mechanics in the appropriate limit. The reader should bear in mind in any case the warning of W. Heisenberg in his introduction to the physical principles of the theory of quanta: '...because without mathematics it is not possible to tackle any physical problem'.

Integrating the equations of motion

The solution of the Schrödinger equation for a given initial condition is studied. For this purpose, the initial condition is first expanded in terms of eigenfunctions of the Hamiltonian operator. This is then 'propagated' to determine the evolution of the wave function. Thus, we are led to consider the Green kernel of the Schrödinger equation.

Integration of the equations of motion is also studied in the Heisenberg picture, with application to the one-dimensional harmonic oscillator.

5.1 Green kernel of the Schrödinger equation

Now we revert to Schrödinger's picture and point out that the main technical problem of wave mechanics is the solution of the Schrödinger equation once an initial condition is given. We are going to see that two key steps emerge for this purpose, i.e.

(i) to find (at least implicitly) eigenvalues and eigenfunctions of the Hamiltonian operator H (taken to be independent of t), once a domain of essential self-adjointness[1] for H has been determined;

(ii) to evaluate the Green kernel, which makes it possible to 'propagate' the initial condition and hence leads to complete knowledge of the wave function at all times.

To begin our analysis it is helpful to consider a simpler problem, i.e. how to solve the linear equation

$$\frac{d\varphi}{dt} = A\varphi \tag{5.1.1}$$

on a finite-dimensional vector space V with initial condition $\varphi(t = 0) = \varphi_0$. If the matrix A is diagonalizable and all eigenvalues are simple (i.e. without degeneracy), the space V has a basis of right eigenvectors $\{v_j\}$:

$$Av_j = \varepsilon_j v_j, \tag{5.1.2}$$

---

[1] An unbounded linear operator is said to be self-adjoint if it is symmetric, and its domain is equal to the domain of its adjoint. Moreover, an unbounded linear operator is said to be essentially self-adjoint if its closure is self-adjoint, which implies that there is a unique self-adjoint extension of the operator. For further details, we refer the reader to the glossary of functional analysis in Esposito *et al.* (2004).

and the initial condition can be expanded in the form

$$\varphi(0) = \sum_{k=1}^{N} b^k v_k, \qquad (5.1.3)$$

where the coefficients of linear combination are obtained by using the left eigenvectors v^j satisfying (no summation over repeated index j on the right-hand side below)

$$v^j A = \varepsilon_j v^j$$

in the form

$$b^j = \langle v^j | \varphi(0) \rangle, \qquad (5.1.4)$$

since left and right eigenvectors belonging to different eigenvalues are orthogonal, i.e.

$$\langle v^j | v_k \rangle = \delta^j_{\ k}. \qquad (5.1.5)$$

The solution of our first-order equation is thus found to be

$$\varphi(t) = e^{tA} \varphi(0) = \sum_{k=1}^{N} b^k v_k e^{\varepsilon_k t}, \qquad (5.1.6)$$

because any power of the matrix A acts as a multiplication operator on the right eigenvectors v_k:

$$A^r v_k = (\varepsilon_k)^r v_k \quad \forall r = 0, 1, \ldots \qquad (5.1.7)$$

by virtue of the eigenvalue equation (5.1.2). The left and right eigenvectors are conjugates of each other if the matrix A is Hermitian: $(A^*)^T = A$.

It is quite remarkable that the expression $\varphi(t) = e^{tA} \varphi(0)$ provides us with a solution of Eq. (5.1.1) even if A cannot be diagonalized. For this to make sense, we only require that the Taylor expansion

$$e^{tA} = I + tA + \frac{t^2 A^2}{2!} + \cdots \qquad (5.1.8)$$

should converge uniformly, which is always the case in finite dimensions, for any fixed value of t. This consideration suggests replacing Eq. (5.1.1) on the vector space with an equation on the space of matrices.

If we consider a fundamental set of solutions, i.e. a maximal independent set of solutions, say $\varphi_1(t), \varphi_2(t), \ldots, \varphi_n(t)$ with corresponding initial conditions $\varphi_1(0), \varphi_2(0), \ldots, \varphi_n(0)$, we may form a matrix $\Phi(t)$ by using solutions $\varphi_j(t)$ as columns, and similarly for $\Phi(0)$. By virtue of the independence requirement, both $\Phi(t)$ and $\Phi(0)$ will be invertible, therefore we may form the product $\Phi(t)\Phi^{-1}(0) \equiv G(t, t_0)$. This matrix is a solution of the equations of motion

$$\frac{\mathrm{d}}{\mathrm{d}t} G(t, t_0) = A G(t, t_0)$$

with initial condition $G(t_0, t_0) = \mathbb{I}$. This equation may be a substitute for Eq. (5.1.1), which is particularly valuable when A cannot be diagonalized. With this matrix we construct the generic solution with the initial condition $\varphi(t_0)$:

$$\varphi(t) = G(t, t_0)\varphi(t_0).$$

Similarly in quantum mechanics, let us assume that H, realized as $L^2(\mathbf{R}^3)$, possesses a complete set of orthonormal eigenvectors $\{u_j\}$:

$$Hu_j(\vec{x}) = E_j u_j(\vec{x}). \tag{5.1.9}$$

Then, the initial condition ψ_0 for the Schrödinger equation can be expanded as

$$\psi_0 \equiv \psi(\vec{x}, 0) = \sum_{j=1}^{\infty} c_j u_j(\vec{x}), \tag{5.1.10}$$

where the Fourier coefficients c_j can be computed by using the formula

$$c_j = (u_j, \psi_0) = \int_{\mathbf{R}^3} \mathrm{d}^3 x'\, u_j^*(\vec{x}')\psi(\vec{x}', 0), \tag{5.1.11}$$

since eigenvectors belonging to different eigenvalues are orthogonal:

$$(u_j, u_l) = \int_{\mathbf{R}^3} \mathrm{d}^3 x\, u_j^*(\vec{x})u_l(\vec{x}) = \delta_{jl}.$$

The solution of the initial-value problem for the Schrödinger equation is thus found to be (bearing in mind that our equation is of first order in the time variable)

$$\psi(\vec{x}, t) = \mathrm{e}^{-\frac{itH}{\hbar}} \psi(\vec{x}, 0) = \sum_{j=1}^{\infty} c_j \sum_{r=0}^{\infty} \frac{(-\mathrm{i}t/\hbar)^r}{r!} H^r u_j(\vec{x})$$

$$= \sum_{j=1}^{\infty} c_j u_j(\vec{x})\mathrm{e}^{-\frac{iE_j t}{\hbar}}, \tag{5.1.12}$$

where we have used the formal Taylor series for $\mathrm{e}^{-\frac{itH}{\hbar}}$, jointly with the eigenvalue equation (5.1.9) and the purely discrete nature of the spectrum of H (for a generalization, see below). In other words, the general solution is expressed as an infinite sum of elementary solutions $u_j(\vec{x})\mathrm{e}^{\frac{-iE_j t}{\hbar}}$, and it is now clear why, to use this approach, it is convenient to solve first the eigenvalue problem for the stationary Schrödinger equation (5.1.9).

For instance, when A represents the Hamiltonian of the free particle, fundamental solutions are not square-integrable. This situation provides an example of an operator which cannot be diagonalized (in the space we are interested in). To handle this situation, we replace the equation on the Hilbert space with an equation on the space of operators. It is precisely to this approach that the equation of motion for the Green kernel or Green function belongs.

5.1.1 Discrete version of the Green kernel by using a fundamental set of solutions

Another useful expression of the solution is obtained after inserting the result (5.1.11) for the coefficients c_j into Eq. (5.1.12), which leads to

$$\psi(\vec{x}, t) = \sum_{j=1}^{\infty} \int_{\mathbf{R}^3} \mathrm{d}^3 x'\, u_j^*(\vec{x}')\psi(\vec{x}', 0)u_j(\vec{x}) \mathrm{e}^{-\frac{iE_j t}{\hbar}}$$

$$= \int_{\mathbf{R}^3} \mathrm{d}^3 x'\, G(\vec{x}, \vec{x}'; t)\psi(\vec{x}', 0), \tag{5.1.13}$$

where $G(\vec{x}, \vec{x}'; t)$ is the standard notation for the Green function (see the comments below):

$$G(\vec{x}, \vec{x}'; t) \equiv \sum_{n=1}^{\infty} u_n^*(\vec{x}')u_n(\vec{x}) \mathrm{e}^{-\frac{iE_n t}{\hbar}}. \tag{5.1.14}$$

In other words, once the initial condition $\psi(\vec{x}, 0)$ is known, the solution at a time $t \neq 0$ is obtained from Eq. (5.1.13), where $G(\vec{x}, \vec{x}'; t)$ is the Green kernel of the operator $\mathrm{e}^{\frac{-itH}{\hbar}}$. This is, by definition, a solution for $t \neq 0$ of the equation

$$\left(i\hbar \frac{\partial}{\partial t} - H_{(x)} \right) G(\vec{x}, \vec{x}'; t) = 0, \tag{5.1.15}$$

subject to the initial condition (where ρ is a suitably smooth function)

$$\lim_{t \to 0} \int_{\mathbf{R}^3} \mathrm{d}^3 x'\, G(\vec{x}, \vec{x}'; t)\rho(\vec{x}') = \rho(\vec{x}). \tag{5.1.16}$$

This is a more precise way to express the distributional behaviour of the Green kernel. In the physics literature, Eqs. (5.1.15) and (5.1.16) are more frequently re-expressed as follows:

$$\mathcal{G}(\vec{x}, \vec{x}', t) \equiv \theta(t) \sum_{n=1}^{\infty} u_n^*(\vec{x}')u_n(\vec{x}) \mathrm{e}^{-\frac{iE_n t}{\hbar}}, \tag{5.1.17}$$

$$\left(i\hbar \frac{\partial}{\partial t} - H_{(x)} \right) \mathcal{G}(\vec{x}, \vec{x}'; t) = \delta(\vec{x}, \vec{x}')\delta(t), \tag{5.1.18}$$

$$\mathcal{G}(\vec{x}, \vec{x}'; 0) = \delta(\vec{x}, \vec{x}'), \tag{5.1.19}$$

where we have multiplied the right-hand side of the definition (5.1.14) by $\theta(t)$ (θ being the step function) to recover the effect of $\delta(t)$ in Eq. (5.1.18). On considering Eqs. (5.1.12) and (5.1.13) we therefore say that $G(\vec{x}, \vec{x}'; t)$ is the Schrödinger kernel for the 1-parameter strongly continuous unitary group $\mathrm{e}^{\frac{-itH}{\hbar}}$. It propagates both forward and backward in time. Some authors prefer to say that the kernel (5.1.14) is the *propagator*, while the kernel (5.1.17), which incorporates the step function, is called the *Green function*.

Note that in Eq. (5.1.13) the initial condition $\psi(\vec{x}', 0)$ can be any vector in the infinite-dimensional Hilbert space of the problem, since we are with an integral. On the other hand, the formal exponentiation used in Eq. (5.1.12) dealing requires, in general, a C^∞ initial condition. However, if the Hilbert space can be decomposed into a direct sum of finite-dimensional subspaces invariant under H, the operator H becomes a Hermitian matrix on every single sub-space, and hence the exponentiation reduces to the familiar operation used in the finite-dimensional case.

We should now stress that the hypothesis of a purely discrete spectrum for H is very restrictive. For example, the stationary Schrödinger equation for a free particle

$$-\frac{\hbar^2}{2m}\Delta u = Eu, \tag{5.1.20}$$

has a continuous spectrum $E = \frac{\hbar^2 k^2}{2m}$, since $E = \hbar\omega$, and the dispersion relation is then $\omega = \frac{\hbar k^2}{2m}$. The corresponding solutions (also called improper eigenfunctions)

$$u_k(\vec{x}) = e^{i\vec{k}\cdot\vec{x}}, \tag{5.1.21}$$

are not, by themselves, normalizable. Therefore, the eigenvalue problem does not have solutions in the space of square-integrable functions, and in this case it is much more convenient to solve for the Green kernel.

The free-particle improper eigenfunctions are, however, of algebraic growth, in that, as $|\vec{x}| \to \infty$, can find a polynomial p such that

$$|u_k(\vec{x})| \le p(x). \tag{5.1.22}$$

Thus a meaningful wave packet for the general solution can be formed by means of a Fourier transform:

$$\psi(\vec{x}, t) = (2\pi)^{-\frac{3}{2}} \int_{\mathbf{R}^3} d^3k \, e^{i\vec{k}\cdot\vec{x}} \, e^{-\frac{i\hbar k^2 t}{2m}} \, \hat{u}(k), \tag{5.1.23}$$

where $\hat{u}(k)$ is the Fourier transform of the initial condition, i.e.

$$\hat{u}(k) = (2\pi)^{-\frac{3}{2}} \int_{\mathbf{R}^3} d^3x \, e^{-i\vec{k}\cdot\vec{x}} \, \psi(\vec{x}, 0). \tag{5.1.24}$$

Remark In general, the spectrum of H, assumed to be self-adjoint, may consist of a discrete part σ_d and a continuous part σ_c, and hence the general solution of the time-dependent Schrödinger equation reads as

$$\psi(\vec{x}, t) = \int_{\sigma(H)} d\mu(E) \sum_\alpha C_\alpha(E)\psi_{E,\alpha}(\vec{x})e^{-\frac{iEt}{\hbar}}. \tag{5.1.25}$$

With our notation, the symbol

$$\int_{\sigma(H)} d\mu(E)$$

is a condensed notation for the summation over discrete eigenvalues and integration over the continuous spectrum. The corresponding spectral representation of the Hamiltonian operator reads as

$$H = \int \lambda d\hat{E}_\lambda = \sum_{\lambda\in\sigma_d} \lambda\hat{P}_\lambda + \int_{\lambda\in\sigma_c} \lambda d\hat{E}_\lambda, \tag{5.1.26}$$

where $\hat{P}_\lambda$ are spectral projectors for eigenvalues belonging to the discrete spectrum, and $d\hat{E}_\lambda$ are their counterparts for the continuous spectrum. This holds by virtue of the Lebesgue decomposition of a measure on $\mathbf{R}$: any measure is the sum of a part $d\mu_{ac} = f(\alpha)d\alpha$, with $f \geq 0$ and locally integrable, which is absolutely continuous with respect to the Lebesgue measure $d\alpha$; a part $d\mu_p$ concentrated on some separate points

$$d\mu_p = d\alpha \sum_n c_n \delta(\alpha - \alpha_n), \quad \alpha_n \in R;$$

and a remainder $d\mu_s$, the singular spectrum. This last part is pathological and will not occur in the problems considered in our manuscript (although there are 1-electron band models with a non-empty singular spectrum). Each of the three pieces of the measure is concentrated on null sets with respect to the others, and $\mathcal{L}^2(\mathbf{R}, d\mu)$ admits an orthogonal decomposition as

$$\mathcal{L}^2(\mathbf{R}, d\mu) = \mathcal{L}^2(\mathbf{R}, d\mu_p) \oplus \mathcal{L}^2(\mathbf{R}, d\mu_{ac}) \oplus \mathcal{L}^2(\mathbf{R}, d\mu_s).$$

5.1.2 General considerations on how we use solutions of the evolution equation

Free particle

A Green-function approach to the Schrödinger equation for a free particle is rather convenient, since in this situation, as we said before, the eigenvalue problem for the free Hamiltonian does not have solutions in the $\mathcal{L}^2$-space. For this purpose, we look for a fundamental solution of Eq. (5.1.15), which is singular at $t - t' = 0$ in such a way that, for every finite region of integration,

$$\lim_{t \to t'} \int_V G(\vec{x} - \vec{x}'; t - t') d^3 x' = 1 \quad \text{if } \vec{x} - \vec{x}' \text{ is in } V, \tag{5.1.27}$$

and 0 otherwise. Since the Schrödinger equation is linear, the desired solution is obtained from the integral (cf. Eq. (5.1.13))

$$\psi(\vec{x}, t) = \int G(\vec{x} - \vec{x}'; t - t') \psi(\vec{x}', t') d^3 x', \tag{5.1.28}$$

where G solves the equation (cf. Eq. (5.1.15))

$$\left(i\hbar \frac{\partial}{\partial t} + \frac{\hbar^2}{2m} \Delta \right) G(\vec{x} - \vec{x}'; t - t') = 0 \quad \text{for } t \neq t', \tag{5.1.29}$$

and satisfies the initial condition (cf. Eq. (5.1.16))

$$G(\vec{x} - \vec{x}'; 0) = \delta(\vec{x} - \vec{x}'). \tag{5.1.30}$$

For a free particle in one dimension, the Green function is given by (we set $x' = 0$ for simplicity)

$$G(x, 0, t) = \frac{C}{\sqrt{t}} e^{\frac{im}{2\hbar} \frac{x^2}{t}}, \tag{5.1.31}$$

which, for all $t \neq 0$, is not square integrable. The calculation shows indeed that, $\forall t \neq 0$, such a $G(x, 0, t)$ solves the equation

$$\frac{\partial G}{\partial t} = \frac{i\hbar}{2m} \frac{\partial^2 G}{\partial x^2}. \tag{5.1.32}$$

Moreover, motivated by Eq. (5.1.27), we consider the integral

$$\int_{x_1}^{x_2} G(x,0,t)\mathrm{d}x = C\sqrt{\frac{2\hbar}{m}} \int_{\sqrt{\frac{m}{2\hbar}}\frac{x_1}{\sqrt{t}}}^{\sqrt{\frac{m}{2\hbar}}\frac{x_2}{\sqrt{t}}} \mathrm{e}^{\mathrm{i}\xi^2}\mathrm{d}\xi. \tag{5.1.33}$$

Now we recall that, if $a \to \infty$ and $b \to \infty$, then

$$\lim_{a,b\to\infty} \int_a^b \mathrm{e}^{\mathrm{i}\xi^2}\mathrm{d}\xi = 0, \tag{5.1.34}$$

whereas, if $a \to -\infty$ and $b \to \infty$,

$$\int_a^b \mathrm{e}^{\mathrm{i}\xi^2}\mathrm{d}\xi \to \int_{-\infty}^{\infty} \mathrm{e}^{\mathrm{i}\xi^2}\mathrm{d}\xi = \sqrt{\pi}\mathrm{e}^{\mathrm{i}\frac{\pi}{4}}. \tag{5.1.35}$$

From Eqs. (5.1.33)–(5.1.35), the property (5.1.27) is indeed satisfied, the origin corresponding to $x = 0$, and the region V to the interval $[x_1, x_2]$, with

$$C = \mathrm{e}^{-\mathrm{i}\frac{\pi}{4}}\sqrt{\frac{m}{2\pi\hbar}}. \tag{5.1.36}$$

In $\mathbf{R}^3$, we can thus write

$$G(\vec{x},\vec{0},t) = G(x_1,0,t)G(x_2,0,t)G(x_3,0,t)$$

$$= \mathrm{e}^{-\mathrm{i}\frac{3\pi}{4}}\left(\frac{m}{2\pi\hbar}\right)^{\frac{3}{2}} t^{-\frac{3}{2}}\mathrm{e}^{\mathrm{i}\frac{m}{2\hbar}\frac{(x_1^2+x_2^2+x_3^2)}{t}}, \tag{5.1.37}$$

and this formula should be used to evaluate the right-hand side of Eq. (5.1.28). As an example, let us consider a one-dimensional wave packet which, at $t = 0$, is given by

$$\psi(x',0) = \widetilde{C}\,\mathrm{e}^{-\left(\frac{x'}{2\triangle_0 x}\right)^2 + \frac{\mathrm{i}mvx'}{\hbar}}, \tag{5.1.38}$$

where $\widetilde{C}$ is a constant, and $\triangle_0 x$ is the mean quadratic deviation at $t = 0$. From Eq. (5.1.28), adapted to our one-dimensional problem, we find (use Eq. (5.1.31) with x replaced by $x-x'$, and Eq. (5.1.36))

$$\psi(x,t) = \gamma\left(\frac{m}{\hbar}t\right)^{1/2}\mathrm{e}^{-\mathrm{i}\frac{\pi}{4}}\mathrm{e}^{\frac{\mathrm{i}mx^2}{2\hbar t}}\int_{-\infty}^{\infty}\exp\left\{\left[\frac{\mathrm{i}m}{2\hbar t} - \frac{1}{(2\triangle_0 x)^2}\right]x'^2\right.$$

$$\left. + \frac{\mathrm{i}m}{\hbar t}(vt-x)x'\right\}\mathrm{d}x'. \tag{5.1.39}$$

In this equation, $\gamma \equiv \frac{\widetilde{C}}{\sqrt{2\pi}}$, and we are dealing with an integral of the kind

$$\int_{-\infty}^{\infty}\mathrm{e}^{-ax^2+2\mathrm{i}bx}\mathrm{d}x = \sqrt{\frac{\pi}{a}}\,\mathrm{e}^{-\frac{b^2}{a}}. \tag{5.1.40}$$

Thus, on defining

$$(2\triangle x)^2 \equiv (2\triangle_0 x)^2\left\{1 + \left[\frac{2\hbar t}{m(2\triangle_0 x)^2}\right]^2\right\}, \tag{5.1.41}$$

we find

$$|\psi(x,t)|^2 = |\widetilde{C}|^2 \frac{(\triangle_0 x)}{(\triangle x)} e^{-\frac{(x-vt)^2}{2(\triangle x)^2}}.$$ (5.1.42)

The physical interpretation is that the centre of the wave packet moves with velocity v, and its spreading in time is described by Eq. (5.1.41). In particular, for $\triangle x$ to become twice as large as $\triangle_0 x$, we have to wait for a time

$$t = 2\sqrt{3}\frac{m}{\hbar}(\triangle_0 x)^2.$$ (5.1.43)

Thus, if $m = 1.7 \times 10^{-24}$ g (as for the hydrogen atom), with $\triangle_0 x = 10^{-8}$ cm, $t = 5.5 \times 10^{-13}$ s. In contrast, if $m = 10^{-3}$ g, with $\triangle_0 x = 10^{-3}$ cm, $t \cong 3.3 \times 10^{18}$ s, i.e. a time of the order of 10^{11} years! Finally, the evolution is time symmetric, so for $t < 0$ there is also a greater spread.

5.2 Integrating the equations of motion in the Heisenberg picture: harmonic oscillator

It can be instructive to integrate the equations of motion in the Heisenberg formalism by using the basic example of the one-dimensional harmonic oscillator. In this case, the Heisenberg equations of motion

$$\frac{d\hat{A}}{dt} = \frac{i}{\hbar}[\hat{H}, \hat{A}]$$ (5.2.1)

yield (hereafter, we use the subscript H for the time-dependent Heisenberg operators, rather than the hat symbol upon the letter denoting the operator)

$$\frac{dx_H}{dt} = \frac{p_H}{m},$$ (5.2.2)

$$\frac{dp_H}{dt} = -m\omega^2 x_H,$$ (5.2.3)

which display a complete *formal* analogy with the structure of the classical equations of motion. In light of the initial conditions

$$\dot{x}_H(0) = \frac{p_H(0)}{m} = \frac{p_0}{m},$$ (5.2.4)

$$\dot{p}_H(0) = -m\omega^2 x_H(0) = -m\omega^2 x_0,$$ (5.2.5)

we find, with the help of matrix notation,

$$\begin{pmatrix} x_H(t) \\ p_H(t) \end{pmatrix} = \begin{pmatrix} \cos\omega t & \frac{\sin\omega t}{m\omega} \\ -m\omega\sin\omega t & \cos\omega t \end{pmatrix} \begin{pmatrix} x_0 \\ p_0 \end{pmatrix},$$ (5.2.6)

i.e. the linear Hamiltonian system on classical phase space and the corresponding quantum system in the Heisenberg picture admit solutions which are formally identical.

Interestingly, the commutator of position operators at different times t_1 and t_2 is not identically vanishing, but it is found, from the basic property of commutators

$$[\lambda_1 \hat{A}_1 + \lambda_2 \hat{A}_2, \hat{B}] = \lambda_1 [\hat{A}_1, \hat{B}] + \lambda_2 [\hat{A}_2, \hat{B}],$$

that

$$\left[x_H(t_1), x_H(t_2) \right] = \frac{(\cos \omega t_1)(\sin \omega t_2)}{m\omega} [x_0, p_0] + \frac{(\sin \omega t_1)(\cos \omega t_2)}{m\omega} [p_0, x_0]$$

$$= \frac{i\hbar}{m\omega} \sin \omega (t_2 - t_1) I, \tag{5.2.7}$$

because $[x_0, p_0] = i\hbar$. With analogous procedure we find

$$\left[x_H(t_1), p_H(t_2) \right] = (\cos \omega t_1)(\cos \omega t_2)[x_0, p_0] - (\sin \omega t_1)(\sin \omega t_2)[p_0, x_0]$$

$$= i\hbar \cos \omega (t_2 - t_1) I, \tag{5.2.8}$$

$$\left[p_H(t_1), p_H(t_2) \right] = -m\omega (\sin \omega t_1)(\cos \omega t_2)[x_0, p_0] - m\omega (\cos \omega t_1)(\sin \omega t_2)[p_0, x_0]$$

$$= i\hbar m\omega \sin \omega (t_2 - t_1) I. \tag{5.2.9}$$

The commutator (5.2.7) can be applied, for example, to the evaluation of the standard inequality

$$(\Delta A)_\psi (\Delta B)_\psi \geq \frac{1}{2} \left| (\psi, [A, B]\psi) \right|, \tag{5.2.10}$$

upon taking $A \equiv x_H(t_1), B \equiv x_H(t_2), (\psi, \psi) = 1$. This yields, for the product of quadratic deviations,

$$(\Delta A)_\psi (\Delta B)_\psi \geq \frac{\hbar}{2m\omega} \left| \sin \omega (t_2 - t_1) \right|. \tag{5.2.11}$$

6 Elementary applications: one-dimensional problems

The boundary conditions for solving the stationary Schrödinger equation in one spatial dimension are derived. At the points where the potential is discontinuous, the stationary states and their first derivatives have left and right limits that coincide, whenever the jump of the potential is finite. Some wave packets for one-dimensional problems are then constructed and studied with the help of the stationary-phase method, including the step-like potential. The chapter ends with the elementary quantum theory of the one-dimensional harmonic oscillator.

6.1 Boundary conditions

The eigenvalue problem for the Hamiltonian operator cannot be solved in general unless we specify the class of potentials we are interested in, with the associated boundary conditions and self-adjointness domain for H. A great variety of potentials can indeed occur in the investigation of physical phenomena, e.g. central potentials behaving as r^{-n} ($n \geq 1$), Yukawa terms $\frac{e^{-\mu r}}{r}$, isotropic and anisotropic harmonic oscillators, polynomials of suitable degree, Laurent series, logarithmic terms and periodic potentials. For simplicity we shall perform a one-dimensional analysis because many problems of interest are either one-dimensional or reducible to one spatial dimension in the radial variable, as is the case in a central potential.

We thus study the eigenvalue problem for the Schrödinger equation (cf. Eq. (5.1.9))

$$\left(-\frac{\hbar^2}{2m} \Delta + V(x) \right) u(x) = Eu(x), \qquad (6.1.1)$$

where $x \in \mathbf{R}$, $\Delta \equiv \frac{d^2}{dx^2}$ and $u(x) = u(x; E) = u_E(x)$. Since the stationary states $u(x)$ may be locally integrable but not globally square-integrable, it is convenient to introduce also the concept of *weak solutions* of Eq. (6.1.1), i.e. those $u(x)$ such that

$$(u, Hq) = E(u, q), \qquad (6.1.2)$$

where q is a C^∞ function on $\mathbf{R}$ with compact support. The condition (6.1.2) reads explicitly, our scalar product being anti-linear in the first argument, as

$$\int_{\mathbf{R}} dx\, u^*(x) \left(-\frac{\hbar^2}{2m} \Delta + V(x) \right) q(x) = E \int_{\mathbf{R}} dx\, u^*(x) q(x). \qquad (6.1.3)$$

Bearing in mind that the potential V might be discontinuous at some points x_i, the real line should be divided into a number of open intervals, in each of which Eq. (6.1.1) holds. Hence, integrating twice by parts,

$$\int_{\mathbf{R}} dx \, u^* \triangle q(x) - \int_{\mathbf{R}} dx (\triangle u^*(x)) q(x)$$

$$= \sum_{i=1}^{N} \lim_{\delta \to 0} \left[u^* \frac{dq}{dx} - q \frac{du^*}{dx} \right]_{x_i - \delta}^{x_i + \delta}$$

$$= \sum_{i=1}^{N} \left[(u_+^* - u_-^*)_i \frac{dq}{dx} - q \left(\frac{du_+^*}{dx} - \frac{du_-^*}{dx} \right)_i \right]. \tag{6.1.4}$$

Moreover, Eq. (6.1.1) leads to

$$\triangle u^* = -\frac{2m}{\hbar^2} (E - V) u^*, \tag{6.1.5}$$

while Eq. (6.1.3) yields

$$\int_{\mathbf{R}} dx \, u^* \triangle q = -\frac{2m}{\hbar^2} \int_{\mathbf{R}} dx \, u^* (E - V) q. \tag{6.1.6}$$

From Eqs. (6.1.5) and (6.1.6), the sum in Eq. (6.1.4) vanishes identically, and the arbitrariness of q and $\frac{dq}{dx}$ at x_i $\forall i = 1, \ldots, N$ yields therefore the boundary (or matching) conditions at the singular points

$$u_+(x_i) = u_-(x_i), \ \forall i = 1, \ldots, N, \tag{6.1.7}$$

$$\frac{du_+}{dx}(x_i) = \frac{du_-}{dx}(x_i), \ \forall i = 1, \ldots, N. \tag{6.1.8}$$

These conditions should be imposed whenever the potential V has a finite jump at x_i, and we will apply them several times from now on. They are connected with the continuity of the probability current at the discontinuities of the potential.

6.1.1 Particle confined by a potential

If the potential has an infinite jump, this may be viewed as the limiting case of a problem where the discontinuity of V is finite (as one takes the limit for the jump tending to infinity), and hence the above rules for the boundary conditions can be applied to solve all quantum-mechanical problems. For example, if a particle of energy $E \in \,]0, V_0[$ and mass m is subject to the potential

$$\widetilde{V}(x) = \begin{cases} V_0 & x \in \,]-\infty, -a[, \\ 0 & x \in \,]-a, a[, \\ V_0 & x \in \,]a, \infty[, \end{cases} \tag{6.1.9}$$

the solutions of the Schrödinger equation for stationary states read, on setting $\Gamma \equiv \frac{\sqrt{2m(V_0 - E)}}{\hbar}, \kappa \equiv \frac{\sqrt{2mE}}{\hbar}$,

$$u_1(x) = A_1 e^{\Gamma x} + A_2 e^{-\Gamma x} = A_1 e^{\Gamma x} \qquad x \in]-\infty, -a[, \qquad (6.1.10)$$

$$u_2(x) = A_3 \cos \kappa x + A_4 \sin \kappa x \qquad x \in]-a, a[, \qquad (6.1.11)$$

$$u_3(x) = A_5 e^{\Gamma x} + A_6 e^{-\Gamma x} = A_6 e^{-\Gamma x} \qquad x \in]a, \infty[, \qquad (6.1.12)$$

where we have set $A_2 = A_5 = 0$ to obtain a wave function $\in \mathcal{L}^2(\mathbf{R}, \mathrm{d}x)$. The continuity conditions at $x = -a$ and at $x = a$:

$$\lim_{x \to -a^-} u_1(x) = \lim_{x \to -a^+} u_2(x), \qquad (6.1.13)$$

$$\lim_{x \to -a^-} u_1'(x) = \lim_{x \to -a^+} u_2'(x), \qquad (6.1.14)$$

$$\lim_{x \to a^-} u_2(x) = \lim_{x \to a^+} u_3(x), \qquad (6.1.15)$$

$$\lim_{x \to a^-} u_2'(x) = \lim_{x \to a^+} u_3'(x), \qquad (6.1.16)$$

lead to the homogeneous system

$$A_1 e^{-\Gamma a} - A_3 \cos \kappa a + A_4 \sin \kappa a = 0, \qquad (6.1.17)$$

$$A_1 e^{-\Gamma a} - \frac{\kappa}{\Gamma} A_3 \sin \kappa a - \frac{\kappa}{\Gamma} A_4 \cos \kappa a = 0, \qquad (6.1.18)$$

$$A_3 \cos \kappa a + A_4 \sin \kappa a - A_6 e^{-\Gamma a} = 0, \qquad (6.1.19)$$

$$-A_3 \sin \kappa a + A_4 \cos \kappa a + \frac{\Gamma}{\kappa} A_6 e^{-\Gamma a} = 0. \qquad (6.1.20)$$

To find a non-trivial solution for A_1, A_3, A_4, A_6, the determinant of the matrix of coefficients should vanish, and this leads to the eigenvalue condition

$$e^{-2\Gamma a} \left[\left(\frac{\Gamma}{\kappa} - \frac{\kappa}{\Gamma} \right) \sin(2\kappa a) + 2 \cos(2\kappa a) \right] = 0. \qquad (6.1.21)$$

The problem is studied first for finite values of V_0, for which the exponential $e^{-2\Gamma a}$ remains finite and hence does not lead to the vanishing of the determinant, which is expressed therefore by

$$\tan(2\kappa a) = -\frac{2\kappa\Gamma}{(\Gamma^2 - \kappa^2)} = -2\frac{\sqrt{E(V_0 - E)}}{(V_0 - 2E)}. \qquad (6.1.22)$$

Now, as V_0 is increased, the right-hand side of this equation is proportional to $\frac{1}{\sqrt{V_0}}$ and hence tends to zero. The $V_0 \to \infty$ limit leads therefore, eventually, to the eigenvalue condition

$$\sin(2\kappa a) = 0, \qquad (6.1.23)$$

an equation solved by $2\kappa a = n\pi$, and hence we obtain the discrete spectrum

$$E_n = n^2 \frac{\hbar^2 \pi^2}{8ma^2} \quad \forall \, n = 1, 2, \ldots \tag{6.1.24}$$

for the particle confined within the closed interval $[-a, a]$ by the infinite-wall potential

$$\widetilde{V}(x) = \begin{cases} \infty & x \in]-\infty, -a[, \\ 0 & x \in]-a, a[, \\ \infty & x \in]a, \infty[. \end{cases} \tag{6.1.25}$$

Note also that, when $V_0 \to \infty$, $u_1(x)$ in Eq. (6.1.10) and $u_3(x)$ in Eq. (6.1.12) tend to zero for any value of x. This is why it is said that, in the potential (6.1.25), the wave function vanishes outside the interval $]-a, a[$.

6.1.2   A closer look at improper eigenfunctions

It is now appropriate to discuss in greater detail a concept already encountered in Chapter 5. We say that $\{u_W(x)\}$ is a family of *improper eigenfunctions* of the Hamiltonian operator if the following conditions hold:

(i) $\{u_W\}$ is continuous, or at least locally integrable, with respect to the parameter W in a subset σ of the real line, with positive measure;

(ii) each function u_W is a generalized solution of Eq. (6.1.1) for the specific value of W (and hence satisfies the regularity properties and the boundary conditions previously derived);

(iii) for any interval $(W, W + \delta W)$ having intersection of positive measure with σ, the *eigendifferential*

$$F_{(W, W+\delta W)}u(x) \equiv \int_W^{W+\delta W} d\gamma \, u_\gamma(x) \tag{6.1.26}$$

belongs to $\mathcal{L}^2(\mathbf{R})$ and is non-vanishing. For example, when $u_\gamma(x) = e^{i\gamma x}$,

$$\int_W^{W+\delta W} d\gamma \, u_\gamma(x) = \frac{2}{x} \sin\left(\frac{x}{2}\delta W\right) e^{i\left(W + \frac{\delta W}{2}\right)x}, \tag{6.1.27}$$

which is square-integrable on $\mathbf{R}$, its squared modulus being majorized by $\frac{4}{x^2}$ as $x \to \pm\infty$, while remaining finite at the origin. This eigendifferential occurs if integration is performed by parts in the free-particle wave packet

$$\int_{-\infty}^{\infty} d\gamma \, A(\gamma) e^{i(\gamma x - \omega t)},$$

and hence the concept of eigendifferential makes precise the condition under which a physically admissible solution of the Schrödinger equation can be constructed.

The improper eigenfunctions are, in general, *tempered distributions*. The class of ordinary functions, which correspond to tempered distributions, can be essentially identified with the set of locally integrable functions of *algebraic growth*. By definition, the latter are functions bounded at infinity by a polynomial: for each u there exists $r \in N$ such that, as $|x| \to \infty$,

$$|u(x)| \le \sum_{s=0}^{r} a_s x^s. \tag{6.1.28}$$

The property of having algebraic growth can be used, for differential operators, as a requisite to select solutions of the (formal) eigenvalue equation among which one should look for eigenfunctions. For example, according to this criterion, solutions of the form (see Eqs. (6.1.10) and (6.1.12))

$$e^{\sqrt{W}|x|} \quad W > 0$$

should be ruled out, whereas, in $\mathcal{L}^2(\mathbf{R})$, wave functions $\psi(x, t)$ can be constructed from the improper eigenfunctions

$$e^{i\sqrt{W}x}, \ e^{-i\sqrt{W}x} \quad W > 0.$$

In $\mathcal{L}^2(\mathbf{R}^p)$, with $p > 1$, some extra care is necessary to select suitable improper eigenfunctions.

6.2 Reflection and transmission

The investigation of reflection and transmission of wave packets elucidates many peculiar properties of the quantum-mechanical analysis of the motion of particles or beams of particles. For example, let a particle of mass m and positive energy, $W > 0$, move in the one-dimensional potential

$$V(x) = \begin{cases} -|V_0| & |x| < b \\ 0 & |x| > b. \end{cases} \tag{6.2.1}$$

Since $W + |V_0|$ is also positive, we find the improper eigenfunctions

$$u_{\mathrm{I}}(x) = A\,e^{ikx} + B\,e^{-ikx}, \tag{6.2.2}$$

$$u_{\mathrm{II}}(x) = C\,e^{i\bar{k}x} + D\,e^{-i\bar{k}x}, \tag{6.2.3}$$

$$u_{\mathrm{III}}(x) = E\,e^{ikx} + F\,e^{-ikx}, \tag{6.2.4}$$

where $k \equiv \sqrt{2mW}/\hbar$, $\bar{k} \equiv \sqrt{2m(W + |V_0|)}/\hbar$, and the labels I, II, III refer to the intervals $x \in]-\infty, -b[, x \in]-b, b[, x \in]b, \infty[$, respectively. Recall from Section 6.1.2 that these stationary 'states' are not in Hilbert space, but are used to build a normalizable wave function. Thus, in this problem, the eigendifferential (see Eq. (6.1.27))

$$\int_{k}^{k+\delta k} d\gamma \ u_\gamma(x)$$

belongs to $\mathcal{L}^2(\mathbf{R})$ for any k and δk and for any choice of the two arbitrary constants in Eqs. (6.2.3) and (6.2.4). This means that the whole interval $W > 0$ belongs to the continuous spectrum of H, and a twofold degeneracy occurs. In the problems studied here, the choice of a particular formal solution of the eigenvalue equation is made in such a way that it is then possible to build a wave function satisfying certain asymptotic conditions. For potentials vanishing at infinity more rapidly than $\frac{1}{|x|}$ the requirement is that it should be possible to build a wave function $\psi(x, t)$ which, for $t \to -\infty$ or $t \to +\infty$ behaves as a free particle, before interacting (or after interacting) with the potential for $t \approx 0$. In one-dimensional problems we can also make statements on the motion of the particle in the presence of potentials that remain non-vanishing at infinity. In such a case an asymptotic condition as $t \to -\infty$ is obtained by imposing the condition that the particle should be located at $x \to +\infty$ or at $x \to -\infty$, and that it should move towards the origin, hence specifying how the linear momentum is directed as $t \to -\infty$. Similarly, as $t \to +\infty$, one can impose that the particle is located at $x \to +\infty$ or at $x \to -\infty$ and that it is moving away from the origin.

We may exploit the arbitrariness in the value of two constants by choosing

$$F = 0, \quad A = \frac{1}{\sqrt{2\pi\hbar}}. \tag{6.2.5}$$

As will be seen shortly, this is the choice of constants that pertains to the case of particles located at $x = -\infty$ as $t \to -\infty$, and hence coming from the left. Of course, we might have chosen, instead of $F = 0$, the alternative condition $E = 0$, describing instead particles coming from the right. The value for A is chosen to make it easier to compare with the case of the free particle, although the crucial role will be played by ratios of coefficients involving A. We set (bearing in mind that $p \equiv \hbar k$)

$$\beta(p) \equiv \frac{B}{A}, \quad \varepsilon(p) \equiv \frac{E}{A}. \tag{6.2.6}$$

The improper eigenfunctions resulting from our choice of parameters are thus

$$u_p^{(+)}(x) = \frac{1}{\sqrt{2\pi\hbar}} \left[e^{\frac{i}{\hbar}px} + \beta(p)e^{-\frac{i}{\hbar}px} \right] \quad \text{if } x < -b, \tag{6.2.7}$$

$$u_p^{(+)}(x) = \frac{1}{\sqrt{2\pi\hbar}} \varepsilon(p)e^{\frac{i}{\hbar}px} \quad \text{if } x > b. \tag{6.2.8}$$

Of course, nothing prevents us from writing the form of $u_p^+(x)$ if $x \in]-b, b[$, but this calculation is not relevant for the analysis of reflection and transmission coefficients, and is hence omitted.

We are eventually interested in a solution $\psi(x, t)$ of the time-dependent Schrödinger equation, which consists of a wave packet built from the above improper eigenfunctions, according to the well-established rule

$$\psi(x, t) = \int_{-\infty}^{\infty} dp \, c(p) u_p^{(+)}(x) e^{-\frac{i}{\hbar}\frac{p^2}{2m}t}. \tag{6.2.9}$$

From Eqs. (6.2.7) and (6.2.8), Eq. (6.2.9) leads to a first non-trivial result:

$$\psi(x, t) = \psi_{\text{in}}(x, t) + \psi_{\text{rifl}}(x, t) = \frac{1}{\sqrt{2\pi\hbar}} \int_{-\infty}^{\infty} dp \, c(p) e^{\frac{i}{\hbar}\left(px - \frac{p^2}{2m}t\right)}$$

$$+ \frac{1}{\sqrt{2\pi\hbar}} \int_{-\infty}^{\infty} dp \, c(p)\beta(p) e^{-\frac{i}{\hbar}\left(px + \frac{p^2}{2m}t\right)} \quad \text{if } x < -b, \tag{6.2.10}$$

$$\psi(x,t) = \psi_{\rm tr}(x,t) = \frac{1}{\sqrt{2\pi\hbar}} \int_{-\infty}^{\infty} dp\; c(p)\varepsilon(p) e^{\frac{i}{\hbar}(px - \frac{p^2}{2m}t)} \quad \text{if } x > b. \tag{6.2.11}$$

If $c(p)$ is sufficiently smooth and takes non-vanishing values only in a small neighbourhood of a particular value p_0 of p, such formulae describe wave packets moving with speed $\frac{p_0}{m}, -\frac{p_0}{m}, \frac{p_0}{m}$, respectively, without a sensible spreading effect. The first and third wave packet move from the left to the right, while the second wave packet moves from the right to the left (see appendix).

In other words, at the beginning of the motion, $\psi(x,t)$ coincides with a solution of the Schrödinger equation for a free particle, and represents a wave packet moving from the left to the right. When such a packet reaches the region where the potential takes a non-vanishing value it splits into two new packets, one of which is transmitted, while the other is reflected. Both packets eventually have the same speed, in magnitude, as the initial packet. From the physical point of view, the solution (6.2.7)–(6.2.9), jointly with the hypothesis that $c(p)$ differs from zero only for p in the neighbourhood of a value p_0, describes a particle which has a certain probability τ of being transmitted, and a probability ρ of being reflected. Unlike classical mechanics, τ is, in general, smaller than 1, and ρ does not vanish.

To complete the calculations of general nature note first that, as $t \to -\infty$,

$$\int_{-\infty}^{\infty} dx |\psi(x,t)|^2 \longrightarrow \int_{-\infty}^{\infty} dx |\psi_{\rm in}(x,t)|^2 = \int_{-\infty}^{\infty} dp |c(p)|^2. \tag{6.2.12}$$

On the other hand, the terms on both sides of Eq. (6.2.12) are essentially independent of time, and hence should also coincide for finite values of the time variable. The normalizability condition of $\psi(x,t)$ therefore implies

$$\int_{-\infty}^{\infty} dp |c(p)|^2 = 1. \tag{6.2.13}$$

Now the probability that, at large t, the particle has been transmitted, is given by ($c(p)$ being of compact support)

$$\tau \equiv \lim_{t \to \infty} \int_b^{\infty} dx |\psi(x,t)|^2 = \int_{-\infty}^{\infty} dx |\psi_{\rm tr}(x,t)|^2$$

$$= \int_{-\infty}^{\infty} dp |c(p)|^2 |\varepsilon(p)|^2 = \int_{p_0 - \frac{\delta}{2}}^{p_0 + \frac{\delta}{2}} dp |c(p)|^2 |\varepsilon(p)|^2$$

$$= \lim_{\delta \to 0} \int_{p_0 - \frac{\delta}{2}}^{p_0 + \frac{\delta}{2}} dp |c(p)|^2 |\varepsilon(p)|^2 = |\varepsilon(p_0)|^2 \lim_{\delta \to 0} \delta |c(p_0)|^2$$

$$\sim |\varepsilon(p_0)|^2 \int_{-\infty}^{\infty} dp |c(p)|^2 = |\varepsilon(p_0)|^2, \tag{6.2.14}$$

where we have used the Parseval lemma to go from the x- to the p-integration, the mean-value theorem and, in the last equality, we have used Eq. (6.2.13). Moreover, the reflection probability can be evaluated as follows:

$$\rho \equiv \lim_{t \to \infty} \int_{-\infty}^{-b} dx |\psi(x,t)|^2 = \int_{-\infty}^{\infty} dx |\psi_{\rm rifl}(x,t)|^2$$

$$= \int_{-\infty}^{\infty} dp |c(p)|^2 |\beta(p)|^2 \sim |\beta(p_0)|^2, \tag{6.2.15}$$

where we have again used the Parseval lemma and the property of c of being a function with compact support, jointly with Eq. (6.2.13), while omitting for simplicity of presentation the limit as $\delta \to 0$. The explicit calculation in the potential (6.2.1), bearing in mind the definitions (6.2.6), shows that

$$\rho + \tau = 1. \tag{6.2.16}$$

Indeed, the continuity conditions at the points $-b$ and b lead to the system

$$A\left(e^{-ikb} + \beta e^{ikb}\right) = Ce^{-i\bar{k}b} + De^{i\bar{k}b}, \tag{6.2.17}$$

$$kA\left(e^{-ikb} - \beta e^{ikb}\right) = \bar{k}\left(Ce^{-i\bar{k}b} - De^{i\bar{k}b}\right), \tag{6.2.18}$$

$$Ce^{i\bar{k}b} + De^{-i\bar{k}b} = A\varepsilon e^{ikb}, \tag{6.2.19}$$

$$\bar{k}\left(Ce^{i\bar{k}b} - De^{-i\bar{k}b}\right) = A\varepsilon k \ e^{ikb}, \tag{6.2.20}$$

which is solved by

$$C = \frac{A}{2}\varepsilon e^{i(k-\bar{k})b}\left(1 + \frac{k}{\bar{k}}\right), \tag{6.2.21}$$

$$D = \frac{A}{2}\varepsilon e^{i(k+\bar{k})b}\left(1 - \frac{k}{\bar{k}}\right), \tag{6.2.22}$$

$$\beta = \frac{i\varepsilon}{2k\bar{k}}\left(\bar{k}^2 - k^2\right)\sin(2\bar{k}b), \tag{6.2.23}$$

$$\varepsilon = \frac{4k\bar{k}e^{-2ikb}}{\left[4k\bar{k}\cos(2\bar{k}b) - 2i(k^2 + \bar{k}^2)\sin(2\bar{k}b)\right]}, \tag{6.2.24}$$

and a patient check shows that $|\beta|^2 + |\varepsilon|^2 = 1$, because $|\beta|^2 + |\varepsilon|^2$ is found to coincide with the ratio $\frac{F(k,\bar{k})}{F(k,\bar{k})}$, where

$$F(k,\bar{k}) \equiv 16k^2\bar{k}^2\cos^2(2\bar{k}b) + 4(k^2 + \bar{k}^2)^2\sin^2(2\bar{k}b). \tag{6.2.25}$$

The functions ρ and τ are called the *reflection* and *transmission coefficient*, respectively. The underlying reason for such names is that, if a sufficiently large number N of particles, all with the same initial value of the momentum, move in the potential (6.2.1), then by virtue of the law of big numbers the number of reflected particles is ρN, and the number of transmitted particles is τN. Indeed, if one considers a *beam of particles*, all of them under the same initial conditions, and in sufficiently large number so that statistical fluctuations can be neglected, then $|\psi|^2 d^3x$ can describe the *percentage of particles observed within the volume* d^3x. Similarly, if a beam of particles of sufficiently high intensity is available, all of them being in the same initial conditions, it is possible to interpret the quantity

$$\vec{j} \cdot \vec{n} d\sigma \, dt$$

as the percentage of particles that cross the surface $d\sigma$ in the time interval dt. This interpretation holds both for charged and neutral particles.

6.3 Step-like potential

We know from Section 6.1 that, whenever the potential in the Schrödinger equation has finite discontinuities at some points, continuity conditions can, correspondingly be imposed on the stationary state and its first derivative. These boundary conditions ensure (essential) self-adjointness of the Hamiltonian operator, under suitable assumptions on the potential (Reed and Simon 1975). According to this scheme, in each open interval of the real line where the potential is continuous, we have to solve the second-order equation

$$\left\{ \frac{d^2}{dx^2} + \frac{2m}{\hbar^2}[E - V_i(x)] \right\} u = 0, \tag{6.3.1}$$

where V_i is the potential in the interval I_i, subject to the continuity conditions

$$\lim_{x \to x_i^-} u(x) = \lim_{x \to x_i^+} u(x), \tag{6.3.2}$$

$$\lim_{x \to x_i^-} \frac{du}{dx} = \lim_{x \to x_i^+} \frac{du}{dx}, \tag{6.3.3}$$

where x_i denote the various discontinuity points of the potential $V(x)$.

A first non-trivial consequence of these properties is that, if a beam of particles with the same mass m and energy $E > W$ enters a region where the potential is step-like and given by

$$V(x) = 0 \quad \text{if } x < 0, \ W \text{ if } x > 0, \tag{6.3.4}$$

we have a solution of the kind (Squires 1995)

$$u(x) = e^{i\kappa x} + Re^{-i\kappa x} \quad \text{if } x < 0, \tag{6.3.5}$$

$$u(x) = Te^{i\gamma x} \quad \text{if } x > 0, \tag{6.3.6}$$

where $\kappa^2 \equiv \frac{2mE}{\hbar^2}, \gamma^2 \equiv \frac{2m(E-W)}{\hbar^2}$, and the application of Eqs. (6.3.2) and (6.3.3) leads to the system

$$1 + R = T, \tag{6.3.7}$$

$$i\kappa(1 - R) = iT\gamma, \tag{6.3.8}$$

which implies

$$R = \frac{(1 - \gamma/\kappa)}{(1 + \gamma/\kappa)}, \quad T = \frac{2}{(1 + \gamma/\kappa)}. \tag{6.3.9}$$

The solution of the time-dependent Schrödinger equation is then found to be (hereafter $p \equiv \hbar\kappa, \overline{p} \equiv \hbar\gamma$)

$$\psi(x,t) = \frac{1}{\sqrt{2\pi\hbar}} \int_{-\infty}^{\infty} c(p) \left[e^{\frac{i}{\hbar}\left(px - \frac{p^2}{2m}t\right)} + R(p)e^{-\frac{i}{\hbar}\left(px + \frac{p^2}{2m}t\right)} \right] dp, \qquad (6.3.10)$$

if $x < 0$, and

$$\psi(x,t) = \frac{1}{\sqrt{2\pi\hbar}} \int_{-\infty}^{\infty} c(p)T(p)e^{\frac{i}{\hbar}[\overline{p}x - E(p)t]} dp \qquad (6.3.11)$$

if $x > 0$. The integration variable is p for both wave packets. However, the latter reads (factors of $\sqrt{2\pi\hbar}$ are here omitted for simplicity)

$$\psi_{\mathrm{II}}(x,t) = \int_{-\infty}^{\infty} f(p,t)e^{\frac{i}{\hbar}\overline{p}x} dp, \qquad (6.3.12)$$

where

$$f(p,t) \equiv c(p)T(p)e^{-\frac{i}{\hbar}E(p)t}, \qquad (6.3.13)$$

whereas, to be able to apply the Parseval lemma, we would like it to be able to express ψ_{II} in the form

$$\psi_{\mathrm{II}}(x,t) = \int_{-\infty}^{\infty} \widetilde{\psi}_{\mathrm{II}}(\overline{p},t)e^{\frac{i}{\hbar}\overline{p}x} d\overline{p}. \qquad (6.3.14)$$

Since the integrals (6.3.12) and (6.3.14) represent the same wave packet, we have to change variables in Eq. (6.3.12), writing

$$\psi_{\mathrm{II}}(x,t) = \int_{-\infty}^{\infty} f(p(\overline{p}),t) \frac{dp}{d\overline{p}} e^{\frac{i}{\hbar}\overline{p}x} d\overline{p}, \qquad (6.3.15)$$

and hence the spatial Fourier transform of the wave packet ψ_{II} reads as

$$\widetilde{\psi}_{\mathrm{II}}(\overline{p},t) = f(p(\overline{p}),t) \frac{dp}{d\overline{p}} = \frac{\overline{p}}{p} c(p)T(p)e^{-\frac{i}{\hbar}E(p)t}, \qquad (6.3.16)$$

having exploited the property

$$\frac{dp}{d\overline{p}} = \frac{d}{d\overline{p}} \left(\overline{p}^2 + 2mW\right)^{\frac{1}{2}} = \frac{\overline{p}}{p}. \qquad (6.3.17)$$

Now the Parseval lemma and the mean-value theorem yield (since $c(p)$ has compact support)

$$\begin{aligned}
(\psi_{\mathrm{II}}, \psi_{\mathrm{II}}) &= \left(\widetilde{\psi}_{\mathrm{II}}, \widetilde{\psi}_{\mathrm{II}}\right) \\
&= \int_{-\infty}^{\infty} |c(p(\overline{p}))|^2 |T(p(\overline{p}))|^2 \frac{\overline{p}}{p} \frac{\overline{p}}{p} d\overline{p} \\
&= \frac{\overline{p}}{p} |T|^2 \int_{-\infty}^{\infty} |c(p(\overline{p}))|^2 \frac{\overline{p}}{p} d\overline{p} \\
&= \frac{\overline{p}}{p} |T|^2 \int_{-\infty}^{\infty} |c(p)|^2 dp = \frac{\overline{p}}{p} |T|^2 \\
&= \frac{\gamma}{\kappa} |T|^2.
\end{aligned} \qquad (6.3.18)$$

This is the transmission coefficient

$$\tau \equiv \lim_{t \to \infty} \int_0^\infty |\psi(x, t)|^2 \mathrm{d}x = \frac{\gamma}{\kappa} |T|^2, \tag{6.3.19}$$

under the assumption that the apparatuses used for preparation and detection refer to the same state.

The evaluation of the reflection coefficient is easier, since no change of measure is necessary when $V(x) = 0$ for $x \in (-\infty, 0)$. We then find

$$\rho \equiv \lim_{t \to \infty} \int_{-\infty}^0 |\psi(x, t)|^2 \mathrm{d}x = |R|^2. \tag{6.3.20}$$

Thus, although all particles have energy greater than W, a non-vanishing fraction is reflected, and is expressed by $|R|^2$. A consistency check shows that

$$1 = |R|^2 + \frac{\gamma}{\kappa} |T|^2 = |R|^2 + \frac{\overline{p}}{p} |T|^2. \tag{6.3.21}$$

The stationary-phase method (see appendix) shows that, as $t \to \infty$, the first of the two terms in Eq. (6.3.10) does not contribute at all, because it corresponds to a wave packet located at $x = +\infty$, i.e. outside the domain of validity of Eq. (6.3.10). Thus, as $t \to \infty$, Eq. (6.3.10) reduces to a reflected wave packet located at $x = -\infty$. Moreover, in the limit $t \to \infty$, Eq. (6.3.11) indeed describes a wave packet located at $x = +\infty$, which is an acceptable asymptotic solution, because it agrees with the domain of definition of Eq. (6.3.11).

In contrast, if all particles have energy $E < W$, the stationary state again takes the form (6.3.5) if $x < 0$, but for $x > 0$ is a decreasing exponential:

$$u(x) = Q \mathrm{e}^{-\Gamma x} \quad \text{if } x > 0, \tag{6.3.22}$$

where $\Gamma^2 \equiv \frac{2m(W-E)}{\hbar^2}$. Thus, the application of Eqs. (6.3.2) and (6.3.3) leads instead to the system

$$1 + R = Q, \tag{6.3.23}$$

$$\mathrm{i}\kappa(1 - R) = -\Gamma Q, \tag{6.3.24}$$

the solution of which is (Squires 1995)

$$R = \frac{(1 - \frac{\mathrm{i}\Gamma}{\kappa})}{(1 + \frac{\mathrm{i}\Gamma}{\kappa})}, \quad Q = \frac{2}{(1 + \frac{\mathrm{i}\Gamma}{\kappa})}. \tag{6.3.25}$$

Interestingly, although all particles have energy smaller than W, a non-vanishing solution of the Schrödinger equation *for stationary states* exists if $x > 0$. However, if $\rho \equiv \frac{\Gamma}{\kappa} \equiv \tan\theta$, from it is found Eq. (6.3.25) that

$$R = \mathrm{e}^{-2\mathrm{i}\theta}, \tag{6.3.26}$$

while the stationary state for $x > 0$ reads (see Eq. (6.3.22))

$$u(x) = 2\mathrm{e}^{-\mathrm{i}\theta} \cos\theta \, \mathrm{e}^{-\Gamma x}. \tag{6.3.27}$$

Interestingly, reflected waves pick up a delay term which has no classical counterpart. It should be stressed that, from Eq. (6.3.26), the definition of the reflection coefficient used so far, jointly with the stationary-phase method, implies that

$$\rho = |R|^2 = 1, \tag{6.3.28}$$

i.e. *all particles are reflected*. This is in agreement with Eq. (6.3.27), because the exponential fall-off of u implies that $\psi(x, t)$ tends to 0 as $t \to \infty$ when $x \to \infty$. Thus, it would be incorrect to claim that the transmission coefficient is $|Q|^2$, with Q expressed by the second equation in (6.3.25). Waves falling off exponentially like Eq. (6.3.27) are called *evanescent waves*, and carry no current or momentum. But the density does not vanish abruptly; the waves are continuous across the boundary and gradually tend to zero far from the boundary.

6.3.1 Tunnelling effect

The above potential can be modified to show that quantum mechanics makes it possible to build a theoretical model for penetration through a potential barrier. This is called the tunnelling effect. The complete scheme, in one-dimensional problems, requires the introduction of a potential V such that

$$V(x) = 0 \text{ if } x < 0 \text{ or } x > a, \ W \text{ if } x \in \,]0, a[, \tag{6.3.29}$$

and a beam of particles of mass m and energy $E < W$ is affected by $V(x)$. From Eq. (6.3.29), we are dealing with three regions. The corresponding forms of the stationary states are

$$u(x) = u_1(x) = e^{i\kappa x} + Re^{-i\kappa x} \quad \text{if } x < 0, \tag{6.3.30}$$

$$u(x) = u_2(x) = Ae^{\Gamma x} + Be^{-\Gamma x} \quad \text{if } x \in \,]0, a[, \tag{6.3.31}$$

$$u(x) = u_3(x) = Te^{i\kappa x} \quad \text{if } x > a. \tag{6.3.32}$$

Of course, since we have three second-order differential equations with four boundary conditions, resulting from the application of Eqs. (6.3.2) and (6.3.3) at $x = 0$ and a, we expect to have in general $6 - 4 = 2$ unknown coefficients. At this stage, physical considerations concerning reflected and transmitted particles, suggested by the simpler examples discussed at the beginning of this section, lead to Eqs. (6.3.30)–(6.3.32), where the number of coefficients, i.e. A, B, R and T, is equal to the number of boundary conditions. Indeed, from Eqs. (6.3.2), (6.3.3) and (6.3.30)–(6.3.32),

$$1 + R = A + B, \tag{6.3.33}$$

$$i\kappa(1 - R) = \Gamma(A - B), \tag{6.3.34}$$

$$Ae^{\Gamma a} + Be^{-\Gamma a} = Te^{i\kappa a}, \tag{6.3.35}$$

$$\Gamma\left(Ae^{\Gamma a} - Be^{-\Gamma a}\right) = iT\kappa e^{i\kappa a}. \tag{6.3.36}$$

It is now convenient to define new parameters

$$\rho \equiv \frac{\Gamma}{\kappa}, \quad \theta \equiv \Gamma a, \quad \delta \equiv \kappa a. \tag{6.3.37}$$

In terms of these, we find (from Eqs. (6.3.35) and (6.3.36))

$$A = \frac{1}{2}\left(1 + \frac{i}{\rho}\right) T e^{i\delta - \theta}, \tag{6.3.38}$$

$$B = \frac{1}{2}\left(1 - \frac{i}{\rho}\right) T e^{i\delta + \theta}. \tag{6.3.39}$$

Equations (6.3.38) and (6.3.39) are now inserted into Eqs. (6.3.33) and (6.3.34). On adding term by term, this yields

$$4e^{-i\delta} = T\left[\left(2 + \frac{i}{\rho} - i\rho\right) e^{-\theta} + \left(2 + i\rho - \frac{i}{\rho}\right) e^{\theta}\right]. \tag{6.3.40}$$

The exponentials on the right-hand side of Eq. (6.3.40) are now re-expressed in terms of the well-known hyperbolic functions, so that the formula for the coefficient T reads

$$T = \frac{e^{-i\delta}}{\left[\cosh\theta + \frac{i}{2}\left(\rho - \frac{1}{\rho}\right)\sinh\theta\right]}. \tag{6.3.41}$$

Thus, the fraction of transmitted particles is

$$\mathcal{F} = |T|^2 = \left[1 + \frac{(\rho^2 + 1)^2}{4\rho^2}\sinh^2\theta\right]^{-1}$$

$$= \left[1 + \frac{1}{4}\frac{W^2}{E(W - E)}\sinh^2\Gamma a\right]^{-1}, \tag{6.3.42}$$

because $\rho^2 = \frac{W - E}{E}$. In particular, in the limit as $W \to \infty$ and $a \to 0$, while $Wa = b =$ constant (this corresponds to the *delta-like* potential), the result (6.3.42) reduces to (Squires 1995)

$$\mathcal{F} = \left(1 + \frac{mb^2}{2\hbar^2 E}\right)^{-1}. \tag{6.3.43}$$

This calculation shows once again that, when the potential has an infinite discontinuity, one can first consider a problem where it takes finite values, and then evaluate the limit on the solution obtained with finite values of V.

6.4 One-dimensional harmonic oscillator

One of the few systems for which the Schrödinger equation can be solved is the harmonic oscillator. From a physical point of view, all cases when equilibrium states are approached can be studied by using harmonic oscillators, and this makes such an example particularly relevant.

The Schrödinger equation for stationary states of a one-dimensional harmonic oscillator reads

$$\left(-\frac{\hbar^2}{2m}\frac{d^2}{dx^2} + \frac{m}{2}\omega^2 x^2 \right) \psi_n = E_n \psi_n. \tag{6.4.1}$$

Here m is the mass, ω is the frequency of oscillations (in radians per unit time). Rather than using an abstract (algebraic) approach for the solution of this equation, for which we refer the reader to Chapters 7 and 8, we here rely upon elementary methods, which nevertheless can also be exploited for other problems. For this purpose, we notice that Eq. (6.4.1) can be factorized in the form

$$-\frac{\hbar^2}{2m}\left(\frac{d}{dx} - \frac{m\omega}{\hbar}x \right)\left(\frac{d}{dx} + \frac{m\omega}{\hbar}x \right)\psi_n = \left(E_n - \frac{\hbar\omega}{2} \right)\psi_n, \tag{6.4.2}$$

because the Leibniz rule implies that

$$\frac{d}{dx}(x\psi_n) = x\frac{d}{dx}\psi_n + \psi_n. \tag{6.4.3}$$

The factorized form (6.4.2) shows that the operator $\frac{d}{dx} - \frac{m\omega}{\hbar}x$ makes it possible to generate a new solution from a given one. Thus, the function ψ_{n+1} defined by

$$\psi_{n+1} \equiv \left(\frac{d}{dx} - \frac{m\omega}{\hbar}x \right)\psi_n, \tag{6.4.4}$$

is a solution of Eq. (6.4.1) with eigenvalue $E_n + \hbar\omega$, and the function

$$\psi_{n-1} \equiv \left(\frac{d}{dx} + \frac{m\omega}{\hbar}x \right)\psi_n \tag{6.4.5}$$

solves Eq. (6.4.1) with eigenvalue $E_n - \hbar\omega$. The latter property follows easily from the factorization

$$-\frac{\hbar^2}{2m}\left(\frac{d}{dx} + \frac{m\omega}{\hbar}x \right)\left(\frac{d}{dx} - \frac{m\omega}{\hbar}x \right)\psi_n = \left(E_n + \frac{\hbar\omega}{2} \right)\psi_n \tag{6.4.6}$$

and from the relation

$$\hat{H}\left(\frac{d}{dx} + \frac{m\omega}{\hbar}x \right)\psi_n = (E_n - \hbar\omega)\left(\frac{d}{dx} + \frac{m\omega}{\hbar}x \right)\psi_n. \tag{6.4.7}$$

Repeated application shows that

$$\hat{H}\left(\frac{d}{dx} + \frac{m\omega}{\hbar}x \right)^r \psi_n = (E_n - r\hbar\omega)\left(\frac{d}{dx} + \frac{m\omega}{\hbar}x \right)^r \psi_n. \tag{6.4.8}$$

Such properties suggest calling the operators $\frac{d}{dx} - \frac{m\omega}{\hbar}x$ and $\frac{d}{dx} + \frac{m\omega}{\hbar}x$ the creation and annihilation operators, respectively.

At this stage, a naturally occurring question is whether negative energy eigenvalues can be obtained by repeated application of the annihilation operator. Indeed, if ψ_0 is a normalized eigenfunction belonging to the eigenvalue E_0 such that $E_0 > 0$, $E_0 - \hbar\omega < 0$, we can define $\psi_{-1} \equiv \left(\frac{d}{dx} + \frac{m\omega}{\hbar}x \right) \psi_0$, and integration by parts, jointly with Eq. (6.4.2), yields

$$\int_{-\infty}^{\infty} \psi_{-1}^* \psi_{-1} \, dx = \int_{-\infty}^{\infty} \psi_0^* \left[\left(-\frac{d}{dx} + \frac{m\omega}{\hbar}x \right) \left(\frac{d}{dx} + \frac{m\omega}{\hbar}x \right) \psi_0 \right] dx$$
$$= \frac{2m}{\hbar^2} \left(E_0 - \frac{\hbar\omega}{2} \right). \tag{6.4.9}$$

If E_0 were smaller than $\frac{\hbar\omega}{2}$ we would obtain a negative norm of ψ_{-1}, which is impossible. If E_0 were larger than $\frac{\hbar\omega}{2}$ (but smaller than $\hbar\omega$), we might iterate the procedure and consider

$$\psi_{-2} \equiv \left(\frac{d}{dx} + \frac{m\omega}{\hbar}x \right) \psi_{-1}, \tag{6.4.10}$$

the 'norm' of which would then be negative by construction. To avoid having such inconsistencies we can only accept that $E_0 = \frac{\hbar\omega}{2}$, which leads to

$$\left(\frac{d}{dx} + \frac{m\omega}{\hbar}x \right) \psi_0 = 0. \tag{6.4.11}$$

This is a first-order equation, which is solved by separation of variables to find

$$\psi_0 = K_0 \, e^{-\frac{m\omega x^2}{2\hbar}}, \tag{6.4.12}$$

where K_0 is a normalization constant obtained by requiring that

$$K_0^2 \int_{-\infty}^{\infty} e^{-\frac{m\omega x^2}{\hbar}} \, dx = 1, \tag{6.4.13}$$

which implies

$$K_0 = \left(\frac{m\omega}{\pi\hbar} \right)^{\frac{1}{4}}. \tag{6.4.14}$$

All the eigenfunctions of the harmonic oscillator can be constructed by starting from the ground state ψ_0 and repeatedly applying the creation operator $\frac{d}{dx} - \frac{m\omega}{\hbar}x$. In such a way we obtain

$$\psi_n(x) = K_n \left(\frac{d}{dx} - \frac{m\omega}{\hbar}x \right)^n \psi_0, \quad n \in \mathcal{N}, \tag{6.4.15}$$

where the normalization constant K_n is such that

$$K_n^2 = \left(\frac{\hbar}{m\omega} \right)^n \sqrt{\frac{m\omega}{\pi\hbar}} \frac{1}{2^n n!}. \tag{6.4.16}$$

In summary, we have found that the solutions of our eigenvalue problem satisfy the boundary conditions $\psi(x) \to 0$, as $x \to \pm\infty$. The problem has a completely discrete spectrum, since $V(x) \to +\infty$ as $x \to \pm\infty$. The ground state has energy $E_0 = \frac{\hbar\omega}{2}$, while in the classical case $E_0 = 0$, i.e. the particle is at rest in the state of lowest energy. According to quantum mechanics the particle cannot be at rest for $x = 0$; there is an unavoidable zero-point energy which represents the energy of fluctuations.

6.4.1 Hermite polynomials

Note now that

$$\left(\frac{\mathrm{d}}{\mathrm{d}x} + \frac{m\omega}{\hbar}x\right)\psi = e^{-\frac{m\omega x^2}{2\hbar}}\frac{\mathrm{d}}{\mathrm{d}x}\left(e^{\frac{m\omega x^2}{2\hbar}}\psi\right). \tag{6.4.17}$$

The generalization to arbitrary powers of the annihilation and creation operators is straightforward:

$$\left(\frac{\mathrm{d}}{\mathrm{d}x} + \frac{m\omega}{\hbar}x\right)^n\psi = e^{-\frac{m\omega x^2}{2\hbar}}\frac{\mathrm{d}^n}{\mathrm{d}x^n}\left(e^{\frac{m\omega x^2}{2\hbar}}\psi\right) \tag{6.4.18}$$

and

$$\left(\frac{\mathrm{d}}{\mathrm{d}x} - \frac{m\omega}{\hbar}x\right)^n\psi = e^{\frac{m\omega x^2}{2\hbar}}\frac{\mathrm{d}^n}{\mathrm{d}x^n}\left(e^{-\frac{m\omega x^2}{2\hbar}}\psi\right). \tag{6.4.19}$$

This remark leads to the introduction of the Hermite polynomials. This means that, on defining the variable

$$\xi \equiv \sqrt{\frac{m\omega}{\hbar}}\,x, \tag{6.4.20}$$

we have

$$H_n(\xi) \equiv (-1)^n e^{\xi^2}\frac{\mathrm{d}^n}{\mathrm{d}\xi^n}e^{-\xi^2}. \tag{6.4.21}$$

From Eq. (6.4.19), the term on the right-hand side of Eq. (6.4.21) can be re-expressed in the form

$$e^{\frac{\xi^2}{2}}\left(\frac{\mathrm{d}}{\mathrm{d}\xi} - \xi\right)^n e^{-\frac{\xi^2}{2}} = e^{\xi^2}\frac{\mathrm{d}^n}{\mathrm{d}\xi^n}e^{-\xi^2}. \tag{6.4.22}$$

The Hermite polynomials therefore make it possible to write, for the harmonic oscillator eigenfunctions,

$$\psi_n(\xi) = \widetilde{K}_n H_n(\xi)e^{-\frac{\xi^2}{2}}, \tag{6.4.23}$$

having set

$$\widetilde{K}_n \equiv K_n\left(\frac{m\omega}{\hbar}\right)^{\frac{n}{2}}(-1)^n. \tag{6.4.24}$$

For example, for the lowest values of the integer n, the Hermite polynomials read as

$$H_0(\xi) = 1, \quad H_1(\xi) = 2\xi, \tag{6.4.25}$$

$$H_2(\xi) = 4\xi^2 - 2, \quad H_3(\xi) = 8\xi^3 - 12\xi. \tag{6.4.26}$$

Thus, the eigenfunctions are even or odd depending on whether n is even or odd, respectively:

$$\psi_{2k}(x) = \psi_{2k}(-x), \tag{6.4.27}$$

$$\psi_{2k+1}(x) = -\psi_{2k+1}(-x). \tag{6.4.28}$$

6.5 Problems

6.P1. Given a free particle moving on the real line, whose position x_1 at time t_1 is known, compute the probability of finding it at the position $x_2 = x_1 + \delta x$ at time $t_2 = t_1 + \delta t$.

6.P2. Consider a one-dimensional quantum system in presence of the potential

$$V(x) = 0 \text{ if } |x| > a, \quad -V_0 + \frac{m}{2}\omega^2 x^2 \text{ if } |x| \leq a, \tag{6.5.1}$$

where m is the mass of the particle and $V_0 > 0$. Consider the stationary states $\psi_\pm(x)$ which, for $|x| \leq a$, read as

$$\psi_\pm(x) = N_\pm \exp\left(\pm \frac{x^2}{2x_0^2}\right), \tag{6.5.2}$$

where $N_\pm$ is a normalization constant.

(i) Prove that $\psi_+(x)$ cannot describe a bound state of the system, whereas $\psi_-(x)$ can, for a particular value of x_0 and provided V_0 and a fulfill a suitable condition.

(ii) Compute the energy eigenvalue of the bound state and its full wave function (up to N).

(iii) If $x_0 >> a$, find to order $O(x_0^{-2})$ the ratio between probabilities of finding the particle inside or, instead, outside the well.

6.P3. A particle of mass m is subject to the one-dimensional potential

$$V(x) = -K\Big[\delta(x + R) + \delta(x - R)\Big], \tag{6.5.3}$$

where the parameter K takes positive values. Prove that the system admits always a symmetric bound state, and find under which condition an antisymmetric bound state can exist.

6.P4. A particle of mass m moves on the real line, where it is subject to a potential $U(x)$. The first excited state is described by the stationary state

$$\psi_1(x) = x\psi_0(x), \tag{6.5.4}$$

belonging to the energy eigenvalue E_1. The function $\psi_0(x)$ describes the ground state and belongs to the eigenvalue E_0.

Find the potential $U(x)$ as a function of E_0, knowing that it vanishes at $x = 0$.

6.P5. Consider a particle of mass m in the one-dimensional real-valued potential

$$V(x) = \lambda^2 x^6 - \frac{3\lambda\hbar}{\sqrt{2m}} x^2. \tag{6.5.5}$$

Prove that the Hamiltonian operator can be written as $H = A^\dagger A$, where

$$A \equiv \frac{p}{\sqrt{2m}} - i\lambda x^3. \tag{6.5.6}$$

Find the eigenvalues of A and $A^\dagger$, and their eigenfunctions in the Hilbert space $L^2(\mathbf{R})$, and the lowest-energy eigenvalue of the system.

6.P6. A particle confined in the closed interval $[0, a]$ by an infinite well is initially in the stationary state

$$\psi_0(x) = N \sin\left(\frac{\pi x}{a}\right)\left[1 + e^{i\alpha} \cos\left(\frac{\pi x}{a}\right)\right], \tag{6.5.7}$$

where α is real and N is a normalization constant. Compute the time-evolution of mean values of position and momentum.

6.P7. A particle confined in the closed interval $[0, a]$ by an infinite well is in the stationary state

$$\psi(x) = N\left(e^{i\frac{2\pi x}{a}} - 1\right). \tag{6.5.8}$$

Find the mean quadratic deviation of position, the probability distribution of energy eigenvalues and the mean value of energy in such a state.

6.P8. A particle confined in the closed interval $[-a, a]$ by an infinite well is, at a certain time, in the state

$$\psi(x) = 0 \text{ if } |x| \geq b, \ \cos\left(\frac{\pi x}{2b}\right) \text{ if } |x| < b, \tag{6.5.9}$$

with $b < a$.

(i) What are the possible values of measurements of energy, and their probabilities?

(ii) Find the mean values of energy and momentum.

(iii) Do these mean values depend on time?

Appendix 6.A Wave-packet behaviour at large time values

The stationary-phase method makes it possible to study integrals of the type

$$I(t) \equiv \int_{-\infty}^{\infty} \varphi(x) e^{itF(x)} dx, \tag{6.A.1}$$

where φ is a function with compact support. At large values of $|t|$, the exponential $e^{itF(x)}$ oscillates rapidly and hence its contributions to the integral tend to cancel each other, with the exception of points where $F'(x)$ vanishes, which correspond to a slow variation of F. Any point x_0 such that

$$F'(x_0) = 0 \tag{6.A.2}$$

is called a *stationary-phase point*. What is crucial is to understand whether or not x_0 belongs to the support of φ. In the affirmative case, the significant contribution to $I(t)$, as $|t| \to +\infty$, is given by those points x in the neighbourhood of x_0. In contrast, if x_0 does not belong to the support of φ, $I(t)$ tends rapidly to zero as $|t| \to +\infty$. More precisely, many relevant applications of the method rely on the following theorem.

Theorem 6.1. Let $\varphi \in C_0^\infty(\mathbf{R})$, and let $F \in C_0^\infty(\mathbf{R})$ be a real-valued function, such that the equation $F'(x) = 0$ has a unique solution x_0 belonging to the support of φ, where $F''(x) \neq 0$. The integral (6.A.1) then has the asymptotic expansion

$$I(t) \sim e^{itF(x_0)} \sum_{j=0}^{n} a_j(\varphi, F) t^{-j-\frac{1}{2}} + O\left(t^{-n-\frac{3}{2}}\right). \tag{6.A.3}$$

In Sections 6.2 and 6.3, the stationary-phase method has been applied to wave packets of the form

$$\psi_{\mathrm{I}}(x,t) = \int_{-\infty}^{\infty} C_{\mathrm{I}}(p) \left[e^{\frac{i}{\hbar}\left(px - \frac{p^2}{2m}t\right)} + R(p)e^{-\frac{i}{\hbar}\left(px + \frac{p^2}{2m}t\right)} \right] dp, \tag{6.A.4}$$

$$\psi_{\mathrm{II}}(x,t) = \int_{-\infty}^{\infty} C_{\mathrm{II}}(p)T(p)e^{\frac{i}{\hbar}\left(\tilde{p}x - \frac{p^2}{2m}t\right)} dp. \tag{6.A.5}$$

With our notation, C_{I} and C_{II} are functions with compact support, and all functions

$$R, T, C_{\mathrm{I}}, C_{\mathrm{II}}$$

have, in general, a phase depending on p:

$$C_{\mathrm{I}}(p) = \chi_1(p)\, e^{i\varphi_1(p)}, \tag{6.A.6}$$

$$R(p) = \chi_2(p)\, e^{i\varphi_2(p)}, \tag{6.A.7}$$

$$C_{\mathrm{II}}(p) = \chi_3(p)\, e^{i\varphi_3(p)}, \tag{6.A.8}$$

$$T(p) = \chi_4(p)\, e^{i\varphi_4(p)}. \tag{6.A.9}$$

Moreover, ψ_{I} is defined for $x \in]-\infty, -a[$, ψ_{II} is defined for $x \in]a, \infty[$, with $a \geq 0$, and $\tilde{p}$ either coincides with p or is a more complicated function of p: $\tilde{p} = \tilde{p}(p)$. Thus, stationarity of the phase implies, for the two parts of ψ_{I}, the conditions

$$\left(\frac{\partial \varphi_1}{\partial p} + \frac{x}{\hbar} - \frac{p}{m}\frac{t}{\hbar} \right)_{p=p_1} = 0, \tag{6.A.10}$$

$$\left(\frac{\partial \varphi_1}{\partial p} + \frac{\partial \varphi_2}{\partial p} - \frac{x}{\hbar} - \frac{p}{m}\frac{t}{\hbar} \right)_{p=p_2} = 0, \tag{6.A.11}$$

whereas, for ψ_{II}, it leads to the equation

$$\left(\frac{\partial \varphi_3}{\partial p} + \frac{\partial \varphi_4}{\partial p} + \frac{x}{\hbar}\frac{\partial \tilde{p}}{\partial p} - \frac{p}{m}\frac{t}{\hbar} \right)_{p=p_3} = 0. \tag{6.A.12}$$

These equations may be re-expressed in the form

$$x = \frac{p_1}{m}t - \hbar \left.\frac{\partial \varphi_1}{\partial p}\right|_{p_1}, \tag{6.A.13}$$

$$x = -\frac{p_2}{m}t + \hbar \left.\frac{\partial(\varphi_1 + \varphi_2)}{\partial p}\right|_{p_2}, \qquad (6.A.14)$$

$$x = \frac{p_3}{m}t - \hbar \left.\frac{\partial(\varphi_3 + \varphi_4)}{\partial p}\right|_{p_3}. \qquad (6.A.15)$$

The following comments are now in order.

(i) As $t \to \infty$, Eq. (6.A.13) implies that $x \to \infty$, which is incompatible with the domain of definition of ψ_{I}. This implies that no incident packet survives after the interaction with a potential of compact support, in agreement with what is expected on physical grounds.

(ii) As $t \to \infty$, Eq. (6.A.14) implies that $x \to -\infty$, which is compatible with the domain of definition of ψ_{I}. Thus, after the interaction with the potential, the only asymptotic state in the negative-x region is a reflected wave packet.

(iii) As $t \to \infty$, Eq. (6.A.15) implies that $x \to \infty$, which is compatible with the domain of definition of ψ_{II}. Thus, after the interaction with the potential, the only asymptotic state in the positive-x region is a transmitted wave packet.

(iv) All the above conclusions hold for a state which, as $t \to -\infty$, describes a free particle located at $x = -\infty$ and moving from the left to the right with velocity $\frac{p_1}{m}$. However, one might equally well require that, as $t \to -\infty$, the initial wave packet is located at $x = +\infty$, and evaluate the probability of detecting an asymptotic state (i.e. as $t \to \infty$) at $x = -\infty$ after the interaction with a short-range potential.

(v) The localization of the wave packet at large times does not contradict the spreading of the wave packet evaluated in Chapter 4, because the former results from an asymptotic calculation, whereas the latter refers to finite time intervals.

7 Elementary applications: multi-dimensional problems

This chapter is devoted to central potentials with the associated theory of angular momentum, harmonic polynomials and spherical harmonics. Applications deal with the hydrogen atom, s-wave bound states in the square-well potential and the isotropic harmonic oscillator.

7.1 The Schrödinger equation in a central potential

When we consider composite systems of elementary particles or quantum systems corresponding to classical ones with a configuration space Q, the wave function will depend on coordinates $(q_1, q_2, \ldots, q_n)$. If the Hamiltonian operator possesses particular symmetries it may be convenient to describe Q by means of appropriate coordinates which take into account the existing symmetries. In this way it may be possible to look for wave functions which are written in a factorized form, e.g. $\psi = f(\xi)g(\eta)$. A particular instance of this general situation is provided by Hamiltonian operators, which are invariant under the rotation group.

Recall now that a central potential in $\mathbf{R}^3$ is a real-valued function that only depends on the magnitude $r \equiv \sqrt{x^2 + y^2 + z^2}$ of the position vector, where x, y, z are Cartesian coordinates in $\mathbf{R}^3$. Our aim is to build a general formalism to analyse the Hamiltonian

$$H = \frac{1}{2m}\left(p_x^2 + p_y^2 + p_z^2\right) + V, \tag{7.1.1}$$

where p_x, p_y, p_z are Cartesian components of the linear momentum, and V is a central potential. By virtue of the hypothesis on the potential, the symmetry group of our problem is the rotation group in three dimensions, since the rotations preserve the length of vectors in $\mathbf{R}^3$ and their real nature. In classical mechanics the angular momentum is a constant of motion, i.e.

$$\{H, L_x\} = \{H, L_y\} = \{H, L_z\} = 0. \tag{7.1.2}$$

If we use two commuting first integrals, e.g. L^2 and L_z, we can reduce our system and find a reduced system having only one degree of freedom. A simple way to see this reduction procedure at work is to write the Hamiltonian in the form

$$H = \frac{1}{2m}\left(p_r^2 + \frac{L^2}{r^2}\right) + V(r). \tag{7.1.3}$$

where $p_r \equiv \frac{\vec{r}}{r} \cdot \vec{p}$ is the radial component of the linear momentum $\vec{p}$. Moreover, the identity

$$L^2 = (\vec{r} \wedge \vec{p})^2 = r^2 p^2 - (\vec{r} \cdot \vec{p})^2$$

yields indeed

$$p^2 = \frac{L^2}{r^2} + p_r^2.$$

The classical system in three dimensions turns out to be reduced to a 1-parameter family of one-dimensional systems, a different one for each value of L^2. The particular value we have to use depends on the initial conditions, for which $L_0^2 = (\vec{r}_0 \wedge \vec{p}_0)^2$. We exploit the fact that it is a constant of motion. This same procedure also applies in quantum mechanics.

The quantum angular momentum is obtained by using the correspondence between symbols and operators and noting that the factors occurring in the expression of each component of the angular momentum are pairwise commuting, and make it possible to use the multiplicative property. We find

$$\hat{L}_x \equiv \hat{y}\hat{p}_z - \hat{z}\hat{p}_y = \frac{\hbar}{i}\left(y\frac{\partial}{\partial z} - z\frac{\partial}{\partial y}\right), \tag{7.1.4}$$

$$\hat{L}_y \equiv \hat{z}\hat{p}_x - \hat{x}\hat{p}_z = \frac{\hbar}{i}\left(z\frac{\partial}{\partial x} - x\frac{\partial}{\partial z}\right), \tag{7.1.5}$$

$$\hat{L}_z \equiv \hat{x}\hat{p}_y - \hat{y}\hat{p}_x = \frac{\hbar}{i}\left(x\frac{\partial}{\partial y} - y\frac{\partial}{\partial x}\right). \tag{7.1.6}$$

These obey the commutation relations

$$\hat{L}_j\hat{L}_k - \hat{L}_k\hat{L}_j = i\hbar\varepsilon_{jks}\hat{L}_s. \tag{7.1.7}$$

It is quite easy to derive

$$[\hat{L}_x, \hat{x}^2 + \hat{y}^2 + \hat{z}^2] = [\hat{L}_y, \hat{x}^2 + \hat{y}^2 + \hat{z}^2] = [\hat{L}_z, \hat{x}^2 + \hat{y}^2 + \hat{z}^2] = 0, \tag{7.1.8}$$

and similarly for $\hat{p}_x^2 + \hat{p}_y^2 + \hat{p}_z^2$. Therefore, the angular momentum commutes with the Hamiltonian operator and it follows that the expectation values of the angular momentum in every state which is a solution of the equations of motion associated with H are constant in time; they are constants of motion.

Now we can rewrite the angular momentum in spherical polar coordinates.[1] We recall that these transformations are given by

$$x = r\sin\theta\cos\varphi, \; y = r\sin\theta\sin\varphi, \; z = r\cos\theta, \tag{7.1.9}$$

---

[1] To appreciate what happens classically when we move from Cartesian coordinates to polar spherical coordinates, let us consider the corresponding equations of motion

$$\dot{\vec{r}} = \vec{v}, \; \dot{\vec{v}} = \vec{r}V'(r).$$

When we want to write them in polar spherical coordinates, we would have

$$\frac{d}{dt}(r\sin\theta\cos\varphi) = v_x, \; \frac{d}{dt}(r\sin\theta\sin\varphi) = v_y, \; \frac{d}{dt}(r\cos\varphi) = v_z.$$

By spelling out the time derivative we would get equations of motion involving $\ddot{r}, \ddot{\theta}, \ddot{\varphi}$ even if the motion is a free motion. We can say that, from a free system, we have obtained an *effective* interacting system. A different, with $\theta \in [0, \pi]$, $\varphi \in [0, 2\pi[$, from which we derive

$$\hat{L}_x = \frac{\hbar}{i} \left(\sin \varphi \frac{\partial}{\partial \theta} + \cos \varphi \cot \theta \frac{\partial}{\partial \varphi} \right), \qquad (7.1.10)$$

$$\hat{L}_y = \frac{\hbar}{i} \left(\cos \varphi \frac{\partial}{\partial \theta} - \sin \varphi \cot \theta \frac{\partial}{\partial \varphi} \right), \qquad (7.1.11)$$

$$\hat{L}_z = \frac{\hbar}{i} \frac{\partial}{\partial \varphi}. \qquad (7.1.12)$$

These formulae will be useful for computations on wave functions expressed in polar coordinates.

To express the Hamiltonian operator in spherical polar coordinates we are obliged to use first the formula in Cartesian coordinates, i.e.

$$\hat{H} = \frac{1}{2m} \left(\hat{p}_x^2 + \hat{p}_y^2 + \hat{p}_z^2 \right) + V \left(\hat{x}^2 + \hat{y}^2 + \hat{z}^2 \right), \qquad (7.1.13)$$

and to express it in terms of $\frac{\partial}{\partial r}, \frac{\partial}{\partial \theta}, \frac{\partial}{\partial \varphi}$. We use $r = \sqrt{x^2 + y^2 + z^2}$ and express $\hat{p}^2 = -\hbar^2 \Delta$ in polar coordinates

$$\hat{p}^2 = -\hbar^2 \left[\frac{1}{r^2} \frac{\partial}{\partial r} \left(r^2 \frac{\partial}{\partial r} \right) + \frac{1}{r^2 \sin \theta} \frac{\partial}{\partial \theta} \left(\sin \theta \frac{\partial}{\partial \theta} \right) + \frac{1}{r^2 \sin^2 \theta} \frac{\partial^2}{\partial \varphi^2} \right]$$

$$= -\hbar^2 \left[\frac{1}{r^2} \frac{\partial}{\partial r} \left(r^2 \frac{\partial}{\partial r} \right) - \frac{\hat{L}^2}{\hbar^2 r^2} \right], \qquad (7.1.14)$$

and eventually

$$\hat{H} = \frac{\hat{p}^2}{2m} + V(r)$$

$$= -\frac{\hbar^2}{2m} \left[\frac{1}{r^2} \frac{\partial}{\partial r} \left(r^2 \frac{\partial}{\partial r} \right) - \frac{\hat{L}^2}{\hbar^2 r^2} \right] + V(r). \qquad (7.1.15)$$

An equivalent way to express the radial part is

$$\frac{\hat{p}_r^2}{2m} = -\frac{\hbar^2}{2mr} \frac{\partial^2}{\partial r^2} (r \cdot)$$

$$= -\frac{\hbar^2}{2m} \left(\frac{\partial^2}{\partial r^2} + \frac{2}{r} \frac{\partial}{\partial r} \right)$$

$$= -\frac{\hbar^2}{2mr^2} \frac{\partial}{\partial r} \left(r^2 \frac{\partial}{\partial r} \right). \qquad (7.1.16)$$

more geometrical way to consider the same situation is as follows. We decompose the motion along the radius $\vec{r}(t)$ and along the orthogonal plane $\vec{r}^{\perp}$, and we find

$$\frac{d}{dt} (\vec{r} \wedge \dot{\vec{r}}) = \vec{r} \wedge \ddot{\vec{r}} = \vec{r} \wedge \vec{r} V'(r) = 0, \text{ whenever } \ddot{\vec{r}} = \vec{r} f(r).$$

Thus, *the motion along the orthogonal plane is always the same, independently of the particular field of central forces*. By using the conservation of angular momentum, the equation along the radius may be expressed as a radial equation, one for each value of the angular momentum.

To integrate the equations of motion associated with the Hamiltonian of a central potential, we need to introduce appropriate 'coordinates' in the Hilbert space. From the Schrödinger equation

$$i\hbar\frac{\mathrm{d}}{\mathrm{d}t}|\psi\rangle = \hat{H}|\psi\rangle \tag{7.1.17}$$

we use a basis of orthonormal vectors for the Hilbert space denoted by $|r, \theta, \varphi\rangle$ and restrict, preliminarly, our analysis to component functions of the following form (see Eq. (4.3.9) and related comments to refresh the memory on 'bra' and 'ket' notation):

$$\langle r, \theta, \varphi|\psi\rangle = R(r)Y(\theta, \varphi). \tag{7.1.18}$$

These are special vectors $|\psi\rangle$, which are called separable in spherical polar coordinates. By restricting our attention to them, we find that the eigenvalue equation associated with $\hat{H}$ gives, for some real parameter λ, the equations

$$\frac{\mathrm{d}}{\mathrm{d}r}\left(r^2\frac{\mathrm{d}R}{\mathrm{d}r}\right) + \frac{2mr^2}{\hbar^2}\Big[E - V(r)\Big]R = \lambda R(r), \tag{7.1.19}$$

$$\frac{1}{\sin\theta}\frac{\partial}{\partial\theta}\left(\sin\theta\frac{\partial Y}{\partial\theta}\right) + \frac{1}{\sin^2\theta}\frac{\partial^2 Y}{\partial\varphi^2} = -\lambda Y(\theta, \varphi). \tag{7.1.20}$$

At this stage, we make a further separation of variables, and look for $Y(\theta, \varphi)$ in the form of a product

$$Y(\theta, \varphi) = \widetilde{Y}(\theta)\sigma(\varphi). \tag{7.1.21}$$

This leads to ordinary differential equations for σ and $\widetilde{Y}$, respectively:

$$\left(\frac{\mathrm{d}^2}{\mathrm{d}\varphi^2} + \mu\right)\sigma = 0, \tag{7.1.22}$$

$$\left[\frac{1}{\sin\theta}\frac{\mathrm{d}}{\mathrm{d}\theta}\left(\sin\theta\frac{\mathrm{d}}{\mathrm{d}\theta}\right) + \left(\lambda - \frac{\mu}{\sin^2\theta}\right)\right]\widetilde{Y} = 0. \tag{7.1.23}$$

Equation (7.1.23) can be solved by Legendre polynomials, but we do not emphasize the special-functions aspect in our introductory treatment, for which we refer the reader to the many textbooks in the literature. If we impose the periodicity conditions

$$\sigma(0) = \sigma(2\pi), \quad \sigma'(0) = \sigma'(2\pi), \tag{7.1.24}$$

these imply that, in the general form of the solution:

$$\sigma(\varphi) = A + B\varphi \quad \text{if } \mu = 0, \tag{7.1.25}$$

$$\sigma(\varphi) = Ae^{i\sqrt{\mu}\varphi} + Be^{-i\sqrt{\mu}\varphi} \quad \text{if } \mu \neq 0, \tag{7.1.26}$$

we must set $B = 0$, while $\sqrt{\mu}$ is an integer. Thus,

$$\sigma_m(\varphi) = A_m e^{im\varphi}, \quad m = 0, \pm 1, \pm 2 \ldots . \tag{7.1.27}$$

The constant $A_m = \frac{1}{\sqrt{2\pi}}$, from the condition

$$\int_0^{2\pi} |\sigma_m|^2 \mathrm{d}\varphi = 1. \tag{7.1.28}$$

To discuss our Hamiltonian operator

$$\hat{H} = \frac{1}{2m} \left(\hat{p}_r^2 + \frac{\widehat{L}^2}{r^2} \right) + V(r) \tag{7.1.29}$$

at a deeper level, it is convenient to discuss more specifically the operators $\hat{p}_r^2$ and $\widehat{L}^2$ with their domains of definition.

We study first $\hat{p}_r$ associated with $\frac{\vec{r}}{r} \cdot \vec{p}$. Indeed, we have the classical formula

$$p_r \equiv \frac{\vec{r}}{r} \cdot \vec{p} = \frac{1}{2} \left(\frac{\vec{r}}{r} \cdot \vec{p} + \vec{p} \cdot \frac{\vec{r}}{r} \right). \tag{7.1.30}$$

However, in quantum mechanics, if we were to consider, naively, the operator

$$\hat{D}_r \equiv \frac{\vec{r}}{r} \cdot \vec{p} = \frac{\hbar}{i} \left(\frac{x}{r} \frac{\partial}{\partial x} + \frac{y}{r} \frac{\partial}{\partial y} + \frac{z}{r} \frac{\partial}{\partial z} \right) = \frac{\hbar}{i} \frac{\partial}{\partial r}, \tag{7.1.31}$$

we would realize that this is not the appropriate choice. What happens is that $\hat{D}_r$ is not symmetric, because

$$\begin{aligned}
\hat{D}_r^\dagger &\equiv \frac{\vec{p} \cdot \vec{r}}{r} = \frac{\hbar}{i} \left(\frac{\partial}{\partial x} \frac{x}{r} + \frac{\partial}{\partial y} \frac{y}{r} + \frac{\partial}{\partial z} \frac{z}{r} \right) \\
&= \frac{\hbar}{i} \left(\frac{3}{r} + x \frac{\partial}{\partial x} r^{-1} + y \frac{\partial}{\partial y} r^{-1} + z \frac{\partial}{\partial z} r^{-1} + \frac{x}{r} \frac{\partial}{\partial x} + \frac{y}{r} \frac{\partial}{\partial y} + \frac{z}{r} \frac{\partial}{\partial z} \right) \\
&= \frac{\hbar}{i} \left[\frac{3}{r} - \frac{(x^2 + y^2 + z^2)}{r^3} + \frac{\partial}{\partial r} \right] = \frac{\hbar}{i} \left(\frac{\partial}{\partial r} + \frac{2}{r} \right).
\end{aligned} \tag{7.1.32}$$

The calculations (7.1.31) and (7.1.32) suggest defining (Dirac 1958)

$$\hat{p}_r \equiv \frac{1}{2} \left(\hat{D}_r + \hat{D}_r^\dagger \right) = \frac{\hbar}{i} \left(\frac{\partial}{\partial r} + \frac{1}{r} \right). \tag{7.1.33}$$

Of course, the formula (7.1.33) can also be derived from the scalar product

$$(f, g) \equiv \int_0^\infty f^*(r) g(r) r^2 \, dr.$$

The operator (7.1.33) is symmetric by construction, and obeys the desired form of the commutation rules:

$$\left[\hat{r}, \hat{p}_r \right] \equiv \hat{r} \hat{p}_r - \hat{p}_r \hat{r} = i\hbar. \tag{7.1.34}$$

Moreover, it leads to the following form of the operator $\hat{p}_r^2$:

$$\begin{aligned}
\hat{p}_r^2 &= -\hbar^2 \left(\frac{\partial^2}{\partial r^2} + \frac{2}{r} \frac{\partial}{\partial r} \right) = -\frac{\hbar^2}{r^2} \frac{\partial}{\partial r} \left(r^2 \frac{\partial}{\partial r} \right) \\
&= -\frac{\hbar^2}{r} \frac{\partial^2}{\partial r^2} (r \cdot).
\end{aligned} \tag{7.1.35}$$

Formula (7.1.29) shows very clearly that the total Hamiltonian decomposes into two operators, acting on the tensor product of two Hilbert spaces $L^2(\mathbf{R}) \otimes L^2(S^2)$. We should first find invariant subspaces of $L^2(S^2)$ on which $\hat{L}^2$ acts as a multiple of the identity, say $\hbar^2 \lambda^2 I$, and then the Hamiltonian operator

$$\left(\frac{1}{2m} \hat{p}_r^2 + V(r) \right) \otimes I + \frac{1}{2mr^2} \otimes \widehat{L^2}$$

becomes, when restricted to the appropriate eigenspace,

$$\left(\frac{1}{2m} \hat{p}_r^2 + V(r) \right) \otimes I + \frac{\lambda^2 \hbar^2}{2mr^2} \otimes I.$$

This means that we can solve first the motion associated with $\widehat{L^2}$ along the 2-sphere and thereafter compose with the motion along the radius. This is a variant of the Newtonian composition of independent motions, indeed the particular Hamiltonian operator we have to use along the radius is dynamically determined by the initial conditions of the motion we want to describe. Let us therefore address the problem of free motion on the 2-sphere.

Note that the conditions of regularity at the origin, and square integrability on $(0, \infty)$, imply that $\hat{p}_r$ is symmetric on

$$C_0^\infty \left(\mathbf{R}^3 - \{0\} \right),$$

whereas $\hat{p}_r^2$ is self-adjoint (recall that, to define a unitary evolution of the quantum system, we need $\hat{H}$ to be self-adjoint; it is not enough that $\hat{H}$ is symmetric). Without giving too many details, we can, however, make contact with what is know in one-dimensional problems concerning these sorts of issues.

Digression Let T be the operator $i\frac{d}{dx}$ on $\mathcal{L}^2(0, 1)$, with domain (for simplicity, we consider a closed interval)

$$D(T) = \{\varphi : \varphi \in AC[0, 1], \varphi(0) = 0 = \varphi(1)\}. \tag{7.1.36}$$

In other words, the domain of T consists of absolutely continuous functions in $[0, 1]$ that vanish at the end points of such an interval. The adjoint of T, here denoted by $T^\dagger$, is again the operator $i\frac{d}{dx}$, and for $\varphi \in D(T)$ and $\psi \in D(T^\dagger)$ we find

$$\left(T\varphi, \psi \right) - \left(\varphi, T^\dagger \psi \right) = -i[\varphi^*(1)\psi(1) - \varphi^*(0)\psi(0)] = 0. \tag{7.1.37}$$

It is hence clear that the boundary conditions on the functions $\in D(T)$ are so 'strong' that no boundary conditions whatsover are necessary for the functions $\in D(T^\dagger)$. This is why T is not self-adjoint (although a family of self-adjoint extensions can be obtained by requiring proportionality of the boundary values at 0 and 1). For instance, the conditions

$$\psi(1) = e^{i\gamma\alpha} \psi(0), \ \varphi(1) = e^{i\gamma\alpha} \varphi(0), \tag{7.1.38}$$

provide the domain of essential self-adjointness for $i\frac{d}{dx}$, each domain being characterized by an element of $U(1)$. The family of all different self-adjoint extensions is characterized by

$$\psi^*(0)\varphi(0) = \psi^*(1)\varphi(1), \tag{7.1.39}$$

therefore

$$U\psi(0) = \psi(1),\tag{7.1.40}$$

where U is a unitary transformation.

However, if we study the operator $A \equiv \frac{d^2}{dx^2}$ on $\mathcal{L}^2(0,1)$, the condition

$$\left(Au, v\right) - \left(u, A^\dagger v\right) = 0 \tag{7.1.41}$$

is fulfilled if and only if both the functions $\in D(A)$ and the functions $\in D(A^\dagger)$ obey the *same* boundary conditions, because integration by parts leads to the condition

$$\left(\frac{du^*}{dx}v - u^*\frac{dv}{dx}\right)_0^1 = 0,\tag{7.1.42}$$

which is satisfied if both $u \in D(A)$ and $v \in D(A^\dagger)$ vanish at 0 and at 1. To find the most general boundary conditions we have to further analyze Eq. (7.1.42), which may be written as

$$\left(\dot{u}^*v - u^*\dot{v}\right)(1) - \left(\dot{u}^*v - u^*\dot{v}\right)(0) = 0,\tag{7.1.43}$$

where the derivative with respect to x, evaluated at the boundary, has been denoted by an overdot. If we define

$$\Phi \equiv \begin{pmatrix} u(0) \\ u(1) \end{pmatrix}, \quad \dot{\Phi} \equiv \begin{pmatrix} \dot{u}(0) \\ \dot{u}(1) \end{pmatrix},\tag{7.1.44}$$

the vanishing condition written before is then equivalent to

$$(\Phi - i\dot{\Phi}) = U(\Phi + i\dot{\Phi}),\tag{7.1.45}$$

and

$$(\Psi - i\dot{\Psi}) = U(\Psi + i\dot{\Psi}),\tag{7.1.46}$$

having set

$$\Psi \equiv \begin{pmatrix} v(0) \\ v(1) \end{pmatrix}, \quad \dot{\Psi} \equiv \begin{pmatrix} \dot{v}(0) \\ \dot{v}(1) \end{pmatrix}.\tag{7.1.47}$$

Condition (7.1.42) appears now as the conservation of the scalar product of Φ with Ψ, say

$$\langle \Phi - i\dot{\Phi} | \Psi - i\dot{\Psi} \rangle = \langle \Phi + i\dot{\Phi} | \Psi + i\dot{\Psi} \rangle,$$

which is equivalent to (7.1.45) and (7.1.46). Thus, all domains on which $\triangle$ is self-adjoint on the interval $[0, 1]$ are in 1-to-1 correspondence with elements of the unitary group $U(2)$. By spelling out the condition, we have

$$\Phi(0) - i\dot{\Phi}(0) = U_{11}(\Phi(0) + i\dot{\Phi}(0)) + U_{12}(\Phi(1) + i\dot{\Phi}(1)),\tag{7.1.48}$$

$$\Phi(1) - i\dot{\Phi}(1) = U_{21}(\Phi(0) + i\dot{\Phi}(0)) + U_{22}(\Phi(1) + i\dot{\Phi}(1)).\tag{7.1.49}$$

It can be easily seen that for $U = I$ Neumann boundary conditions are obtained, while for $U = -I$ Dirichlet boundary conditions (Asorey *et al.* 2005) are recovered.

7.1.1 Use of symmetries and geometrical interpretation

The equations of motion associated with the Hamiltonian function (7.1.1) may be considered in spherical polar coordinates in the form associated with (7.1.3). Since H and L^2 commute, it is possible to solve the equations of motion for L^2 and H separately for each initial condition $(\vec{r}_0, \vec{p}_0)$. We would have the equations for H given by

$$H = \frac{1}{2m}\left(p_r^2 + \frac{L_0^2}{r^2}\right) + V(r),$$

along the radius r with initial conditions

$$r_0 = \sqrt{\vec{r}_0 \cdot \vec{r}_0}, \; p_r(0) = \frac{\vec{r}_0 \cdot \vec{p}_0}{r_0}.$$

Similarly, we might solve the equations of motion for L^2 with initial conditions $\vec{r}_0, \vec{p}_0$. Eventually, we find the solution of the equations of motion associated with H with initial conditions $(\vec{r}_0, \vec{p}_0)$ by simply combining the solution found for $H(r, p_r)$ and that for L^2.

The usual Lagrangian procedure is to introduce coordinates adapted to the decomposition of $\mathbf{R}^3 - \{0\}$ into $S^2 \times \mathbf{R}_+$. We use spherical coordinates to find the Lagrangian for the motion along the sphere of radius r given by

$$L_S = \frac{mr^2}{2}\left(\dot{\theta}^2 + \sin^2\theta\,\dot{\varphi}^2\right),$$

mr^2 playing the role of effective mass of a particle, and one along the radius

$$L_r = \frac{m\dot{r}^2}{2} - V(r) - \frac{L_0^2}{r^2}.$$

Notice the change in sign we have to introduce by hand with respect to the term in H. For this motion, it should be clear that, along the radius, we get a 1-parameter family of Lagrangians, depending on the initial conditions $\vec{r}_0 \wedge \vec{p}_0$. Thus, in composing the two motions, to find the evolution of $\vec{r}(t), \vec{p}(t)$ the motion associated with appropriate corresponding Lagrangians identified by $\vec{r}_0$ and $\vec{p}_0$ should be used.

For the motion along the sphere we further notice that φ is a cyclic coordinate, with associated constant of motion $l_\varphi = \sin^2\theta\,\dot{\varphi}$, therefore for any initial condition $(\varphi, \theta; \dot{\varphi}, \dot{\theta})$ we might solve separately the motion along the radius and the motion associated with

$$L_\theta = \frac{m}{2}r^2\left(\dot{\theta}^2 + \frac{l_\varphi^2}{\sin^2\theta}\right),$$

for each l_φ fixed by the initial conditions. This reduction by *stages* is commonly used to solve problems in the presence of a large group of symmetries. However, there is another way to solve the problems, which requires almost no computations and fully relies on symmetries and geometrical interpretation. The equations of motion defined by L^2 on the three-dimensional space possess as constants of motion $L_x, L_y, L_z, \vec{r} \cdot \vec{r}, \vec{r} \cdot \vec{p}, \vec{p} \cdot \vec{p}$. The invariance of $\vec{L}$ implies that the motion, for a fixed value of $\vec{L}$ identified by the initial conditions, takes place along a plane passing through the origin and orthogonal to $\vec{L}$. The invariance of $\vec{r} \cdot \vec{r} = \vec{r}_0 \cdot \vec{r}_0$ implies that the motion should also be along such a sphere;

therefore, the motion will be given by intersecting the sphere with a previously identified plane. The resulting effect is a motion along great circles. Thus, free motion along the sphere has trajectories provided by geodesic motions along great circles. In conclusion, we get an overall picture of the trajectories without any computation. Of course, to have an explicit analytical expression we need to introduce coordinates; now we may choose between $(\vec{r}, \vec{p})$ and $(r, p_r, \theta, p_\theta, \varphi, p_\varphi)$.

7.1.2 Angular momentum operators and spherical harmonics

The factorization of the operator gives rise to a separation of variables for the components of wave functions, which read as $\psi(r, \theta, \varphi) = R(r)Y(\theta, \varphi)$. Thus, we are entitled to study separately square-integrable functions on $\mathbf{R}_+$ with respect to the measure $r^2 dr$ (for three-dimensional problems) and square-integrable functions on the two-dimensional sphere S^2, $\mathcal{L}^2(S^2)$, with respect to the measure on S^2 induced by the Euclidean metric on $\mathbf{R}^3$, i.e. $\sin\theta \, d\theta \, d\varphi$.

We first consider complex-valued polynomials on $\mathbf{R}^3$, with coordinates (x_1, x_2, x_3). This infinite-dimensional vector space may be given the structure of a Hilbert space if we define the following inner product (Balachandran *et al.* 2010 and references therein):

$$\langle \mathcal{P}_1 | \mathcal{P}_2 \rangle = \int_{\mathbf{R}^3} \mathcal{P}_1^* \mathcal{P}_2 \exp\left(-(x_1^2 + x_2^2 + x_3^2)\right) dx_1 \, dx_2 \, dx_3. \tag{7.1.50}$$

This space is decomposed into orthogonal sub-spaces of homogeneous polynomials, i.e. $\mathcal{P} = \oplus_k \mathcal{P}_k$. Each differential operator $\hat{L}_x, \hat{L}_y, \hat{L}_z$ preserves the degree of homogeneous polynomials, therefore their corresponding matrix in $\mathcal{P}$ has a block-diagonal form.

For instance, x_1, x_2, x_3, polynomials of degree 1, are mapped into themselves. The quadratic polynomials, generated by $x_1^2, x_2^2, x_3^2, x_1 x_2, x_2 x_3, x_3 x_1$, are mapped into themselves and the corresponding linear transformation acts by preserving the one-dimensional subspace generated by $x_1^2 + x_2^2 + x_3^2$ and the five-dimensional complementary subspace generated by $x_1^2 - x_2^2, x_2^2 - x_3^2, x_1 x_2, x_2 x_3, x_3 x_1$. The corresponding 6×6 matrix of each infinitesimal generator of rotations splits into a 5×5 matrix and a 1×1 one.

Similarly, for cubic polynomials, generated by

$$x_1^3, x_2^3, x_3^3, x_1^2 x_2, x_1 x_2^2, x_2^2 x_3, x_2 x_3^2, x_3^2 x_1, x_3 x_1^2, x_1 x_2 x_3,$$

the space is ten-dimensional and splits into two invariant subspaces of dimension three and seven, respectively. They are generated by $x_1 r^2, x_2 r^2, x_3 r^2, r^2 = x_1^2 + x_2^2 + x_3^2$ and a complementary one, orthogonal to it for the scalar product above.

Each subspace $\mathcal{P}_{2k}$ of homogeneous polynomials of even degree $2k$ and each subspace $\mathcal{P}_{2k+1}$ of homogeneous polynomials of odd degree $2k + 1$ decompose into irreducible orthogonal invariant subspaces, i.e.

$$\mathcal{P}_{2k} = H_0 r^{2k} \oplus H_2 r^{2(k-1)} \oplus \dots \oplus H_{2(k-1)} r^2 \oplus H_{2k}, \tag{7.1.51}$$

$$\mathcal{P}_{2k+1} = H_1 r^{2k} \oplus H_3 r^{2(k-1)} \oplus \dots \oplus H_{2(k-1)+1} r^2 \oplus H_{2k+1}. \tag{7.1.52}$$

In this decomposition, the subspaces of homogeneous polynomials H_l are characterized by the property that they contain no element having r^2 as a factor.

By virtue of the invariance of any polynomial of r^2 under rotations, each H_k is invariant under rotations, and hence is preserved under the action of the angular momentum operators. Moreover, the action of the generators of the angular momentum on H_k does not possess diagonal blocks. By using the expression of the Laplacian as div grad,

$$\triangle(\vec{r} \cdot \vec{a})^l = \text{div grad}(\vec{r} \cdot \vec{a})^l = \text{div}(l\vec{a}(\vec{r} \cdot \vec{a})^{l-1})$$
$$= l(l-1)\vec{a} \cdot \vec{a}(\vec{r} \cdot \vec{a})^{l-2}. \tag{7.1.53}$$

In particular, if the vector $\vec{a}$ is chosen with complex coefficients, e.g.

$$\vec{a} = \left(i(t^2+1), t^2-1, 2t\right), \tag{7.1.54}$$

which has vanishing pseudo-norm:

$$\vec{a} \cdot \vec{a} = -(t^2+1)^2 + (t^2-1)^2 + 4t^2 = 0,$$

we obtain

$$\left(i(t^2+1)x + (t^2-1)y + 2tz\right)^l = \left(t^2(y+ix) - (y-ix) + 2tz\right)^l$$
$$= t^l \sum_{m=-l}^{l} t^m H_{lm}(\vec{r}), \tag{7.1.55}$$

with H_{lm} being the generators of homogeneous polynomials of degree l which generate the spaces H_l of our previous decomposition. They are $2l+1$ in number and they are in the kernel of $\triangle$, i.e. they are annihilated by $\triangle$. They are called harmonic polynomials. They possess the property of being invariant under the action of the generators of the angular momentum, with the resulting matrix possessing no further block-diagonal decomposition. In addition, they have the same dimension of the subspaces appearing in Eqs. (7.1.51) and (7.1.52). Thus, we consider them as a practical way to exhibit the decomposition of homogeneous polynomials. When each of these harmonic polynomials is restricted to the unit sphere in $\mathbf{R}^3$, we obtain the spherical harmonics $Y_{l,m}(\theta, \varphi)$. By using the familiar expression of the vector $\vec{r}$ in radial and polar coordinates, we find

$$\left(i(t^2+1)\sin\theta\cos\varphi + (t^2-1)\sin\theta\sin\varphi + 2t\cos\theta\right)^l = t^l \sum_{m=-l}^{l} t^m Y_{l,m}(\theta, \varphi). \tag{7.1.56}$$

The whole set of spherical harmonics constitute a complete orthogonal system on the unit sphere. This means that any polynomial in x, y, z, when restricted to the sphere, can be written as a sum of spherical harmonics. To sum up, the harmonic polynomials, which solve the Laplace equation in $\mathbf{R}^3$ and hence are a proper subset of the general set of homogeneous polynomials, have the property of decomposing the Hilbert space of complex-valued polynomials in $\mathbf{R}^3$ into irreducible subspaces under the action of the angular momentum operator (i.e. such subspaces cannot be 'broken' into smaller building blocks). When we restrict them to the two-dimensional sphere, they provide a decomposition of $\mathcal{L}^2(S^2)$ into irreducible subspaces, called spherical harmonics.

Some formulae for spherical harmonics (which are orthonormal) that are frequently used in the applications are as follows:

$$Y_{0,0} = \frac{1}{\sqrt{4\pi}}, \tag{7.1.57}$$

$$Y_{1,0} = \sqrt{\frac{3}{4\pi}} \cos\theta, \tag{7.1.58}$$

$$Y_{1,\pm 1} = \mp\sqrt{\frac{3}{8\pi}} \sin\theta\ e^{\pm i\varphi}, \tag{7.1.59}$$

$$Y_{2,0} = \sqrt{\frac{5}{16\pi}}\left(3\cos^2\theta - 1\right), \tag{7.1.60}$$

$$Y_{2,\pm 1} = \mp\sqrt{\frac{15}{8\pi}} (\cos\theta)(\sin\theta)e^{\pm i\varphi}, \tag{7.1.61}$$

$$Y_{2,\pm 2} = \sqrt{\frac{15}{32\pi}} \sin^2\theta\ e^{\pm 2i\varphi}. \tag{7.1.62}$$

They can all be obtained from the generating formula

$$Y_{l,m}(\theta,\varphi) = \frac{e^{im\varphi}}{\sqrt{2\pi}}\Theta_{lm}(\theta), \tag{7.1.63}$$

where

$$\Theta_{lm}(\theta) \equiv \sqrt{\frac{(2l+1)}{2}}\sqrt{\frac{(l\pm m)!}{(l\mp m)!}}(\pm\sin\theta)^{\mp m}\left[\frac{\mathrm{d}}{\mathrm{d}\cos\theta}\right]^{l\mp m}\frac{(\cos^2\theta - 1)^l}{2^l l!}, \tag{7.1.64}$$

where $\pm$ signs occur because

$$Y_{l,m}(\theta,\varphi) = Y_{l,-m}(-\theta,-\varphi). \tag{7.1.65}$$

By virtue of the generating formula, spherical harmonics have a simple behaviour under complex conjugation, i.e.

$$Y_{l,m}^*(\theta,\varphi) = Y_{l,m}(\theta,-\varphi). \tag{7.1.66}$$

Moreover, they obey the orthogonality condition

$$\int_0^{2\pi} \mathrm{d}\varphi \int_0^{\pi} \sin\theta\,\mathrm{d}\theta\, Y_{l,m}^*(\theta,\varphi) Y_{l',m'}(\theta,\varphi) = \delta_{ll'}\,\delta_{mm'}. \tag{7.1.67}$$

Since the harmonic polynomials belong to the kernel of the Laplacian,

$$\Delta\left(r^l Y_{l,m}(\theta,\varphi)\right) = 0, \tag{7.1.68}$$

which implies that

$$\Delta_S Y_{l,m}(\theta,\varphi) = -l(l+1)Y_{l,m}(\theta,\varphi), \tag{7.1.69}$$

where Δ_S is the Laplacian on the 2-sphere, and follows directly from

$$\left(\frac{\partial^2}{\partial r^2} + \frac{2}{r}\frac{\partial}{\partial r} - \frac{\Delta_S}{r^2}\right)\left(r^l Y_{l,m}(\theta,\varphi)\right) = 0. \tag{7.1.70}$$

Usually, l and m are called the *azimuthal quantum number* and the *magnetic quantum number*, respectively. If we now consider square-integrable functions on $\mathbf{R}^3$, they can all be expanded in a series of terms of the form $\Phi(r)Y_l(\theta,\varphi)$. If we choose a complete orthogonal system $\phi_1(r), \phi_2(r), \ldots$, i.e. such that

$$\int_0^\infty r^2 \phi_n^*(r)\phi_m(r)\mathrm{d}r = \delta_{mn},$$

we obtain a $(2l+1)$-dimensional space by means of functions of the form $\phi_n(r)Y_l(\theta,\varphi)$. These spaces can be denoted by $\mathcal{H}_{nl}$ and are mutually orthogonal.

7.1.3 Angular momentum eigenvalues: algebraic treatment

Let us consider the self-adjoint operators associated to the angular momenta in three dimensions, hereafter denoted with $\widehat{L}_i$. As already stated they satisfy the following algebra:

$$\left[\widehat{L}_j,\widehat{L}_k\right] = i\hbar\varepsilon_{jkl}\widehat{L}_l. \tag{7.1.71}$$

As can easily be seen, from Eq. (7.1.71) it follows that the operator $\widehat{L}^2 \equiv \sum_{i=1}^3 \widehat{L}_i^2$ commutes with all components of the angular momentum, i.e.

$$\left[\widehat{L}^2,\widehat{L}_i\right] = 0 \quad \forall i. \tag{7.1.72}$$

From this relation, it is possible to find a basis of common eigenvectors for $\widehat{L}^2$ and an arbitrary component of angular momentum; let us choose for example $\widehat{L}_3$. Denoting the elements of such a basis with $\psi_{l,m}$, they satisfy by definition both eigenvalue equations

$$\begin{cases} \widehat{L}^2\,\psi_{l,m} = l(l+1)\,\hbar^2\,\psi_{l,m}, \\ \widehat{L}_3\,\psi_{l,m} = m\,\hbar\,\psi_{l,m}, \end{cases} \tag{7.1.73}$$

where the prefactors $\hbar^2$ and $\hbar$ have been placed for dimensional reasons. Since both $\widehat{L}^2$ and $\widehat{L}_3$ are self-adjoint, this implies that l and m are real; moreover, since $\widehat{L}^2$ is positive-definite, it can be assumed that $l \geq 0$. A stronger condition concerning m and l can be found by defining the angular-momentum raising and lowering operators, i.e. $\widehat{L}_+ \equiv \widehat{L}_1 + i\widehat{L}_2$ and $\widehat{L}_- \equiv \widehat{L}_1 - i\widehat{L}_2 = \widehat{L}_+^\dagger$, respectively. From the definition of these operators it can easily be proved that

$$\left[\widehat{L}_3,\widehat{L}_\pm\right] = \pm\hbar\widehat{L}_\pm,$$
$$\left[\widehat{L}_+,\widehat{L}_-\right] = 2\hbar\widehat{L}_3. \tag{7.1.74}$$

From Eq. (7.1.74),

$$\widehat{L}_-\widehat{L}_+ = \widehat{L}^2 - \widehat{L}_3^2 - \hbar\widehat{L}_3,$$
$$\widehat{L}_+\widehat{L}_- = \widehat{L}^2 - \widehat{L}_3^2 + \hbar\widehat{L}_3. \tag{7.1.75}$$

By computing the expectation value of the previous operators on the state $\psi_{l,m}$,

$$0 \leq ||\widehat{L}_+\psi_{l,m}||^2 = (\widehat{L}_+\psi_{l,m},\widehat{L}_+\psi_{l,m}) = (\psi_{l,m},\widehat{L}_-\widehat{L}_+\psi_{l,m})$$

$$= (\psi_{l,m},\left[\widehat{L}^2 - \widehat{L}_3^2 - \hbar\widehat{L}_3\right]\psi_{l,m}) = \hbar^2[l(l+1) - m(m+1)], \tag{7.1.76}$$

and analogously

$$0 \leq ||\widehat{L}_-\psi_{l,m}||^2 = (\widehat{L}_-\psi_{l,m}, \widehat{L}_-\psi_{l,m}) = (\psi_{l,m}, \widehat{L}_+\widehat{L}_-\psi_{l,m})$$

$$= (\psi_{l,m}, \left[\widehat{L}^2 - \widehat{L}_3^2 + \hbar\widehat{L}_3\right]\psi_{l,m}) = \hbar^2[l(l+1) - m(m-1)]. \tag{7.1.77}$$

This means that we must satisfy at the same time the inequalities

$$\begin{cases} l(l+1) - m(m+1) \geq 0, \\ l(l+1) - m(m-1) \geq 0, \end{cases} \tag{7.1.78}$$

which are fulfilled for $-l \leq m \leq l$.

The raising and lowering operators have another important property, which can be found by applying them to the general eigenvector $\psi_{l,m}$. In this case in fact, by using the first of Eqs. (7.1.74) we have

$$\widehat{L}_3\widehat{L}_+\psi_{l,m} = \widehat{L}_+\widehat{L}_3\psi_{l,m} + \hbar\widehat{L}_+\psi_{l,m} = \hbar(m+1)\widehat{L}_+\psi_{l,m}, \tag{7.1.79}$$

$$\widehat{L}_3\widehat{L}_-\psi_{l,m} = \widehat{L}_-\widehat{L}_3\psi_{l,m} - \hbar\widehat{L}_-\psi_{l,m} = \hbar(m-1)\widehat{L}_-\psi_{l,m}. \tag{7.1.80}$$

This means that

$$\widehat{L}_\pm\psi_{l,m} = N_\pm\psi_{l,m\pm1}. \tag{7.1.81}$$

The normalization constants $N_\pm$ can be fixed by comparing the above expressions with Eqs. (7.1.76) and (7.1.77), thus obtaining

$$N_\pm = \hbar\sqrt{l(l+1) - m(m\pm1)}, \tag{7.1.82}$$

up to a phase. However, the relation Eq. (7.1.81) leads to a paradox. In fact, starting from any $\psi_{l,m}$ and acting sufficiently many times with $\widehat{L}_+$ the necessary condition $m \leq l$ might easily be overcome. This would occur unless, by iterating the application of $\widehat{L}_+$, the conditions to have a vanishing N_+ are not reached exactly, which would prevent a further increase of m. Of course, the analogous result should hold for $-l \leq m$ and the iterative application of $\widehat{L}_-$. These considerations imply that, for a given m, there should be an integer number q such that $m + q = l$, which is the condition to have a vanishing $\psi_{l,m+q+1}$. At the same time, there should be a negative integer $-p$ such that $m - p = -l$, which means a vanishing $\psi_{l,m-p-1}$. To sum up, we have for any m and l

$$l = m + q, \tag{7.1.83}$$

$$-l = m - p, \tag{7.1.84}$$

with p and q positive integers or vanishing numbers. From this it follows that $l = (q+p)/2$, i.e. the eigenvalue of $\widehat{L}^2$ can be any integer or half-odd positive number (including 0). Correspondingly, for integer or half-odd l, we have m, assuming all integer or half-odd values in the interval $[-l, l]$, respectively.

7.1.4 Radial part of the eigenvalue problem in a central potential

We are now ready to resume our starting Hamiltonian operator and the associated eigenvalue equation for a particle of mass m in a central potential. In Eq. (7.1.19) it is now convenient to set

$$R(r) \equiv \frac{u(r)}{r}. \tag{7.1.85}$$

The function R is square-integrable with respect to the measure $r^2 dr$, whereas the function u is square-integrable with respect to the measure dr. Hence we obtain the eigenvalue equation (where E denotes the desired eigenvalues)

$$\left\{ \frac{d^2}{dr^2} + \frac{2m}{\hbar^2} \left[E - V(r) - \frac{l(l+1)\hbar^2}{2mr^2} \right] \right\} u = 0. \tag{7.1.86}$$

Equation (7.1.86) is supplemented by the $\mathcal{L}^2$ condition

$$\int_0^\infty |u(r)|^2 dr < \infty, \tag{7.1.87}$$

and the boundary condition ensuring essential self-adjointness of the Hamiltonian operator:

$$u(0) = 0. \tag{7.1.88}$$

In other words, Eq. (7.1.86) holds in the sense of differential operators on $\mathcal{L}^2(0, \infty)$ with the boundary condition (7.1.88), rather than in the sense of classical differential equations. If the potential V is C^∞ with compact support, a theorem ensures that the solutions of Eq. (7.1.86) are C^∞ (this goes under the name of the *elliptic regularity* theorem). In the applications, we will see that regularity of u, for given values of l and E, is specified, more precisely, by the conditions

$$u_l(0, E) = 0, \tag{7.1.89}$$

$$\lim_{r \to 0} r^{-l-1} u_l(r, E) = 1. \tag{7.1.90}$$

The definition (7.1.85) is a particular case of a more general setting. Indeed, in q space dimensions, the radial part $\chi_{nl}(r)$ of the stationary states obeys the equation (cf. (7.1.86))

$$\left[\frac{d^2}{dr^2} + \frac{(q-1)}{r} \frac{d}{dr} - \frac{l(l+q-2)}{r^2} + \frac{2mE}{\hbar^2} \right] \chi_{nl}(r) = \frac{2m}{\hbar^2} V(r) \chi_{nl}(r). \tag{7.1.91}$$

At this stage, if we set (cf. Eq. (7.1.85))

$$\chi_{nl}(r) = r^\alpha y_{nl}(r), \tag{7.1.92}$$

we can choose α in such a way that the coefficient multiplying $\frac{dy_{nl}}{dr}$ is vanishing. The desired α is found from an algebraic equation to take the value

$$\alpha = -\frac{(q-1)}{2}, \tag{7.1.93}$$

and, on defining

$$\lambda \equiv l + \frac{(q-2)}{2}, \tag{7.1.94}$$

we find

$$\left[\frac{d^2}{dr^2} + \frac{2m}{\hbar^2} (E - V(r)) - \frac{\left(\lambda^2 - \frac{1}{4}\right)}{r^2} \right] y_{nl}(r) = 0. \tag{7.1.95}$$

Interestingly, the radial part of the stationary states therefore obeys a second-order ordinary differential equation *having the same form in all dimensions. The dimension only affects the value taken by the parameter* λ defined in Eq. (7.1.94).

7.2 Hydrogen atom

The investigation of the hydrogen atom is a two-body problem in quantum mechanics. Before quantization, we have a classical Hamiltonian

$$H = \frac{p_n^2}{2m_n} + \frac{p_e^2}{2m_e} - \frac{q_e^2}{|\vec{x}_n - \vec{x}_e|}, \tag{7.2.1}$$

where the subscripts 'n' and 'e' refer to the nucleus and the electron, respectively. What we are really interested in is the relative motion of the electron and the nucleus. For this purpose, we pass to new variables according to the standard formulae for 2-body problems:

$$\vec{P} \equiv \vec{p}_n + \vec{p}_e, \quad \vec{X} \equiv \frac{\left(m_e \vec{x}_e + m_n \vec{x}_n\right)}{(m_e + m_n)}, \tag{7.2.2}$$

$$\vec{p}_{en} \equiv \frac{\left(m_n \vec{p}_e - m_e \vec{p}_n\right)}{(m_e + m_n)}, \quad \vec{r} \equiv \vec{x}_e - \vec{x}_n. \tag{7.2.3}$$

These formulae are obtained by requiring that a canonical transformation relates $(\vec{x}_e, \vec{x}_n, \vec{p}_e, \vec{p}_n)$ to $(\vec{r}, \vec{X}, \vec{p}_{en}, \vec{P})$. For simplicity, these formulae can be derived in one space dimension, starting from two generic coordinates (x_1, x_2) and their conjugate momenta (p_1, p_2). We can thus write

$$X = \frac{m_1 x_1 + m_2 x_2}{(m_1 + m_2)}, \quad P_X = p_1 + p_2, \tag{7.2.4}$$

$$r = x_1 - x_2, \quad p = A p_1 + B p_2, \tag{7.2.5}$$

with coefficients A and B to be determined from the requirement that the change of coordinates should be a canonical transformation, hence preserving the Poisson-bracket structure. Then,

$$\{r, X\} = \{p, P_X\} = \{r, P_X\} = 0, \tag{7.2.6}$$

while

$$\{X, P_X\} = 1, \tag{7.2.7}$$

$$\{X, p\} = \frac{m_1 A + m_2 B}{(m_1 + m_2)} = 0, \tag{7.2.8}$$

$$\{r, p\} = A - B = 1. \tag{7.2.9}$$

Equations (7.2.8) and (7.2.9) are solved by

$$A = \frac{m_2}{(m_1 + m_2)}, \quad B = -\frac{m_1}{(m_1 + m_2)}, \tag{7.2.10}$$

and hence

$$p = \frac{m_2 p_1 - m_1 p_2}{(m_1 + m_2)}. \tag{7.2.11}$$

On reverting now to the three-dimensional case, the resulting form of the Hamiltonian reads as

$$H = \frac{P^2}{2M} + \frac{p_{en}^2}{2\mu} - \frac{q_e^2}{r}, \tag{7.2.12}$$

where $M \equiv m_e + m_n$ and μ is the reduced mass

$$\mu \equiv \frac{m_e m_n}{(m_e + m_n)}. \tag{7.2.13}$$

At this stage, we quantize in the coordinate representation, and factorize the stationary state as

$$\psi(\vec{X}, r) = \Omega(\vec{X})\psi(\vec{r}). \tag{7.2.14}$$

The centre-of-mass degrees of freedom represent a free particle and are encoded into $\Omega(\vec{X})$. The motion of the electron with respect to the nucleus is instead associated to $\psi(\vec{r})$. In the frame where the centre of mass is at rest, $\psi(\vec{r})$ obeys a stationary Schrödinger equation in a central field, which can be written as in Eq. (7.1.86). In this equation, with m replaced by μ, we look for solutions in the form

$$u(r) = e^{-wr} f(r), \tag{7.2.15}$$

where $w^2 \equiv -\varepsilon \equiv \frac{2\mu|E|}{\hbar^2}$, because we are studying bound states, for which the allowed energies here are negative (as in the classical case). Of course, one only takes into account a decreasing exponential, to ensure that $u \in \mathcal{L}^2(0, \infty)$. At this stage, we write

$$f(r) = r^s \sum_{k=0}^{\infty} a_k r^k, \tag{7.2.16}$$

where the parameter s is derived by imposing regularity at the origin. Differentiation of u with respect to r yields

$$u'' = \left(f'' - 2wf' + w^2 f \right) e^{-wr}, \tag{7.2.17}$$

and hence

$$a_0 \left[s(s-1) - l(l+1) \right] r^{s-2}$$

$$+ \sum_{k=0}^{\infty} a_{k+1} \left[(k+s+1)(k+s) - l(l+1) \right] r^{k+s-1}$$

$$+ \sum_{k=0}^{\infty} a_k \left[\frac{2\mu q_e^2}{\hbar^2} - 2w(k+s) \right] r^{k+s-1} = 0. \tag{7.2.18}$$

The idea is, of course, to use the property according to which, if

$$\sum_{k=0}^{\infty} b_k r^k = 0 \quad \forall r,$$

then $b_k = 0$ $\forall k$. The coefficient a_0 should not vanish, and hence what is multiplied by a_0 must be set to zero, i.e.

$$s(s-1) - l(l+1) = 0. \tag{7.2.19}$$

Among the two roots of Eq. (7.2.19): $s_1 = l+1, s_2 = -l$; only s_1 is compatible with having a regular solution at $r = 0$.

From the remaining part of Eq. (7.2.18), we find recurrence relations among a_{k+1} and a_k. If the algorithm could be implemented for all integer values of k, the series expressing $f(r)$ would be divergent because, at large k, the ratio $\frac{a_{k+1}}{a_k}$ would be proportional to $\frac{1}{k}$. This would lead, in turn, to a solution $u(r)$ that is not square-integrable on $(0, \infty)$. Thus, it is found that a value k^* of k exists such that

$$a_{k^*} \neq 0, \tag{7.2.20}$$

but

$$a_{k^*+1} = 0. \tag{7.2.21}$$

In other words, the series in Eq. (7.2.16) actually reduces to a polynomial, and (recall that $s_1 = l+1$)

$$\frac{\mu q_e^2}{\hbar^2} - w\left(k^* + s_1\right) = 0. \tag{7.2.22}$$

The comparison with the definition of w leads to

$$\varepsilon = -\frac{\mu q_e^4}{2\hbar^2} \frac{1}{n^2}, \tag{7.2.23}$$

which is the Balmer formula for bound states of the hydrogen atom (we have set $n \equiv k^* + s_1 \equiv k^* + l + 1$). The number n is called the principal quantum number, and is ≥ 1. For a given value of n, there exist n^2 linearly independent states with the *same* energy. The quantum number l takes all integer values from 0 to $n - 1$, while m ranges over all values from $-l$ to $+l$, including 0. For example, the values $l = 0, 1, 2$ correspond to the so-called s-, p- and d-wave sectors, respectively (sharp, principal, diffuse). From the allowed ranges, it is clear that there is one $1s$ state ($n = 1, l = m = 0$), one $2s$ state ($n = 2, l = m = 0$), three $2p$ states ($n = 2, l = 1, m = -1, 0, 1$), one $3s$ state ($n = 3, l = m = 0$), three $3p$ states ($n = 3, l = 1, m = -1, 0, 1$), five $3d$ states ($n = 3, l = 2, m = -2, -1, 0, 1, 2$), and so on.

For the first few values of the quantum numbers n, l, for example, for hydrogen-like atoms (see Eq. (7.1.85), bearing in mind that Z is the atomic number), and defining the Bohr radius

$$a_B \equiv \frac{\hbar^2}{m_e q_e^2} \approx 5.28 \cdot 10^{-2} \text{ nm}, \tag{7.2.24}$$

and the linear function

$$\rho_n(r) \equiv \frac{2Zr}{na_B}, \tag{7.2.25}$$

the following radial parts of stationary states:

$$R_{10}(r) = (Z/a_B)^{\frac{3}{2}} 2 e^{-\frac{1}{2}\rho_1}, \tag{7.2.26}$$

$$R_{20}(r) = (Z/a_B)^{\frac{3}{2}} \frac{1}{2\sqrt{2}} (2 - \rho_2) e^{-\frac{1}{2}\rho_2}, \tag{7.2.27}$$

$$R_{21}(r) = (Z/a_B)^{\frac{3}{3}} \frac{1}{2\sqrt{6}} \rho_2 e^{-\frac{1}{2}\rho_2}, \tag{7.2.28}$$

$$R_{30}(r) = (Z/a_B)^{\frac{3}{2}} \frac{1}{9\sqrt{3}} \left(6 - 6\rho_3 + \rho_3^2\right) e^{-\frac{1}{2}\rho_3}, \tag{7.2.29}$$

$$R_{31}(r) = (Z/a_B)^{\frac{3}{2}} \frac{1}{9\sqrt{6}} (4 - \rho_3) \rho_3 e^{-\frac{1}{2}\rho_3}, \tag{7.2.30}$$

$$R_{32}(r) = (Z/a_B)^{\frac{3}{2}} \frac{1}{9\sqrt{30}} \rho_3^2 e^{-\frac{1}{2}\rho_3}. \tag{7.2.31}$$

The occurrence of factors which decrease exponentially implies that the probability of finding the electron at large distances becomes very small indeed. Here, very large means a few times the Bohr radius.

7.2.1 Runge–Lenz vector

The Hamiltonian of the hydrogen atom has a large symmetry algebra. In addition to the conserved angular momentum: $[H, \vec{L}] = 0$, there is also the Runge–Lenz vector $\vec{R}$, i.e. the Hermitian operator

$$\vec{R} \equiv \frac{1}{2m_e} \left(\vec{p} \times \vec{L} - \vec{L} \times \vec{p}\right) - \frac{q_e^2}{r} \vec{r}. \tag{7.2.32}$$

Thus,

$$R_k = \frac{1}{m_e} \left(x_k p_l p_l - x_l p_l p_k + i\hbar p_k\right) - q_e^2 \frac{x_k}{r}, \tag{7.2.33}$$

where we have used the identities

$$\left(\vec{A} \times \vec{B}\right)_k = \varepsilon_{kij} A_i B_j, \tag{7.2.34}$$

$$\varepsilon_{ijk} \varepsilon_{kln} = \delta_{il} \delta_{jn} - \delta_{in} \delta_{jl}, \tag{7.2.35}$$

and the canonical commutation relations

$$x_k p_l - p_l x_k = i\hbar \delta_{kl}, \tag{7.2.36}$$

with $x_1 = x, x_2 = y, x_3 = z$. The explicit calculation shows that $[H, \vec{R}] = 0$. Note that, in classical mechanics, the electron would follow an elliptic orbit with the origin at one focus, and the vector $\vec{R}$ would point from the origin to the nearer vertex of the ellipse. Its length would be proportional to the eccentricity of the orbit.

Other useful identities involving the Runge–Lenz vector are as follows (hereafter, we set $\hbar = 1$):

$$\vec{R} \cdot \vec{L} = \vec{L} \cdot \vec{R} = 0, \tag{7.2.37}$$

$$\left[L_j, R_k \right] = i\varepsilon_{jkl}R_l, \tag{7.2.38}$$

$$|\vec{R}|^2 = q_e^4 I + \frac{2}{m_e}H\left(|\vec{L}|^2 + I\right), \tag{7.2.39}$$

$$\left[R_j, R_k \right] = -\frac{2i}{m_e}H\varepsilon_{jkl}L_l. \tag{7.2.40}$$

Now, denoting by $D \subset \mathcal{L}^2(\mathbf{R}^3)$ the span of the eigenvectors of H having negative eigenvalues, an Hermitian operator $\vec{K}$ can be defined with domain D by setting

$$\vec{K} \equiv \sqrt{-\frac{m_e}{2H}}\,\vec{R}. \tag{7.2.41}$$

At this stage, we can also define the operators

$$\vec{M} \equiv \frac{1}{2}(\vec{L} + \vec{K}), \quad \vec{N} \equiv \frac{1}{2}(\vec{L} - \vec{K}), \tag{7.2.42}$$

which obey the commutation relations

$$\left[M_j, M_k \right] = i\varepsilon_{jkl}M_l, \tag{7.2.43}$$

$$\left[N_j, N_k \right] = i\varepsilon_{jkl}N_l, \tag{7.2.44}$$

$$\left[M_j, N_k \right] = 0, \tag{7.2.45}$$

and the identities

$$|\vec{M}|^2 = |\vec{N}|^2 = \frac{1}{4}\left(|\vec{L}|^2 + |\vec{K}|^2\right), \tag{7.2.46}$$

$$H = -\frac{m_e q_e^4}{2(4|\vec{M}|^2 + I)}. \tag{7.2.47}$$

This means that the operators $\vec{M}$ and $\vec{N}$ acting in D generate a representation of the Lie algebra $su(2) \times su(2)$. Suppose now we decompose D as the direct sum of irreducible representations of this algebra. The operators $|\vec{M}|^2$ and $|\vec{N}|^2$ commute with the generators:

$$[M^2, M_k] = [M^2, N_k] = [N^2, N_k] = [N^2, M_k] = 0, \tag{7.2.48}$$

and are hence constant on each irreducible representation, by virtue of the Schur lemma. Thus, H is also constant on each irreducible representation, so these are the eigenspaces of H. Only those irreducible representations having $|\vec{M}|^2 = |\vec{N}|^2$ occur, and the value of each of these operators is $j(j+1)$ in the $(2j+1)$-dimensional irreducible representation of $su(2)$ having highest weight $j = 0, \frac{1}{2}, 1, \frac{3}{2}, \ldots$.

7.3 s-Wave bound states in the square-well potential

The central potential

$$V(r) = -V_0 \text{ if } r < a, \ 0 \text{ if } r > a, \tag{7.3.1}$$

can be used to approximate the neutron–proton interaction in nuclear physics. Here, we are aiming to find under which conditions on V_0 it is possible to find s-wave bound states. Recall from Section 7.1 that, for a central potential in $\mathbf{R}^3$, bound states, if they exist, solve the problem expressed by the differential equation (here $\lambda \equiv l + \frac{1}{2}$)

$$\left[\frac{d^2}{dr^2} + \frac{2m}{\hbar^2}(E - V(r)) - \frac{\left(\lambda^2 - \frac{1}{4}\right)}{r^2} \right] y_{nl}(r) = 0, \tag{7.3.2}$$

jointly with regularity at the origin:

$$\lim_{r \to 0} r^{-l-1} y_{nl}(r; E) = \text{constant} \implies y_{nl}(r = 0; E) = 0, \tag{7.3.3}$$

and square-integrability on the positive half-line $\mathbf{R}_+$:

$$\int_0^\infty |y_{nl}(r)|^2 dr < \infty. \tag{7.3.4}$$

In our case, Eqs. (7.3.3) and (7.3.4) imply that s-wave bound states, if they exist, must have $E < 0$, to avoid violating the condition (7.3.4). On defining

$$k_1 \equiv \frac{\sqrt{2m(V_0 - |E|)}}{\hbar}, \ k_2 \equiv \frac{\sqrt{2m|E|}}{\hbar}, \tag{7.3.5}$$

we therefore have

$$y_{n0}(r) = y_{n0}^I(r) \text{ if } r < a, \ y_{n0}^{II}(r) \text{ if } r > a, \tag{7.3.6}$$

where

$$\left(\frac{d^2}{dr^2} + k_1^2 \right) y_{n0}^I(r) = 0, \text{ if } r \in]0, a[, \tag{7.3.7}$$

$$\left(\frac{d^2}{dr^2} - k_2^2 \right) y_{n0}^{II}(r) = 0, \text{ if } r > a, \tag{7.3.8}$$

subject to the continuity conditions

$$\lim_{r \to a^-} y_{n0}(r) = \lim_{r \to a^+} y_{n0}(r), \tag{7.3.9}$$

$$\lim_{r \to a^-} y_{n0}'(r) = \lim_{r \to a^+} y_{n0}'(r). \tag{7.3.10}$$

Regularity at the origin according to Eq. (7.3.3) implies that

$$y_{n0}^I(r) = A(E) \sin k_1 r, \tag{7.3.11}$$

while square-integrability according to Eq. (7.3.4) yields

$$y_{n0}^{II}(r) = B(E) e^{-k_2 r}. \tag{7.3.12}$$

The continuity conditions (7.3.9) and (7.3.10) lead therefore to the linear homogeneous system

$$A \sin k_1 a - B e^{-k_2 a} = 0, \tag{7.3.13}$$

$$A k_1 \cos k_1 a + B k_2 e^{-k_2 a} = 0. \tag{7.3.14}$$

To find non-trivial solutions of Eqs. (7.3.13) and (7.3.14) we have therefore to set to zero the resulting determinant of the matrix of coefficients, which engenders the equation

$$k_1 a \cot k_1 a = -k_2 a. \tag{7.3.15}$$

Now Eq. (7.3.15) suggests defining the dimensionless variables

$$\xi \equiv k_1 a, \quad \omega \equiv k_2 a, \tag{7.3.16}$$

so that Eqs. (7.3.5) and (7.3.15) lead to the system

$$\xi^2 + \omega^2 = \frac{2m V_0 a^2}{\hbar^2}, \tag{7.3.17}$$

$$\xi \cot \xi = -\omega. \tag{7.3.18}$$

Equations (7.3.17) and (7.3.18) yield, implicitly, the desired energy eigenvalues via graphical and numerical methods, but the conditions upon V_0 can be worked out by squaring up Eq. (7.3.18) and then exploiting Eq. (7.3.17), i.e.

$$\xi^2 \cot^2 \xi = \omega^2 = \frac{2m V_0 a^2}{\hbar^2} - \xi^2. \tag{7.3.19}$$

We are therefore studying whether the graphs of the functions

$$f_1 : \xi \to f_1(\xi) \equiv \xi^2 (1 + \cot^2 \xi) = \frac{\xi^2}{(\sin \xi)^2}, \tag{7.3.20}$$

$$f_2 : \xi \to f_2(\xi) \equiv \frac{2m V_0 a^2}{\hbar^2}, \tag{7.3.21}$$

intersect at one or more points. Indeed, f_2 describes a 1-parameter family of lines (V_0 being the parameter), all parallel to the ξ-axis, while f_1, studied in the $\left[\frac{\pi}{2}, \frac{5}{2}\pi \right]$ interval, is a function having an absolute minimum at $\xi = \frac{\pi}{2}$, where $f_1\left(\frac{\pi}{2}\right) = \frac{\pi^2}{4}$, vertical asymptotes at $\xi = \pi$ and $\xi = 2\pi$, and a relative minimum at $\xi = y = 4.49341$ such that

$$y = \tan y, \tag{7.3.22}$$

since

$$f_1'(\xi) = \frac{2\xi}{(\sin \xi)^3} (\sin \xi - \xi \cos \xi), \tag{7.3.23}$$

$$f_1''(\xi) = \frac{2}{(\sin \xi)^2} \left[1 + \xi^2 + 3\xi^2 \cot^2 \xi - 4\xi \cot \xi \right], \tag{7.3.24}$$

and hence

$$f_1''(y) = \frac{2y^2}{(\sin y)^2} > 0. \tag{7.3.25}$$

To find at least one bound state we have therefore to impose that

$$f_2(\xi) = \frac{2mV_0a^2}{\hbar^2} \geq f_1\left(\frac{\pi}{2}\right). \tag{7.3.26}$$

Moreover, there will be exactly one bound state provided that Eq. (7.3.26) holds jointly with

$$f_2(\xi) = \frac{2mV_0a^2}{\hbar^2} < f_1(y) < f_1\left(\frac{3}{2}\pi\right). \tag{7.3.27}$$

The above conditions lead eventually to

$$V_0 \geq \frac{\pi^2}{8}\frac{\hbar^2}{ma^2}, \tag{7.3.28}$$

and

$$\frac{\pi^2}{8}\frac{\hbar^2}{ma^2} < V_0 < \frac{1}{2(\cos y)^2}\frac{\hbar^2}{ma^2} < \frac{9}{8}\pi^2\frac{\hbar^2}{ma^2}, \tag{7.3.29}$$

respectively. Unfortunately, some treatments in the literature are over-simplified and identify our y with $\frac{3}{2}\pi$, which is incorrect.

7.4 Isotropic harmonic oscillator in three dimensions

As a further example of problem in a central potential studied through elementary power-series methods, we now consider the isotropic harmonic oscillator in three dimensions. The angular part of the stationary states is given by the spherical harmonics as we just mentioned, while Eq. (7.1.95) yields in this case, passing to dimensionless variables

$$\alpha \equiv \sqrt{\frac{m\omega}{\hbar}}, \; \xi \equiv \alpha r, \tag{7.4.1}$$

the differential equation

$$\left[\frac{d^2}{d\xi^2} - \frac{l(l+1)}{\xi^2} + \varepsilon_n - \xi^2\right]y_{nl} = 0, \tag{7.4.2}$$

where $\varepsilon = 2n + 3$, as can be proved by a method entirely analogous to the hydrogen atom case (see below). Indeed, regularity at the origin and square-integrability on the positive half-line lead to the ansatz

$$y_{nl} = e^{-\frac{\xi^2}{2}}\xi^\rho \sum_{p=0}^{\infty} a_p\xi^p, \tag{7.4.3}$$

and its substitution into the differential equation for y_{nl} yields

$$\sum_{p=0}^{\infty}\left\{\left[(p+\rho)(p+\rho-1) - l(l+1)\right]a_p\xi^{p+\rho-2}\right.$$
$$\left. + \left[\varepsilon_n - 1 - 2(p+\rho)\right]a_p\xi^{p+\rho}\right\} = 0. \tag{7.4.4}$$

This infinite sum contains only two negative powers of ξ, and hence it is convenient to re-express it in the form

$$\frac{a_0}{\xi^2}[\rho(\rho - 1) - l(l+1)] + \frac{a_1}{\xi}[(\rho+1)\rho - l(l+1)]$$

$$+ \sum_{p=0}^{\infty} \left\{ \left[\varepsilon_n - 1 - 2(p+\rho) \right] a_p \right.$$

$$+ \left. \left[(p+\rho+2)(p+\rho+1) - l(l+1) \right] a_{p+2} \right\} \xi^p = 0. \tag{7.4.5}$$

Here we should set to zero the coefficients of all powers of ξ. Indeed, if $a_0 \neq 0$, the coefficient of ξ^{-2} vanishes if and only if

$$\rho^2 - \rho - l(l+1) = 0, \tag{7.4.6}$$

and the root of this equation compatible with regularity at the origin for y_{nl} is

$$\rho_+ = l+1. \tag{7.4.7}$$

Moreover, if $a_1 \neq 0$, the coefficient of ξ^{-1} vanishes if and only if

$$\rho^2 + \rho - l(l+1) = 0, \tag{7.4.8}$$

and the root of this equation leading to a form of y_{nl} regular at the origin is

$$\widetilde{\rho}_+ = l. \tag{7.4.9}$$

Last, but not least, the coefficients of positive powers of ξ vanish if and only if the following recursion relation holds:

$$a_{p+2} = \frac{(1 + 2(p+\rho) - \varepsilon_n)}{[(p+\rho+2)(p+\rho+1) - l(l+1)]} a_p. \tag{7.4.10}$$

At large p, the right-hand side of this formula tends to $\frac{2}{p} a_p$, which implies that the ratio $\frac{a_{p+2}}{a_p}$ tends to zero too slowly for the series to be convergent. This implies that a particular value $\overline{p}$ exists such that

$$a_{\overline{p}} \neq 0, \quad a_{\overline{p}+2} = 0, \tag{7.4.11}$$

so that the series reduces to a polynomial. The corresponding numerator in the previous equation should vanish, i.e.

$$1 + 2(\overline{p} + \rho) - 2n - 3 = 0. \tag{7.4.12}$$

At this stage, we insert $\rho = \rho_+ = l+1$ and find

$$l = n - \overline{p}, \quad \overline{p} = 0, 2, 4, \ldots . \tag{7.4.13}$$

Similarly, insertion of $\rho = \widetilde{\rho}_+ = l$ yields instead

$$l = n - (\overline{p} - 1), \quad \overline{p} = 1, 3, 5, \ldots . \tag{7.4.14}$$

To sum up, the quantum number l is found to take the values

$$l = n, n - 2, \ldots, 0 \text{ if } n \text{ is even}, \tag{7.4.15}$$

$$l = n, n - 2, \ldots, 1 \text{ if } n \text{ is odd}, \tag{7.4.16}$$

which is a peculiar property of the isotropic harmonic oscillator in three dimensions.

7.5 Multi-dimensional harmonic oscillator: algebraic treatment

Let us consider an harmonic oscillator in D dimensions ($\vec{x} \equiv (x_1, x_2, \ldots, x_D)$). In this case, the potential reads as

$$U(x_1, \ldots, x_D) = \frac{m}{2} \sum_{i=1}^{D} \omega_i^2 x_i^2. \tag{7.5.1}$$

The results obtained in Section 6.4 allow for a straightforward generalization. In fact, in this case the corresponding eigenvalue equation for the Hamiltonian operator is

$$\sum_{i=1}^{D} \left[-\frac{\hbar^2}{2m} \left(\frac{\partial^2}{\partial x_i^2} \right) + \frac{m}{2} \omega_i^2 x_i^2 \right] \psi_E = E \psi_E. \tag{7.5.2}$$

Such an equation can be solved by separation of variables, i.e. by writing

$$\psi_E(\vec{x}) = \prod_i \psi_{n_i}(x_i), \tag{7.5.3}$$

where, $\forall i = 1, \ldots, D$, ψ_{n_i} are eigenfunctions of a one-dimensional harmonic oscillator. One thus obtains in general

$$E = \sum_{i=1}^{D} \hbar \omega_i \left(n_i + \frac{1}{2} \right). \tag{7.5.4}$$

In particular, if all frequencies coincide (isotropic harmonic oscillator): $\omega_i = \omega \; \forall i$,

$$E = \hbar \omega \left(n + \frac{D}{2} \right), \tag{7.5.5}$$

where $n = \sum_{i=1}^{D} n_i$. In this case the resulting energy levels are degenerate, and the degeneracy N_{deg} is the number of independent choices $n_1, n_2, \ldots, n_D$ satisfying the condition

$$\sum_{i=1}^{D} n_i = n. \tag{7.5.6}$$

This can be easily computed by imagining the separation of n identical white balls by $(D-1)$ black ones. In this case, for each configuration of black balls the set of white ones has been partitioned in D subsets corresponding to n_i. Hence it is obvious that

$$N_{\text{deg}} = \left(\begin{array}{c} n + D - 1 \\ D - 1 \end{array} \right), \tag{7.5.7}$$

which for $D = 2$ gives $N_{\deg} = n + 1$, and for $D = 3$ yields $N_{\deg} = (n + 1)(n + 2)/2$.

Moreover, some degeneracy may survive even if the frequencies are not equal. This means that there are at least two different choices $\{n_i\}$ and $\{n'_i\}$ producing the same level of energy E. Hence, subtracting the two different expressions (7.5.4),

$$\sum_{i=1}^{D} \omega_i \left(n_i - n'_i \right) = 0, \tag{7.5.8}$$

which implies a *rational* relation among the ω_i.

7.5.1 An example: two-dimensional isotropic harmonic oscillator

As discussed in the previous section, in two dimensions an isotropic harmonic oscillator has the Hamiltonian

$$H = H_x + H_y = \frac{p_x^2}{2m} + \frac{p_y^2}{2m} + \frac{1}{2}m\omega^2 x^2 + \frac{1}{2}m\omega^2 y^2, \tag{7.5.9}$$

where the resulting operators $\hat{H}_x$ and $\hat{H}_y$ obtained upon quantization mutually commute. For simplicity it is convenient to introduce the dimensionless Hamiltonian coordinates

$$Q_x = \sqrt{\frac{m\omega}{\hbar}}x, \qquad Q_y = \sqrt{\frac{m\omega}{\hbar}}y, \tag{7.5.10}$$

$$P_x = \frac{p_x}{\sqrt{m\omega\hbar}}, \qquad P_y = \frac{p_y}{\sqrt{m\omega\hbar}}, \tag{7.5.11}$$

which obey the algebra $[Q_x, Q_y] = [Q_x, P_y] = [P_x, Q_y] = [P_x, P_y] = 0$ and $[Q_x, P_x] = [Q_y, P_y] = iI$. In terms of such quantities the Hamiltonian (7.5.9) becomes

$$H = \frac{\hbar\omega}{2}\left(P_x^2 + P_y^2 + Q_x^2 + Q_y^2\right) = \hbar\omega\left(a_x^\dagger a_x + a_y^\dagger a_y + I\right), \tag{7.5.12}$$

with

$$a_x = \frac{Q_x + iP_x}{\sqrt{2}}, \qquad a_x^\dagger = \frac{Q_x - iP_x}{\sqrt{2}}, \tag{7.5.13}$$

and similarly for the y coordinate. The annihilation operators a_x, a_y and the creation ones obey the algebra $[a_x, a_y] = [a_x^\dagger, a_y^\dagger] = [a_x, a_y^\dagger] = [a_x^\dagger, a_y] = 0$ whereas $[a_x, a_x^\dagger] = [a_y, a_y^\dagger] = I$. The expressions $N_x \equiv a_x^\dagger a_x$ and $N_y \equiv a_y^\dagger a_y$ are commonly denoted as *number* operators.

As already discussed in the previous section a straightforward solution of the eigenvalue problem is provided by the expression (7.5.3) for the eigenvectors, where in this case $n_1 = n_x$ and $n_2 = n_y$ and the energy eigenvalues are

$$E = \hbar\omega(n_x + n_y + 1) = \hbar\omega(n + 1), \tag{7.5.14}$$

with $n = n_x + n_y$. By using the definition (7.5.13) and the expressions (7.5.10) and (7.5.11) the only component of angular momentum can easily be computed, i.e. $\hat{L}_z$ turns out to be given by

$$\hat{L}_z = i\hbar \left(a_y^\dagger a_x - a_x^\dagger a_y \right). \tag{7.5.15}$$

Such an operator commutes with the Hamiltonian $[\hat{H}, \hat{L}_z] = 0$; hence it is possible to find a basis of eigenfunctions for both operators, but this is not the two-dimensional version of (7.5.3), because on this function $\hat{L}_z$ is not diagonal. The simultaneous diagonalization of $\hat{H}$ and $\hat{L}_z$ is performed *via* a change of coordinate, i.e. by defining $x_+ \equiv (x - iy)/\sqrt{2}$ and $x_- \equiv (x + iy)/\sqrt{2}$. In correspondence of this new set of coordinates we can define two new pairs of annihilation and creation operators:

$$A_\pm \equiv \frac{1}{\sqrt{2}} \left(a_x \mp i a_y \right), \qquad A_\pm^\dagger \equiv \frac{1}{\sqrt{2}} \left(a_x^\dagger \pm i a_y^\dagger \right). \tag{7.5.16}$$

In terms of Eq. (7.5.16) it is possible to recast $\hat{H}$ and $\hat{L}_z$ in the following form:

$$\hat{H} = \hbar\omega \left(A_+^\dagger A_+ + A_-^\dagger A_- + I \right) = \hbar\omega \left(N_+ + N_- + I \right), \tag{7.5.17}$$

$$\hat{L}_z = \hbar \left(N_+ - N_- \right), \tag{7.5.18}$$

which are mutually diagonalized on the basis of eigenvectors

$$\psi_E(x_+, x_-) = \psi_{n_+}(x_+) \psi_{n_-}(x_-), \tag{7.5.19}$$

where the single factors are constructed by using the same approach as in the case of a one-dimensional harmonic oscillator. On this basis the eigenvalues of $\hat{H}$ and $\hat{L}_z$ are

$$E = \hbar\omega(n_+ + n_- + 1) = \hbar\omega(n + 1), \tag{7.5.20}$$

$$l_z = \hbar(n_+ - n_-) = \hbar m. \tag{7.5.21}$$

It is easy to see that once the value of E is fixed, i.e. $n = n_+ + n_-$, the $n+1$ different choices of pairs (n_+, n_-) corresponding to the same n yield different values of m. The simultaneous diagonalization of $\hat{H}$ and $\hat{L}_z$ for the two-dimensional isotropic harmonic oscillator removes completely the degeneracy in the eigenspaces. In particular one can prove that, for fixed n, the allowed values of m are $n, n - 2, \ldots, -n + 2, -n$.

The generalization of the above results to the three-dimensional isotropic harmonic oscillator is almost straightforward by using cylindrical coordinates, i.e. x_+, x_- and z and the corresponding Hamiltonians. However, since in this case $U(\vec{r}) = m\omega^2 r^2/2$ such a problem can be regarded as a particular case of central potential and hence treated as previously discussed.

7.6 Problems

7.P1. Two particles of masses m_1 and m_2 interact in three-dimensional space through the potential

$$V(\vec{r}_1, \vec{r}_2) = A|\vec{r}_1 - \vec{r}_2|^2 + \frac{B}{|\vec{r}_1 - \vec{r}_2|^2}, \qquad (7.6.1)$$

where A and B are positive. Find the allowed energy levels and their degeneracy.

7.P2. The Hamiltonian of a particle in two dimensions, which is bound to move on the strip $x \in [0, l]$, $y \in]-\infty, \infty[$, reads as

$$H = \frac{1}{2m}\left[(p_x + \alpha y)^2 + p_y^2\right] + V(y), \qquad (7.6.2)$$

where α is real and the potential V is linear in y. Find eigenvalues and eigenfunctions of H, and discuss their degeneracy when periodic boundary conditions are imposed.

7.P3. A particle moves freely in the plane with polar coordinates ρ, ϕ, and is described by the stationary state

$$\psi(\rho, \phi) = N\theta(\rho - \lambda)\left(\frac{\rho}{\lambda} + \cos\phi\right), \qquad (7.6.3)$$

where N is a normalization constant, λ is positive and θ is the Heaviside step-function. Find the possible values if angular momentum is measured in such a state, and their probabilities.

7.P4. A particle of mass m is in the stationary state

$$\psi(x, y, z) = Nre^{-\frac{r}{a}}, \qquad (7.6.4)$$

where r is the distance from the origin and a has dimension length.

(i) Find the mean quadratic deviations for the components of position and linear momentum.

(ii) Find the most probable values of r and of the kinetic energy.

7.P5. Consider the commutation relations obeyed by the angular momentum operators.

(i) If ψ is eigenvector of J_z, find the expectation value of J_x and J_y in such a state.

(ii) If A is an operator commuting with J_x and J_z, prove that it also commutes with J^2.

7.P6. A particle of mass m is in the stationary state

$$\psi(x, y, z) = (x + y + z)e^{-\alpha r}, \qquad (7.6.5)$$

where α has the dimension of inverse length. What are the possible values of a measurement of L^2 and L_z in such a state? Find their probabilities and the mean values of L^2 and L_z in such a state.

7.P7. A beam of particles possessing orbital angular momentum is subject to a pair of devices affected by a magnetic field. Assume that the directions of the magnetic field in the two devices form an angle α, and that the beam is originally propagating along the y-axis, while the magnetic field is along the z-axis in the first device. If the beam is split by the first device into three beams, the lowest of which enters the second device, how many beams are eventually seen and what are their relative intensities?

7.P8. A particle of mass m and charge q has velocity $\vec{v}$ and is subject to a magnetic field $\vec{B}$. Bearing in mind the classical Poisson brackets derived at the end of Section 3.2, compute the commutator among components of the orbital angular momentum $\vec{L}$. Moreover, if

$$\vec{B} = g\frac{\vec{r}}{r^3}, \tag{7.6.6}$$

which is the magnetic field of a monopole of magnetic charge g, prove that the operators

$$J_k \equiv L_k - qg\frac{x_k}{r} \tag{7.6.7}$$

obey the commutation relations of angular momentum operators.

7.P9. For a two-dimensional isotropic harmonic oscillator, find which operators among the ones listed below form a complete set of observables:

$$\hat{H} = \hbar\omega\left(\hat{a}_1^\dagger \hat{a}_1 + \hat{a}_2^\dagger \hat{a}_2 + I\right), \tag{7.6.8}$$

$$\hat{L} = -i\left(\hat{a}_2^\dagger \hat{a}_1 - \hat{a}_1^\dagger \hat{a}_2\right), \tag{7.6.9}$$

$$\hat{N}_{12} = \hat{a}_1^\dagger \hat{a}_2, \tag{7.6.10}$$

$$\hat{N}_1 = \hat{a}_1^\dagger \hat{a}_1, \tag{7.6.11}$$

$$\hat{L}_+ = \left(\hat{a}_1^\dagger + i\hat{a}_2^\dagger\right)(\hat{a}_1 + i\hat{a}_2). \tag{7.6.12}$$

For each complete set of observables classify the stationary states.

7.P10. In dimensionless variables, a two-dimensional isotropic harmonic oscillator is in the stationary state

$$\psi(x,y) = N \exp\left\{-\frac{(x^2 + y^2)}{2} + x + iy\right\}. \tag{7.6.13}$$

Find the probability distribution of energy levels and angular momentum.

7.P11. A two-dimensional isotropic harmonic oscillator is described, at a certain time, by the following wave function in polar coordinates:

$$\psi(\rho,\theta) = A(\rho^2 - a^2) \text{ if } \rho \leq a, \ 0 \text{ if } \rho > a. \tag{7.6.14}$$

Find the probability of obtaining, in the measurement of energy in such a state, the three lowest-energy eigenvalues, and the time-evolution of such a probability.

7.P12. A particle of mass m moves in a plane and is in the ground state of the Hamiltonian

$$H = \frac{p^2}{2m} + \frac{k}{2}(x^2 + y^2). \tag{7.6.15}$$

Suppose that we are able to obtain suddenly an elastic constant which is four times larger. Compute the probability of finding the particle in each of the three lowest levels of the new Hamiltonian.

7.P13. A three-dimensional isotropic harmonic oscillator is in an energy eigenstate with quantum numbers $n_x = 1, n_y = n_z = 0$. Is it possible to measure angular momentum without affecting the energy? In the affirmative case, compute the possible values of a measurement of L_z in such a state, and their probabilities.

Coherent states and related formalism

The theory of quantum harmonic oscillators is studied in greater detail; this chapter deals with, among other factors, the representation of states in the space of entire functions, coherent states and the Ehrenfest picture.

8.1 General considerations on harmonic oscillators and coherent states

In previous chapters we have often used the abstract setting of Hilbert spaces and operators to make general considerations or to derive general results. However, in concrete problems, to extract numerical results, we have introduced specific 'coordinate systems' in the Hilbert space, specifically we have considered realizations of the Hilbert space either in the position or the momentum representation.

When dealing with probability distributions we have already argued that an important tool to go from a coordinate system to another one is provided by a decomposition of the identity, in the form

$$\int |a\rangle \, \mathrm{d}a \langle a| = I,$$

or

$$\sum_n |\phi_n\rangle \langle \phi_n| = I$$

in terms of orthonormal bases, associated with continuous or discrete eigenvalues, respectively. Each partition of the identity is associated with the eigenvectors of a maximal set of pairwise commuting self-adjoint operators. Thus, position operators give rise to $\int |x\rangle \, \mathrm{d}x \langle x|$, momentum operators to $\int |p\rangle \, \mathrm{d}p \langle p|$, while L_z and L^2 make it possible to derive a partition of the identity in terms of spherical harmonics for the Hilbert space of square-integrable functions on the sphere.

By using partitions of the identity we introduce coordinate systems, e.g. we have

$$|\psi\rangle = \int |x\rangle \, \mathrm{d}x \langle x|\psi\rangle,$$

or

$$|\psi\rangle = \int |p\rangle \, \mathrm{d}p \langle p|\psi\rangle,$$

or also

$$|\psi\rangle = \sum_n |\phi_n\rangle\langle\phi_n|\psi\rangle,$$

where the components $\langle x|\psi\rangle$, $\langle p|\psi\rangle$ or $\langle\phi_n|\psi\rangle$ represent the coordinate system associated with each partition of the identity. They are also very useful to take us from one coordinate system to another.

We have already noticed that in both cases, position or momentum coordinate systems, we are obliged to introduce eigenvectors which are not associated with square-integrable functions, but only locally square-integrable. However, it is possible to introduce a resolution of the identity associated with a continuous basis, which identifies wave functions on phase space rather than as functions of the coordinates or the momenta, and moreover saturates the uncertainty relations, so as to be as close as possible to the classical setting. These particular states are called coherent states and they emerge in the study of the harmonic oscillator Hamiltonian, when it is factorized as a product of creation and annihilation operators, very much in the spirit of the complex coordinates we introduced in Chapter 3 for the classical harmonic oscillator. We saw therein how the introduction of complex coordinates makes it possible to introduce first-order differential equations in one complex variable, instead of a second-order differential equation in one real variable. We dealt with symmetries and constants of motion, describing the role that the unitary group $U(2)$ plays for the two-dimensional isotropic harmonic oscillator. We recall that, in the Heisenberg picture, we have found that the evolution of position and momentum operators is ruled by the equations

$$\begin{pmatrix} x_H(t) \\ p_H(t) \end{pmatrix} = \begin{pmatrix} \cos\omega t & \frac{\sin\omega t}{m\omega} \\ -m\omega\sin\omega t & \cos\omega t \end{pmatrix} \begin{pmatrix} x_0 \\ p_0 \end{pmatrix}.$$

This expression does not distinguish between classical and quantum evolution. Therefore, as far as linear transformations are concerned, we should expect that conclusions which hold in the classical setting will also have a counterpart at the quantum level. It is only when products of various quantities come into play that the non-commutative nature of quantum mechanics makes the difference. For instance, we expect also that the analogue of complex coordinates should play a relevant role also in the quantum setting as was anticipated briefly in Section 7.5.1. In this chapter we assess the consequences of these similarities at the level of states; by doing this we shall encounter coherent states, which were introduced by Schrödinger in 1926 as those states that describe a maximal kind of classical behaviour. In particular, they are states with minimum (balanced) uncertainty and such that their time evolution is concentrated along the classical trajectories. The term *coherent* originates from quantum optics (coherent radiation). As already remarked for the harmonic oscillator, coherent states arise in the quantum description of a wide range of physical systems, most notably in the quantum theory of light (quantum electrodynamics) and other bosonic quantum fields. Thus, in this chapter we shall study in greater detail the harmonic oscillator, focusing our attention on those aspects which play a relevant role for the description of coherent states. There have been a lot of review papers and books devoted to the subject, and we refer to them for further mathematical aspects and applications (Klauder and Skagerstam 1985, Perelomov 1986, Zhang *et al.* 1990, Syad *et al.* 2000, Gazeau 2009).

8.2 Quantum harmonic oscillator: a brief summary

Starting with the classical Hamiltonian function $H = \frac{p^2}{2m} + \frac{m}{2}\omega^2 q^2$, it is possible to redefine position and momentum according to

$$Q \equiv q\sqrt{m\omega},\ P \equiv \sqrt{\frac{p}{m\omega}},$$

so that the Hamiltonian function may be rewritten in the form

$$H = \frac{\omega}{2}(P^2 + Q^2),$$

in such a way that P and Q appear absolutely on the same footing, and therefore making the introduction of complex coordinates straightforward, leading to

$$H = \omega z z^*.$$

With this choice

$$z = \frac{1}{\sqrt{2}}(Q + iP),\ z^* = \frac{1}{\sqrt{2}}(Q - iP) \tag{8.2.1}$$

position Q and momentum P have dimension of square root of an action. We recall, from Chapter 3, that the evolution in time is described by

$$z(t) = e^{-i\omega t}z(0),\ z^*(t) = e^{i\omega t}z^*(0).$$

As we have already seen partially, the quantum harmonic oscillator can be described by means of the operator

$$\widehat{H} = \frac{\omega}{2}(\widehat{P}^2 + \widehat{Q}^2), \tag{8.2.2}$$

which is obtained from the classical Hamiltonian by using the multiplicative property of the symbol map. Note that $\widehat{P}$ and $\widehat{Q}$ have the dimension of $\sqrt{\hbar}$. It is possible to solve the associated equations of motion in the position, the momentum, the energy or number, and the phase-space or complex or Bargmann–Fock representations.

In the position representation, which amounts to a coordinate system associated with the (improper) eigenvectors $|x\rangle$ such that $Q|x\rangle = x|x\rangle$, each vector is replaced by its components $\langle x|\psi\rangle = \psi(x)$, and the Hamiltonian operator gets replaced by

$$\widehat{H} = \frac{\omega}{2}\left(-\hbar^2 \frac{d^2}{dQ^2} + Q^2\right). \tag{8.2.3}$$

In this coordinate description of the Hilbert space,

$$\widehat{Q}\psi(x,t) = x\psi(x,t),\ \widehat{P}\psi(x,t) = -i\hbar\frac{\partial}{\partial x}\psi(x,t). \tag{8.2.4}$$

Time evolution associated with $\widehat{H}$ is given by

$$\psi(x,t) = e^{-i\frac{\widehat{H}(t-t_0)}{\hbar}}\psi(x,t_0). \tag{8.2.5}$$

The abstract equation $\widehat{H}|\psi\rangle = E|\psi\rangle$ may be written in the position coordinate system, and the eigenvectors will be represented by eigenfunctions $\psi_n(x)$. Using of such eigenfunctions $\psi_n(x)$ we find a countable infinity of elementary solutions

$$\psi_n(x, t) = e^{-i\frac{E_n t}{\hbar}}\psi_n(x), \tag{8.2.6}$$

where the energy eigenvalues are equally spaced on the positive half-line, since

$$E_n = \hbar\omega\left(n + \frac{1}{2}\right), \quad n = 0, 1, 2, \ldots, \infty.$$

Eigenstates have been already derived in Chapter 6 and have the form (6.4.15). We now revert to the units used therein and note, as a spin-off, that $\sqrt{\frac{2\hbar}{m\omega}}$ has the dimension of a characteristic length to make the argument of the exponential dimensionless. If we introduce the specific characteristic length

$$\langle\psi_0|\widehat{Q}^2|\psi_0\rangle = \frac{\hbar}{2m\omega} = l_c^2, \tag{8.2.7}$$

it is indeed found that $\frac{x^2}{4l_c^2} = \frac{m\omega x^2}{2\hbar}$.

Because of our specific choice, in the momentum representation the situation is very similar, i.e.

$$\widehat{P}\psi(p, t) = p\psi(p, t), \quad \widehat{Q}\psi(p, t) = i\hbar\frac{\partial}{\partial p}\psi(p, t), \tag{8.2.8}$$

and the transformation from one representation to the other is given by the unitary transformation represented by the Fourier transform at each fixed time t, i.e.

$$\psi(p, t) = \frac{1}{\sqrt{2\pi\hbar}}\int_{-\infty}^{\infty} e^{-\frac{ipx}{\hbar}}\psi(x, t)dx, \tag{8.2.9}$$

$$\psi(x, t) = \frac{1}{\sqrt{2\pi\hbar}}\int_{-\infty}^{\infty} e^{\frac{ipx}{\hbar}}\psi(p, t)dp. \tag{8.2.10}$$

This unitary transformation takes us from eigenvectors of the position operators to eigenvectors of the momentum operators. This time, to form the argument of the exponential we introduce a characteristic momentum, so that

$$p_c^2 = \frac{\hbar m\omega}{2} = \langle\psi_0|\widehat{P}^2|\psi_0\rangle. \tag{8.2.11}$$

Eigenvectors of the Hamiltonian in this coordinate system have the form

$$\psi_n(p) = \left(\frac{1}{2\pi p_c^2}\right)^{\frac{1}{4}}\frac{1}{\sqrt{2^n n!}}e^{-\frac{p^2}{4p_c^2}}H_n\left(\frac{p}{\sqrt{2}p_c}\right), \tag{8.2.12}$$

where H_n denotes the Hermite polynomial.

In the energy or number representation we introduce the quantum analogue of z and z^* of Section 3.6 to obtain

$$\hat{a} = \frac{1}{\sqrt{2\hbar}}(\widehat{Q} + i\widehat{P}), \quad \hat{a}^\dagger = \frac{1}{\sqrt{2\hbar}}(\widehat{Q} - i\widehat{P}). \tag{8.2.13}$$

With this choice, $\hat{a}$ and $\hat{a}^\dagger$ are dimensionless. Therefore

$$\hat{a}\hat{a}^\dagger = \frac{1}{2}\frac{(\widehat{P^2} + \widehat{Q^2})}{\hbar} + \frac{1}{2}I, \hat{a}^\dagger\hat{a} = \frac{1}{2}\frac{(\widehat{P^2} + \widehat{Q^2})}{\hbar} - \frac{1}{2}I, \qquad (8.2.14)$$

$$\widehat{H} = \hbar\omega\left(\hat{a}\hat{a}^\dagger - \frac{1}{2}I\right) = \hbar\omega\left(\widehat{N} + \frac{1}{2}I\right), \qquad (8.2.15)$$

where we have introduced the number operator $\widehat{N} = \hat{a}^\dagger\hat{a}$, which is Hermitian, and, along with $\hat{a}, \hat{a}^\dagger, I$ closes on the complex Lie algebra

$$[\widehat{N}, \hat{a}] = -\hat{a}, \ [\widehat{N}, \hat{a}^\dagger] = \hat{a}^\dagger, \ [\hat{a}, \hat{a}^\dagger] = I. \qquad (8.2.16)$$

The energy basis or the number basis for the Hilbert space is defined by

$$\widehat{H}|\psi_n\rangle = \hbar\omega\left(n + \frac{1}{2}\right)|\psi_n\rangle, \qquad (8.2.17)$$

or, equivalently,

$$\widehat{N}|\psi_n\rangle = n|\psi_n\rangle, \qquad (8.2.18)$$

with the spectrum given by all positive integers. The equations of motion in the Heisenberg picture are given by

$$\frac{\mathrm{d}}{\mathrm{d}t}\hat{a} = \frac{\mathrm{i}}{\hbar}[\hat{H}, \hat{a}] = -\mathrm{i}\omega\hat{a}, \qquad (8.2.19)$$

$$\frac{\mathrm{d}}{\mathrm{d}t}\hat{a}^\dagger = \frac{\mathrm{i}}{\hbar}[\hat{H}, \hat{a}^\dagger] = \mathrm{i}\omega\hat{a}^\dagger, \qquad (8.2.20)$$

whose solutions are

$$\hat{a}(t) = \mathrm{e}^{-\mathrm{i}\omega t}\hat{a}_0, \ \hat{a}^\dagger(t) = \mathrm{e}^{\mathrm{i}\omega t}\hat{a}_0^\dagger. \qquad (8.2.21)$$

These equations of evolution have the same form as those we derived for the classical case by means of complex coordinates.

It is now possible to derive also that, for the isotropic quantum harmonic oscillator in higher dimensions, the quadratic expressions $\hat{a}_j^\dagger A^{jk}\hat{a}_k$ are constants of motion for any matrix A, while the constants of motion are real, i.e. represented by Hermitian operators, for all Hermitian matrices. Moreover, all complex linear transformations map solutions into solutions even if they are not unitary. We notice that, in the energy basis,

$$\hat{a}^\dagger|\psi_n\rangle = \sqrt{n+1}|\psi_{n+1}\rangle, \ \hat{a}|\psi_n\rangle = \sqrt{n}|\psi_{n-1}\rangle, \ \widehat{N}|\psi_n\rangle = n|\psi_n\rangle, \qquad (8.2.22)$$

and

$$|\psi_n\rangle = \frac{(\hat{a}^\dagger)^n}{\sqrt{n!}}|\psi_0\rangle. \qquad (8.2.23)$$

Remark   We have stressed the similarity between the classical and the quantum oscillator; let us now stress a striking difference between the classical and the quantum case. If we write the solution $x(t) = A\sin(\omega t + \varphi)$ for the classical case, with A the amplitude of oscillations and φ the initial phase, we find

$$E = \frac{\dot{x}^2}{2} + \frac{\omega^2 x^2}{2} = \frac{\omega^2 A^2}{2}. \tag{8.2.24}$$

For a given value of the energy, say E_n, the classical motion is bounded by $|x| < A = \frac{1}{\omega}\sqrt{2E}$. On the other hand, $|\psi_n(x)|^2 \neq 0$ if $|x| > A$ shows that, when the quantum oscillator is in the stationary state $|\psi_n\rangle$, there is a non-vanishing probability of finding the system in a position forbidden by the classical motion. In particular,

$$\int_{|x|>\sqrt{2}\lambda_c} \psi_0^* \psi_0(x)\mathrm{d}x \approx 0.157 \tag{8.2.25}$$

gives the probability of finding the particle outside the classically allowed region.

8.3 Operators in the number operator basis

By using the number operator basis $|\psi_n\rangle$ it is possible to introduce a particular coordinate system for the Hilbert space, in which the fundamental operators like position, momentum, creation or annihilation operators become infinite-dimensional matrices. By using the orthonormal basis $|\psi_n\rangle$ we can represent any vector $|\varphi\rangle$ as a complex combination

$$|\varphi\rangle = \sum_n C_n(\varphi)|\psi_n\rangle, \tag{8.3.1}$$

where the expansion coefficients $C_n(\varphi) = \langle\psi_n|\varphi\rangle$. By using the partition of the identity $I = \sum_n |\psi_n\rangle\langle\psi_n|$, we find for any two vectors $|\varphi\rangle, |\psi\rangle$

$$\langle\psi|\varphi\rangle = \sum_n C_n^*(\psi)C_n(\varphi). \tag{8.3.2}$$

In this basis, the Hilbert space $\mathcal{H}$ is coordinatized by means of infinite-dimensional sequences, and we say it is realized as l_2. In this basis, operators are represented by infinite-dimensional matrices. In particular, the annihilation operator has matrix elements

$$a_{nm} = \langle\psi_n|\hat{a}|\psi_m\rangle = \sqrt{m}\delta_{n,m-1}, \tag{8.3.3}$$

which corresponds to an infinite matrix having all zeros on the main diagonal, with an adjacent sequence, on the right of the zeros, of entries $\sqrt{1}, \sqrt{2}, \sqrt{3}, \ldots$, while the creation operator has matrix elements

$$a_{nm}^\dagger = \langle\psi_n|\hat{a}^\dagger|\psi_m\rangle, \tag{8.3.4}$$

which corresponds to an infinite matrix having all zeros on the main diagonal, with an adjacent sequence, below the zeros, of entries $\sqrt{1}, \sqrt{2}, \sqrt{3}, \ldots$. The operators $\hat{P}$ and $\hat{Q}$

may be derived from $(\hat{a} - \hat{a}^\dagger)$ and $(\hat{a} + \hat{a}^\dagger)$, respectively. Of course, the infinite-dimensional matrices associated with $\widehat{Q}$ and $\widehat{P}$ satisfy the canonical commutation relations

$$\left[\widehat{Q}, \widehat{P}\right] = i\hbar I. \tag{8.3.5}$$

Note that, since $\widehat{Q}$ and $\widehat{P}$ are represented by infinite-dimensional matrices, it is no longer true that the trace of a commutator is vanishing.

8.4 Representation of states on phase space, the Bargmann–Fock representation

We have seen that there is a correspondence between complex coordinates and creation and annihilation operators. By considering the phase-space $T^*\mathbf{R} = \mathbf{R}^2 = \mathbf{C}$, we can build the Hilbert space $\mathcal{E}$ of entire functions (i.e. analytic in the whole complex plane) defined on phase space and given by

$$f(z) = \sum_n c_n \frac{z^n}{\sqrt{n!}}, \quad \sum_n |c_n|^2 < \infty, \tag{8.4.1}$$

endowed with the scalar product

$$\langle f_1 | f_2 \rangle \equiv \frac{1}{\pi} \int e^{-|z|^2} f_1^*(z) f_2(z) d^2 z. \tag{8.4.2}$$

The entire functions are being introduced by analogy with the correspondence

$$|\psi\rangle = \sum_n c_n(\psi) \frac{(\hat{a}^\dagger)^n}{\sqrt{n!}} |\psi_0\rangle \Longleftrightarrow f(z) = \sum_n c_n \frac{z^n}{\sqrt{n!}}. \tag{8.4.3}$$

We claim that $u_n(z) = \frac{z^n}{\sqrt{n!}}$ form an orthonormal basis of $\mathcal{E}$. This is easily shown when the Hilbert space is considered with respect to the measure $e^{-|z|^2} d^2 z$. Indeed,

$$\langle u_n | u_m \rangle = \frac{1}{\pi} \int \frac{z^m (z^*)^n}{\sqrt{n!m!}} e^{-|z|^2} d^2 z. \tag{8.4.4}$$

This is easily evaluated by using the Argand–Gauss representation of complex numbers $z = \rho e^{i\varphi}$. The integral becomes then

$$\frac{1}{\pi} \frac{1}{\sqrt{n!m!}} \int_0^\infty \rho d\rho \int_0^{2\pi} d\varphi \, \rho^{n+m} e^{i\varphi(m-n)} e^{-\rho^2}.$$

This vanishes if $n \neq m$, while for $n = m$,

$$\langle u_n | u_n \rangle = \frac{2}{n!} \int_0^\infty \rho^{2n+1} e^{-\rho^2} d\rho = \frac{1}{n!} \int_0^\infty t^n e^{-t} dt = 1. \tag{8.4.5}$$

Clearly, the coefficients $\{c_n(\psi)\}$ make it possible to set up a 1-to-1 correspondence between entire functions and states written in the basis of the number operator. One should pay attention to the fact that the measures we use for the definition of the scalar product are not both translationally invariant.

To write vectors as functions on phase space we need an appropriate family of vectors defining a partition of the unity. We introduce eigenvectors defined by the equation

$$\hat{a}|z\rangle = z|z\rangle \quad z \in \mathbf{C}, \tag{8.4.6}$$

where, since $\hat{a}$ is not self-adjoint, the eigenvalue z is not real. A solution of this eigenvalue problem is provided by

$$|z\rangle = e^{-\frac{1}{2}|z|^2} \sum_{n=0}^{\infty} \frac{z^n}{\sqrt{n!}} |n\rangle, \tag{8.4.7}$$

where we recall that $\hat{a}|n\rangle = \sqrt{n}|n-1\rangle$. We are solving an equation by resorting to a coordinate system where the action of $\hat{a}$ is known and simple. By letting $\hat{a}$ act on each term of the series, the result follows immediately. Thus, the evaluation of the bra $\langle\psi_n|$ on the ket $|z\rangle$ yields

$$\langle\psi_n|z\rangle = e^{-\frac{1}{2}|z|^2} \frac{z^n}{\sqrt{n!}}. \tag{8.4.8}$$

It is seen that the Gaussian factor which belongs to the measure in the space of entire functions is transferred to the state, because the relevant measure is the Lebesgue measure, which has to be translationally invariant. The states $|z\rangle$ constructed as above are called (standard) coherent states (Glauber 1963, Sudarshan 1963). They have quite a few interesting properties, which are as follows.

  (i) The map $z \in \mathbf{C} \to |z\rangle \in \mathcal{L}^2(\mathbf{R})$ is continuous.
 (ii) $|z\rangle$ is defined to be an eigenvector of the annihilation operator, i.e. $\hat{a}|z\rangle = z|z\rangle$.
(iii) Coherent states provide a decomposition of the identity, i.e.

$$\frac{1}{\pi} \int_{\mathbf{C}} |z\rangle\langle z| d^2 z = I. \tag{8.4.9}$$

(iv) Coherent states saturate the uncertainty relation

$$(\Delta\widehat{Q})(\Delta\widehat{P}) = \frac{\hbar}{2}. \tag{8.4.10}$$

 (v) The family of coherent states is equivariant with respect to the classical evolution, i.e.

$$e^{-i\frac{\widehat{H}t}{\hbar}}|z\rangle = e^{-i\frac{t\omega}{2}}|e^{-i\omega t}z\rangle, \tag{8.4.11}$$

where $\widehat{H}$ is the harmonic oscillator Hamiltonian and $e^{-i\omega t}z_0 = z(t)$ represents the classical dynamics.
(vi) The family of coherent states is the orbit of the ground state of the harmonic oscillator under the action of the Weyl displacement operator $D(z)$, i.e.

$$D(z) \equiv e^{(z\hat{a}^\dagger - z^*\hat{a})}, \tag{8.4.12}$$

$$|z\rangle = D(z)|0\rangle. \tag{8.4.13}$$

The continuity of the map $z \to |z\rangle$ is understood in terms of the metric topologies on $\mathbf{C}$ and $\mathcal{H}$, respectively. It follows from computing

$$\||z_1\rangle - |z_2\rangle\|^2 = 2\left(1 - \mathrm{Re}\langle z_1|z_2\rangle\right)$$

$$= 2\left[1 - e^{-\frac{|z_1 - z_2|^2}{2}} \cos\left(\frac{q_1 p_2 - p_1 q_2}{2}\right)\right], \tag{8.4.14}$$

which tends to 0 as $z_1 \to z_2$, having set $z \equiv \frac{1}{\sqrt{2}}(q + ip)$.

As for the resolution of the identity, we find from the expansion of coherent states on the harmonic oscillator basis, and from the orthonormality of the basis of entire functions,

$$\int_{\mathbf{C}} |z\rangle\langle z| \frac{d^2 z}{\pi} = \sum_{n,m} |n\rangle\langle m| \int_{\mathbf{C}} \frac{z^n}{\sqrt{n!}} \frac{(z^*)^m}{\sqrt{m!}} e^{-|z|^2} \frac{d^2 z}{\pi}$$

$$= \sum_{n,m=0}^{\infty} |n\rangle\langle m|\delta_{nm} = \sum_{n=0}^{\infty} |n\rangle\langle n| = I. \tag{8.4.15}$$

The saturation of uncertainty relations follows easily from $|\psi_n\rangle = \frac{(\hat{a}^\dagger)^n}{\sqrt{n!}}|\psi_0\rangle$, and the expression of $\widehat{Q}$ and $\widehat{P}$ in terms of creation and annihilation operators. We leave it as an exercise.

The equivariance property follows from what we have already observed about equations of motion in the Heisenberg picture, i.e.

$$\frac{d}{dt}\hat{a} = \frac{i}{\hbar}[\widehat{H}, \hat{a}] = -i\omega\hat{a}, \tag{8.4.16}$$

$$\frac{d}{dt}\hat{a}^\dagger = \frac{i}{\hbar}[\widehat{H}, \hat{a}^\dagger] = i\omega\hat{a}^\dagger, \tag{8.4.17}$$

with solutions $\hat{a}(t) = e^{-i\omega t}\hat{a}_0$, $\hat{a}^\dagger(t) = e^{i\omega t}\hat{a}_0^\dagger$, while for the classical case

$$\frac{d}{dt}z = \{z, H\} = -i\omega z, \tag{8.4.18}$$

$$\frac{d}{dt}z^* = \{z^*, H\} = i\omega z^*, \tag{8.4.19}$$

with

$$z(t) = e^{-i\omega t}z_0, \quad z^*(t) = e^{i\omega t}z_0^*. \tag{8.4.20}$$

The relations we have written follow from setting $\widehat{H} = \hbar\omega\left(\hat{a}^\dagger\hat{a} + \frac{1}{2}I\right)$ and using the previous equations.

8.4.1 The Weyl displacement operator

From the Weyl displacement operator defined in Eq. (8.4.12),

$$D(z^*) = D^\dagger(z) = e^{z^*\hat{a} - z\hat{a}^\dagger}, \tag{8.4.21}$$

and

$$D(z_1)D(z_2) = D(z_1 + z_2)e^{i\varphi(z_1,z_2)}. \qquad (8.4.22)$$

To find the form of the phase $\varphi(z_1, z_2)$ we may proceed as follows. For a real parameter s,

$$D(sz) = e^{sz\hat{a}^\dagger - sz^*\hat{a}} \qquad (8.4.23)$$

is a 1-parameter group of unitary transformations and solves the equation

$$\frac{\mathrm{d}}{\mathrm{d}s}D(sz)|_{s=0} = z\hat{a}^\dagger - z^*\hat{a}. \qquad (8.4.24)$$

The expression

$$D_1(sz) = e^{-\frac{1}{2}s^2|z|^2}\, e^{sz\hat{a}^\dagger}\, e^{-sz^*\hat{a}} \qquad (8.4.25)$$

is again a 1-parameter group of unitary transformations and solves the same equation (8.4.24). Thus, by exploiting a uniqueness argument for solutions of differential equations we find

$$D(sz) = D_1(sz). \qquad (8.4.26)$$

In the argument, it is crucial that both expressions define 1-parameter groups of transformations. To prove the statement expressed by Eq. (8.4.12) it is now enough to use

$$D(z) = e^{-\frac{1}{2}|z|^2}\, e^{z\hat{a}^\dagger}\, e^{-z^*\hat{a}}. \qquad (8.4.27)$$

To find the form of the phase φ, we use the identity

$$e^{A+B} = e^{-\frac{1}{2}[A,B]}e^A e^B,$$

which holds provided that both A and B commute with their commutator, i.e.

$$[A,[A,B]] = [B,[A,B]] = 0.$$

By explicit computation we find

$$i\varphi(z_1, z_2) = \frac{1}{2}(z_1 z_2^* - z_1^* z_2), \qquad (8.4.28)$$

or, equivalently,

$$D(z_1)D(z_2) = e^{i\mathrm{Im}(z_1 z_2^*)} D(z_1 + z_2). \qquad (8.4.29)$$

If we introduce real variables for $\mathbf{C} = \mathbf{R}^2 = T^*\mathbf{R}$, say $z = \frac{1}{\sqrt{2}}(x + i\alpha)$, and write the annihilation operator in the form $\hat{a} = \frac{1}{\sqrt{2\hbar}}(\hat{Q} + i\hat{P})$, we find

$$z\hat{a}^\dagger - z^*\hat{a} = i\frac{(\alpha\hat{Q} - x\hat{P})}{\sqrt{\hbar}}, \qquad (8.4.30)$$

and hence the Weyl operator reads as

$$D(x,\alpha) = e^{\frac{i}{\sqrt{\hbar}}(\alpha\hat{Q} - x\hat{P})} = e^{i\frac{\alpha\hat{Q}}{\sqrt{\hbar}}}e^{-i\frac{x\hat{P}}{\sqrt{\hbar}}}e^{-i\frac{\alpha x}{2}}. \qquad (8.4.31)$$

The relation

$$D(z_1)D(z_2)D^{-1}(z_1)D^{-1}(z_2) = e^{(z_1 z_2^* - z_1^* z_2)} \, I \qquad (8.4.32)$$

represents the integrated version of the canonical commutation relations $[\widehat{Q}, \widehat{P}] = i\hbar I$. The association $z \rightarrow D(z)$ defines a projective unitary representation of $\mathbf{C}$, since the composition of two operators $D(z_1)$ and $D(z_2)$ fails to commute by a factor proportional to the phase $\varphi(z_1, z_2)$.

8.5   Basic operators in the coherent states' basis

We now consider a realization of the creation and annihilation operators as differential operators on the Hilbert space of entire functions. We first write their action on the basis functions $u_n = \frac{z^n}{\sqrt{n!}}$. From the properties

$$\hat{a}|\psi\rangle = \sum_{n=0}^{\infty} C_n \frac{n(\hat{a}^\dagger)^{n-1}}{\sqrt{n!}}|\psi_0\rangle, \; \hat{a}^\dagger|\psi\rangle = \sum_{n=0}^{\infty} C_n \frac{(n+1)(\hat{a}^\dagger)^{n+1}}{\sqrt{(n+1)!}}|\psi_0\rangle, \qquad (8.5.1)$$

$$f(z) = \sum_{n=0}^{\infty} C_n \frac{z^n}{\sqrt{n!}}, \qquad (8.5.2)$$

we derive

$$\hat{a}f(z) = \sum_{n=0}^{\infty} C_n \frac{z^{n-1}}{\sqrt{(n-1)!}} = \frac{\mathrm{d}}{\mathrm{d}z}f(z), \qquad (8.5.3)$$

$$\hat{a}^\dagger f(z) = \sum_{n=0}^{\infty} \frac{C_n z^{n+1}}{\sqrt{n!}} = zf(z). \qquad (8.5.4)$$

Thus, a representation for the operators $\hat{a}$ and $\hat{a}^\dagger$ in terms of differential operators is given by

$$\hat{a} = \frac{\mathrm{d}}{\mathrm{d}z}, \; \hat{a}^\dagger = z. \qquad (8.5.5)$$

For the Hamiltonian, position and momentum of the harmonic oscillator,

$$\widehat{H} = \hbar\omega\left(z\frac{\mathrm{d}}{\mathrm{d}z} + \frac{1}{2}\right), \; \widehat{Q} = \sqrt{\frac{\hbar}{2}}\left(\frac{\mathrm{d}}{\mathrm{d}z} + z\right), \; \widehat{P} = -i\sqrt{\frac{\hbar}{2}}\left(\frac{\mathrm{d}}{\mathrm{d}z} - z\right). \qquad (8.5.6)$$

This particular realization is called the holomorphic or Bargmann–Fock representation, with the correspondence from the number basis to the basis of entire functions being an isometry of Hilbert spaces. In advanced quantum theory the anti-holomorphic representation also turns out to be useful, which consists of replacing z with z^*. In this representation

$$\hat{a}^\dagger = z^*, \; \hat{a} = \frac{\mathrm{d}}{\mathrm{d}z^*}. \qquad (8.5.7)$$

8.6 Uncertainty relations

It is possible to compute the variance of position and momentum. We start with

$$e_a(z) = \frac{\langle z|\hat{a}|z\rangle}{\langle z|z\rangle} = z, \; e_{a^\dagger}(z) = \frac{\langle z|\hat{a}^\dagger|z\rangle}{\langle z|z\rangle} = z^*, \tag{8.6.1}$$

and for the number operator $\widehat{N} = \hat{a}^\dagger \hat{a}$

$$e_N(z) = \frac{\langle z|\widehat{N}|z\rangle}{\langle z|z\rangle} = z^* z. \tag{8.6.2}$$

For the position operator $\widehat{Q} = \sqrt{\frac{\hbar}{2}}(\hat{a} + \hat{a}^\dagger)$ and momentum operator $\widehat{P} = -i\sqrt{\frac{\hbar}{2}}(\hat{a} - \hat{a}^\dagger)$,

$$e_Q(z) = q, \; e_P(z) = p, \tag{8.6.3}$$

where q and p have both the dimension of $\sqrt{\hbar}$, i.e.

$$q = \sqrt{\frac{\hbar}{2}}(z + z^*), \; p = -i\sqrt{\frac{\hbar}{2}}(z - z^*).$$

Moreover

$$e_{Q^2}(z) = \frac{\langle z|\widehat{Q}^2|z\rangle}{\langle z|z\rangle} = q^2 + \frac{\hbar}{2}, \; e_{P^2}(z) = p^2 + \frac{\hbar}{2}, \tag{8.6.4}$$

and

$$e_H(z) = \frac{\langle z|\widehat{H}|z\rangle}{\langle z|z\rangle} = \frac{\omega}{2}(q^2 + p^2) + \frac{1}{2}\omega\hbar, \tag{8.6.5}$$

for $\widehat{H} = \frac{\omega}{2}(\widehat{P}^2 + \widehat{Q}^2)$. In the course of performing this calculation, the formula

$$\widehat{Q}^2 = \frac{\hbar}{2}(\hat{a} + \hat{a}^\dagger)(\hat{a} + \hat{a}^\dagger)$$

leads to

$$e_{Q^2}(z) = \frac{\hbar}{2}[(z + z^*)^2 + 1], \tag{8.6.6}$$

and similarly for $\widehat{P}^2$. It is interesting that the expectation-value functions are not multiplied pointwise, i.e.

$$\left(e_Q(z)\right)^2 \neq e_{Q^2}(z), \tag{8.6.7}$$

showing that the product on expectation-value functions is not local. Moreover, the variance $e_{Q^2}(z) - \left(e_Q(z)\right)^2$ is exactly the measure of their difference, i.e. it measures the deviation from locality.

By computing

$$(\Delta Q)^2 \equiv \langle z|\widehat{Q}^2|z\rangle - (\langle z|\widehat{Q}|z\rangle)^2 = \frac{\hbar}{2}, \tag{8.6.8}$$

$$(\Delta P)^2 \equiv \langle z|\widehat{P}^2|z\rangle - (\langle z|\widehat{P}|z\rangle)^2 = \frac{\hbar}{2}, \tag{8.6.9}$$

we find that

$$(\Delta Q)(\Delta P) = \frac{\hbar}{2}. \tag{8.6.10}$$

Thus, by saturating the uncertainty relation, coherent states are the closest possible to their classical counterpart.

8.7 Ehrenfest picture

In Chapter 4, when dealing with mean values, we derived the equations of motion

$$\frac{\mathrm{d}}{\mathrm{d}t}\langle x_l \rangle_\psi = \frac{1}{m}\langle p_l \rangle_\psi = \left\langle \frac{\partial H}{\partial p_l} \right\rangle_\psi,$$

$$\frac{\mathrm{d}}{\mathrm{d}t}\langle p_l \rangle_\psi = -\left\langle \frac{\partial H}{\partial x_l} \right\rangle_\psi.$$

We also remarked that the first correction to the formula expressing complete analogy between equations of motion for mean values and classical equations of motion for dynamical variables involves corrections depending on the third derivatives and higher, evaluated at a point equal to the mean value of position. It follows that for quadratic potentials, like the harmonic oscillator, there will be no corrections. Let us introduce therefore the expectation-value functions

$$e_Q(\psi_z) \equiv \frac{\langle \psi_z | Q | \psi_z \rangle}{\langle \psi_z | \psi_z \rangle}, \tag{8.7.1}$$

$$e_P(\psi_z) \equiv \frac{\langle \psi_z | P | \psi_z \rangle}{\langle \psi_z | \psi_z \rangle}. \tag{8.7.2}$$

We point out that from now on we are going to use the name of expectation-value function instead of mean value. The equations of motion are given by the Hamiltonian form

$$\frac{\mathrm{d}}{\mathrm{d}t}e_Q(\psi_z) = \left\{ e_H(\psi_z), e_Q(\psi_z) \right\} = e_{-\frac{\mathrm{i}}{\hbar}[H,Q]}(\psi_z) = \omega e_P(\psi_z), \tag{8.7.3}$$

$$\frac{\mathrm{d}}{\mathrm{d}t}e_P(\psi_z) = \left\{ e_H(\psi_z), e_P(\psi_z) \right\} = e_{-\frac{\mathrm{i}}{\hbar}[H,P]}(\psi_z) = -\omega e_Q(\psi_z). \tag{8.7.4}$$

It should be stressed that the Poisson bracket appearing in Eqs. (8.7.3) and (8.7.4) is actually defined by these formulae, i.e.

$$\left\{ e_A(\psi_z), e_B(\psi_z) \right\} \equiv e_{-\frac{\mathrm{i}}{\hbar}[A,B]}(\psi_z), \tag{8.7.5}$$

because it involves the fundamental variables e_P and e_Q. As solutions of the previous equations of motion we find

$$e_Q(\psi_z)(t) = \langle e_Q(\psi_z) \rangle_0 \cos \omega t + \langle e_P(\psi_z) \rangle_0 \sin \omega t, \tag{8.7.6}$$

$$e_P(\psi_z)(t) = \langle e_P(\psi_z) \rangle_0 \cos \omega t - \langle e_Q(\psi_z) \rangle_0 \sin \omega t. \tag{8.7.7}$$

If we use $z = \rho e^{i\theta}$ we also find

$$e_Q(\psi_z) = q = \sqrt{2\hbar}\rho\cos\theta, \tag{8.7.8}$$

$$e_P(\psi_z) = p = \sqrt{2\hbar}\rho\sin\theta, \tag{8.7.9}$$

$$e_{P^2}(\psi_z) = p^2 + \frac{\hbar}{2} = \left(2\rho^2\sin^2\theta + \frac{1}{2}\right)\hbar, \tag{8.7.10}$$

$$e_{Q^2}(\psi_z) = q^2 + \frac{\hbar}{2} = \left(2\rho^2\cos^2\theta + \frac{1}{2}\right)\hbar. \tag{8.7.11}$$

We notice explicitly that $e_{P^2}(\psi_z) \neq \left(e_P(\psi_z)\right)^2$, indeed the product on expectation-value functions is not local and we should always interpret

$$e_{A^2}(\psi_z) = \frac{\langle\psi_z|A^2|\psi_z\rangle}{\langle\psi_z|\psi_z\rangle}, \tag{8.7.12}$$

which differs from

$$\left(e_A(\psi_z)\right)^2 = \left(\frac{\langle\psi_z|A|\psi_z\rangle}{\langle\psi_z|\psi_z\rangle}\right)^2 \tag{8.7.13}$$

unless the state is a dispersion-free state for A! If we write q for $e_Q(\psi_z)$ and p for $e_P(\psi_z)$, we have a skew-symmetric product

$$q\star p - p\star q = \hbar,$$

and a symmetric product

$$q\star q = q^2 + \frac{\hbar}{2}, \quad p\star p = p^2 + \frac{\hbar}{2},$$

$$q\star p + p\star q = 2qp.$$

It is not difficult to derive the Poisson bracket in terms of (ρ, θ), indeed

$$2\hbar\{\rho\cos\theta, \rho\sin\theta\} = 1,$$

corresponding to

$$\{e_P, e_Q\}(\psi_z) = e_{-\frac{i}{\hbar}[P,Q]}(\psi_z)$$

and $[Q, P] = i\hbar$, gives

$$\{\rho, \theta\} = \frac{1}{2\rho}. \tag{8.7.14}$$

Thus, we may use the classical-like bracket but we should bear in mind that the expectation-value functions, considered as classical-like functions, do not satisfy the condition $e_A \cdot e_B = e_{AB}$ for the pointwise product, moreover $e_A \cdot e_B \neq e_B \cdot e_A$. Therefore, in this classical-like description of quantum mechanics, we may use classical-like structures, but the product we need to use on expectation-value functions is a non-local and non-commutative product. These two aspects characterize the quantum nature of our description with respect to the classical pointwise and commutative product.

It is interesting to compute the probability distribution associated with $|\psi_z\rangle$ in the resolution of the identity provided by the eigenstates of the number operator. Then,

$$P_{z,N}(n) = \langle \psi_z | \psi_n \rangle \langle \psi_n | \psi_z \rangle = e^{-|z|^2} \frac{|z|^{2n}}{n!} = e^{-\rho^2} \frac{\rho^{2n}}{n!}. \tag{8.7.15}$$

8.8 Problems

8.P1. At time $t = 0$, a one-dimensional harmonic oscillator is in the quantum state

$$|\psi\rangle_0 = \frac{1}{\sqrt{5}} (2a^\dagger + I)|0\rangle. \tag{8.8.1}$$

Find the time-evolution of mean values of position and momentum, and compare them with classical analogues; do the same for total energy and discuss the meaning of your results.

8.P2. A one-dimensional harmonic oscillator is in a state which is linear combination of the first two eigenstates of the Hamiltonian, i.e. $|0\rangle$ and $|1\rangle$. Find the maximum of the mean value $\langle x \rangle$ as a function of the mean value of energy.

8.P3. Consider the Hamiltonian operator

$$\hat{H} = \frac{\hat{p}^2}{2m} + a\hat{x}^2 + b\hat{x} + cI, \tag{8.8.2}$$

with a, b, c real-valued.

(i) Bearing in mind that a, b, c can be positive or negative, in which cases does $\hat{H}$ admit bound states? What physical system is described by the various sign choices?

(ii) Find eigenvalues and eigenfunctions when the parameters make it possible to obtain bound states.

(iii) Find the time-evolution of the mean value of position when the initial quantum state is the ground state, or the nth excited state, or their normalized linear combination.

8.P4. Find the mean quadratic deviation of a one-dimensional harmonic oscillator which is at thermal equilibrium at temperature T. Compare with the result from classical physics.

8.P5. Prove that the quantum state

$$|\psi\rangle = e^{\lambda \hat{a}^\dagger} |0\rangle \tag{8.8.3}$$

is an eigenstate of the annihilation operator $\hat{a}$, normalizable for all complex values of $\lambda \in C$. Compute the probability distribution, mean value and mean quadratic deviation for position and energy in the quantum state $|\psi\rangle$.

9 Introduction to spin

First, the experimental foundations for the existence of a new dynamical variable, the spin of particles, are presented. The Pauli equation is then derived in detail, two applications are given and the energy levels of a particle with spin in a constant magnetic field are studied. This is the analysis of Landau levels, which can be performed by using the known results on the spectrum of harmonic oscillators. In this chapter we also study in detail the addition of orbital angular momentum and spin, as well as the spin–orbit interaction with the resulting Thomas precession.

9.1 Stern–Gerlach experiment and electron spin

The hypothesis that the electron has an intrinsic magnetic moment and an angular momentum, in short a *spin*, was first suggested in Uhlenbeck and Goudsmit (1926). They noticed, even before the discovery of quantum mechanics, that a complete description of spectra was not possible unless a magnetic moment and a mechanical moment were ascribed to the electron, and hence the concept of an electron as a point charge was insufficient. First, let us therefore try to understand how a magnetic moment can be associated to an atomic system. For this purpose, consider for simplicity the Bohr model of an hydrogen atom, where the electron moves along a circular orbit and rotates with orbital angular momentum $\vec{L}$. A moving charge is equivalent to an electric current, hence an electron along a circular orbit can be treated as a loop along which an electric current flows, and such a loop has a magnetic moment. Starting from the magnetic moment associated to each individual electron, a magnetic moment for the whole atom can be derived. Indeed, a current i along a loop enclosing a small area S_δ gives rise to a dipole magnetic moment

$$\vec{M} = \vec{n}\frac{i}{c}S_\delta, \tag{9.1.1}$$

where $\vec{n}$ is the normal to the plane containing the loop. If the loop has a uniform charge density ρ_e with magnitude $-\frac{q_e}{2\pi r}$, the modulus of $\vec{M}$ therefore reads as ($q_e = -|q_e|$ being the negative charge of the electron)

$$M = -\frac{q_e v}{2\pi r}\frac{\pi r^2}{c} = -\frac{q_e v r}{2c} = -\frac{q_e L}{2m_e c}. \tag{9.1.2}$$

Since the rotation of the electron is opposite to the current, the corresponding vectors are related by

$$\vec{M} = \frac{q_e}{2m_e c} \vec{L}. \tag{9.1.3}$$

The Bohr quantization rules suggest replacing the orbital angular momentum by $\frac{\vec{L}}{\hbar}$, and hence we write eventually

$$\vec{M} = \frac{q_e \hbar}{2m_e c} \frac{\vec{L}}{\hbar} = -\mu_B \frac{\vec{L}}{\hbar}. \tag{9.1.4}$$

The quantity

$$\mu_B \equiv \frac{|q_e| \hbar}{2m_e c} \tag{9.1.5}$$

has, of course, the dimensions of a magnetic moment and is called the Bohr magneton (equal to 0.9274×10^{-20} erg gauss^{-1}. In general, a system of electrons with total angular momentum $\vec{J}$ has a magnetic moment $\vec{M}$ antiparallel to $\vec{J}$ and we usually write

$$\vec{M} = -g\mu_B \frac{\vec{J}}{\hbar}, \tag{9.1.6}$$

where g is called the gyromagnetic ratio.

If an atom with magnetic moment $\vec{M}$ is affected by a magnetic field $\vec{B}$, the interaction potential is

$$V_I = -\vec{M} \cdot \vec{B} \equiv H_I, \tag{9.1.7}$$

so that the resulting Hamilton equations read as

$$\frac{d\vec{x}}{dt} = \frac{\vec{p}}{m}, \tag{9.1.8}$$

$$\frac{d\vec{p}}{dt} = -\frac{\partial H_I}{\partial \vec{x}} = \vec{M} \cdot \frac{\partial \vec{B}}{\partial \vec{x}}. \tag{9.1.9}$$

These formulae make it clear that it is the gradient of the magnetic field that really plays the key role in the equations of motion. In particular, if the magnetic field is uniform, the total force on the magnetic dipole is vanishing.

We now describe in some detail the key steps in the experimental detection of spin. First, in Gerlach and Stern (1922) the authors measured the possible values of the *magnetic dipole moment* for silver atoms by sending a beam of these atoms throughout a non-uniform magnetic field. In this experiment, a beam of neutral atoms is formed by evaporating silver from an oven. The beam is collimated by a diaphragm and it enters a magnet (see Figure 9.1). The cross-sectional view of the magnet shows that it produces a field that increases in intensity in the z-direction (third axis), which is also the direction of the magnetic field itself in the region of the beam. Since the atoms are neutral overall, the only net force acting on them is a force proportional to M_3 and the gradient of the external magnetic field. Each atom is hence deflected, in passing through the non-uniform magnetic field, by an amount that is proportional to M_3. This means that the beam is analysed into components,

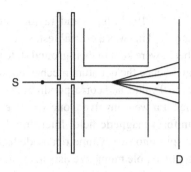

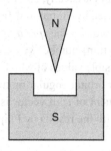

Fig. 9.1   Experimental apparatus used in the Stern–Gerlach experiment.

depending on the various values of M_3. Lastly, deflected atoms strike a metallic plate, upon which they condense and leave a visible trace. According to the general properties of angular momentum operators, M_3 can only take discrete values

$$M_3 = -g_l\, \mu_B\, m_3, \tag{9.1.10}$$

where m_3 ranges from $-l$ to $+l$, including 0. Thus, according to quantum mechanics, the deflected beam should be split into several discrete components, and for all orientations of the analysing magnet. In other words, the experimental setting consists of an oven, a collimator, a magnet and eventually a detector plate. The magnet acts essentially as a measuring device, which investigates the quantization of the component of the magnetic dipole moment along a z-axis. Such an axis is defined by the direction in which its field increases in intensity.

Stern and Gerlach found that the beam of silver atoms is split into *two* discrete components, one component being bent in the positive z-direction and the other bent in the negative z-direction. Moreover, they found that these results hold independently of the choice of the z-direction. The experiment was repeated using several other species of atoms, and in all cases it was found that the deflected beam is split into two, or more, discrete components. However, these experimental data were puzzling for a simple but fundamental reason: since l is an integer, the number of possible values of M_3 is always odd. This is incompatible with the beam of silver atoms being split into *only two* components, both of which are deflected.

Besides all this in 1927 Phipps and Taylor used the Stern–Gerlach technique on a beam of hydrogen atoms. This was a crucial test, since these atoms contain a single electron. The atoms in the beam were kept in their ground state by virtue of the relatively low temperature of the oven, and hence the quantum mechanical theory predicts that the quantum number l can only take the 0 value, and correspondingly m_3 vanishes as well. Thus, if only the orbital angular momentum were involved, one would expect that the beam should be unaffected by the non-uniform magnetic field. Interestingly, however, Phipps and Taylor found that the beam was split into two symmetrically deflected components. This implies in turn that a sort of magnetic dipole moment exists in the atom, though not quite of the sort suggested by the theory of orbital angular momentum. Note that we are treating the translational degree of freedom classically, and using the quantum spin in the external magnetic field.

The data obtained by Phipps and Taylor can indeed be understood if the angular momentum formalism holds for the *spin angular momentum*. In other words, one is led to assume that a new (quantum) number s_3 exists, which implies the existence of a new degree of freedom, the possible values of which range from $-s$ to $+s$, as is true of the quantum numbers m_3 and l for orbital angular momentum. Note that the choice to label the states using the third component of spin vector is just arbitrary as for the orbital angular momentum case. To agree with the results by Phipps and Taylor, the two possible values of m_s are

$$s_3 = -\frac{1}{2}, +\frac{1}{2},$$ (9.1.11)

and hence s can only take the value

$$s = \frac{1}{2}.$$ (9.1.12)

As far as the very idea of *electron spin* is concerned, credit is given, appropriately, to Goudsmit and Uhlenbeck, as was stated at the beginning of this section. In 1925 they were young graduate students who were trying to understand why certain lines of the optical spectra of the hydrogen atom and alkali atoms consist of a closely spaced pair of lines. They were thus dealing with the *fine structure* of these atoms, and proposed that the electron has an intrinsic angular momentum and magnetic dipole moment with the properties outlined above. The non-trivial step was to assign a fourth quantum number to the electron, rather than the three that would be obtained from the (usual) Schrödinger theory. They tried to understand the electron spin in terms of a model where the electron is rotating. However, Lorentz studied the electromagnetic properties of rotating electrons, and was able to show that serious inconsistencies would result from such a model. In particular, the magnetic energy would be so large that, by the equivalence of mass and energy, the electron would have a larger mass than the proton, and would be bigger than the whole atom! None of the people concerned, including Wolfgang Pauli, were aware of Elie Cartan's discovery of spinors (Cartan 1938) and their properties. Thus, the understanding of the underlying reasons for the existence of electron spin was completely lacking when Goudsmit and Uhlenbeck first brought their idea to the attention of the scientific community (thanks to the enthusiastic support of Ehrenfest).

In 'classical physics', the Hamiltonian description of a non-relativistic particle with spin can be obtained by considering the vectors $\vec{x}, \vec{p}$ and $\vec{s}$, and requiring the fundamental

Poisson brackets

$$\{x_i, x_j\} = \{p_i, p_j\} = 0, \tag{9.1.13}$$

$$\{x_i, p_j\} = \delta_{ij}, \tag{9.1.14}$$

$$\{s_j, x_k\} = \{s_j, p_k\} = 0, \quad \{s_i, s_j\} = \varepsilon_{ijk} s_k. \tag{9.1.15}$$

On using the Hamiltonian $H \equiv H_0 + \mu \vec{s} \cdot \vec{B}$, the resulting Hamilton equations of motion read as

$$\dot{x}_i = \frac{p_i}{m}, \tag{9.1.16}$$

$$\dot{p}_i = -\mu s_j \partial_i B_j, \tag{9.1.17}$$

$$\dot{s}_i = \mu \varepsilon_{ijk} B_j s_k. \tag{9.1.18}$$

Remark On linear functions, there is no difference between classical and quantum. It is only when we consider products that the difference shows up. In the quantum case the product is non-commutative and non-local.

9.2 Wave functions with spin

Wave functions for particles with spin s belongs to the Hilbert space $\mathcal{L}^2(\mathbf{R}^3) \otimes \mathbf{C}^{2s+1}$. In other words, each state can be represented as a $(2s + 1)$-dimensional column vector of square-integrable functions on $\mathbf{R}^3$, with norm given by (in this section we study the stationary theory)[1]

$$\|\psi\|^2 = (\psi, \psi) = \int d^3x \psi^\dagger(\vec{x}) \psi(\vec{x})$$

$$= \int d^3x \left(\psi_1^* \psi_1 + \psi_2^* \psi_2 + \cdots + \psi_{2s+1}^* \psi_{2s+1} \right). \tag{9.2.1}$$

Let us now consider in more detail this non-relativistic formalism for the electron, which is a particle of spin $\frac{1}{2}$. Its wave function reads as

$$\psi(\vec{x}) = \begin{pmatrix} \psi_+(\vec{x}) \\ \psi_-(\vec{x}) \end{pmatrix} = \psi_+(\vec{x}) \begin{pmatrix} 1 \\ 0 \end{pmatrix} + \psi_-(\vec{x}) \begin{pmatrix} 0 \\ 1 \end{pmatrix}. \tag{9.2.2}$$

Thus, on defining

$$\chi_+ \equiv \begin{pmatrix} 1 \\ 0 \end{pmatrix}, \tag{9.2.3}$$

$$\chi_- \equiv \begin{pmatrix} 0 \\ 1 \end{pmatrix}, \tag{9.2.4}$$

---

[1] Hereafter, to simplify the notation we often omit the $\otimes$ symbol assuming it implicitly.

we can write

$$\psi(\vec{x}) = \psi_+(\vec{x})\chi_+ + \psi_-(\vec{x})\chi_-. \tag{9.2.5}$$

The column vectors χ_+ and χ_- are a basis in the space $\mathbf{C}^2$ of spin-states for a spin-$\frac{1}{2}$ particle. They are orthonormal, because

$$(\chi_+, \chi_+) = (\chi_-, \chi_-) = 1, \tag{9.2.6}$$

$$(\chi_+, \chi_-) = 0. \tag{9.2.7}$$

Moreover, every element χ of $\mathbf{C}^2$ can be written as a linear combination of χ_+ and χ_-, i.e.

$$\chi = C_+\chi_+ + C_-\chi_-, \tag{9.2.8}$$

where

$$C_+ = (\chi_+, \chi), \tag{9.2.9}$$

$$C_- = (\chi_-, \chi). \tag{9.2.10}$$

Our basis vectors χ_+ and χ_- therefore form a complete orthonormal system in $\mathbf{C}^2$.

Given now the spin operators acting on $\mathbf{C}^2$

$$\hat{S}_1 \equiv \frac{\hbar}{2}\sigma_1, \quad \hat{S}_2 \equiv \frac{\hbar}{2}\sigma_2, \quad \hat{S}_3 \equiv \frac{\hbar}{2}\sigma_3, \tag{9.2.11}$$

where σ_i with $i = 1, 2, 3$ stand for 2×2 Pauli matrices, we find that these operators obey the commutation relations of angular momentum:

$$\left[\hat{S}_k, \hat{S}_l\right] = i\hbar\, \varepsilon_{klm}\, \hat{S}_m. \tag{9.2.12}$$

Moreover,

$$\hat{S}^2\chi_\pm = \frac{3}{4}\hbar^2\chi_\pm, \tag{9.2.13}$$

$$\hat{S}_3\chi_\pm = \pm\frac{\hbar}{2}\chi_\pm. \tag{9.2.14}$$

The *raising* and *lowering* operators can also be defined:

$$\hat{S}_+ \equiv \hat{S}_1 + i\hat{S}_2, \tag{9.2.15}$$

$$\hat{S}_- \equiv \hat{S}_1 - i\hat{S}_2, \tag{9.2.16}$$

for which

$$\hat{S}_+\chi_+ = 0, \quad \hat{S}_+\chi_- = \hbar\chi_+, \quad \hat{S}_-\chi_+ = \hbar\chi_-, \quad \hat{S}_-\chi_- = 0. \tag{9.2.17}$$

As a corollary, $\hat{S}_+$ and $\hat{S}_-$ are nilpotent, in that $\hat{S}_+^2 = \hat{S}_-^2 = 0$.

Furthermore, bearing in mind the split (9.2.2),

$$\hat{S}_1\psi(\vec{x}) = \frac{\hbar}{2}\left[\psi_-(\vec{x})\chi_+ + \psi_+(\vec{x})\chi_-\right], \tag{9.2.18}$$

$$\hat{S}_2 \psi(\vec{x}) = -\frac{i\hbar}{2} \Big[\psi_-(\vec{x}) \chi_+ - \psi_+(\vec{x}) \chi_- \Big], \tag{9.2.19}$$

$$\hat{S}_3 \psi(\vec{x}) = \frac{\hbar}{2} \Big[\psi_+(\vec{x}) \chi_+ - \psi_-(\vec{x}) \chi_- \Big], \tag{9.2.20}$$

$$\hat{S}^2 \psi(\vec{x}) = \frac{3}{4} \hbar^2 \psi(\vec{x}). \tag{9.2.21}$$

Several examples concerning spin in quantum mechanics can be found in Cohen-Tannoudji *et al.* (1977a,b).

9.3 Addition of orbital and spin angular momenta

On considering the quantum theory of the electron which is described in the Hilbert space $\mathcal{L}^2(\mathbf{R}^3) \otimes \mathbf{C}^2$, the total angular momentum $\hat{J}$ can be defined as

$$\hat{J}_i \equiv \hat{L}_i \otimes I_2 + I_\infty \otimes \hat{S}_i, \tag{9.3.1}$$

where I_2 and I_∞ are the identity operators on $\mathbf{C}^2$ and $\mathcal{L}^2(\mathbf{R}^3)$, respectively. By previous expression the two terms on the right-hand side of Eq. (9.3.1) define implicitly the orbital and spin angular momentum in the enlarged space $\mathcal{L}^2(\mathbf{R}^3) \otimes \mathbf{C}^2$, respectively. This simply implements the natural assumption that both angular momenta mutually commute. The action of $\hat{J}_i$ on an arbitrary state is then given by

$$\hat{J}_i(\psi_-(\vec{x}) \chi_- + \psi_+(\vec{x}) \chi_+) \equiv \Big[(\hat{L}_i \psi_-(\vec{x})) \chi_- + (\hat{L}_i \psi_+(\vec{x})) \chi_+ \Big]$$
$$+ \Big[\psi_-(\vec{x}) \hat{S}_i \chi_- + \psi_+(\vec{x}) \hat{S}_i \chi_+ \Big]. \tag{9.3.2}$$

From Eq. (9.3.1) we find that $\hat{J}_i$ obey the commutation relations of angular momentum

$$\Big[\hat{J}_k, \hat{J}_l \Big] = \Big[\hat{L}_k, \hat{L}_l \Big] \otimes I_2 + I_\infty \otimes \Big[\hat{S}_k, \hat{S}_l \Big]$$
$$= i\hbar \varepsilon_{klm} \Big(\hat{L}_m \otimes I_2 + I_\infty \otimes \hat{S}_m \Big) = i\hbar \varepsilon_{klm} \hat{J}_m. \tag{9.3.3}$$

Hence, in complete analogy to the angular momentum case, we find that $\hat{J}^2$ and any $\hat{J}_k$ form a pair of compatible self-adjoint operators. For this reason it is possible to show that $\hat{L}^2, \hat{L}_3, \hat{S}^2$ and $\hat{S}_3$, form a complete set of observables as well as $\hat{L}^2, \hat{S}^2, \hat{J}^2$ and $\hat{J}_3$. Note that the choice of the third component in the previous sets is just arbitrary and thus any other component can be alternatively adopted. In order to prove that $\hat{L}^2, \hat{S}^2, \hat{J}^2$ and $\hat{J}_3$ represent a complete set of observables, it is only necessary to check that $[\hat{L}^2, \hat{J}^2] = [\hat{S}^2, \hat{J}^2] = 0$ since all other commutation relations are trivially vanishing. Indeed,

$$\Big[\hat{L}^2, \hat{J}^2 \Big] = \Big[\hat{L}^2, \hat{L}^2 + \hat{S}^2 + 2 \hat{\vec{L}} \cdot \hat{\vec{S}} \Big] = 2 \Big[\hat{L}^2, \hat{\vec{L}} \cdot \hat{\vec{S}} \Big] = 2 \sum_i \Big[\hat{L}^2, \hat{L}_i \Big] \hat{S}_i = 0. \tag{9.3.4}$$

The analogous expression holds for $[\hat{S}^2, \hat{J}^2]$.

Since both sets are complete, we can decide to find a basis by simultaneously diagonalizing one of them. Let us denote by $|l, m_3, s, s_3\rangle$ a basis of eigenvectors such that

$$\hat{L}^2 |l, m_3, s, s_3\rangle = \hbar^2 \, l(l+1) \, |l, m_3, s, s_3\rangle, \tag{9.3.5}$$

$$\hat{L}_3 |l, m_3, s, s_3\rangle = \hbar \, m_3 \, |l, m_3, s, s_3\rangle, \tag{9.3.6}$$

$$\hat{S}^2 |l, m_3, s, s_3\rangle = \hbar^2 \, s(s+1) \, |l, m_3, s, s_3\rangle, \tag{9.3.7}$$

$$\hat{S}_3 |l, m_3, s, s_3\rangle = \hbar \, s_3 \, |l, m_3, s, s_3\rangle, \tag{9.3.8}$$

whereas alternatively, we denote by $|l, s, j, j_3\rangle$ a basis of eigenvectors such that

$$\hat{L}^2 |l, s, j, j_3\rangle = \hbar^2 \, l(l+1) \, |l, s, j, j_3\rangle, \tag{9.3.9}$$

$$\hat{S}^2 |l, s, j, j_3\rangle = \hbar^2 \, s(s+1) \, |l, s, j, j_3\rangle, \tag{9.3.10}$$

$$\hat{J}^2 |l, s, j, j_3\rangle = \hbar^2 \, j(j+1) \, |l, s, j, j_3\rangle, \tag{9.3.11}$$

$$\hat{J}_3 |l, s, j, j_3\rangle = \hbar j_3 \, |l, s, j, j_3\rangle. \tag{9.3.12}$$

Once l and s are fixed the space spanned by $|l, m_3, s, s_3\rangle$ has dimension $(2l+1)(2s+1)$, which should be recovered by using $|l, s, j, j_3\rangle$ as well. This means that for fixed l and s, a range of possible values are obtained for j, i.e. $j_{\min} \leq j \leq j_{\max}$. To find $j_{\max}$ we can simply observe that $j_3 = m_3 + s_3 \leq l + s$; this implies that a multiplet characterized by $j \geq l + s$ must exist since it has to contain the state with $j_3 = l + s$. On the other hand, if it were possible to find a multiplet characterized by $j > l + s$, this would imply the presence of states with $j_3 > l + s$, which is not the case. For this reason we obtain that $j_{\max} = l + s$. In particular, for the extremal values $j_3 = l + s$ and $j = l + s$ it is easy to prove the coincidence between two vectors of the two bases, i.e.

$$|l, l, s, s\rangle = |l, s, l+s, l+s\rangle, \tag{9.3.13}$$

where, of course, the vector on the left-hand side of previous equation belongs to the set of vectors $|l, m_3, s, s_3\rangle$, and the one on the right-hand side to $|l, s, j, j_3\rangle$, respectively. Acting on both sides of Eq. (9.3.13) with $\hat{J}_-$ we get

$$\hat{J}_- |l, l, s, s\rangle = (\hat{L}_- + \hat{S}_-) |l, l, s, s\rangle$$

$$= \hbar \left(\sqrt{2l} |l, l-1, s, s\rangle + \sqrt{2s} |l, l, s, s-1\rangle \right)$$

$$= \hat{J}_- |l, s, l+s, l+s\rangle = \hbar \sqrt{2(l+s)} \, |l, s, l+s, l+s-1\rangle, \tag{9.3.14}$$

which leads to

$$|l, s, l+s, l+s-1\rangle = \sqrt{\frac{l}{(l+s)}} \, |l, l-1, s, s\rangle + \sqrt{\frac{s}{(l+s)}} \, |l, l, s, s-1\rangle. \tag{9.3.15}$$

On the right-hand side of this equation there is a combination of two independent states, which admits an orthogonal one, i.e.

$$|\psi\rangle \equiv -\sqrt{\frac{s}{(l+s)}} \, |l, l-1, s, s\rangle + \sqrt{\frac{l}{(l+s)}} \, |l, l, s, s-1\rangle, \tag{9.3.16}$$

which is still an eigenvector of $\hat{J}_3$ with eigenvalue $l + s - 1$. It is easy to see that it is an eigenvector of $\hat{J}^2$ too, of eigenvalue $l + s - 1$ as well. This can easily be proved by recalling that $\hat{J}^2 = \hat{L}^2 + \hat{S}^2 + 2\hat{\vec{L}} \cdot \hat{\vec{S}} = \hat{L}^2 + \hat{S}^2 + 2\hat{L}_3\hat{S}_3 + \hat{L}_+\hat{S}_- + \hat{L}_-\hat{S}_+$. In fact,

$$\hat{J}^2|\psi\rangle = \hbar^2 \, (l + s - 1)(l + s)|\psi\rangle. \tag{9.3.17}$$

This implies that

$$-\sqrt{\frac{s}{(l+s)}} \, |l, l-1, s, s\rangle + \sqrt{\frac{l}{(l+s)}} \, |l, l, s, s-1\rangle = |l, s, l+s-1, l+s-1\rangle. \tag{9.3.18}$$

This process can easily be iterated. Acting on Eqs. (9.3.18) and (9.3.15) with $\hat{J}_-$, the expressions of $|l, s, l+s-1, l+s-2\rangle$ and $|l, s, l+s, l+s-2\rangle$ are obtained in terms of three elements of the basis $|l, m_3, s, s_3\rangle$, i.e. $|l, l, s, s-2\rangle$, $|l, l-1, s, s-1\rangle$ and $|l, l-2, s, s\rangle$. Hence a third combination of these three eigenvectors orthogonal simultaneously to $|l, s, l+s-1, l+s-2\rangle$ and $|l, s, l+s, l+s-2\rangle$ can be found. Such a combination turns out to be the eigenvector $|l, s, l+s-2, l+s-2\rangle$. Hence, at the end of the process, are obtained all multiplets with $j_{\min} \leq j \leq l+s$ are obtained. The total dimension of such a subspace should recover the value $(2l + 1)(2s + 1)$, i.e. (recall that $\sum_{k=0}^{n} k = \frac{n(n+1)}{2}$)

$$(2l + 1)(2s + 1) = \sum_{j=j_{\min}}^{l+s} (2j + 1) = \sum_{j=0}^{l+s} (2j + 1) - \sum_{j=0}^{j_{\min}-1} (2j + 1)$$
$$= (l+s)(l+s+1) - j_{\min}(j_{\min} - 1)$$
$$+ (l+s-j_{\min}+1), \tag{9.3.19}$$

which implies $j_{\min} = |l - s|$. Summarizing the results of this section we can simply state that summing an orbital angular momentum l with a spin s the total angular momentum j obtained assumes all values $|l-s|, |l-s|+1, |l-s|+2, \ldots, l+s$. In the case of an electron ($s = \frac{1}{2}$) the addition between orbital and spin angular momenta is particularly simple since we only have $j = l \pm \frac{1}{2}$.

9.4 The Pauli equation

The analysis of an electron in an external electromagnetic field is an important problem in quantum mechanics. As the reader knows, the electromagnetic field can be described in terms of the potential A_μ, with components (ϕ, A_x, A_y, A_z). With our notation, ϕ corresponds to the temporal component (the so-called scalar potential), and A_x, A_y and A_z are the components of the vector potential in Cartesian coordinates. When relativistic effects are negligible, the Lagrangian is thus found to be

$$\mathcal{L} = \frac{m_e}{2}\left(\dot{x}^2 + \dot{y}^2 + \dot{z}^2\right) + \frac{q_e}{c}\left(\dot{x}A_x + \dot{y}A_y + \dot{z}A_z\right) - q_e\phi. \tag{9.4.1}$$

Interestingly, the canonical momenta $p_i \equiv \frac{\partial \mathcal{L}}{\partial \dot{q}^i}$ do not coincide with the kinematic momenta (i.e. the components of linear momentum) $P_x \equiv m_e \dot{x}, P_y \equiv m_e \dot{y}, P_z \equiv m_e \dot{z}$, but

$$p_k = m_e \dot{x}_k + \frac{q_e}{c} A_k = P_k + \frac{q_e}{c} A_k. \tag{9.4.2}$$

The energy of the electron is a quadratic function of $\dot{x}, \dot{y}$ and $\dot{z}$, without linear terms, because

$$E = \dot{x} p_x + \dot{y} p_y + \dot{z} p_z - \mathcal{L} = \frac{m_e}{2} \left(\dot{x}^2 + \dot{y}^2 + \dot{z}^2 \right) + q_e \phi. \tag{9.4.3}$$

Thus, on using (again) Eq. (9.4.2), the Hamiltonian can eventually be expressed in terms of the canonical momenta as

$$H = \frac{1}{2m_e} \left[\left(p_x - \frac{q_e}{c} A_x \right)^2 + \left(p_y - \frac{q_e}{c} A_y \right)^2 + \left(p_z - \frac{q_e}{c} A_z \right)^2 \right] + q_e \phi. \tag{9.4.4}$$

Note that the terms linear in the velocities disappear in the Hamiltonian, by virtue of the Euler theorem on homogeneous functions. If the external magnetic field vanishes, a gauge choice makes it possible to write

$$H = \frac{1}{2m_e} \left(p_x^2 + p_y^2 + p_z^2 \right) + q_e \phi. \tag{9.4.5}$$

Because of the existence of spin, the quantum Hamiltonian operator has to act on wave functions

$$\psi : \mathbf{R}^3 \to \mathbf{C}^2.$$

Thus, our three momentum operators should be combined into a 2×2 matrix-valued operator. Our experience with Pauli matrices suggests considering the operator (with summation over repeated indices)

$$\sigma_k p_k = \frac{\hbar}{i} \begin{pmatrix} \frac{\partial}{\partial z} & \frac{\partial}{\partial x} - i \frac{\partial}{\partial y} \\ \frac{\partial}{\partial x} + i \frac{\partial}{\partial y} & -\frac{\partial}{\partial z} \end{pmatrix}. \tag{9.4.6}$$

Thus, on taking into account the electron spin, the Hamiltonian operator is taken to be (σ_0 being the 2×2 identity matrix)

$$H_0 = \frac{1}{2m_e} \left(\sigma_k p_k \right)^2 + U(x, y, z) \sigma_0. \tag{9.4.7}$$

If the external magnetic field vanishes, however, no new effect can be appreciated, and the operator H_0 acts in a diagonal way on the wave function.

In contrast, when an external magnetic field $\vec{B}$ is switched on, it is convenient to use the gauge-invariant kinematic momenta P_x, P_y and P_z. From Eq. (9.4.2)

$$\frac{i}{\hbar} (P_y P_z - P_z P_y) = -\frac{q_e}{c} \left(\frac{\partial A_z}{\partial y} - \frac{\partial A_y}{\partial z} \right) = -\frac{q_e}{c} B_x, \tag{9.4.8}$$

$$\frac{i}{\hbar} (P_z P_x - P_x P_z) = -\frac{q_e}{c} \left(\frac{\partial A_x}{\partial z} - \frac{\partial A_z}{\partial x} \right) = -\frac{q_e}{c} B_y, \tag{9.4.9}$$

$$\frac{i}{\hbar} (P_x P_y - P_y P_x) = -\frac{q_e}{c} \left(\frac{\partial A_y}{\partial x} - \frac{\partial A_x}{\partial y} \right) = -\frac{q_e}{c} B_z. \tag{9.4.10}$$

The crucial step in building the Hamiltonian operator with spin is now to use the gauge-invariant kinematic momenta by writing (cf. Eq. (9.4.4))

$$H = \frac{1}{2m_e}\left(\sigma_k P_k\right)^2 + U(x,y,z)\sigma_0. \tag{9.4.11}$$

On using the labels 1, 2 and 3 for the components along the x-, y- and z-axes, respectively, this leads to the lengthy but useful formula

$$H = \frac{1}{2m_e}\left(\sigma_1^2 P_1^2 + \sigma_2^2 P_2^2 + \sigma_3^2 P_3^2 + \sigma_1\sigma_2 P_1 P_2 + \sigma_2\sigma_1 P_2 P_1\right.$$
$$\left. + \sigma_1\sigma_3 P_1 P_3 + \sigma_3\sigma_1 P_3 P_1 + \sigma_2\sigma_3 P_2 P_3 + \sigma_3\sigma_2 P_3 P_2\right)$$
$$+ U(x,y,z)\sigma_0. \tag{9.4.12}$$

The cross-terms in Eq. (9.4.12) are conveniently rearranged with the help of the identity

$$\sigma_j\sigma_k = \delta_{jk}\sigma_0 + i\varepsilon_{jkl}\sigma_l,$$

and hence

$$H = \widetilde{H}_0 + \frac{1}{2m_e}\left[\sigma_1\sigma_2(P_1 P_2 - P_2 P_1) + \sigma_2\sigma_3(P_2 P_3 - P_3 P_2)\right.$$
$$\left. + \sigma_3\sigma_1(P_3 P_1 - P_1 P_3)\right]$$
$$= \widetilde{H}_0 + \frac{1}{2m_e}\left(i\sigma_k i\frac{\hbar}{c}q_e B_k\right) = \widetilde{H}_0 + \mu_B\vec{\sigma}\cdot\vec{B}, \tag{9.4.13}$$

where

$$\widetilde{H}_0 \equiv \frac{1}{2m_e}\left(\sigma_1^2 P_1^2 + \sigma_2^2 P_2^2 + \sigma_3^2 P_3^2\right) + U(x,y,z)\sigma_0. \tag{9.4.14}$$

The Pauli equation for a generic charged particle with spin in an external magnetic field is therefore

$$i\hbar\frac{\partial\psi}{\partial t} = H\psi, \tag{9.4.15}$$

with the Hamiltonian H given by Eq. (9.4.13), and having replaced in the expression (9.1.5) for μ_B the charge and mass of the electron with the analogous quantities of the particle under study. The coupling term $\vec{\sigma}\cdot\vec{B}$ was derived by Pauli in a highly original way, at a time when the relativistic quantum theory of the electron was not yet developed. It should be stressed, however, that such a theory, due to Dirac (1928, 1958), can be used to put on firmer ground the *ad hoc* non-relativistic derivation of Pauli, in which spin degrees of freedom are added by hand.

9.5 Solutions of the Pauli equation

Let us consider for simplicity the equation (9.4.15) for an electron. In this case only the term $\mu_B\vec{\sigma}\cdot\vec{B}$ in the Hamiltonian depends on the spin operators. Thus, on solving Eq. (9.4.15) by separation of variables, with $\hat{X},\hat{Y},\hat{Z},\hat{S}^2$ and $\hat{S}_z$ as fundamental operators, the time evolution of that part of the wave function depending on spin variables is ruled by the equation

$$i\hbar\frac{\partial\chi}{\partial t} = \frac{2\mu_{\mathrm{B}}}{\hbar}\vec{B}\cdot\widehat{\vec{S}}\,\chi, \tag{9.5.1}$$

where

$$\chi(t) = \begin{pmatrix}\xi_+(t)\\ \xi_-(t)\end{pmatrix} = \xi_+(t)\chi_+ + \xi_-(t)\chi_-. \tag{9.5.2}$$

Two important features are hence found to emerge:

(i) the Pauli equation leads to a coupled system of first-order differential equations, to be solved for given initial conditions;

(ii) the operator on the right-hand side of Eq. (9.5.1) may be time-dependent. Thus, some care is necessary to map the problem into one for which the well-known exponentiation of time-independent matrices can be applied.

Following Landau and Lifshitz (1958), we first consider the Pauli equation for a neutral particle of spin $\frac{1}{2}$ in a magnetic field $\vec{B}$ for which the only non-vanishing component is directed along the z-axis: $B_z = B(t)$. Its explicit form is not specified. Equations (9.5.1) and (9.5.2) therefore lead to a first-order system which, in this particular case, is decoupled:

$$\frac{\partial\xi_+}{\partial t} = -\frac{i\mu_{\mathrm{B}}B}{\hbar}\xi_+, \tag{9.5.3}$$

$$\frac{\partial\xi_-}{\partial t} = \frac{i\mu_{\mathrm{B}}B}{\hbar}\xi_-. \tag{9.5.4}$$

The solution is hence given by

$$\xi_+(t) = C_+\,e^{-\frac{i\mu_{\mathrm{B}}}{\hbar}\int_{t_0}^{t}B(t')\mathrm{d}t'}, \tag{9.5.5}$$

$$\xi_-(t) = C_-\,e^{\frac{i\mu_{\mathrm{B}}}{\hbar}\int_{t_0}^{t}B(t')\mathrm{d}t'}, \tag{9.5.6}$$

where the constants C_+ and C_- can be determined once the initial conditions are given.

In a second example, we consider instead a magnetic field $\vec{B}$ with components (Landau and Lifshitz 1958)

$$B_x = B\sin\theta\cos\omega t, \tag{9.5.7}$$

$$B_y = B\sin\theta\sin\omega t, \tag{9.5.8}$$

$$B_z = B\cos\theta. \tag{9.5.9}$$

In this case, the modulus of $\vec{B}$ is constant in time, whereas all its components change smoothly according to Eqs. (9.5.7)–(9.5.9). Thus, the Pauli equation can be expressed by the coupled system

$$i\hbar\frac{\partial}{\partial t}\begin{pmatrix}\xi_+\\ \xi_-\end{pmatrix} = \mu_{\mathrm{B}}B\begin{pmatrix}\cos\theta & \sin\theta\,e^{-i\omega t}\\ \sin\theta\,e^{i\omega t} & -\cos\theta\end{pmatrix}\begin{pmatrix}\xi_+\\ \xi_-\end{pmatrix}. \tag{9.5.10}$$

The matrix on the right-hand side of Eq. (9.5.10) depends explicitly on time. We can map our equation into an equivalent equation where the time dependence of the matrix disappears on setting

$$\phi_+ \equiv e^{\frac{i}{2}\omega t}\,\xi_+,$$

(9.5.11)

$$\phi_- \equiv e^{-\frac{i}{2}\omega t}\,\xi_-.$$

(9.5.12)

On using Eqs. (9.5.10)–(9.5.12) and the Leibniz rule we then find the first-order equation

$$\frac{\partial}{\partial t}\begin{pmatrix}\phi_+ \\ \phi_-\end{pmatrix} = \frac{i}{2}\begin{pmatrix}(\omega + 2\omega_B\cos\theta) & 2\omega_B\sin\theta \\ 2\omega_B\sin\theta & -(\omega + 2\omega_B\cos\theta)\end{pmatrix}\begin{pmatrix}\phi_+ \\ \phi_-\end{pmatrix},$$

(9.5.13)

where $\omega_B \equiv \mu_B B/\hbar$. The system (9.5.13) can be decoupled by acting with $\frac{\partial}{\partial t}$ on both sides and then using the original first-order equations again. Then,

$$\phi_+(t) = A_1 e^{\frac{i}{2}\Omega t} + A_2 e^{-\frac{i}{2}\Omega t},$$

(9.5.14)

$$\phi_-(t) = C_1 e^{\frac{i}{2}\Omega t} + C_2 e^{-\frac{i}{2}\Omega t},$$

(9.5.15)

where

$$\Omega \equiv \sqrt{(\omega + 2\omega_B\cos\theta)^2 + 4\omega_B^2\sin^2\theta},$$

(9.5.16)

$$C_1 \equiv \frac{2\omega_B\sin\theta}{(\Omega + \omega + 2\omega_B\cos\theta)}A_1,$$

(9.5.17)

$$C_2 \equiv \frac{-2\omega_B\sin\theta}{(\Omega - \omega - 2\omega_B\cos\theta)}A_2.$$

(9.5.18)

9.5.1 Another simple application of the Pauli equation

Suppose that a particle with magnetic moment $\vec{\mu}$ is placed in a constant magnetic field pointing along the x-axis. At the initial time $t = 0$, the particle is found to have $s_3 = \frac{1}{2}$. We are aiming to evaluate the probabilities, at any later time, of finding the particle with $s_2 = \pm\frac{1}{2}$ (Lim et al. 1998 problem 3024). The spin wave function is a column vector with rows ξ_+ and ξ_-, say. The resulting Pauli equation reads as

$$\left[i\hbar I\frac{d}{dt} + \mu_B B\begin{pmatrix}0 & 1 \\ 1 & 0\end{pmatrix}\right]\begin{pmatrix}\xi_+ \\ \xi_-\end{pmatrix} = 0.$$

(9.5.19)

This yields a coupled system of first-order ordinary differential equations. They can be decoupled by differentiating with respect to time and then re-inserting the original first-order equations. Eventually we get

$$\left[\frac{d^2}{dt^2} + \frac{\mu_B^2 B^2}{\hbar^2}\right]\xi(t) = 0,$$

(9.5.20)

where $\xi(t)$ equals $\xi_+(t)$ or $\xi_-(t)$. Thus, on defining

$$\omega_B \equiv \frac{\mu_B B}{\hbar};$$ (9.5.21)

this equation has the familiar oscillating solutions

$$\xi_\pm = A_\pm e^{i\omega_B t} + B_\pm e^{-i\omega_B t},$$ (9.5.22)

where $A_\pm$ and $B_\pm$ are determined from the initial conditions. In our case, the initial spin wave function is $\begin{pmatrix} 1 \\ 0 \end{pmatrix}$, which satisfies the initial condition

$$\hat{S}_3 \begin{pmatrix} 1 \\ 0 \end{pmatrix} = \frac{1}{2} \begin{pmatrix} 1 \\ 0 \end{pmatrix},$$ (9.5.23)

from which $\xi_+(0) = 1, \xi_-(0) = 0$. The other pair of initial conditions are obtained from the Pauli equation, i.e.

$$\dot{\xi}_+(0) = 0, \ \dot{\xi}_-(0) = i\omega.$$ (9.5.24)

The full set of four initial conditions yields a linear system of algebraic equations for the integration constants, i.e.

$$A_+ + B_+ = 1,$$ (9.5.25)

$$A_- + B_- = 0,$$ (9.5.26)

$$\omega(A_+ - B_+) = 0,$$ (9.5.27)

$$\omega(A_- - B_-) = \omega,$$ (9.5.28)

which is solved by

$$A_+ = A_- = B_+ = -B_- = \frac{1}{2}.$$ (9.5.29)

As a function of time, the spin wave function reads therefore as

$$\begin{pmatrix} \cos \omega_B t \\ i \sin \omega_B t \end{pmatrix}.$$

On the other hand, the eigenstate of $\hat{S}_2$ with eigenvalue $\frac{1}{2}$ can be written, up to a phase factor, in the form

$$|s_2(+)\rangle = \frac{1}{\sqrt{2}} \begin{pmatrix} 1 \\ i \end{pmatrix},$$ (9.5.30)

while the eigenstate of $\hat{S}_2$ with eigenvalue $-\frac{1}{2}$ can be expressed by the column vector

$$|s_2(-)\rangle = \frac{1}{\sqrt{2}} \begin{pmatrix} 1 \\ -i \end{pmatrix}.$$ (9.5.31)

Thus, the desired probabilities are obtained by squaring up the modulus of the projection of the spin wave function upon these eigenstates, i.e.

$$P_+ = |\langle s_2(+)|\psi(t)\rangle|^2 = \frac{1}{2}(1 + \sin 2\omega_B t), \tag{9.5.32}$$

$$P_- = |\langle s_2(-)|\psi(t)\rangle|^2 = \frac{1}{2}(1 - \sin 2\omega_B t). \tag{9.5.33}$$

9.6 Landau levels

Here we are interested in the energy levels of a particle of mass m and charge q with spin in a time-independent constant magnetic field. We will see that a profound link exists between this problem and the spectrum of harmonic oscillators (Landau 1930).

It is convenient to take the vector potential in the form

$$A_x = -By, \tag{9.6.1}$$

$$A_y = A_z = 0, \tag{9.6.2}$$

if the magnetic field is directed along the z-axis. Denoting by μ the intrinsic magnetic moment of the particle, the Hamiltonian operator reads (Landau and Lifshitz 1958)

$$\hat{H} = \hat{H}_0 + \hat{H}_S \tag{9.6.3}$$

where

$$\hat{H}_0 = \frac{1}{2m}\left(\hat{p}_x + \frac{qB}{c}\hat{y}\right)^2 + \frac{\hat{p}_y^2}{2m} + \frac{\hat{p}_z^2}{2m},$$
$$\hat{H}_S = -\frac{\mu}{\hbar}B\hat{S}_3, \tag{9.6.4}$$

where $\hat{p}_i \equiv m\hat{v}_i + \frac{q}{c}\hat{A}_i$, $\forall i = x, y, z$. Since the only contribution in the Hamiltonian depending on the spin, i.e. $\hat{H}_S$, does not depend on spatial coordinates, this implies that $[\hat{H}_0, \hat{H}_S] = 0$, hence the two Hamiltonians can be diagonalized separately: $\hat{H}_S$ in $\mathbf{C}^2$ and $\hat{H}_0$ in $\mathcal{L}_2(R^3)$, respectively. The diagonalization of $\hat{H}_S$ is straightforward and its eigenvalues can be expressed in terms of the eigenvalue $\hbar \lambda_3$ of $\hat{S}_3$; for this reason we focus our attention on the diagonalization of $\hat{H}_0$ only. In this case the equation to solve reads

$$\frac{1}{2m}\left[\left(\hat{p}_x + \frac{qB}{c}\hat{y}\right)^2 + \hat{p}_y^2 + \hat{p}_z^2\right]\varphi(\vec{r}) - \mu\,\lambda_3\,B\,\varphi(\vec{r}) = E\varphi(\vec{r}). \tag{9.6.5}$$

The Hamiltonian of this form of the stationary Schrödinger equation does not depend explicitly on x and z. This implies that p_x and p_z are conserved quantities of our problem, and φ may be expressed in the form

$$\varphi(\vec{r}) = e^{\frac{i}{\hbar}(p_x x + p_z z)}\,Y(y). \tag{9.6.6}$$

Note that the eigenvalues p_x and p_z take all values from $-\infty$ to $+\infty$. Moreover, since $A_z = 0$, the p_z component of the canonical momentum coincides with the z component of the linear momentum, and hence the component v_z of the velocity of the particle may take arbitrary values. This peculiar property is expressed by saying that the motion along the field *is not quantized*.

On requiring that

$$\hat{p}_x \equiv \frac{\hbar}{i} \frac{\partial}{\partial x}, \quad \hat{p}_y \equiv \frac{\hbar}{i} \frac{\partial}{\partial y}, \quad \hat{p}_z \equiv \frac{\hbar}{i} \frac{\partial}{\partial z},$$

as is always the case for canonical momenta in Cartesian coordinates, the insertion of Eq. (9.6.6) into (9.6.5) leads to

$$Y''(y) + \frac{2m}{\hbar^2}\left[\left(E + \frac{\mu \lambda_3}{s}B - \frac{p_z^2}{2m}\right) - \frac{m}{2}\omega_B^2(y - y_0)^2\right]Y(y) = 0, \tag{9.6.7}$$

where we have defined

$$y_0 \equiv -\frac{cp_x}{qB}, \tag{9.6.8}$$

$$\omega_B \equiv \frac{|q|B}{mc}. \tag{9.6.9}$$

Note that Eq. (9.6.7) is the stationary Schrödinger equation for a one-dimensional harmonic oscillator of frequency ω_B, and we know that the energy eigenvalues are $\left(n + \frac{1}{2}\right)\hbar\omega_B$. In other words,

$$E_n = \left(n + \frac{1}{2}\right)\hbar\omega_B + \frac{p_z^2}{2m} - \mu\,\lambda_3\,B. \tag{9.6.10}$$

For an electron, $\mu = -\mu_B$, so that the energy spectrum becomes

$$E_n = \left(n + \frac{1}{2} + \lambda_3\right)\hbar\omega_B + \frac{p_z^2}{2m}. \tag{9.6.11}$$

The corresponding eigenfunctions are expressed through the Hermite polynomials in the standard way

$$Y_n(y) = \frac{1}{\pi^{\frac{1}{4}}a_B^{\frac{1}{2}}\sqrt{2^n n!}}\exp\left[-\frac{(y - y_0)^2}{2a_B^2}\right]H_n\left(\frac{y - y_0}{a_B}\right), \tag{9.6.12}$$

where $a_B \equiv \sqrt{\frac{\hbar}{m\omega_B}}$.

A striking consequence of Eq. (9.6.10) is that, since p_x does not contribute to the energy spectrum, while it ranges continuously over all values from $-\infty$ to $+\infty$, the energy levels have an *infinite degeneracy*. It is usually suggested that such an infinite degeneracy may be removed by confining the motion in the xy plane to a large but finite area. The problem remains, however, to understand what is the deeper underlying reason for the occurrence of such an infinite degeneracy, from the point of view of the general formalism of quantum mechanics. For this purpose a more advanced treatment is needed, which can be found, for example, in our previous work (Esposito *et al.* 2004).

9.7 Spin–orbit interaction: Thomas precession

The above description of the spin of the electron as an *empirical angular momentum* with giromagnetic factor $g = 2$ prepared the conceptual ground for L.H. Thomas to discover the presence of a relativistic kinematical correction to atomic energy. Such an effect, better known as spin–orbit interaction, reproduces in fact the anomalous Zeeman effect instead of the normal one for atoms in magnetic fields (as discussed in the following chapters), and the correct fine-structure splitting of energy multiplets (Jackson 1975).

Under the Uhlenbeck–Goudsmit hypothesis an electron possesses a spin angular momentum $\vec{S}$ which can assume quantized levels $\pm\frac{\hbar}{2}$, and hence a magnetic moment $\vec{\mu} = -\frac{|q_e|}{(m_e c)}\vec{S}$. The equation of motion of the angular momentum in the rest frame of the particle reads

$$\frac{d\vec{S}}{dt} = \vec{\mu} \wedge \vec{B}', \tag{9.7.1}$$

with the *prime* symbol on the magnetic field denoting that it is calculated in the particle rest frame. It is possible to show, by using the proper Lorentz transformation, that $\vec{B}'$ is given in terms of the electric and magnetic fields $\vec{E}$ and $\vec{B}$ in the laboratory frame as

$$\vec{B}' \approx \left(\vec{B} - \frac{\vec{v}}{c} \wedge \vec{E}\right), \tag{9.7.2}$$

up to the first order in $\frac{v}{c}$ expansion. The equation of motion (9.7.1) is obtained by the potential energy

$$U' = -\vec{\mu} \cdot \left(\vec{B} - \frac{\vec{v}}{c} \wedge \vec{E}\right). \tag{9.7.3}$$

In an atom the electric field can be approximated as the one produced by the nucleus, i.e.

$$q_e\vec{E} = -\frac{\vec{r}}{r}\frac{dV}{dr}; \tag{9.7.4}$$

hence, substituting in Eq. (9.7.3),

$$U' = \frac{|q_e|}{m_e c}\vec{S} \cdot \vec{B} + \frac{1}{m_e^2 c^2}\left(\vec{S} \cdot \vec{L}\right)\frac{1}{r}\frac{dV}{dr}. \tag{9.7.5}$$

Such a result is not completely correct, since as pointed out by Thomas the electron rest frame will in general be rotating with respect to the laboratory frame. The rate of change of a vector $\vec{P}$ in the lab frame is generically given by

$$\frac{d\vec{P}}{dt} = \left(\frac{d\vec{P}}{dt}\right)_{\text{non-rot}} - \vec{\omega} \wedge \vec{P}, \tag{9.7.6}$$

where $\vec{\omega}$ is the angular velocity of rotation. By virtue of this additional term, the new potential energy becomes

$$U = U' - \vec{S} \cdot \vec{\omega}. \tag{9.7.7}$$

As shown in Jackson (1975) $\vec{\omega}$ can be evaluated by computing the Lorentz transformation between laboratory and particle rest frame up to first order in the $\frac{v}{c}$ expansion parameter. In this case,

$$\vec{\omega} \approx \frac{1}{2c^2}\vec{v} \wedge \vec{a}. \tag{9.7.8}$$

For electrons in an atom the acceleration $\vec{a}$ is produced by the screened Coulomb field, thus

$$\vec{\omega} \approx \frac{1}{2c^2} \frac{\vec{r} \wedge \vec{v}}{m_e} \frac{1}{r} \frac{dV}{dr} = \frac{1}{2m_e^2 c^2} \frac{1}{r} \frac{dV}{dr} \vec{L}. \tag{9.7.9}$$

Thus, substituting Eq. (9.7.9) in Eq. (9.7.7) we get the final expression

$$U = \frac{|q_e|}{m_e c} \vec{S} \cdot \vec{B} + \frac{1}{2m_e^2 c^2} \left(\vec{S} \cdot \vec{L} \right) \frac{1}{r} \frac{dV}{dr}, \tag{9.7.10}$$

which differs from Eq. (9.7.5) by a factor two at the denominator of the second contribution. In the absence of an external magnetic field, i.e. $\vec{B} = 0$, the only remaining term is

$$U = \frac{1}{2m_e^2 c^2} \left(\vec{S} \cdot \vec{L} \right) \frac{1}{r} \frac{dV}{dr}, \tag{9.7.11}$$

which is typically known as the spin–orbit interaction term. In this case the contribution (9.7.11) has to be added to the usual Hamiltonian of the hydrogen atom which, as shown in the previous chapters, is diagonalized in the basis $|n, l, m, s, s_z\rangle$. Such a basis does not diagonalize U too, but rather it would require using as complete set of observables $\mathcal{H}, \hat{L}^2$, $\hat{S}^2, \hat{J}^2$ and $\hat{J}_3^2$; in fact, one can easily prove that $\hat{\vec{S}} \cdot \hat{\vec{L}} = \frac{(\hat{J}^2 - \hat{L}^2 - \hat{S}^2)}{2}$. By exploiting such an observation we get

$$U|l, s, j, j_z\rangle = \frac{1}{2m_e^2 c^2} \frac{1}{r} \frac{dV}{dr} \left(\hat{\vec{S}} \cdot \hat{\vec{L}} \right) |l, s, j, j_z\rangle$$

$$= \frac{1}{2m_e^2 c^2} \frac{1}{r} \frac{dV}{dr} \frac{\hbar^2}{2} \left[j(j+1) - l(l+1) - \frac{3}{4} \right] |l, s, j, j_z\rangle. \tag{9.7.12}$$

When an external magnetic moment is present, the term $\frac{|q_e|}{m_e c} \vec{S} \cdot \vec{B}$ must be considered, and hence even the basis $|l, s, j, j_z\rangle$ does not diagonalize the complete Hamiltonian. In this case, approximation methods are necessary, as discussed in Chapter 11.

9.8 Problems

9.P1. The Hamiltonian of a system of two spin-$\frac{1}{2}$ particles reads as

$$H = a\vec{s}_1 \cdot \vec{s}_2. \tag{9.8.1}$$

At the initial time $t = 0$ the z-components of the spin of the two particles are measured simultaneously, and the values $\frac{\hbar}{2}$ and $-\frac{\hbar}{2}$ are found, respectively. What is the probability of finding the same result if the measurement is repeated at time $t = \frac{\pi}{2\hbar a}$?

9.P2. A particle of mass m, spin $\frac{1}{2}$ and magnetic moment $\vec{\mu} = g\vec{s}$, subject to an isotropic elastic force, is subject to a uniform magnetic field along the z-axis. The particle is initially in the state

$$|\psi_0\rangle = \exp\left\{ -\frac{i}{\hbar} \vec{a} \cdot \vec{P} \right\} |\psi_0\rangle |\chi_+^x\rangle, \tag{9.8.2}$$

where $|\psi_0\rangle$ is the ground state of the elastic interaction, $\vec{P}$ is the linear momentum, $\vec{a}$ is an arbitrary vector and $|\chi_+^x\rangle$ is the eigenstate of s_x belonging to the eigenvalue $\frac{\hbar}{2}$. Compute the mean values of position, energy and x-component of the total angular momentum, and their time-evolution.

9.P3. A spin-$\frac{1}{2}$ particle with magnetic moment μ is immersed in a uniform magnetic field directed along the z-axis and has modulus B. Suppose that, at time $t = 0$, the particle is in the eigenstate of S_x belonging to the eigenvalue $-\frac{\hbar}{2}$. Find, at a generic time t, the probability that the particle is in the eigenstate of S_x, or S_y, or S_z, belonging to the eigenvalue $\frac{\hbar}{2}$.

9.P4. Three distinguishable spin-$\frac{1}{2}$ particles are described by the Hamiltonian

$$H = A\left(\vec{S}_1 \cdot \vec{S}_2 + \vec{S}_2 \cdot \vec{S}_3 + \vec{S}_3 \cdot \vec{S}_1\right), \tag{9.8.3}$$

where A is a constant. Find

(i) The dimension of the Hilbert system of this physical system.
(ii) A basis for the system.
(iii) The energy levels of the system and their degeneracies.
(iv) The energy levels and their degeneracies if the particles are taken to be identical.

9.P5. Consider a spin-$\frac{1}{2}$ particle moving on the real line and described by the following Hamiltonian:

$$H = \frac{1}{2m}\left(p^2 + W^2 + \hbar\sigma_3\frac{dW}{dx}\right), \tag{9.8.4}$$

where W is an arbitrary C^1-function and σ_i is the standard notation for Pauli matrices.

(i) Prove that the following quantities are constants of motion:

$$N = \frac{1}{2}(I - \sigma_3), \tag{9.8.5}$$

$$Q_1 = \frac{1}{2}(\sigma_1 p + \sigma_2 W), \tag{9.8.6}$$

$$Q_2 = \frac{1}{2}(\sigma_2 p - \sigma_1 W). \tag{9.8.7}$$

Symmetries in quantum mechanics

The meaning of symmetries is studied in some detail, and a study is performed of transformations which preserve the description, of transformations of frames and the corresponding quantum symmetries. The second part of the chapter outlines Galilei transformations, time translations, spatial reflection and time reversal.

10.1 Meaning of symmetries

Symmetries and constants of motion have already appeared here and there in this book, most notably in Appendix 3.A. Recall that, therein, we have defined symmetries as those transformations that generate new solutions out of known solutions; more generally, that permute solutions among themselves. Constants of motion are functions on the carrier space which, when restricted to solutions, acquire a constant value depending on the solution considered. This point of view may be viewed as a dynamical point of view, i.e. depending on the specific system under consideration.

There is, however, another meaning of symmetry, which deals with the symmetry of the description that ensures the covariance of the equations of motion, e.g. transformations that map equations of motion in Hamiltonian form into other equations still in Hamiltonian form, or transformations that map second-order differential equations into second-order ones, i.e. equations admitting a Lagrangian description into new ones also admitting a Lagrangian description. Transformations of the dynamical type emerge out of the previous ones when we further require that the Hamiltonian is preserved or the Lagrangian is preserved up to a total time derivative. As Wigner advocates, 'symmetries are used, in physics, in two distinct manners. First, they are used as superlaws of nature in that, once their validity has been suggested by their consistency with the known laws of nature, they serve as guides in our search for as yet unknown laws of nature. Second, they serve as tools for obtaining properties of the solutions of the equations provided by the laws of nature. It is desirable for the first use to give a formulation of symmetries directly in terms of the primitive concepts of physical theory, i.e. in terms of observations, or measurements, and their results' (Wightman 1995).

In our previous usage of symmetries in quantum mechanics we have implicitly assumed that, by virtue of the probabilistic interpretation, our transformations are required to be unitary. Moreover, we have taken the active point of view for the discussion of symmetries. In this chapter we would like to discuss further, at a deeper level, both the symmetries of the description and the difference between the active and passive points of view. This last discussion is particularly relevant in quantum mechanics because of the important role played by the observer. To be concrete, we consider the Schrödinger–Dirac picture of quantum mechanics. With every physical system S we associate a Hilbert space $\mathcal{H}_S$, considered as a primary object. Observables are identified as real operators acting on $\mathcal{H}_S$, and are a derived concept. Expectation-value functions, associated with an operator A and a vector $|\psi\rangle \in \mathcal{H}_S$, are defined as

$$e_A(\psi) \equiv \frac{\langle \psi | A | \psi \rangle}{\langle \psi | \psi \rangle}.$$

These functions capture the probabilistic aspects of quantum mechanics. In this manner we realize the minimal properties that the description of a physical system should possess (see Section 4.7):

  (i) a space of states $\mathcal{S}$;
  (ii) a space of observables $\mathcal{O}$;
  (iii) a real-valued pairing $\mu : \mathcal{O} \times \mathcal{S} \to \mathbf{R}$, associating a real number with any state and any observable.

In addition, we need an identification of physical variables. Along with these aspects, we should pay attention to the quantum-to-classical transition, i.e., under some appropriate conditions (normally, pictorially expressed as $\hbar \to 0$), we should recover classical dynamics. A possible way to achieve this last requirement is to realize the Hilbert space associated with our quantum system by means of square-integrable functions on some configuration space. We usually refer to this realization as the 'position representation', which differs from the 'momentum representation' that is associated with the realization of the Hilbert space $\mathcal{H}$ by means of square-integrable functions on momentum space. An implicit connection with the classical limit is contained in the observation that the 'direct product' of the configuration space times the momentum space reproduces the classical phase space. Within a purely quantum setting, these spaces would be obtained separately as the joint spectrum of maximal sets of commuting operators, each joint spectrum being a differentiable manifold. Notice, however, that in the coherent-state picture we are dealing with 'wave functions' defined on phase space.

In particular, when our quantum system consists of elementary particles, we may assume our configuration space to be space–time or suitable copies of it, specifically $\mathbf{R}^3 \times \mathbf{R}$ or $\mathbf{R}^3 \times \cdots \times \mathbf{R}^3 \times \mathbf{R}$. With this identification we may give a meaning to the notion of reference frame and observer that we have encountered already in a classical setting.

Since our formulation is non-relativistic, we assume that we are dealing with a given splitting of space–time into time and space. The time function will be assumed to be the same for all reference frames, and space will be assumed to be an affine space. Each observer will introduce a Cartesian coordinate system to describe its space part, by selecting an orthonormal basis in the vector space obtained after an origin has been fixed in the affine space, and two different Cartesian coordinate systems will be connected by a rotation plus a translation, an element of the Euclidean group. The translation connects the two selected 'origins' in the affine space, while the rotation takes us from one orthonormal basis to another in the model vector space of the affine space. From an observational point of view, we shall assume that any two reference frames connected by a transformation belonging to the Euclidean group are equivalent. This equivalence is usually motivated to implement the observed isotropy and homogeneity of space. From a more concrete point of view, the equivalence implies that there is a prescription to compare the experimental results of an observer with those of another, when they are performing experiments on the same physical system.

This point of view is usually called the *passive point of view*, i.e. the same system is observed by two different observers and we compare the descriptions they provide by means of well-established rules.

The *active point of view*, on the contrary, considers a given observer who makes observations on a system and on a transformed one, i.e. another copy of the same system but located in a different part of space and oriented otherwise. If two classically equivalent reference frames, e.g. connected by a Galilei transformation, describe a quantum system, how should each of these reference frames compare their experimental results so that they might also be considered as equivalent from the point of view of quantum mechanics? Or, stated differently, if two frames are classically equivalent, are they also quantum mechanically equivalent? In Section 4.4.1 we already took a direct approach to implement the Galilei transformations (Eq. (4.4.7)) on the differential operator occurring in the Schrödinger equation; there we showed that in two different frames connected by a Galilei transformation the Schrödinger equation has the same form if we allow the wave function in one frame to gain a phase with respect to the one considered in the other frame. Now we take up this problem in more general terms.

A transformation from one frame to another is associated with a transformation on the space of states that preserves the transition probabilities. Similarly to the classical situation, for quantum mechanics there should also be no quantum experiment that would allow the establishment the state of the uniform motion of a frame.

10.1.1 Transformations that preserve the description

As we have already seen, by virtue of the probabilistic interpretation, physical states, are not described by vectors in $\mathcal{H}_S$ but rather by rays (equivalence classes of vectors differing by multiplication by a non-vanishing complex number). We denote the space of rays by $R(\mathcal{H}_S)$. A convenient way to parametrize rays is by means of rank-1 projectors

$$\rho \equiv \frac{|\psi\rangle\langle\psi|}{\langle\psi|\psi\rangle}. \tag{10.1.1}$$

Here $|\psi\rangle$ is any representative in the ray to which $|\psi\rangle$ belongs any other vector $|\psi'\rangle = \lambda e^{i\varphi}|\psi\rangle$, in the same equivalence class, gives the same pure state, i.e.

$$\rho = \frac{|\psi'\rangle\langle\psi'|}{\langle\psi'|\psi'\rangle}. \tag{10.1.2}$$

Having stated that pure states are rays, elements of $R(\mathcal{H}_S)$, it is natural to require that transformations that preserve the description of our theory should be invertible 1-to-1 maps from $R(\mathcal{H}_S)$ to $R(\mathcal{H}_S)$, with the requirement that they preserve the transition probabilities between pure states, i.e. they should preserve the scalar product between rank-1 projectors:

$$\langle \rho_1 | \rho_2 \rangle = \text{Tr}(\rho_1 \rho_2) = \frac{\langle \psi_1 | \psi_2 \rangle \langle \psi_2 | \psi_1 \rangle}{\langle \psi_1 | \psi_1 \rangle \langle \psi_2 | \psi_2 \rangle}. \tag{10.1.3}$$

This expression makes it possible to conclude by inspection that unitary transformations on $\mathcal{H}_S$ will induce transition-probability preserving maps; indeed, if

$$U : |\psi\rangle \to U|\psi\rangle = |\psi'\rangle$$

is a unitary map, then $T(U) : \rho \to T(U)\rho$, given by

$$T(U)\rho \equiv \frac{U|\psi\rangle\langle\psi|U^\dagger}{\langle U\psi | U\psi \rangle}, \tag{10.1.4}$$

has the required property

$$\langle \rho_1 | \rho_2 \rangle = \langle T(U)\rho_1 | T(U)\rho_2 \rangle. \tag{10.1.5}$$

Is it possible to get all maps we are interested in by using this procedure? An important theorem by Wigner answers this question almost positively: any transition-probability preserving map arises from a unitary or anti-unitary transformation on the Hilbert space (see Esposito *et al.* 2004). We are disregarding a possible multiplication by a positive real number by requiring that the transformation on $\mathcal{H}_S$ should not alter the norm of the vector.

As we have already said in Appendix 3.A, a reference frame identifies a splitting of space–time into time and space. In our setting, space is an affine three-dimensional space. An observer in a frame introduces coordinate systems by selecting an origin, so that points of the affine space are in one-to-one correspondence with vectors. Then the observer selects an orthonormal basis, say $|e_1\rangle|e_2\rangle|e_3\rangle$, so that coordinates of any vector $|v\rangle$ are defined by $\langle e_j | v \rangle = x_j(v)$. Here, we are using Dirac notation of bra-ket for our Euclidean vector space which models the three-dimensional affine space. The selection of this space is equivalent to the choice of a set of orthogonal Cartesian axes.

Two different choices of origin and the subsequent basis of orthonormal vectors may be ascribed to two different observers, and a transformation from one description to another will be represented by a translation (connecting the two origins) and a rotation, transforming one orthonormal basis into the other. Usually, this approach is dealt with as a transformation connecting one coordinate system with the other; it is thus considered to represent the passive point of view. If we consider, instead, the transformation mapping one point of the affine space into another point of the same space, we say we are dealing with the active point of view.

The experimental observation of homogeneity and isotropy of space requires that frames connected by rototranslations are equivalent; if we detect a difference between one location and the other homogeneity is lost, and if we detect a difference between one orientation and the other isotropy is lost. To better explain the difference between active and passive transformations we shall make first our considerations on the Hilbert space $\mathcal{H}_s$.

We have already seen that, to introduce coordinate systems on $\mathcal{H}_S$, each observer selects an orthonormal basis of vectors

$$\{|e\rangle\} = \{|e_1\rangle, |e_2\rangle, \ldots, |e_n\rangle, \ldots\}$$

with $\sum_n |e_n\rangle\langle e_n| = I$, and defines linear coordinate functions on vectors by setting

$$\langle e_j|\psi\rangle = \psi_e(j),$$

also denoted by ψ_j for simplicity. In the continuum case we use

$$\int_{\mathbf{R}} |a\rangle \, \mathrm{d}a \langle a| = I$$

with $\langle a|a'\rangle = \delta(a - a')$, and introduce our coordinate system by means of

$$\langle a|\psi\rangle = \psi(a).$$

By analogy with the position or momentum representation, very often in the literature the components of $|\psi\rangle$, in the discrete case ψ_j or $\psi(a)$ in the continuum case, are called wave functions. By selecting a different orthonormal basis, say a different observer, for which

$$\{|f\rangle\} = \{|f_1\rangle, |f_2\rangle, \ldots, |f_n\rangle, \ldots\}, \quad \sum_n |f_n\rangle\langle f_n| = I,$$

we define a new coordinate system with $\psi_f(j) \equiv \langle f_j|\psi\rangle$. It is possible to pass from one coordinate system to the other by exploiting Dirac notation in terms of bra and kets along with the decomposition of the identity. We have

$$|\psi\rangle = \sum_n |e_n\rangle\langle e_n|\psi\rangle,$$

and acting on both sides by $\langle f_j|$ we get

$$\langle f_j|\psi\rangle = \sum_n \langle f_j|e_n\rangle\langle e_n|\psi\rangle,$$

or

$$\psi_f(j) = \sum_n \langle f_j|e_n\rangle\psi_e(n). \tag{10.1.6}$$

The matrix $(\langle f_j|e_n\rangle)$ taking us from one coordinate system to another is a unitary matrix $U(e, f)$ because the bases have been chosen to be orthonormal. In this picture the vector $|\psi\rangle$ remains unaltered and we only transform wave functions. Again, this is called the passive point of view. A different possibility is also available, i.e. we define a new vector $|\psi'\rangle$ in $\mathcal{H}_S$, which is obtained by setting

$$|\psi'\rangle \equiv \sum_n \psi_e(n)|f_n\rangle. \tag{10.1.7}$$

In other words, we consider the components of $|\psi\rangle$ in the $\{|e\rangle\}$-basis and use them as coefficients in the $\{|f\rangle\}$-basis to reconstruct a vector $|\psi'\rangle$. In this way we define a unitary transformation in $\mathcal{H}_S$, mapping $|\psi\rangle$ into $|\psi'\rangle$, which is said to represent the active counterpart of previously constructed coordinate transformations. By using the duality relation between states and observables, this same construction can be considered on operators. This time, the coordinate representation of the operator will be a matrix and, in the passive point of view, we transform from one matrix representation to another while the operator is unchanged. To define the associated active transformation we consider the associated matrix in the $\{|e\rangle\}$-basis and turn it into an operator by using the $\{|f\rangle\}$-basis, i.e.

$$A' = \sum_{j,k} |f_j\rangle \langle e_j|A|e_k\rangle \langle f_k|. \tag{10.1.8}$$

We have therefore

$$A' = U(e,f)AU^\dagger(e,f), \tag{10.1.9}$$

an active transform on the observables.

Remark   In the reconstruction of a vector from its components in a given basis, a different procedure is possible. We might define

$$|\psi''\rangle = \sum_n \psi_e^*(n)|f_n\rangle, \tag{10.1.10}$$

by using the complex conjugate of the components in the $\{|e\rangle\}$-basis. The induced transformation from $|\psi\rangle$ to $|\psi''\rangle$ is no longer unitary, but is anti-unitary. Nevertheless,

$$|\langle\psi|\varphi\rangle|^2 = |\langle\psi'|\varphi'\rangle|^2 = |\langle\psi''|\varphi''\rangle|^2, \tag{10.1.11}$$

i.e. at the level of transition probabilities the transformation could be either unitary or anti-unitary, and the transition probability would still be preserved. It is possible to show that any anti-unitary transformation may be written as the product of a unitary transformation times a complex conjugation. The product of two anti-unitary transformations is unitary, so that the group of unitary transforms is a subgroup of index 2 of the group of unitary or anti-unitary transformations. In conclusion, we have found that the group of transition-probability preserving transformations acting on $R(\mathcal{H}_S)$ is the image of the group made of unitary or anti-unitary transformations acting on $\mathcal{H}_S$.

Let us show how to use the *symmetry of the description* in the case of a Galilei transformation so as to recover what we have already found in Chapter 4. We consider two frames Σ and Σ' with coordinates $(\vec{x}, t)$ and $(\vec{x}', t')$, respectively. We assume the coordinate transformation

$$\vec{x}' = \vec{x} - \vec{v}t, \ t' = t. \tag{10.1.12}$$

The frame Σ' is moving uniformly with velocity $\vec{v}$ relative to Σ, then we find

$$\frac{\partial}{\partial t'} = \frac{\partial}{\partial t} + \vec{v} \cdot \vec{\nabla}, \ \frac{\partial}{\partial \vec{x}'} = \frac{\partial}{\partial \vec{x}}. \tag{10.1.13}$$

The potential energy $V(\vec{x}, t)$ in Σ is $V'(\vec{x}', t')$ in Σ', with

$$V(\vec{x}, t) = V'(\vec{x}', t'). \tag{10.1.14}$$

In Σ' the Schrödinger equation should have the form

$$\left[-\frac{\hbar^2}{2m} \Delta' + V(\vec{x}', t') \right] \psi'(\vec{x}', t') = i\hbar \frac{\partial}{\partial t'} \psi'(\vec{x}', t'), \tag{10.1.15}$$

and the identical form

$$\left[-\frac{\hbar^2}{2m} \Delta + V(\vec{x}, t) \right] \psi(\vec{x}, t) = i\hbar \frac{\partial}{\partial t} \psi(\vec{x}, t) \tag{10.1.16}$$

in the Σ frame; i.e. the two equations have the same form, they provide the same description, independently of the reference frame.

The probability density at a point in space–time should be the same both in Σ and Σ'; since the volume is the same in both frames,

$$\psi^*(\vec{x}, t)\psi(\vec{x}, t) = \psi'^*(\vec{x}', t')\psi'(\vec{x}', t'), \tag{10.1.17}$$

and hence the wave function should transform according to

$$\psi(\vec{x}, t) = e^{if(\vec{x}, t)} \psi'(\vec{x}', t'). \tag{10.1.18}$$

Writing the differential equation in Σ' in terms of the differential operators in Σ yields

$$\left[-\frac{\hbar^2}{2m} \Delta + V + i\hbar \left(\frac{\hbar}{m} \vec{\nabla} f - \vec{v} \right) \cdot \vec{\nabla} \right] \psi$$

$$+ \left[\frac{i\hbar^2}{2m} \Delta f + \frac{\hbar^2}{2m} \vec{\nabla} f \cdot \vec{\nabla} f - \hbar \vec{v} \cdot \vec{\nabla} f - \hbar \frac{\partial f}{\partial t} \right] \psi$$

$$= i\hbar \frac{\partial \psi}{\partial t}. \tag{10.1.19}$$

This equation will reduce to the Schrödinger equation only if f is such that all extra terms identically vanish. Thus,

$$\frac{\hbar}{m} \vec{\nabla} f - \vec{v} = 0, \quad \Delta f = 0, \tag{10.1.20}$$

$$\frac{\hbar}{2m} \vec{\nabla} f \cdot \vec{\nabla} f - \vec{v} \cdot \vec{\nabla} f - \frac{\partial f}{\partial t} = 0. \tag{10.1.21}$$

The first two equations for f require $f = \frac{m\vec{v} \cdot \vec{x}}{\hbar} + g(t)$ with $g(t)$ an arbitrary function. The last condition requires

$$f(\vec{x}, t) = \frac{m\vec{v} \cdot \vec{x} - \frac{1}{2} m\vec{v}^2 t}{\hbar}, \tag{10.1.22}$$

neglecting an additive constant term. Thus, by requiring the covariance, or the form invariance, of the equation when going from one frame to another implies that the wave function transforms with an additional phase factor in agreement with what we have already derived in the discussion of the transformation properties of the wave function (Chapter 4).

Before moving to specific transformations arising from space-time transformations, let us make a few comments on constants of motion. We have seen that physical significance is ascribed to the expectation-value functions rather than to operators and vectors. If an observable is represented by an operator A, then

$$e_A(\rho_t) = \text{Tr}(\rho(t)A), \tag{10.1.23}$$

where the state evolves according to

$$\rho(t) = U(t, t_0)\rho(t_0)U^\dagger(t, t_0). \tag{10.1.24}$$

By using the cyclic property of the trace we also find

$$e_{A_t}(\rho) = \text{Tr}\Big(\rho_0 U^\dagger(t, t_0)AU(t, t_0)\Big), \tag{10.1.25}$$

having set $\rho(t_0) = \rho_0$. The rate of change of the expectation-value function is

$$\frac{\mathrm{d}}{\mathrm{d}t}e_{A_t}(\rho) = \frac{\mathrm{d}}{\mathrm{d}t}e_A(\rho_t) = \text{Tr}\left(\frac{\partial\rho}{\partial t}A + \rho\frac{\partial A}{\partial t}\right)$$

$$= \text{Tr}\left[-\frac{i}{\hbar}(H\rho A - \rho HA) + \rho\frac{\partial A}{\partial t}\right]$$

$$= \text{Tr}\left[-\frac{i}{\hbar}(\rho AH - \rho HA) + \rho\frac{\partial A}{\partial t}\right]. \tag{10.1.26}$$

Therefore,

$$\frac{\mathrm{d}}{\mathrm{d}t}e_A(\rho_t) = \text{Tr}\left[\frac{i}{\hbar}\rho(t)[H, A] + \rho(t)\frac{\partial A}{\partial t}\right], \tag{10.1.27}$$

while on the other hand

$$\frac{\mathrm{d}}{\mathrm{d}t}e_{A_t}(\rho) = \text{Tr}\left[\rho_0\frac{\mathrm{d}A}{\mathrm{d}t}\right] = \text{Tr}\left[\frac{i}{\hbar}\rho_0[H, A(t)] + \rho_0\frac{\partial A}{\partial t}\right]. \tag{10.1.28}$$

We may conclude that the expectation-value function is constant in time if both $\frac{\partial A}{\partial t}$ and $[H, A]$ vanish. This is the closest analogue of the classical counterpart.

We may consider the relation between (dynamical) symmetries and constants of motion by considering a continuous 1-parameter group of unitary transformations with generator $A = A^\dagger$, say $U(s) = e^{isA}$. To say that the Hamiltonian operator H is invariant under this transformation means that

$$U(s)HU^{-1}(s) = H, \tag{10.1.29}$$

or, equivalently, that $[H, U(s)] = 0$. By letting the parameters be infinitesimally small,

$$U(s) = I + isA + O(s^2).$$

The condition for invariance reduces to $[H, A] = 0$. Thus, if the generator of symmetry transformations does not depend explicitly on time, $\frac{\mathrm{d}}{\mathrm{d}t}e_{A_t}(\rho) = 0$. This is the constant of motion corresponding to the symmetry transformation. It is clear that, by reversing the derivation stepwise, it is also found that, from $\frac{\mathrm{d}}{\mathrm{d}t}e_{A_t}(\rho) = 0$, there follows $[H, A] = 0$.

Remark   The notion of constant of motion should not be confused with the notion of stationary state. Indeed, assume that the Hamiltonian operator does not depend on time, and that the initial state vector is an eigenvector of H, $|\psi(0)\rangle = |E_n\rangle$ with $H|E_n\rangle = E_n|E_n\rangle$. The evolution of this state is $|\psi(t)\rangle = e^{-iE_nt/\hbar}|E_n\rangle$. It follows that the expectation value of any dynamical variable A, i.e. $e_A(\psi(t)) = \langle E_n|A|E_n\rangle$, does not depend on time. Thus, in a stationary state, the expectation-value functions of all dynamical variables are independent of time, whereas a constant of motion has an expectation value independent of time for all states.

10.2 Transformations of frames and corresponding quantum symmetries

If we denote by $\text{Aut}(R(\mathcal{H}_S))$ the group of probability preserving transformations on the space of pure states, to show that classical equivalence, defined by the action of the Galilei group from one frame to another, also remains true at the quantum level we will have to define a homomorphism from the Galilei group to $\text{Aut}(R(\mathcal{H}_S))$. In particular, we will have to construct this homomorphism for rototranslations and for Galilei boosts. We will also show how to implement discrete transformations like parity and time reversal.

By exploiting Wigner's theorem we may define our homomorphism by associating either unitary transformations or anti-unitary ones. The essential idea to build this homomorphism is to identify the vector space we associate with the affine space, once an origin has been selected, with the spectrum of a maximal set of independent commuting position operators, say $\vec{X} = (\widehat{X}_1, \widehat{X}_2, \widehat{X}_3)$. The association

$$\vec{X}|\vec{x}\rangle = \vec{x}|\vec{x}\rangle \tag{10.2.1}$$

allows the selection of a basis of (improper) vectors in $\mathcal{H}_S$ with the properties:

(i) one-to-one correspondence between eigenvalues and eigenvectors, up to a phase;
(ii) the partition of unity $\int_{\mathbf{R}^3} |\vec{x}\rangle\langle\vec{x}|\mathrm{d}^3x$ makes it possible to reconstruct the vector operator $\vec{X}$ (for simplicity of notation, we do not use the vector symbol for the vector operators of position and momentum) by setting

$$\vec{X} = \int |\vec{x}\rangle\vec{x}\langle\vec{x}|\mathrm{d}^3x. \tag{10.2.2}$$

10.2.1 Rototranslations

If we consider a translation on $\mathbf{R}^3$, say $\vec{x} \to \vec{x} + \vec{a}$, we can reconstruct the operator

$$\vec{X} + \vec{a}I = \int |\vec{x}\rangle(\vec{x} + \vec{a})\langle\vec{x}|\mathrm{d}^3x \tag{10.2.3}$$

such that

$$(\vec{X} + \vec{a}I)|\vec{x}\rangle = (\vec{x} + \vec{a})|\vec{x}\rangle. \tag{10.2.4}$$

It is possible to define a vector denoted by $|\vec{x} + \vec{a}\rangle$, eigenvector of $\vec{X}$ according to

$$\vec{X}|\vec{x} + \vec{a}\rangle = (\vec{x} + \vec{a})|\vec{x} + \vec{a}\rangle. \tag{10.2.5}$$

The correspondence $|\vec{x}\rangle \rightarrow |\vec{x} + \vec{a}\rangle$ defines a unitary operator

$$U_{\vec{a}}|\vec{x}\rangle = |\vec{x} + \vec{a}\rangle. \tag{10.2.6}$$

Note that, since $|\vec{x} + \vec{a}\rangle$ is defined up to a phase, the same is true for $U_{\vec{a}}$. Now, starting with the equation obeyed by $|\vec{x} + \vec{a}\rangle$,

$$\vec{X}U_{\vec{a}}|\vec{x}\rangle = (\vec{x} + \vec{a})U_{\vec{a}}|\vec{x}\rangle, \tag{10.2.7}$$

from which we derive

$$U_{\vec{a}}^{\dagger}\vec{X}U_{\vec{a}}|\vec{x}\rangle = (\vec{x} + \vec{a})|\vec{x}\rangle = (\vec{X} + \vec{a}I)|\vec{x}\rangle. \tag{10.2.8}$$

Since $|\vec{x}\rangle$ is a basis, we get eventually

$$U_{\vec{a}}^{\dagger}\vec{X}U_{\vec{a}} = \vec{X} + \vec{a}I. \tag{10.2.9}$$

This relation holds for any vector $\vec{a}$, therefore we may assume it to be infinitesimal, say $\varepsilon\vec{a}$, then up to infinitesimal terms of second order in ε we have

$$U_{\varepsilon\vec{a}} = I + i\varepsilon\widehat{G} + O(\varepsilon^2). \tag{10.2.10}$$

Since $U_{\varepsilon\vec{a}}$ is taken to be unitary, we have

$$U_{\varepsilon\vec{a}}^{\dagger}U_{\varepsilon\vec{a}} = I. \tag{10.2.11}$$

This implies that $\widehat{G}^{\dagger} = \widehat{G}$. From the condition

$$U_{\varepsilon\vec{a}}^{\dagger}\vec{X}U_{\varepsilon\vec{a}} = \vec{X} + \varepsilon\vec{a}I \tag{10.2.12}$$

we find

$$[\widehat{G}, \vec{X}] = iI, \tag{10.2.13}$$

which implies in turn

$$\widehat{G} = \frac{\vec{P}}{\hbar} + F(\vec{X}), \tag{10.2.14}$$

where $\vec{P}$ denotes the triplet of momentum operators $(\widehat{P}_1, \widehat{P}_2, \widehat{P}_3)$. Very much as it occurs in classical mechanics, the implementation of translations by means of canonical transformations is uniquely fixed by requiring the invariance of the symplectic potential $p_i dx^i$, which is a stronger requirement than just requiring the invariance of the Poisson bracket. Here we could also require

$$U_{\varepsilon\vec{a}}^{\dagger}\vec{P}U_{\varepsilon\vec{a}} = \vec{P}; \tag{10.2.15}$$

in this way we would find that $F(\vec{X})$ should be a multiple of the identity, which we can set equal to zero. Thus, as expected, the implementation of translations is achieved by means of the momentum operators. By composing infinitesimal operators, say

$$U_{\vec{a}} = \lim_{j \to \infty} \left(I + \frac{i\vec{a} \cdot \vec{P}}{j\hbar}\right)^j = e^{\frac{i\vec{a} \cdot \vec{P}}{\hbar}}, \tag{10.2.16}$$

in the particular coordinate system we have selected for the Hilbert space $\mathcal{H}_S$, i.e. the Q-basis, we find

$$U_{\vec{a}} = e^{\vec{a}\cdot\frac{\partial}{\partial x}}. \tag{10.2.17}$$

We also find that

$$\langle x|(U_{\vec{a}}|\psi\rangle) = (\langle x|U_{\vec{a}}^{\dagger})|\psi\rangle = \langle \vec{x} - \vec{a}|\psi\rangle = \psi(\vec{x} - \vec{a}). \tag{10.2.18}$$

The same procedure may be applied to rotations. On the spectrum of the vector operator

$$\vec{X}|x_1,x_2,x_3\rangle = \vec{x}|x_1,x_2,x_3\rangle \tag{10.2.19}$$

we consider the transformation $\vec{x}' = R(s\vec{n})\vec{x}$, where $\vec{n}$ is the unit vector identifying the rotation axis and s is the amount of rotation, i.e. the angle of rotation. We can consider the transformed vector operator $R(s\vec{n})\cdot\vec{X}$ and find

$$R(s\vec{n})\cdot\vec{X}|\vec{x}\rangle = R(s\vec{n})\vec{x}|\vec{x}\rangle = s(\vec{n}\wedge\vec{x})|\vec{x}\rangle. \tag{10.2.20}$$

We consider the eigenvector corresponding to the eigenvalues, say $|R(s\vec{n})\vec{x}\rangle$, and define the unitary matrix

$$U_R|\vec{x}\rangle = |\mathcal{R}\vec{x}\rangle. \tag{10.2.21}$$

Again we consider

$$\vec{X}U_R|\vec{x}\rangle = \mathcal{R}\vec{x}U_R|\vec{x}\rangle, \tag{10.2.22}$$

and derive

$$U_R^{\dagger}\vec{X}U_R|\vec{x}\rangle = \mathcal{R}\vec{x}|\vec{x}\rangle = (\mathcal{R}\vec{X})|\vec{x}\rangle; \tag{10.2.23}$$

therefore,

$$U_R^{\dagger}\vec{X}U_R = \mathcal{R}\vec{X}. \tag{10.2.24}$$

Now we fix s to be infinitesimal and derive that

$$(I - is\widehat{G})\vec{X}(I + is\widehat{G}) = (I\vec{X} + s\vec{n}\wedge\vec{X}), \tag{10.2.25}$$

which implies

$$-is[\widehat{G},\vec{X}] = s\vec{n}\wedge\vec{X}. \tag{10.2.26}$$

By using the canonical commutation relations we find

$$\widehat{G} = \frac{1}{\hbar}\vec{n}\cdot\vec{L} + \hat{r}(\vec{X}), \tag{10.2.27}$$

where $\vec{L} = (\widehat{L}_1,\widehat{L}_2,\widehat{L}_3)$ is the angular momentum. Again we have also to select the action on $\vec{P}$, which may be required to be

$$U_R^{\dagger}\vec{P}U_R = \mathcal{R}\vec{P}; \tag{10.2.28}$$

in this way $F(\vec{X}) = \lambda I$.

By repeating the infinitesimal transformations j times, and in the limit as $j \to \infty$, we get the finite transformation

$$U_{R(s\vec{n})} = \lim_{j \to \infty} \left(I + \frac{1}{\hbar}\frac{\vec{n}}{j} \cdot \vec{L} \right) = \exp\left(\frac{1}{\hbar}\vec{n} \cdot \vec{L} \right), \qquad (10.2.29)$$

and likewise for many particles, replacing $\vec{L}$ with $\sum_k \vec{L}_k$. These computations might have been carried out in the active point of view. Let us show how this could be done with respect to time translations associated with time evolution. The equivalence stated in terms of

$$\mathrm{Tr}(\rho_1 \rho_2) = \mathrm{Tr}((\widehat{S}\rho_1)(s\rho_2)), \qquad (10.2.30)$$

required to hold also during time evolution, assuming that $\widehat{S}$ is the transformation connecting the two frames Σ and Σ', would give

$$\left| \langle \psi | \widehat{U}(t) | \varphi \rangle \right| = \left| \langle \widehat{S}^\dagger \psi | \widehat{U}(t) | \widehat{S}^\dagger \varphi \rangle \right|, \qquad (10.2.31)$$

where $\widehat{U}(t)$ is the time evolution operator of the system. We can also re-express the previous condition in the form

$$\left| \langle \psi | \widehat{U}(t) | \varphi \rangle \right| = \left| \langle \psi | \widehat{S}\widehat{U}(t)\widehat{S}^\dagger | \varphi \rangle \right|. \qquad (10.2.32)$$

From the above expression we deduce that the two vectors $\widehat{U}(t)|\varphi\rangle$ and $\widehat{S}\widehat{U}(t)\widehat{S}^\dagger|\varphi\rangle$ can differ at most by a phase factor, i.e. (see e.g. Messiah 2014)

$$\widehat{S}\widehat{U}(t)\widehat{S}^\dagger|\varphi\rangle = e^{i\gamma}\widehat{U}(t)|\varphi\rangle. \qquad (10.2.33)$$

Since $\widehat{S}\widehat{U}(t)\widehat{S}^\dagger$ is a linear operator, γ cannot depend on the particular state $|\varphi\rangle$, which implies

$$\widehat{S}\widehat{U}(t)\widehat{S}^\dagger = e^{i\gamma}\widehat{U}(t). \qquad (10.2.34)$$

By differentiating the previous expression with respect to time, and bearing in mind that

$$i\hbar\frac{\mathrm{d}}{\mathrm{d}t}\widehat{U}(t) = \widehat{H}\widehat{U}(t), \qquad (10.2.35)$$

we get

$$\widehat{S}\widehat{H}\widehat{S}^\dagger = -\hbar\frac{\mathrm{d}\gamma}{\mathrm{d}t}I + \widehat{H}. \qquad (10.2.36)$$

Since the two operators $\widehat{S}\widehat{H}\widehat{S}^\dagger$ and $\widehat{H}$ should have the same spectrum (they are related by the unitary transformation $\widehat{S}$) $\frac{\mathrm{d}\gamma}{\mathrm{d}t} = 0$, and since $\gamma(0) = 0$, this implies $\gamma(t) = 0$. From these considerations, the equivalence of the two observers $\mathcal{O}$ in Σ and $\mathcal{O}'$ in Σ' is expressed by the relation

$$\widehat{S}\widehat{H}\widehat{S}^\dagger = \widehat{H}. \qquad (10.2.37)$$

In the case of a one-particle system, the expression (10.2.37) becomes

$$\widehat{S}\widehat{H}\left(\vec{X}, \vec{P}\right)\widehat{S}^\dagger = \widehat{H}\left(\widehat{S}\vec{X}\widehat{S}^\dagger, \widehat{S}\vec{P}\widehat{S}^\dagger\right),$$

$$\widehat{H}\left(R^{-1}(s\vec{n})(\vec{X} + \vec{a}), R^{-1}(s\vec{n})\vec{P}\right) = \widehat{H}\left(\vec{X}, \vec{P}\right), \qquad (10.2.38)$$

which implies

$$\left[\vec{P} \cdot \vec{a}, \widehat{H} \right] = \left[\vec{L} \cdot \vec{n}, \widehat{H} \right] = 0. \tag{10.2.39}$$

The previous considerations make it possible to prove the equivalence of the following statements:

(i) All observers related to each other through a rototranslation are equivalent.

(ii) The form of the Hamiltonian $\widehat{H}$ is invariant under a rototranslation.

(iii) The components of $\vec{P}$ and $\vec{L}$ along $\vec{a}$ and $\vec{\omega}$, respectively, are constants of motion.

10.3 Galilei transformations

Let us consider two observers $\mathcal{O}$ in Σ and $\mathcal{O}'$ in Σ' in uniform motion with respect to each other. The coordinates of a point of space described by $\mathcal{O}$ and by $\mathcal{O}'$, respectively, are related by the simple expression

$$\vec{x}' = \vec{x} - \vec{v}\, t, \tag{10.3.1}$$

where $\vec{v}$ is the relative velocity of $\mathcal{O}'$ with respect to $\mathcal{O}$. This is what we call a Galilei transformation (already encountered when we studied the invariance properties of the Schrödinger equation) and, since for $\vec{v} = \vec{0}$ it should correspond to the identical transformation in the Hilbert space $\mathcal{H}$, this ensures that, for $\vec{v} \neq \vec{0}$, the transformation in $\mathcal{H}$ has to be unitary.

The way in which the classical position and momentum change under a Galilei transformation can be used to derive, along the lines of previous treatment, a behaviour for the corresponding self-adjoint operators in quantum mechanics, i.e.

$$\widehat{S}_{\vec{v}}\, \vec{X}\, \widehat{S}_{\vec{v}}^{\dagger} = \vec{X} + I\vec{v}\, t,$$

$$\widehat{S}_{\vec{v}}\, \vec{P}\, \widehat{S}_{\vec{v}}^{\dagger} = \vec{P} + Im\vec{v}. \tag{10.3.2}$$

If we consider an infinitesimal Galilei transformation, $\widehat{S}_{\varepsilon\vec{v}}$ can be written as

$$\widehat{S}_{\varepsilon\vec{v}} = I + \frac{i}{\hbar}\, \varepsilon\vec{v} \cdot \vec{K}(t) + O(\varepsilon^2), \tag{10.3.3}$$

and using such expression in Eqs. (10.3.2), they lead to the following commutation relations:

$$\frac{i}{\hbar} \left[\widehat{K}_j, \widehat{X}_l \right] = t\, \delta_{jl} I,$$

$$\frac{i}{\hbar} \left[\widehat{K}_j, \widehat{P}_l \right] = m\, \delta_{jl} I. \tag{10.3.4}$$

The solution of Eqs. (10.3.4) is, up to a multiple of the identity,

$$\vec{K}(t) = - \left(m\vec{X} - \vec{P}\, t \right). \tag{10.3.5}$$

The generalization of $\vec{K}(t)$ to the case of N particles is straightforward, i.e.

$$\vec{K}(t) = - \sum_{j=1}^{N} \left(m_j \vec{X}_j - \vec{P}_j\, t \right) = -M\vec{X} + \vec{P}t, \tag{10.3.6}$$

where $M = \sum_{j=1}^{N} m_j$, $\vec{X}$ is the self-adjoint operator associated the position of the centre of mass of the system, and $\vec{P}$ is the total[1] momentum operator. The expression of $\widehat{S}_{\vec{v}}$ is obtained by the exponentiation of Eqs. (10.3.4), i.e.

$$\widehat{S}_{\vec{v}} = \exp\left\{ \frac{i}{\hbar} \vec{v} \cdot \vec{K}(t) \right\} = \exp\left\{ -\frac{i}{\hbar} \vec{v} \cdot \left(M\vec{X} - \vec{P}t \right) \right\}. \tag{10.3.7}$$

By using the Baker–Campbell–Hausdorff formula, i.e.

$$e^{\alpha \widehat{A} + \beta \widehat{B}} = e^{-\frac{1}{2}\alpha\beta[\widehat{A},\widehat{B}]}\, e^{\alpha \widehat{A}}\, e^{\beta \widehat{B}}, \tag{10.3.8}$$

which, we recall, is valid when both $\widehat{A}$ and $\widehat{B}$ commute with the commutator $[\widehat{A}, \widehat{B}]$, from Eq. (10.3.7).

$$\widehat{S}_{\vec{v}}^{\dagger} = e^{-\frac{i}{\hbar}\frac{1}{2}M\vec{v}^2 t}\, e^{\frac{i}{\hbar}M\vec{v}\cdot\vec{X}}\, e^{-\frac{i}{\hbar}\vec{v}\cdot\vec{P}t}. \tag{10.3.9}$$

From the above expression we can compute the transformed wave function as

$$\langle \vec{x}_1, \ldots, \vec{x}_N | \widehat{S}_{\vec{v}}^{\dagger} | \psi \rangle = e^{-\frac{i}{\hbar}\frac{1}{2}M\vec{v}^2 t} e^{\frac{i}{\hbar}M\vec{v}\cdot\vec{X}} \langle \vec{x}_1 - \vec{v}_1 t, \ldots, \vec{x}_N - \vec{v}_N t | \psi \rangle. \tag{10.3.10}$$

By defining $\widehat{S}_{\vec{v}}^{\dagger} |\psi\rangle (\vec{x}_1, \ldots, \vec{x}_N) \equiv \langle \vec{x}_1, \ldots, \vec{x}_N | \widehat{S}_{\vec{v}}^{\dagger} | \psi \rangle$, Eq. (10.3.10) can be re-expressed in the more familiar form

$$\widehat{S}_{\vec{v}}^{\dagger} \psi(\vec{x}_1, \ldots, \vec{x}_N) = e^{-\frac{i}{\hbar}\frac{1}{2}M\vec{v}^2 t} e^{\frac{i}{\hbar}M\vec{v}\cdot\vec{X}} \psi(\vec{x}_1 - \vec{v}_1 t, \ldots, \vec{x}_N - \vec{v}_N t). \tag{10.3.11}$$

Let us apply the expression (10.3.11) to the simple case of one free particle only with mass m, i.e.

$$
\widehat{S}_{\vec{v}}^{\dagger} \int \frac{d^3 p}{(2\pi\hbar)^{\frac{3}{2}}}\, c(\vec{p})\, \exp\left\{ \frac{i}{\hbar} \left(\vec{p}\cdot\vec{x} - \frac{\vec{p}^2}{2m}t \right) \right\}
$$

$$
= \exp\left\{ -\frac{i}{\hbar}\frac{1}{2}m\vec{v}^2 t + \frac{i}{\hbar}m\vec{v}\cdot\vec{x} \right\}
$$

$$
\times \int \frac{d^3 p}{(2\pi\hbar)^{\frac{3}{2}}}\, c(\vec{p})\, \exp\left\{ \frac{i}{\hbar} \left(\vec{p}\cdot(\vec{x} - \vec{v}t) - \frac{\vec{p}^2}{2m}t \right) \right\}
$$

$$
= \int \frac{d^3 p}{(2\pi\hbar)^{\frac{3}{2}}}\, c(\vec{p} - m\vec{v})\, \exp\left\{ \frac{i}{\hbar} \left(\vec{p}\cdot\vec{x} - \frac{\vec{p}^2}{2m}t \right) \right\}. \tag{10.3.12}
$$

As can be easily seen from the previous expression, the Galilei transformation just shifts the wave function in the momentum space, $c(\vec{p})$, of the fixed quantity $-m\vec{v}$.

The same approach applied to reference frames related by a rototranslation can now be used for a Galilei transformation. The relation which ensures that the time evolution of the

---

[1] This standard notation for centre of mass and total momentum accounts for our choice of reverting to the $\vec{X}_j$, $\vec{P}_l$ operators for the individual particles.

system is observed in equivalent ways by $\mathcal{O}$ in Σ and $\mathcal{O}'$ in Σ', related to each other by a Galilei transform, reads as

$$\left| \langle \psi | \widehat{U}(t) | \varphi \rangle \right| = \left| \langle \psi | \widehat{S}_{\vec{v}}(t) \widehat{U}(t) \widehat{S}_{\vec{v}}^{\dagger}(0) | \varphi \rangle \right|, \qquad (10.3.13)$$

or analogously

$$\widehat{S}_{\vec{v}}(t)\, \widehat{U}(t)\, \widehat{S}_{\vec{v}}^{\dagger}(0) = e^{i\gamma_{\vec{v}}}\, \widehat{U}(t). \qquad (10.3.14)$$

By differentiating Eq. (10.3.14) with respect to time we obtain

$$i\hbar \frac{d\widehat{S}_{\vec{v}}(t)}{dt}\, \widehat{U}(t)\, \widehat{S}_{\vec{v}}^{\dagger}(0) + \widehat{S}_{\vec{v}}(t)\widehat{H}\widehat{U}(t)\widehat{S}_{\vec{v}}^{\dagger}(0)$$
$$= -\hbar \frac{d\gamma_{\vec{v}}}{dt}\, e^{i\gamma_{\vec{v}}}\widehat{U}(t) + e^{i\gamma_{\vec{v}}}\widehat{H}\widehat{U}(t), \qquad (10.3.15)$$

which can be cast in the form

$$\widehat{S}_{\vec{v}}(t)\widehat{H}\widehat{S}_{\vec{v}}^{\dagger}(t) + i\hbar \frac{d\widehat{S}_{\vec{v}}(t)}{dt}\, \widehat{S}_{\vec{v}}^{\dagger}(t) = \widehat{H} - \hbar \frac{d\gamma_{\vec{v}}}{dt}I, \qquad (10.3.16)$$

once we point out that, from Eq. (10.3.14),

$$\widehat{U}(t)\, \widehat{S}_{\vec{v}}^{\dagger}(0) = e^{i\gamma_{\vec{v}}}\, \widehat{S}_{\vec{v}}^{\dagger}(t)\, \widehat{U}(t). \qquad (10.3.17)$$

Let us consider the infinitesimal Galilei transformation written in Eq. (10.3.3) and substitute such an expression into Eq. (10.3.16). In this case one obtains

$$\frac{1}{i\hbar}\left[\vec{K}(t), \widehat{H} \right] + \frac{d\vec{K}(t)}{dt} = -\vec{b}(t)I, \qquad (10.3.18)$$

where by definition $\hbar d\gamma_{\vec{v}}/dt = \delta\vec{v}\cdot\vec{b}(t)$. By using for $\vec{\widehat{K}}$ the expression (10.3.5) and inserting it into Eq. (10.3.18) we have

$$\frac{m}{i\hbar}\left[\vec{X}, \widehat{H} \right] - \frac{1}{i\hbar}\left[\vec{P}, \widehat{H} \right] t - \vec{P} = \vec{b}(t)I. \qquad (10.3.19)$$

In case $\widehat{H}$ does not depend on time, $\vec{b}(t) = \vec{\alpha} + \vec{\beta} t$ can be parametrized with $\vec{\alpha}$ and $\vec{\beta}$ vector constants. Hence we obtain the set of equations which implement the condition (10.3.13), i.e.

$$\frac{m}{i\hbar}\left[\vec{X}, \widehat{H} \right] = \vec{\alpha}I + \vec{P}, \qquad (10.3.20)$$

$$\frac{1}{i\hbar}\left[\widehat{H}, \vec{P} \right] = \vec{\beta}I. \qquad (10.3.21)$$

The general expression for the Hamiltonian of a single particle satisfying both Eqs. (10.3.20) and (10.3.21) is the following:

$$\widehat{H} = \frac{\vec{P}^2}{2m} + \frac{1}{m}\vec{\alpha}\cdot\vec{P} + \vec{\beta}\cdot\vec{X} + \lambda I, \qquad (10.3.22)$$

whereas, in case of N particles,

$$\widehat{H} = \sum_{j=1}^{N} \frac{\vec{P}_j^2}{2m_j} + \frac{1}{M}\vec{\alpha}\cdot\vec{P} + \vec{\beta}\cdot\vec{X} + \widehat{V}(\vec{X}_1, \ldots, \vec{X}_N, \vec{P}_1, \ldots, \vec{P}_N) + \mu I, \qquad (10.3.23)$$

where the potential $\widehat{V}$ must satisfy the conditions

$$\left[\vec{X},\widehat{V}\right]=\left[\vec{P},\widehat{V}\right]=0. \tag{10.3.24}$$

This occurs if the potential $\widehat{U}$ depends on relative coordinates and momenta only, i.e. $\vec{r}_{ij}=\vec{x}_i-\vec{x}_j$ and $\vec{q}_{ij}=(m_i\vec{x}_j-m_j\vec{x}_i)/M$. If one further requires the invariance under rototranslations as well, in the case of a single particle $\vec{\alpha}=\vec{\beta}=0$. Thus, the general expression for the Hamiltonian of a single particle satisfying both Eqs. (10.3.20) and (10.3.21) is the following:

$$\widehat{H}=\frac{\vec{P}^2}{2\,m}+\lambda I, \tag{10.3.25}$$

and we may put $\lambda=0$. On the other hand, in case of N particles of such a requirement implies

$$\widehat{H}=\sum_{j=1}^{N}\frac{\vec{P}_j^2}{2\,m_j}+\widehat{V}(\vec{r}_{ij},\vec{q}_{ij}). \tag{10.3.26}$$

The assumption of vanishing $\vec{\alpha}$ and $\vec{\beta}$ implies $\vec{b}(t)=0$, hence Eq. (10.3.18) becomes

$$\frac{1}{i\hbar}\left[\vec{K}(t),\widehat{H}\right]+\frac{d\vec{K}(t)}{dt}=0, \tag{10.3.27}$$

which states that, for the combined rototranslation and Galilei transformation invariance, $\vec{K}(t)$ is a constant of motion, i.e. it coincides with the position of centre of mass at the initial time $(t=0)$.

10.4 Time translation

Let us assume that two reference frames Σ and Σ' coincide spatially but just differ for a choice of the time origin, i.e.

$$\vec{x}'=\vec{x}, \qquad t'=t-\tau. \tag{10.4.1}$$

In this case also the transformation in the Hilbert space, hereafter denoted with $\widehat{S}_\tau$, is unitary, and such that

$$\widehat{S}_\tau^\dagger\,|\psi(t)\rangle=|\psi(t-\tau)\rangle. \tag{10.4.2}$$

By virtue of the above relation $\widehat{S}_\tau^\dagger$ can be taken as coinciding with the time evolution operator $\widehat{U}(t_i,t_f)=\widehat{U}(t-\tau,t)$, but with the sign of the generator having changed. Hence, under an infinitesimal transformation,

$$\widehat{S}_{\delta\tau}^\dagger=I-\frac{i}{\hbar}\,\delta\tau\widehat{H}. \tag{10.4.3}$$

If the Hamiltonian depends explicitly on time the invariance condition becomes

$$\widehat{S}_\tau(t)\widehat{H}(t)\widehat{S}_\tau^\dagger(t)+i\hbar\frac{d\widehat{S}_\tau(t)}{dt}\,\widehat{S}_\tau^\dagger(t)=\widehat{H}(t)-\hbar\frac{d\gamma_\tau}{dt}I. \tag{10.4.4}$$

Under an infinitesimal transformation, Eq. (10.4.4) provides the condition

$$\frac{d\widehat{H}(t)}{dt} = c(t), \qquad \text{where} \quad \hbar \frac{d\gamma_{\delta\tau}}{dt} = -c(t)\,\delta\tau. \tag{10.4.5}$$

Hence the invariance condition is satisfied by a Hamiltonian that depends on time at most for an additive function, which may be reabsorbed in a redefinition of $\widehat{H}$. For this reason, without loss of general form, we may assume that $\widehat{H}$ does not depend on time.

10.5 Spatial reflection

Let us consider two reference frames Σ and Σ' related by spatial reflection, i.e.

$$\vec{x}' = -\vec{x},$$
$$\vec{p}' = -\vec{p}. \tag{10.5.1}$$

If $\widehat{\Pi}$, by virtue of Wigner's theorem, is the corresponding transformation in the Hilbert space, it should fulfill the relations

$$\widehat{\Pi}\,\vec{X}_j\,\widehat{\Pi}^{-1} = -\vec{X}_j,$$
$$\widehat{\Pi}\,\vec{P}_j\,\widehat{\Pi}^{-1} = -\vec{P}_j, \tag{10.5.2}$$

where we are assuming in general a system of N particles. Starting from the commutation relations between a generic pair of momentum and coordinate components, i.e.

$$\left[\vec{X}_{jk}, \vec{P}_{j'_{k'}}\right] = i\,\hbar\,\delta_{jj'}\delta_{kk'}, \tag{10.5.3}$$

and acting on such a relation with $\widehat{\Pi}$ from the left, and with $\widehat{\Pi}^{-1}$ from the right, we get

$$\left[\widehat{\Pi}\vec{X}_{jk}\widehat{\Pi}^{-1}, \widehat{\Pi}\vec{P}_{j'_{k'}}\widehat{\Pi}^{-1}\right] = \pm i\hbar\delta_{jj'}\delta_{kk'}, \tag{10.5.4}$$

where on the right-hand side of Eq. (10.5.4) the sign is positive for a unitary $\widehat{\Pi}$, and negative in the anti-unitary case. Since, as a consequence of Eq. (10.5.2), the sign must be positive, this implies that also for spatial reflection the Wigner transformation $\widehat{\Pi}$ must be a unitary one. Note that in this case, since spatial reflection is a discrete transformation, the continuity criterion cannot be used to prove the unitary nature of $\widehat{\Pi}$.

Starting from the definition of a position eigenvector for a system of N particles, i.e. $\vec{X}_j\,|\vec{x}_1,\ldots,\vec{x}_N\rangle = \vec{x}_j\,|\vec{x}_1,\ldots,\vec{x}_N\rangle\ \forall j = 1,\ldots,N$, and acting on the left with $\widehat{\Pi}$ we get

$$\vec{X}_j\,|\vec{x}_1,\ldots,\vec{x}_N\rangle = \vec{x}_j\,|\vec{x}_1,\ldots,\vec{x}_N\rangle,$$
$$\widehat{\Pi}\,\vec{X}_j\,\widehat{\Pi}^{-1}\,\widehat{\Pi}|\vec{x}_1,\ldots,\vec{x}_N\rangle = \vec{x}_j\,\widehat{\Pi}\,|\vec{x}_1,\ldots,\vec{x}_N\rangle,$$
$$\vec{X}_j\,\widehat{\Pi}|\vec{x}_1,\ldots,\vec{x}_N\rangle = -\vec{x}_j\,\widehat{\Pi}\,|\vec{x}_1,\ldots,\vec{x}_N\rangle. \tag{10.5.5}$$

Hence,

$$\widehat{\Pi}\,|\vec{x}_1,\ldots,\vec{x}_N\rangle = |-\vec{x}_1,\ldots,-\vec{x}_N\rangle. \tag{10.5.6}$$

In analogous way, by neglecting a possible phase factor,

$$\widehat{\Pi} \, |\vec{p}_1, \dots, \vec{p}_N\rangle = |-\vec{p}_1, \dots, -\vec{p}_N\rangle \qquad (10.5.7)$$

for the momenta eigenvectors. It is worth observing that possible phase factors on the momentum and position eigenvectors are related to each other. This can easily be seen by recalling that

$$|\vec{p}_1, \dots, \vec{p}_N\rangle = \int \prod_{l=1}^{N} \frac{\mathrm{d}^3 x_l}{(2\pi\hbar)^{\frac{3}{2}}} \, \exp\left\{ \frac{\mathrm{i}}{\hbar} \left(\sum_{j=1}^{N} \vec{p}_j \cdot \vec{x}_j \right) \right\} \, |\vec{x}_1, \dots, \vec{x}_N\rangle. \qquad (10.5.8)$$

Thus, by acting with $\widehat{\Pi}$ from the left on the previous relation one gets

$$\widehat{\Pi} \, |\vec{p}_1, \dots, \vec{p}_N\rangle = \int \prod_{l=1}^{N} \frac{\mathrm{d}^3 x_l}{(2\pi\hbar)^{\frac{3}{2}}} \, \exp\left\{ \frac{\mathrm{i}}{\hbar} \left(\sum_{j=1}^{N} \vec{p}_j \cdot \vec{x}_j \right) \right\} \widehat{\Pi} \, |\vec{x}_1, \dots, \vec{x}_N\rangle$$

$$= \int \prod_{l=1}^{N} \frac{\mathrm{d}^3 x_l}{(2\pi\hbar)^{\frac{3}{2}}} \, \exp\left\{ \frac{\mathrm{i}}{\hbar} \left(-\sum_{j=1}^{N} \vec{p}_j \cdot \vec{x}_j \right) \right\} \, |\vec{x}_1, \dots, \vec{x}_N\rangle$$

$$= |-\vec{p}_1, \dots, -\vec{p}_N\rangle, \qquad (10.5.9)$$

which shows that the absence of a phase factor in Eq. (10.5.6) implies a vanishing phase factor in Eq. (10.5.7) as well.

From the expressions (10.5.6) and (10.5.7) it is possible to obtain the action of the spatial reflection operator on the configuration space, i.e.

$$\widehat{\mathcal{P}} \, \psi(\vec{x}_1, \dots, \vec{x}_N) \equiv \langle \vec{x}_1, \dots, \vec{x}_N | \widehat{\Pi} | \psi \rangle = \psi(-\vec{x}_1, \dots, -\vec{x}_N). \qquad (10.5.10)$$

By using such a definition the eigenvalue problem can be studied for spatial reflection which, for one particle, reads as

$$\widehat{\mathcal{P}} \, \psi(\vec{x}) = \varepsilon \, \psi(\vec{x}) \Rightarrow \psi(-\vec{x}) = \varepsilon \, \psi(\vec{x}) \Rightarrow \psi(\vec{x}) = \varepsilon^2 \, \psi(\vec{x}); \qquad (10.5.11)$$

hence, $\varepsilon = \pm 1$, which implies $\widehat{\mathcal{P}}^2 = I$. The two possible eigenvalues, ± 1 correspond to *even* and *odd* functions, respectively. The operator $\widehat{\mathcal{P}}$ is self-adjoint, and each wavefunction can be decomposed as $\psi(\vec{x}) = \frac{1}{2} (\psi(\vec{x}) + \psi(-\vec{x})) + \frac{1}{2} (\psi(\vec{x}) - \psi(-\vec{x})) = \psi_E(\vec{x}) + \psi_O(\vec{x})$ where $\psi_E(\vec{x})$ and $\psi_O(\vec{x})$ are eigenvectors belonging to the eigenvalues $+1$ and -1, respectively.

The condition of invariance of time evolution is given in this case by the simple expression

$$\widehat{\Pi} \, \widehat{H} \, \widehat{\Pi}^{-1} = \widehat{H} \quad \Leftrightarrow \quad [\widehat{\Pi}, \widehat{H}] = 0, \qquad (10.5.12)$$

which is equivalent to requiring that the potential should be an even function, i.e.

$$V(-\vec{x}_1, \dots, -\vec{x}_N) = V(\vec{x}_1, \dots, \vec{x}_N).$$

10.6 Time reversal

In case of time reversal, Σ and Σ' are related by

$$\vec{x}' = \vec{x},$$
$$\vec{p}' = -\vec{p}. \tag{10.6.1}$$

By using the above conditions we look for a transformation $\widehat{T}$, unitary or anti-unitary, such that

$$\widehat{T}\,\vec{X}_j\,\widehat{T}^{-1} = \vec{X}_j,$$
$$\widehat{T}\,\vec{P}_j\,\widehat{T}^{-1} = -\vec{P}_j. \tag{10.6.2}$$

The result that the operator $\widehat{T}$ must be anti-unitary is a consequence of the following two relations, which are just implied by Eqs. (10.6.2), i.e.

$$\widehat{T}\,|\vec{x}_1,\dots,\vec{x}_N\rangle = |\vec{x}_1,\dots,\vec{x}_N\rangle,$$
$$\widehat{T}\,|\vec{p}_1,\dots,\vec{p}_N\rangle = |-\vec{p}_1,\dots,-\vec{p}_N\rangle. \tag{10.6.3}$$

In fact, recalling the expression (10.5.8), and acting on the left of this expression by means of $\widehat{T}$, by virtue of Eqs. (10.6.2) $\widehat{T}$ should satisfy the relation

$$\widehat{T}\,\exp\left\{\frac{\mathrm{i}}{\hbar}\left(\sum_{j=1}^{N}\vec{p}_j\cdot\vec{x}_j\right)\right\}|\vec{x}_1,\dots,\vec{x}_N\rangle$$

$$= \exp\left\{-\frac{\mathrm{i}}{\hbar}\left(\sum_{j=1}^{N}\vec{p}_j\cdot\vec{x}_j\right)\right\}|\vec{x}_1,\dots,\vec{x}_N\rangle, \tag{10.6.4}$$

hence, $\widehat{T}$ is anti-linear and consequently it is anti-unitary.

If $\widehat{H}$ is an even function of $\widehat{\vec{p}}$ and $\widehat{\vec{x}}$ it follows that

$$\widehat{T}\widehat{H}\widehat{T}^{-1} = \widehat{H} \;\Rightarrow\; \widehat{T}^{-1}\exp\left\{\frac{\mathrm{i}}{\hbar}\widehat{H}t\right\}|\psi_0\rangle = \exp\left\{-\frac{\mathrm{i}}{\hbar}\widehat{H}t\right\}\widehat{T}^{-1}|\psi_0\rangle. \tag{10.6.5}$$

In this case, since $\langle\vec{x}_1,\dots,\vec{x}_N|\widehat{T}|\psi\rangle = \psi^*(\vec{x}_1,\dots,\vec{x}_N)$, this means that, if

$$\psi(\vec{x}_1,\dots,\vec{x}_N;t)$$

solves the Schrödinger equation, the function $\psi^*(\vec{x}_1,\dots,\vec{x}_N;-t)$ is a solution as well.

10.7 Problems

10.P1. Let Π be the parity operator defined by

$$\Pi\Psi(\vec{x}) = \Psi(-\vec{x}). \tag{10.7.1}$$

(i) Say whether Π commutes or anti-commutes with the position, linear momentum and orbital angular momentum operators.

(ii) Prove that spherical harmonics are also eigenvectors of Π, with eigenvalues depending only on the orbital angular momentum quantum number, i.e.

$$\Pi Y_{lm} = \lambda_l Y_{lm}. \tag{10.7.2}$$

Find the explicit form of λ_l.

10.P2. Consider the one-dimensional harmonic oscillator Hamiltonian H in dimensionless form, and the parity operator Π previously defined. Express Π in terms of H.

10.P3. Prove that, if a system of particles of mass m_i is in a parity eigenstate, the position of the centre of mass

$$\vec{R} = \frac{\sum_{i=1}^{N} m_i \vec{x}_i}{\sum_{i=1}^{N} m_i} \tag{10.7.3}$$

has vanishing mean value.

11 Approximation methods

Since in both classical and quantum mechanics we are rarely dealing with explicitly solvable equations, this chapter is devoted to an elementary introduction to perturbation theory, Jeffreys–Wentzel–Kramers–Brillouin (hereafter JWKB) method and scattering theory, which are approximation methods that play a key role in all applications of quantum mechanics to atomic and nuclear physics.

11A.1 Approximation of eigenvalues and eigenvectors

Many realistic problems that arise in theoretical physics lead to equations which cannot be given exact general solutions. Therefore, many attempts have been made to produce satisfactory methods to provide approximate solutions. The most common ones search for solutions in terms of power series and in terms of asymptotic expansions. The main idea behind approximation methods is to try to find approximate solutions of a given system by means of exact solutions of an approximate system (a 'nearby' system). We start with the eigenvalue equation associated with the Hamiltonian H and consider an approximating one, say H_0, whose equations of motion can be explicitly solved. We introduce an interpolating family of Hamiltonians $H_\varepsilon = H_0 + \varepsilon H_I$ such that, for a given value of ε, say $\bar\varepsilon$, $H_{\bar\varepsilon} \equiv H$. Both eigenvalues and eigenvectors for H_ε may be expanded as a series in ε and solved order by order. The simplest example to illustrate what we have in mind is provided by a second-order algebraic equation, e.g.

$$x^2 - 3.03x + 2.02 = 0. \tag{11A.1.1}$$

This equation may be written by means of an interpolating family

$$x^2 - 3(1 + \varepsilon)x + 2(1 + \varepsilon) = 0, \tag{11A.1.2}$$

which yields the equation we want to solve when $\varepsilon = 10^{-2}$. One can consider the approximate equation to be given by $\varepsilon = 0$, i.e.

$$x^2 - 3x + 2 = 0. \tag{11A.1.3}$$

This equation has two solutions $x_0 = +2, +1$. We can look for approximate solutions by setting

$$x(\varepsilon) = \sum_{n=0}^{n} x_n \varepsilon^n. \tag{11A.1.4}$$

The expansion is around exact solutions of the approximate equation. We get

$$x^2(\varepsilon) - 3(1 + \varepsilon)x(\varepsilon) + 2(1 + \varepsilon) = 0, \tag{11A.1.5}$$

and solve this equation order by order. To zeroth order we find of course $x_0 = +2, +1$. To first order we find

$$2x_1 x_0 - 3x_1 - 3x_0 + 2 = 0, \tag{11A.1.6}$$

which is solved by

$$x_1 = \frac{(3x_0 - 2)}{(2x_0 - 3)}, \tag{11A.1.7}$$

and leads to $x_0 = 2, x_1 = 4$ or $x_0 = 1, x_1 = -1$. Thus,

$$x = 2 + 4 \cdot 10^{-2} + O(\varepsilon^2), \tag{11A.1.8}$$

$$x = 1 - 10^{-2} + O(\varepsilon^2). \tag{11A.1.9}$$

It is possible to continue the approximation to the order ε^2, which means with accuracy 10^{-4}.

As a second example, we may consider the eigenvalue equation

$$A|n\rangle = a_n|n\rangle = \lambda|n\rangle. \tag{11A.1.10}$$

We assume again that it is possible to decompose A according to

$$A = A^{(0)} + \varepsilon A^{(1)}, \tag{11A.1.11}$$

with $A^{(0)}$ representing the approximated system, which has exact solutions, i.e.

$$A^{(0)}|n(0)\rangle = \lambda_n^{(0)}|n(0)\rangle, \tag{11A.1.12}$$

completely soluble. Then we search for eigenvalues

$$\lambda_n(\varepsilon) = \lambda_n^{(0)} + \varepsilon\lambda_n^{(1)} + \varepsilon^2\lambda_n^{(2)} + \cdots, \tag{11A.1.13}$$

i.e. corrections to the exact eigenvalue $\lambda_n^{(0)}$, and eigenvectors which are corrections to the exact eigenvector of the approximated operator. The eigenvectors are expanded according to

$$|n(\varepsilon)\rangle = |n(0)\rangle + \varepsilon|n(1)\rangle + \varepsilon^2|n(2)\rangle + \cdots, \tag{11A.1.14}$$

with vectors in the expansion pairwise orthogonal, and $\lambda_n^{(0)}$ with $|n(0)\rangle$ the exact eigenvalue and eigenvector, respectively, of the approximated equation.

The strategy is to find the first-order correction of the eigenvector by finding its components along the basis of the approximated operator $A^{(0)}$. To first order in ε, the eigenvalue equation

$$(A^{(0)} + \varepsilon A^{(1)})(|n(0)\rangle + \varepsilon|n(1)\rangle) = (\lambda_n^{(0)} + \varepsilon\lambda_n^{(1)})(|n(0) + \varepsilon|n(1)\rangle) \tag{11A.1.15}$$

gives

$$A^{(0)}|n(0)\rangle = \lambda_n^{(0)}|n(0)\rangle, \tag{11A.1.16}$$

$$A^{(1)}|n(0)\rangle + A^{(0)}|n(1)\rangle = \lambda_n^{(1)}|n(0)\rangle + \lambda_n^{(0)}|n(1)\rangle. \tag{11A.1.17}$$

On taking the product on the left with $\langle n(0)|$,

$$\langle n(0)|A^{(1)}|n(0)\rangle + \lambda_n^{(0)}\langle n(0)|n(1)\rangle = \lambda_n^{(1)}\langle n(0)|n(0)\rangle + \lambda_n^{(0)}\langle n(0)|n(1)\rangle, \tag{11A.1.18}$$

from which

$$\lambda_n^{(1)} = \frac{\langle n(0)|A^{(1)}|n(0)\rangle}{\langle n(0)|n(0)\rangle}. \tag{11A.1.19}$$

At this stage we multiply by an auto-bra of $A^{(0)}$ belonging to a different eigenvalue; thus, using the orthogonality property

$$\langle m(0)|n(0)\rangle = 0, \ m \neq n, \tag{11A.1.20}$$

we find that

$$\langle m(0)|A^{(1)}|n(0)\rangle + \lambda_m^{(0)}\langle m(0)|n(1)\rangle = \lambda_n^{(0)}\langle m(0)|n(1)\rangle, \tag{11A.1.21}$$

giving in turn

$$\langle m(0)|n(1)\rangle = \frac{\langle m(0)|A^{(1)}|n(0)\rangle}{(\lambda_n^{(0)} - \lambda_m^{(0)})}. \tag{11A.1.22}$$

It is now possible to take into account eigenvectors; to first order,

$$|n\rangle = |n(0)\rangle + \varepsilon|n(1)\rangle, \tag{11A.1.23}$$

and the procedure can be iterated.

Example. Let us consider

$$A = \begin{pmatrix} 1 & \varepsilon & \varepsilon \\ \varepsilon & 2 & \varepsilon \\ \varepsilon & \varepsilon & 3 \end{pmatrix} = \begin{pmatrix} 1 & 0 & 0 \\ 0 & 2 & 0 \\ 0 & 0 & 3 \end{pmatrix} + \varepsilon \begin{pmatrix} 0 & 1 & 1 \\ 1 & 0 & 1 \\ 1 & 1 & 0 \end{pmatrix}. \tag{11A.1.24}$$

In principle, it is possible to diagonalize the matrix by solving the secular equation

$$\det(A - \lambda I) = 0. \tag{11A.1.25}$$

The approximation procedure, instead, considers

$$\lambda(\varepsilon) = \lambda^{(0)} + \varepsilon\lambda^{(1)} + \varepsilon^2\lambda^{(2)} + \cdots, \tag{11A.1.26}$$

and analogously

$$|n(\varepsilon)\rangle = |n(0)\rangle + \varepsilon|n(1)\rangle + \varepsilon^2|n(2)\rangle + \cdots, \tag{11A.1.27}$$

with vectors on the right-hand side pairwise orthogonal. To first order in ε,

$$(A^{(0)} + \varepsilon A^{(1)})(|n(0)\rangle + \varepsilon|n(1)\rangle) = (\lambda_n^{(0)} + \varepsilon\lambda_n^{(1)})(|n(0)\rangle + \varepsilon|n(1)\rangle), \tag{11A.1.28}$$

which implies

$$A^{(0)}|n(0)\rangle = \lambda_n^{(0)}|n(0)\rangle, \tag{11A.1.29}$$

$$A^{(1)}|n(0)\rangle + A^{(0)}|n(1)\rangle = \lambda_n^{(1)}|n(0)\rangle + \lambda_n^{(0)}|n(1)\rangle. \tag{11A.1.30}$$

We project both members on the basis of $A^{(0)}$, say $|n(0)\rangle$ and $|m(0)\rangle$, with

$$\sum_{m \neq n} |m(0)\rangle\langle m(0)| = I - |n(0)\rangle\langle n(0)|. \tag{11A.1.31}$$

The eigenvalues for $A^{(0)}$ are simple. We find to first order

$$
\begin{aligned}
|n\rangle &= |n(0)\rangle + \varepsilon \sum_m |m(0)\rangle \langle m(0)|n(1)\rangle \\
&= |n(0)\rangle + \varepsilon |n(0)\rangle \langle n(0)|n(1)\rangle + \varepsilon \sum_{m \neq n} |m(0)\rangle \langle m(0)|n(1)\rangle \\
&= |n(0)\rangle + \varepsilon \sum_{m \neq n} \frac{\langle m(0)|A^{(1)}|n(1)\rangle}{(\lambda_n^{(0)} - \lambda_m^{(0)})} |m(0)\rangle.
\end{aligned}
\tag{11A.1.32}
$$

For the eigenvalues, we find

$$
\begin{aligned}
\lambda_n(\varepsilon) = \lambda_n^{(0)} &+ \varepsilon \langle n(0)|A^{(1)}|n(0)\rangle \\
&+ \varepsilon^2 \sum_{m \neq n} \frac{\langle n(0)|A^{(1)}|m(0)\rangle \langle m(0)|A^{(1)}|n(0)\rangle}{(\lambda_n^{(0)} - \lambda_m^{(0)})}.
\end{aligned}
\tag{11A.1.33}
$$

In our example

$$
\lambda_1^{(0)} = 1, \ \lambda_2^{(0)} = 2, \ \lambda_3^{(0)} = 3,
\tag{11A.1.34}
$$

with eigenvectors

$$
|\phi_1(0)\rangle = \begin{pmatrix} 1 \\ 0 \\ 0 \end{pmatrix}, \ |\phi_2(0)\rangle = \begin{pmatrix} 0 \\ 1 \\ 0 \end{pmatrix}, \ |\phi_3(0)\rangle = \begin{pmatrix} 0 \\ 0 \\ 1 \end{pmatrix},
\tag{11A.1.35}
$$

$$
(A^{(1)})_{mn} = 1 - \delta_{mn}.
\tag{11A.1.36}
$$

The first-order correction for eigenvalues is

$$
\lambda_1^{(1)} = (A^{(1)})_{11} = 0,
\tag{11A.1.37}
$$

$$
\lambda_2^{(1)} = (A^{(1)})_{22} = 0,
\tag{11A.1.38}
$$

$$
\lambda_3^{(1)} = (A^{(1)})_{33} = 0.
\tag{11A.1.39}
$$

The corresponding correction for eigenvectors is provided by

$$
\begin{aligned}
|\phi_1(1)\rangle &= \sum_{m \neq 1} \frac{(A^{(1)})_{m1}|\phi_m(0)\rangle}{(\lambda_1^{(0)} - \lambda_m^{(0)})} \\
&= \frac{(A^{(1)})_{21}|\phi_2(0)\rangle}{(\lambda_1^{(0)} - \lambda_2^{(0)})} + \frac{(A^{(1)})_{31}|\phi_3(0)\rangle}{(\lambda_1^{(0)} - \lambda_3^{(0)})} \\
&= \frac{1}{(1-2)} \begin{pmatrix} 0 \\ 1 \\ 0 \end{pmatrix} + \frac{1}{(1-3)} \begin{pmatrix} 0 \\ 0 \\ 1 \end{pmatrix} = \begin{pmatrix} 0 \\ -1 \\ -\frac{1}{2} \end{pmatrix},
\end{aligned}
\tag{11A.1.40}
$$

$$|\phi_2(1)\rangle = \sum_{m \neq 2} \frac{(A^{(1)})_{m2}|\phi_m(0)\rangle}{(\lambda_2^{(0)} - \lambda_m^{(0)})}$$

$$= \frac{(A^{(1)})_{12}|\phi_1(0)\rangle}{(\lambda_2^{(0)} - \lambda_1^{(0)})} + \frac{(A^{(1)})_{32}|\phi_3(0)\rangle}{(\lambda_2^{(0)} - \lambda_3^{(0)})}$$

$$= \frac{1}{(2-1)} \begin{pmatrix} 1 \\ 0 \\ 0 \end{pmatrix} + \frac{1}{(2-3)} \begin{pmatrix} 0 \\ 0 \\ 1 \end{pmatrix} = \begin{pmatrix} 1 \\ 0 \\ -1 \end{pmatrix}, \tag{11A.1.41}$$

$$|\phi_3(1)\rangle = \sum_{m \neq 3} \frac{(A^{(1)})_{m3}|\phi_m(0)\rangle}{(\lambda_3^{(0)} - \lambda_m^{(0)})}$$

$$= \frac{(A^{(1)})_{13}|\phi_1(0)\rangle}{(\lambda_3^{(0)} - \lambda_1^{(0)})} + \frac{(A^{(1)})_{23}|\phi_2(0)\rangle}{(\lambda_3^{(0)} - \lambda_2^{(0)})}$$

$$= \frac{1}{(3-1)} \begin{pmatrix} 1 \\ 0 \\ 0 \end{pmatrix} + \frac{1}{(3-2)} \begin{pmatrix} 0 \\ 1 \\ 0 \end{pmatrix} = \begin{pmatrix} \frac{1}{2} \\ 1 \\ 0 \end{pmatrix}. \tag{11A.1.42}$$

Thus, eigenvectors to first order are

$$|\phi_1\rangle = \begin{pmatrix} 1 \\ 0 \\ 0 \end{pmatrix} + \varepsilon \begin{pmatrix} 0 \\ -1 \\ -\frac{1}{2} \end{pmatrix}, \ |\phi_2\rangle = \begin{pmatrix} 0 \\ 1 \\ 0 \end{pmatrix} + \varepsilon \begin{pmatrix} 1 \\ 0 \\ -1 \end{pmatrix}, \ |\phi_3\rangle = \begin{pmatrix} 0 \\ 0 \\ 1 \end{pmatrix} + \varepsilon \begin{pmatrix} \frac{1}{2} \\ 1 \\ 0 \end{pmatrix}. \tag{11A.1.43}$$

It is now possible to compute eigenvalues to second order in the form

$$\lambda_1^{(2)} = \varepsilon^2 \sum_{m \neq 1} \frac{(A^{(1)})_{1m}(A^{(1)})_{m1}}{(\lambda_1^{(0)} - \lambda_m^{(0)})}$$

$$= \varepsilon^2 \left[\frac{(A^{(1)})_{12}(A^{(1)})_{21}}{(\lambda_1^{(0)} - \lambda_2^{(0)})} + \frac{(A^{(1)})_{13}(A^{(1)})_{31}}{(\lambda_1^{(0)} - \lambda_3^{(0)})} \right]$$

$$= \varepsilon^2 \left(\frac{1}{(1-2)} + \frac{1}{(1-3)} \right) = -\frac{3}{2}\varepsilon^2, \tag{11A.1.44}$$

and similarly for the others.

11A.2 Hellmann–Feynman theorem

This theorem is a useful tool in solid-state physics and in atomic and molecular physics. It was used by Feynman (1939) in a very interesting way, and holds when the Hamiltonian operator depends on parameters; we consider a variation of these parameters. Let us take the eigenvalue equation

$$H_\epsilon |E_n(\epsilon)\rangle = E_n(\epsilon)|E_n(\epsilon)\rangle. \tag{11A.2.1}$$

Assuming the differentiability with respect to ϵ of E_n, H and $|E_n(\epsilon)\rangle$, we find

$$\frac{\partial H_\epsilon}{\partial \epsilon}|E_n(\epsilon)\rangle + H_\epsilon \frac{\partial}{\partial \epsilon}|E_n(\epsilon)\rangle$$
$$= \frac{\partial E_n(\epsilon)}{\partial \epsilon}|E_n(\epsilon)\rangle + E_n(\epsilon)\frac{\partial}{\partial \epsilon}|E_n(\epsilon)\rangle. \tag{11A.2.2}$$

Multiplying by $\langle E_n(\epsilon)|$ and $\langle E_m(\epsilon)|$, with $n \neq m$,

$$\langle E_m(\epsilon)|\frac{\partial H_\epsilon}{\partial \epsilon}|E_n(\epsilon)\rangle + \langle E_m(\epsilon)|\frac{\partial}{\partial \epsilon}|E_n(\epsilon)\rangle E_m(\epsilon)$$
$$= \frac{\partial E_n(\epsilon)}{\partial \epsilon}\langle E_m(\epsilon)|E_n(\epsilon)\rangle + E_n(\epsilon)\langle E_m(\epsilon)|\frac{\partial}{\partial \epsilon}|E_n(\epsilon)\rangle. \tag{11A.2.3}$$

For $m = n$, from $\frac{\partial}{\partial \epsilon}\langle E_n(\epsilon)|E_n(\epsilon)\rangle = 0$,

$$\langle E_n(\epsilon)|\frac{\partial}{\partial \epsilon}|E_n(\epsilon)\rangle = 0, \tag{11A.2.4}$$

and we derive

$$\frac{\partial E_n(\epsilon)}{\partial \epsilon} = \langle E_n(\epsilon)|\frac{\partial H_\epsilon}{\partial \epsilon}|E_n(\epsilon)\rangle, \tag{11A.2.5}$$

while, for $m \neq n$,

$$\langle E_m(\epsilon)|\frac{\partial H_\epsilon}{\partial \epsilon}|E_n(\epsilon)\rangle = (E_n(\epsilon) - E_m(\epsilon))\langle E_m(\epsilon)|\frac{\partial}{\partial \epsilon}|E_n(\epsilon)\rangle. \tag{11A.2.6}$$

We may also derive an integrated form, i.e.

$$E_n(\epsilon_1) - E_n(\epsilon_2) = \frac{\langle E_n(\epsilon_2)|(H(\epsilon_1) - H(\epsilon_2))|E_n(\epsilon_1)\rangle}{\langle E_n(\epsilon_2)|E_n(\epsilon_1)\rangle}. \tag{11A.2.7}$$

The off-diagonal terms are

$$(E_m(\epsilon) - E_n(\epsilon))\langle E_n(\epsilon)|\frac{\partial}{\partial \epsilon}|E_m(\epsilon)\rangle = \langle E_n(\epsilon)|\frac{\partial H_\epsilon}{\partial \epsilon}|E_m(\epsilon)\rangle. \tag{11A.2.8}$$

In this derivation we have assumed that the domain of self-adjointness of H_ϵ, say $\mathcal{D}_\epsilon \subset \mathcal{H}$, does not change with ϵ. However, if $\partial_\epsilon|E_n(\epsilon)\rangle \notin \mathcal{D}_\epsilon$, there will be a correction to the Hellmann–Feynman formula (Esteve et $al.$ 2010).

As an application, we consider the Hamiltonian of the harmonic oscillator when the mass m and the frequency ω are thought of as parameters. The eigenvalue equation is

$$H|n\rangle = \hbar\omega\left(n + \frac{1}{2}\right)|n\rangle, \tag{11A.2.9}$$

with $H = \frac{1}{2m}P^2 + \frac{1}{2}m\omega^2 X^2$ and $E_n = \hbar\omega(n + \frac{1}{2})$. It is easy to show that $m\frac{\partial E_n}{\partial m} = 0$, which implies that

$$-\left\langle\frac{1}{2m}P^2\right\rangle + \left\langle\frac{1}{2}m\omega^2 X^2\right\rangle = 0, \tag{11A.2.10}$$

i.e. the potential energy and kinetic energy are equal. Moreover,

$$\omega\frac{\partial E_n}{\partial \omega} = E_n = 2\left\langle\frac{1}{2}m\omega^2 X^2\right\rangle = 2E_{\text{pot}}, \tag{11A.2.11}$$

and thus

$$E_{\text{kin}} = E_{\text{pot}} = \frac{1}{2}E_n. \tag{11A.2.12}$$

We might also apply this theorem to the hydrogen-like atoms, for which the Hamiltonian H is the sum of $T = \frac{P^2}{2\mu}$ and $V = -\frac{Zq_e^2}{r}$, while the energy eigenvalues read as

$$E = -\frac{Z^2 q_e^4 \mu}{2n^2 \hbar^2}. \tag{11A.2.13}$$

By differentiating with respect to μ,

$$\mu \frac{\partial H}{\partial \mu} = -T, \ \mu \frac{\partial E}{\partial \mu} = E, \tag{11A.2.14}$$

so that $\langle T \rangle = -E$, while differentiation with respect to the atomic number Z yields

$$Z \frac{\partial H}{\partial Z} = V, \ Z \frac{\partial E}{\partial Z} = 2E, \tag{11A.2.15}$$

so that $\langle V \rangle = 2E$, and we obtain the correct sum

$$\langle T \rangle + \langle V \rangle = -E + 2E = E. \tag{11A.2.16}$$

By writing in explicit form,

$$-Zq_e^2 \left\langle \frac{1}{r} \right\rangle = \frac{Z^2 q_e^2}{n^2 a_B}, \tag{11A.2.17}$$

or

$$\left\langle \frac{1}{r} \right\rangle = \frac{Z}{n^2 a_B}. \tag{11A.2.18}$$

Remark Applying this theorem to $(A^{(0)} + \epsilon A^{(1)})|n(\epsilon)\rangle = E(\epsilon)|n(\epsilon)\rangle$, we immediately find the first-order correction on the eigenvalue.

11A.3 Virial theorem

The virial theorem has a statistical nature; it concerns temporal averages of various mechanical quantities. Let us consider first the classical situation. We are dealing with a system of point particles of mass m_j with positions $\vec{r}_j$ and total applied force $\vec{F}_j$. The equations of motion are

$$\frac{\mathrm{d}}{\mathrm{d}t}\vec{P}_j = \vec{F}_j. \tag{11A.3.1}$$

Let us introduce the dynamical variable

$$G \equiv \sum_j \vec{r}_j \cdot \vec{P}_j = \frac{\mathrm{d}}{\mathrm{d}t} \sum_j \frac{m}{2}\vec{r}_j \cdot \vec{r}_j. \tag{11A.3.2}$$

We find

$$\frac{\mathrm{d}}{\mathrm{d}t}G = \sum_j \left(\frac{\mathrm{d}\vec{r}_j}{\mathrm{d}t} \cdot \vec{P}_j + \vec{r}_j \frac{\mathrm{d}\vec{P}_j}{\mathrm{d}t} \right) = \sum_j (m_j v_j^2 + \vec{r}_j \cdot \vec{F}_j), \qquad (11\mathrm{A}.3.3)$$

where we have used $m\frac{\mathrm{d}}{\mathrm{d}t}\vec{r}_j = \vec{P}_j$. By introducing the kinetic energy $T = \sum_j \frac{1}{2} m_j v_j^2$, we get

$$\frac{\mathrm{d}}{\mathrm{d}t}G - 2T = \sum_j \vec{r}_j \cdot \vec{F}_j. \qquad (11\mathrm{A}.3.4)$$

If we average along a trajectory for a time interval τ we obtain

$$\frac{1}{\tau}\int_0^\tau \frac{\mathrm{d}G}{\mathrm{d}t}\mathrm{d}t = \frac{1}{\tau}(G(\tau) - G(0))$$
$$= 2\overline{T} + \sum_j \overline{\vec{r}_j \cdot \vec{F}_j}, \qquad (11\mathrm{A}.3.5)$$

where the bar denotes time average. When the motion is periodic and we integrate on a period, we find the virial theorem

$$\overline{T} = -\frac{1}{2}\sum_j \overline{\vec{F}_j \cdot \vec{r}_j}. \qquad (11\mathrm{A}.3.6)$$

This result holds also when $G(r)$ remains bounded for any value of r. Usually, the expression $\sum_j \vec{F}_j \cdot \vec{r}_j$ is called the *Clausius virial*. If we further introduce the quantity

$$I = \frac{1}{2}\sum_j m_j |\vec{r}_j|^2, \qquad (11\mathrm{A}.3.7)$$

along with

$$T = \frac{1}{2}\sum_j \frac{P_j^2}{m_j}, \qquad (11\mathrm{A}.3.8)$$

we find

$$\frac{\mathrm{d}}{\mathrm{d}t}I = G. \qquad (11\mathrm{A}.3.9)$$

In terms of Poisson brackets, we can write

$$\{\dot{I}, T\} = 2T, \qquad (11\mathrm{A}.3.10)$$

and on considering

$$V = -\sum_{j \neq k} \frac{m_j m_k}{|\vec{r}_j - \vec{r}_k|}, \qquad (11\mathrm{A}.3.11)$$

we can also write

$$\{\dot{I}, V\} = V. \qquad (11\mathrm{A}.3.12)$$

Therefore

$$\ddot{I} = (E + T), \qquad (11\mathrm{A}.3.13)$$

where E is the energy.

Let us consider now the quantum case, with Hamiltonian

$$H = \frac{1}{2m}P^2 + V(\vec{r}). \tag{11A.3.14}$$

We first notice that if H has eigenvalues and eigenvectors which are normalizable vectors, from $H|E\rangle = E|E\rangle$,

$$\langle E|[H,A]|E\rangle = \langle E|(HA - AH)|E\rangle = 0, \tag{11A.3.15}$$

for all A admitting the eigenvectors in its domain. Let us consider $A = \vec{r} \cdot \vec{P}$ (we might introduce $A = \vec{r} \cdot \vec{P} + \vec{P} \cdot \vec{r}$ to make it Hermitian). From the equation

$$\begin{aligned}
[A,H] &= i\hbar\frac{d\vec{r}}{dt} \cdot \vec{P} + i\hbar\vec{r} \cdot \frac{d\vec{P}}{dt} \\
&= i\hbar\frac{\vec{P}}{m} \cdot \vec{P} + i\hbar\vec{r} \cdot (-\vec{\nabla}V(\vec{r}))
\end{aligned} \tag{11A.3.16}$$

we find

$$\left\langle \frac{1}{m}P^2 \right\rangle_{|E\rangle} = \langle \vec{r} \cdot \vec{\nabla}V(\vec{r}) \rangle_{|E\rangle}. \tag{11A.3.17}$$

For every eigenstate of H we have derived

$$E_{\text{kin}} = \frac{1}{2}\langle \vec{r} \cdot \vec{\nabla}V(\vec{r}) \rangle_{|E\rangle}. \tag{11A.3.18}$$

When the potential is a power of $r = |\vec{r}|$

$$V(\vec{r}) = V(r) = kr^n, \tag{11A.3.19}$$

we have

$$r\frac{\partial}{\partial r}V(r) = nV(r), \tag{11A.3.20}$$

i.e. $E_{\text{kin}} = \frac{n}{2}E_{\text{pot}}$. By using $E = E_{\text{kin}} + E_{\text{pot}}$, we get

$$E_{\text{kin}} = \frac{n}{(n+2)}E, \ E_{\text{pot}} = \frac{2}{(n+2)}E. \tag{11A.3.21}$$

In particular, we find

$$n = -1, \ E_{\text{kin}} = -E, \ E_{\text{pot}} = 2E, \ E_{\text{kin}} = -\frac{1}{2}E_{\text{pot}}, \tag{11A.3.22}$$

$$n = 2, \ E_{\text{kin}} = \frac{1}{2}E, \ E_{\text{pot}} = \frac{1}{2}E, \ E_{\text{kin}} = E_{\text{pot}}. \tag{11A.3.23}$$

There is an alternative way to deal with the virial theorem. We consider a perturbation such as

$$|E(\lambda)\rangle = e^{i\lambda G}|E\rangle, \tag{11A.3.24}$$

along with

$$\langle E(\lambda)| = \langle E|e^{-i\lambda G}, \tag{11A.3.25}$$

i.e. G is Hermitian, $G = G^\dagger$. By expanding the unitary operator, we get

$$e^{i\lambda G} = I + i\lambda G - \frac{1}{2}\lambda^2 G^2 + \cdots, \tag{11A.3.26}$$

$$e^{-i\lambda G} = I - i\lambda G - \frac{1}{2}\lambda^2 G^2 + \cdots, \tag{11A.3.27}$$

$$H(\lambda) = e^{-i\lambda G} H_0 e^{i\lambda G}$$

$$\sim (I - i\lambda G - \frac{1}{2}\lambda^2 G^2 + \cdots) H_0 (I + i\lambda G - \frac{1}{2}\lambda^2 G^2 + \cdots)$$

$$\sim H_0 + i\lambda[H_0, G] - \frac{1}{2}\lambda^2 (G^2 H_0 - 2G H_0 G + H_0 G^2). \tag{11A.3.28}$$

Now we consider expectation values in an eigenstate of H_0 with eigenvalue E. We get

$$\langle H(\lambda) \rangle = E + \frac{1}{2}\lambda^2 \langle (2G H_0 G - H_0 G^2 - G^2 H_0) \rangle + O(\lambda^3), \tag{11A.3.29}$$

where the linear term is missing because

$$\langle E | i[H_0, G] | E \rangle = 0. \tag{11A.3.30}$$

In conclusion

$$\frac{\partial}{\partial \lambda} \langle H(\lambda) \rangle |_{\lambda=0} = 0. \tag{11A.3.31}$$

A particular choice of G, say a squeezing transformation

$$\vec{P} \to e^{-i\lambda G} \vec{P} e^{i\lambda G}, \tag{11A.3.32}$$

$$\vec{r} \to e^{-i\lambda G} \vec{r} e^{i\lambda G}, \tag{11A.3.33}$$

with

$$G = \frac{1}{2\hbar}(\vec{r} \cdot \vec{P} + \vec{P} \cdot \vec{r}) = \frac{1}{\hbar}\vec{r} \cdot \vec{P} - \frac{3}{2}i, \tag{11A.3.34}$$

gives

$$\vec{P} \to e^{\lambda}\vec{P}, \ \vec{r} \to e^{-\lambda}\vec{r}. \tag{11A.3.35}$$

Then

$$e^{-i\lambda G} H_0 e^{i\lambda G} = \frac{1}{2m}(e^{\lambda}\vec{P})^2 + V(e^{-\lambda}\vec{r}), \tag{11A.3.36}$$

with average

$$\langle H(\lambda) \rangle = e^{2\lambda} \left\langle \frac{1}{2m}|\vec{P}|^2 \right\rangle + \langle V(e^{-\lambda}\vec{r}) \rangle. \tag{11A.3.37}$$

By differentiating with respect to λ,

$$\frac{\partial}{\partial \lambda} \langle H(\lambda) \rangle |_{\lambda=0} = 2E_{\text{kin}} - \left\langle \vec{r} \cdot \vec{\nabla} V(\vec{r}) \right\rangle = 0, \tag{11A.3.38}$$

and we recover the virial theorem

$$E_{\text{kin}} = \frac{1}{2}\left\langle \vec{r} \cdot \vec{\nabla} V(\vec{r}) \right\rangle = -\frac{1}{2}\langle \vec{r} \cdot \vec{F}(r) \rangle, \tag{11A.3.39}$$

where the force operator is given by

$$\vec{F} = -\vec{\nabla} V(\vec{r}).$$

(11A.3.40)

11A.4 Anharmonic oscillator

Given a one-dimensional harmonic oscillator, here we consider a perturbed potential (Grechko *et al.* 1977)

$$U(\hat{x}) = \frac{1}{2}m\omega^2\hat{x}^2 + \varepsilon_1\hat{x}^3 + \varepsilon_2\hat{x}^4,$$

(11A.4.1)

where ε_1 and ε_2 are two 'small' parameters. Our aim is to evaluate how the energy spectrum and the eigenfunctions are affected by the addition of $\varepsilon_1\hat{x}^3 + \varepsilon_2\hat{x}^4$ to the unperturbed potential $\frac{1}{2}m\omega^2\hat{x}^2$.

Our problem provides a non-trivial application of time-independent perturbation theory in the non-degenerate case, and we can apply the standard formulae of Section 11A.1:

$$E_n \sim E_n^0 + \langle n^{(0)}|\hat{W}|n^{(0)}\rangle + \sum_{k \neq n} \frac{|\langle n^{(0)}|\hat{W}|k^{(0)}\rangle|^2}{(E_n^0 - E_k^0)},$$

(11A.4.2)

$$|n\rangle \sim |n^{(0)}\rangle + \sum_{k \neq n} \frac{\langle k^{(0)}|\hat{W}|n^{(0)}\rangle}{(E_n^0 - E_k^0)}|k^{(0)}\rangle,$$

(11A.4.3)

where, defining $\xi \equiv x\sqrt{\frac{m\omega}{\hbar}}$,

$$\langle \xi|n^{(0)}\rangle = C_n e^{\frac{-\xi^2}{2}} H_n(\xi),$$

(11A.4.4)

$$\langle k^{(0)}|\hat{W}|n^{(0)}\rangle = \langle k^{(0)}|\hat{W}_1|n^{(0)}\rangle + \langle k^0|\hat{W}_2|n^{(0)}\rangle,$$

(11A.4.5)

$$\hat{W}_1 \equiv \varepsilon_1\hat{x}^3,$$

(11A.4.6)

$$\hat{W}_2 \equiv \varepsilon_2\hat{x}^4.$$

(11A.4.7)

The building blocks of our calculation for the anharmonic oscillator are the matrix elements

$$\langle n^{(0)}|\hat{x}|k^{(0)}\rangle = \sqrt{\frac{\hbar}{2m\omega}}(\sqrt{n}\delta_{n,k+1} + \sqrt{n+1}\ \delta_{n,k-1}).$$

(11A.4.8)

Consider first the effect of $\hat{W}_1$. Its matrix elements are

$$\langle n^{(0)}|\hat{W}_1|k^{(0)}\rangle = \varepsilon_1\langle n^{(0)}|\hat{x}^3|k^{(0)}\rangle = \varepsilon_1\sum_l \langle n^{(0)}|\hat{x}^2|l^{(0)}\rangle\langle l^{(0)}|\hat{x}|k^{(0)}\rangle,$$

(11A.4.9)

where we have inserted a resolution of the identity. In an analogous way, the matrix elements $\langle n^{(0)}|\hat{x}^2|l^{(0)}\rangle$ can be evaluated as follows:

$$\langle n^{(0)}|\hat{x}^2|l^{(0)}\rangle = \sum_p \langle n^{(0)}|\hat{x}|p^{(0)}\rangle \langle p^{(0)}|\hat{x}|l^{(0)}\rangle$$

$$= \frac{\hbar}{2m\omega} \sum_p (\sqrt{n}\delta_{n,p+1} + \sqrt{n+1}\,\delta_{n+1,p})$$

$$\times (\sqrt{p}\delta_{p,l+1} + \sqrt{p+1}\,\delta_{l,p+1})$$

$$= \frac{\hbar}{2m\omega} \left[\sqrt{n(l+1)} \sum_p \delta_{n-1,p}\delta_{p,l+1} \right.$$

$$+ \sqrt{(n+1)(l+1)} \sum_p \delta_{n+1,p}\delta_{p,l+1}$$

$$+ \sqrt{nl} \sum_p \delta_{n-1,p}\delta_{p,l-1}$$

$$\left. + \sqrt{(n+1)l} \sum_p \delta_{n+1,p}\delta_{p,l-1} \right]. \tag{11A.4.10}$$

At this stage, we remark that

$$\sqrt{n(l+1)} \sum_p \delta_{n-1,p}\delta_{p,l+1} = \sqrt{n(l+1)}\delta_{n-1,l+1}$$

$$= \sqrt{n(n-1)}\delta_{n,l+2}, \tag{11A.4.11}$$

and, similarly,

$$\sqrt{(n+1)(l+1)} \sum_p \delta_{n+1,p}\delta_{p,l+1} = (n+1)\delta_{n,l}, \tag{11A.4.12}$$

$$\sqrt{nl} \sum_p \delta_{n-1,p}\delta_{p,l-1} = n\delta_{n,l}, \tag{11A.4.13}$$

$$\sqrt{(n+1)l} \sum_p \delta_{n+1,p}\delta_{p,l-1} = \sqrt{(n+1)(n+2)}\delta_{n,l-2}. \tag{11A.4.14}$$

This leads to

$$\langle n^{(0)}|\hat{x}^2|l^{(0)}\rangle = \frac{\hbar}{2m\omega} \left[\sqrt{n(n-1)}\delta_{n,l+2} + (2n+1)\delta_{n,l} \right.$$

$$\left. + \sqrt{(n+1)(n+2)}\delta_{n,l-2} \right], \tag{11A.4.15}$$

which implies (see Eqs. (11A.4.8) and (11A.4.9))

$$\langle n^{(0)}|\hat{x}^3|k^{(0)}\rangle = \left(\frac{\hbar}{2m\omega}\right)^{\frac{3}{2}}\left[\sqrt{n(n-1)(n-2)}\delta_{n,k+3} + 3n^{\frac{3}{2}}\delta_{n,k+1}\right.$$
$$+ 3(n+1)^{\frac{3}{2}}\delta_{n,k-1}$$
$$\left. + \sqrt{(n+1)(n+2)(n+3)}\delta_{n,k-3}\right]. \qquad (11A.4.16)$$

This implies that $(\psi_n^0, \hat{W}_1\psi_n^0) = 0$, and hence $\hat{W}_1$ only contributes to second-order corrections of the energy spectrum, when k equals $n \pm 1$ and $n \pm 3$. The complete formula for first-order corrections to the eigenfunctions is therefore (see Eq. (11A.4.3))

$$|n\rangle \sim |n^{(0)}\rangle + \varepsilon_1\left[|(n-3)^{(0)}\rangle\frac{\sqrt{n(n-1)(n-2)}}{3\hbar\omega} + |(n-1)^{(0)}\rangle\frac{3n^{\frac{3}{2}}}{\hbar\omega}\right.$$
$$+ |(n+1)^{(0)}\rangle\frac{3(n+1)^{\frac{3}{2}}}{-\hbar\omega}$$
$$\left. + |(n+3)^{(0)}\rangle\frac{\sqrt{(n+1)(n+2)(n+3)}}{-3\hbar\omega}\right]\left(\frac{\hbar}{2m\omega}\right)^{\frac{3}{2}}. \qquad (11A.4.17)$$

It should be stressed that $\hat{W}_2$ does not contribute to (11A.4.17), since ε_2 is of higher order with respect to ε_1, in that

$$\frac{\varepsilon_2}{(\varepsilon_1)^2} = O\left((m\omega^2)^{-1}\right). \qquad (11A.4.18)$$

The term $\hat{W}_2$, however, affects the energy spectrum (see the expansion (11A.4.2)). Indeed,

$$\langle n^{(0)}|\hat{W}_2|n^{(0)}\rangle = \varepsilon_2\langle n^{(0)}|\hat{x}^4|n^{(0)}\rangle, \qquad (11A.4.19)$$

where

$$\langle n^{(0)}|\hat{x}^4|n^{(0)}\rangle = \sum_k\langle n^{(0)}|\hat{x}^3|k^{(0)}\rangle\langle k^{(0)}|\hat{x}|n^{(0)}\rangle$$
$$= \left(\frac{\hbar}{2m\omega}\right)^2\left[3n^{\frac{3}{2}}\sqrt{n}\delta_{n-1,n-1} + 3(n+1)^{\frac{3}{2}}\sqrt{n+1}\delta_{n+1,n+1}\right]$$
$$= 3\left(2n^2 + 2n + 1\right)\left(\frac{\hbar}{2m\omega}\right)^2, \qquad (11A.4.20)$$

which implies

$$E_n \sim \left(n + \frac{1}{2}\right)\hbar\omega + \varepsilon_2\langle n^{(0)}|\hat{x}^4|n^{(0)}\rangle$$
$$+ (\varepsilon_1)^2\left[\frac{|\langle n^{(0)}|\hat{x}^3|(n-3)^{(0)}\rangle|^2}{3\hbar\omega} + \frac{|\langle n^{(0)}|\hat{x}^3|(n-1)^{(0)}\rangle|^2}{\hbar\omega}\right.$$

$$\left. + \frac{|\langle n^{(0)}|\hat{x}^3|(n+1)^{(0)}\rangle|^2}{-\hbar\omega} + \frac{|\langle n^{(0)}|\hat{x}^3|(n+3)^{(0)}\rangle|^2}{-3\hbar\omega}\right]$$

$$= \left(n + \frac{1}{2}\right)\hbar\omega + 3\varepsilon_2\left(\frac{\hbar}{2m\omega}\right)^2\left(2n^2 + 2n + 1\right)$$

$$- \frac{(\varepsilon_1)^2}{\hbar\omega}\left(\frac{\hbar}{2m\omega}\right)^3\left(30n^2 + 30n + 11\right). \tag{11A.4.21}$$

On defining

$$\sigma \equiv \sqrt{\frac{\hbar}{2m\omega}}, \tag{11A.4.22}$$

$$A \equiv 6\varepsilon_2 - 15\frac{\varepsilon_1^2}{m\omega^2}, \quad B \equiv 3\varepsilon_2 - \frac{11}{2}\frac{\varepsilon_1^2}{m\omega^2}, \tag{11A.4.23}$$

the result (11A.4.21) may be cast in the very convenient form

$$E_n \sim \left(n + \frac{1}{2}\right)\hbar\omega + \sigma^4\left[An(n+1) + B\right], \tag{11A.4.24}$$

which implies that, to avoid level-crossing (i.e. a sign change of perturbed eigenvalues for sufficiently large values of n), the following inequality should be imposed:

$$\varepsilon_2 > \frac{5}{2}\frac{\varepsilon_1^2}{m\omega^2}. \tag{11A.4.25}$$

11A.5 Secular equation for problems with degeneracy

Let us assume that the level with energy E_l^0 has a degeneracy of multiplicity f. This is indeed what happens in the majority of physical problems (e.g. the hydrogen atom or the harmonic oscillator in dimension greater than 1). As a zeroth-order approximation one may consider the linear combination

$$\psi_l = \sum_{k=1}^{f} a_k \varphi_{lk}, \tag{11A.5.1}$$

where φ_{lk} are taken to be solutions of the equation

$$(H_0 - E_l^0 I)\varphi_{lk} = 0. \tag{11A.5.2}$$

The insertion of the combination (11A.5.1) into the eigenvalue equation, jointly with the scalar product with the unperturbed eigenfunctions, leads to a homogeneous linear system of f equations:

$$\sum_{k=1}^{f}\left(H_{mk} - E_l\delta_{mk}\right)a_k = 0 \quad \forall m = 1, 2, \ldots, f. \tag{11A.5.3}$$

The condition for finding non-trivial solutions is, of course, that the determinant of the matrix of coefficients should vanish:

$$\det\left(H_{mk} - E_l \delta_{mk}\right) = 0. \tag{11A.5.4}$$

This yields an algebraic equation of degree f in the unknown E_l. Such an equation is called the *secular* equation. In particular, it may happen that all roots are distinct. If this turns out to be the case, the effect of the perturbation is to remove the degeneracy completely, and the level with energy E_l^0 splits into f distinct levels, for each of which a different eigenfunction exists:

$$\psi_{lk} = \sum_m a_{mk} \varphi_m. \tag{11A.5.5}$$

In Eq. (11A.5.5), the coefficients a_{mk} are evaluated by replacing E_l with E_{lk} in the system (11A.5.3).

If some roots of the secular equation (11A.5.4) coincide, however, the perturbation does not completely remove the degeneracy, and the following section describes a first concrete example of how this method works.

11A.6 Stark effect

The Stark effect consists in the splitting of the energy levels of the hydrogen atom (or other atoms) resulting from the application of an electric field (Stark 1914). If an electric field is applied, its effect is viewed as a perturbation, and is evaluated within the framework of time-independent perturbation theory for degenerate levels.

Here we are interested in the first-order Stark effect on the $n = 2$ states of the hydrogen atom. Indeed, we know from Chapter 7 that, when the quantum number $n = 2$ for the hydrogen atom, the quantum number l takes the values 0 and 1, and the quantum number m takes the values $-1, 0, 1$. Thus, there are four bound states with $n = 2$: $u_{2,0,0}, u_{2,1,-1}, u_{2,1,0}$ and $u_{2,1,1}$, and they all have the same energy E^0. The perturbation Hamiltonian $H^{(1)}$ for an electron of charge q_e in an electric field $\vec{E}$ directed along the z-axis is $H^{(1)} = -q_e E z$. Note that the first-order correction of the energy for any bound state $u_{n,l,m}$ vanishes (hereafter, dV is the integration measure in spherical coordinates):

$$E^{(1)} = -q_e E \int_{\mathbf{R}^3} u_{n,l,m}^* \, z \, u_{n,l,m} \, dV = 0. \tag{11A.6.1}$$

This happens because the state $u_{n,l,m}$ has either positive parity (if l is even) or negative parity (if l is odd), while $z = r \cos\theta$ is an odd function, so that the integral (11A.6.1), evaluated over the whole of $\mathbf{R}^3$, vanishes by construction (the integrand being an odd function, just as z is).

Non-vanishing values for first-order corrections of the energy levels can only be obtained from states that are neither even nor odd, and hence are linear combinations of states of opposite parity. The selection rule for matrix elements of the perturbation $H^{(1)}$ is obtained from the general formula

$$\left(u_{n',l',m'}, \hat{z}u_{n,l,m}\right) = \left(R_{n',l'}, \hat{r}R_{n,l}\right)\delta_{m,m'}\left(C^A\delta_{l',l+1} + C^B\delta_{l',l-1}\right), \qquad (11A.6.2)$$

where C^A and C^B are parameters depending on l and m. This implies that $H^{(1)}$ has non-vanishing matrix elements if and only if $m' - m = 0$ and $l' - l = \pm 1$.

As a first step, we compute the matrix elements of $H^{(1)}$ on the basis given by the states $u_{2,0,0}, u_{2,1,-1}, u_{2,1,0}$ and $u_{2,1,1}$. From Eq. (11A.6.1), all diagonal matrix elements vanish in this case: $H_{ii}^{(1)} = 0$. The orthogonality properties of the eigenfunctions with different values of m lead to a further simplification of the calculation, and the only non-vanishing matrix elements of $H^{(1)}$ turn out to be those particular off-diagonal elements corresponding to bound states with the same value of m: $u_{2,1,0}$ and $u_{2,0,0}$. Recall now that

$$u_{2,0,0}(r,\theta,\varphi) = R_{2,0}(r)Y_{0,0}(\theta,\varphi), \qquad (11A.6.3)$$

$$u_{2,1,0}(r,\theta,\varphi) = R_{2,1}(r)Y_{1,0}(\theta,\varphi). \qquad (11A.6.4)$$

This leads to (Squires 1995)

$$H_{34}^{(1)} = H_{43}^{(1)} = -q_e E \int u_{2,1,0}^* \, z \, u_{2,0,0} \, dV = -\varepsilon, \qquad (11A.6.5)$$

where

$$\varepsilon \equiv -3q_e E \frac{a_B}{Z}. \qquad (11A.6.6)$$

Thus, the matrix of $H^{(1)}$ on the basis of the four states with $n = 2$ is found to be

$$H_{ij}^{(1)} = \begin{pmatrix} 0 & 0 & 0 & 0 \\ 0 & 0 & 0 & 0 \\ 0 & 0 & 0 & -\varepsilon \\ 0 & 0 & -\varepsilon & 0 \end{pmatrix}. \qquad (11A.6.7)$$

As we mentioned before, since the four unperturbed states have the same energy, we are dealing with perturbation theory in the degenerate case. It is clear from Eq. (11A.6.7) that two eigenvalues of $H_{ij}^{(1)}$ vanish, and their eigenvectors are $u_{2,1,1}$ and $u_{2,1,-1}$, respectively. The non-trivial contribution to the energy spectrum results from the 2×2 sub-matrix $\begin{pmatrix} 0 & -\varepsilon \\ -\varepsilon & 0 \end{pmatrix}$. Its eigenvalues are $\lambda_+ = +\varepsilon$ and $\lambda_- = -\varepsilon$. In the former case, the eigenvector solves the equation

$$\begin{pmatrix} 0 & -\varepsilon \\ -\varepsilon & 0 \end{pmatrix}\begin{pmatrix} u_1 \\ u_2 \end{pmatrix} = \begin{pmatrix} \varepsilon u_1 \\ \varepsilon u_2 \end{pmatrix},$$

and hence is proportional to $\begin{pmatrix} 1 \\ -1 \end{pmatrix}$, while in the latter case it solves the equation

$$\begin{pmatrix} 0 & -\varepsilon \\ -\varepsilon & 0 \end{pmatrix}\begin{pmatrix} v_1 \\ v_2 \end{pmatrix} = \begin{pmatrix} -\varepsilon v_1 \\ -\varepsilon v_2 \end{pmatrix},$$

and is therefore proportional to $\begin{pmatrix} 1 \\ 1 \end{pmatrix}$. The corresponding normalized eigenvectors to first order are

$$u_{\lambda_+} = \frac{1}{\sqrt{2}} \begin{pmatrix} 1 \\ -1 \end{pmatrix} (u_{2,1,0}, u_{2,0,0}) = \frac{1}{\sqrt{2}} (u_{2,1,0} - u_{2,0,0}), \tag{11A.6.8}$$

$$u_{\lambda_-} = \frac{1}{\sqrt{2}} \begin{pmatrix} 1 \\ 1 \end{pmatrix} (u_{2,1,0}, u_{2,0,0}) = \frac{1}{\sqrt{2}} (u_{2,1,0} + u_{2,0,0}). \tag{11A.6.9}$$

Note that $u_{2,1,0}$ has negative parity, while $u_{2,0,0}$ has positive parity. Thus, in agreement with our initial remarks, the eigenfunctions of the problem turn out to have mixed parity. To first order in perturbation theory, the energies are then found to be

$$E^0, \ E^0, \ E^0 - 3q_e E \frac{a_{\rm B}}{Z}, \ E^0 + 3q_e E \frac{a_{\rm B}}{Z}.$$

Interestingly, this simple calculation shows that there can only be a first-order Stark effect when degenerate states exist with different values of the quantum number l. By virtue of the first-order Stark effect, the hydrogen atom behaves as if it had a permanent electric dipole moment of magnitude $-3q_e a_{\rm B}$. This dipole moment can be parallel, or anti-parallel, or orthogonal to the external electric field. In general, however, the ground states of atoms and nuclei are non-degenerate, and hence do not possess such a permanent electric dipole moment (Squires 1995). To second order in perturbation theory, the Stark effect provides a correction to the energy levels proportional to the square of the magnitude of the electric field. This term corresponds to an induced electric dipole moment.

Before concluding this section we should emphasize a crucial point. The perturbation occurring in the Stark effect is unbounded from below and unbounded with respect to the comparison Hamiltonian, and this is a source of non-trivial features. Indeed, in suitable units, the Stark effect on a hydrogen-like atom is described by the Hamiltonian operator

$$H(E) = -\Delta - \frac{Z}{r} + 2Ex_3 \tag{11A.6.10}$$

acting on $\mathcal{L}^2(\mathbf{R}^3)$, where $2E > 0$ is the uniform electric field directed along the x_3-axis (either x or y or z), Z is the atomic number and $r \equiv \sqrt{x^2 + y^2 + z^2}$. The spectrum of $H(E)$ is *absolutely continuous* in $(-\infty, \infty)$, while the spectrum of the unperturbed operator $H(0)$ for the hydrogen atom is *discrete* along $(-\infty, 0)$.

11A.7 Zeeman effect

Another good example of how an external field may remove the degeneracy of the original problem is provided by the Zeeman effect (Zeeman 1897a,b). In this case, the lines of the emission spectrum of an atomic system are split into several nearby components, when a magnetic field of sufficiently high intensity (e.g. of the order of 10^3 G) is switched on. This phenomenon may occur in hydrogen-like atoms and alkali metals. The quantum-mechanical interpretation requires that the electron for which the Schrödinger equation in an external electromagnetic field is written is affected by the potential $U(r)$ describing the effects of the nucleus and of the 'interior electrons'. Moreover, an external magnetic field is applied. The resulting canonical momenta contain the kinematic momenta and a term proportional to the vector potential as we know from Chapter 9, so that the Hamiltonian operator for an electron of charge q_e reads as

$$\hat{H} = \sum_{k=1}^{3} \frac{1}{2m_e} \left(\hat{p}_k - \frac{q_e}{c} \hat{A}_k \right)^2 + q_e \hat{V} + \hat{U}, \qquad (11A.7.1)$$

where $\hat{V}$ is the operator corresponding to the scalar potential of the classical theory. We now use the operator identity (no summation over k)

$$\left(\hat{p}_k - \frac{q_e}{c} \hat{A}_k \right)^2 = \hat{p}_k^2 - \frac{q_e}{c} \left(\hat{p}_k \hat{A}_k + \hat{A}_k \hat{p}_k \right) + \frac{q_e^2}{c^2} \hat{A}_k^2, \qquad (11A.7.2)$$

and another basic result:

$$\hat{\vec{p}} \cdot \hat{\vec{A}} = \frac{\hbar}{i} \sum_{k=1}^{3} \frac{\partial}{\partial x^k} \hat{A}_k = \hat{\vec{A}} \cdot \hat{\vec{p}} - i\hbar \, \vec{\nabla} \cdot \hat{\vec{A}}. \qquad (11A.7.3)$$

The Hamiltonian operator (11A.7.1) is thus found to read as

$$\hat{H} = \frac{\hat{\vec{p}}^2}{2m_e} - \frac{q_e}{m_e c} \hat{\vec{A}} \cdot \hat{\vec{p}} + \frac{iq_e \hbar}{2m_e c} \vec{\nabla} \cdot \hat{\vec{A}}$$

$$+ \frac{q_e^2}{2m_e c^2} \hat{\vec{A}}^2 + q_e \hat{V} + \hat{U}. \qquad (11A.7.4)$$

Now we assume, for simplicity, that the scalar potential vanishes:

$$\hat{V} = 0. \qquad (11A.7.5)$$

Moreover, since the external magnetic field is generated by bodies of macroscopic dimensions, it can be taken to be uniform, in a first approximation, over distances of the order of atomic dimensions: $\vec{B}(\vec{x}, t) = \vec{B}_0$. The form (11A.7.4) suggests choosing the Coulomb gauge to further simplify the calculations:

$$\vec{\nabla} \cdot \hat{\vec{A}} = 0. \qquad (11A.7.6)$$

A possible choice for the vector potential compatible with the above assumptions is thus found to be

$$\vec{A} = \frac{1}{2} \vec{B}_0 \wedge \vec{x}. \qquad (11A.7.7)$$

Its insertion into Eq. (11A.7.4) makes it necessary to derive the identity

$$\left(\vec{B}_0 \wedge \vec{x} \right) \cdot \vec{p} = \varepsilon_{klm} B_l x_m p_k = -B_l \varepsilon_{lkm} x_m p_k$$

$$= B_l \varepsilon_{lmk} x_m p_k = B_l \left(\vec{x} \wedge \vec{p} \right)_l = \vec{B}_0 \cdot \vec{L}. \qquad (11A.7.8)$$

The desired Hamiltonian operator is therefore

$$\hat{H} = \frac{\hat{\vec{p}}^2}{2m_e} - \frac{q_e}{2m_e c} \vec{B}_0 \cdot \hat{\vec{L}} + \frac{q_e^2}{8m_e c^2} \left(\vec{B}_0 \wedge \vec{x} \right)^2 + \hat{U}(r). \qquad (11A.7.9)$$

In a first approximation we now neglect the term quadratic in $\vec{B}_0$ and we choose

$$\vec{B}_0 = (0, 0, B_0),$$ (11A.7.10)

i.e. an external magnetic field directed along the z-axis. By virtue of our assumptions, the Hamiltonian operator reduces to

$$\hat{H} = \frac{\hat{p}^2}{2m_e} + \hat{U}(r) - \frac{q_e}{2m_e c}B_0\hat{L}_z = \hat{H}_0 - \frac{q_e}{2m_e c}B_0\hat{L}_z.$$ (11A.7.11)

The operators $\hat{H}_0, \hat{L}^2$ and $\hat{L}_z$ have common eigenvectors, here denoted by u_{nlm}, with eigenvalue equations

$$\hat{H}_0 u_{nlm} = W_{nl}^{(0)} u_{nlm},$$ (11A.7.12)

$$\hat{L}^2 u_{nlm} = l(l+1)\hbar^2 u_{nlm},$$ (11A.7.13)

$$\hat{L}_z u_{nlm} = m\hbar u_{nlm}.$$ (11A.7.14)

The resulting eigenvalue equation for the full Hamiltonian is

$$\hat{H} u_{nlm} = \left(W_{nl}^{(0)} + m\mu_B B_0\right)u_{nlm}.$$ (11A.7.15)

Thus, on neglecting contributions quadratic in B_0, the external magnetic field does not affect the eigenfunctions but modifies the eigenvalues, which are given by

$$W_{nlm} = W_{nl}^{(0)} + m\mu_B B_0.$$ (11A.7.16)

This means that the original invariance under rotations has been spoiled by the magnetic field, which has introduced a privileged direction. Each unperturbed energy level $W_{nl}^{(0)}$ is therefore split into $2l + 1$ distinct components. For example, in the transition $(n, l, m) \rightarrow (n', l', m')$ we now have the frequency

$$\nu_{nl \rightarrow n'l'} = \frac{W_{nlm} - W_{n'l'm'}}{h}$$

$$= \nu_{nl \rightarrow n'l'}^{(0)} - (\delta m)\frac{\mu_B B_0}{h}.$$ (11A.7.17)

For $(\delta m) = 0, \pm 1$ this leads to the three spectral lines

$$\nu_{nl \rightarrow n'l'}^{(0)} - \frac{\mu_B B_0}{h}, \quad \nu_{nl \rightarrow n'l'}^{(0)}, \quad \nu_{nl \rightarrow n'l'}^{(0)} + \frac{\mu_B B_0}{h}.$$

Such a theoretical model provides results in good agreement with observation only if the intensity of the magnetic field is so high that fine-structure corrections are negligible. The latter result from a relativistic evaluation of the Hamiltonian operator in a central potential, and lead to energy levels depending both on n and l. For example, fine-structure corrections cannot be neglected in the analysis of the D_1 and D_2 spectral lines of sodium, which are split into six and four components, respectively.

11A.8 Anomalous Zeeman effect

From an historical point of view we distinguish between the *normal* and an *anomalous* Zeeman effect that appears on transitions where the net spin of the electrons is not vanishing. In particular, in the anomalous case the number of Zeeman sub-levels becomes even instead of odd if an uneven number of electrons are involved. It was called anomalous because the electron spin had not yet been discovered, and hence there was no good explanation for it at the time that Zeeman observed the effect.

The anomalous Zeeman effect is observed in alkali metals under the influence of weak magnetic fields, and does not admit a classical interpretation. The transition from the anomalous to the normal Zeeman effect as the intensity of the external magnetic field is increased is called the *Paschen–Back effect* (Paschen and Back 1912, 1913).

Let us consider for simplicity a hydrogen atom and apply to it a constant magnetic field $\vec{B}$, which for convenience we imagine to be oriented along the third axis. In this case, if we are in the so-called weak-field approximation, i.e. we can neglect quadratic terms in the magnetic field, we have the following Hamiltonian for the system:

$$
\begin{aligned}
\hat{H} &= \frac{\hat{\vec{p}}^{\,2}}{2m_e} - \frac{q_e^2}{r} + \mu_B \vec{B} \cdot \left(\frac{\hat{\vec{L}} + 2\hat{\vec{S}}}{\hbar} \right) + 2\mu_B^2 \frac{1}{r^3} \left(\frac{\vec{S} \cdot \vec{L}}{\hbar^2} \right) \\
&= \hat{H}_0 + \mu_B \vec{B} \cdot \left(\frac{\hat{\vec{L}} + 2\hat{\vec{S}}}{\hbar} \right) + 2\mu_B^2 \frac{1}{r^3} \left(\frac{\hat{\vec{S}} \cdot \hat{\vec{L}}}{\hbar^2} \right) \\
&= \hat{H}_0 + \mu_B B \left(\frac{\hat{L}_3 + 2\hat{S}_3}{\hbar} \right) + 2\mu_B^2 \frac{1}{r^3} \left(\frac{\hat{\vec{S}} \cdot \hat{\vec{L}}}{\hbar^2} \right),
\end{aligned}
\tag{11A.8.1}
$$

where the spin–orbit term is taken from Eq. (9.7.11) and $\hat{H}_0$ denotes the Hamiltonian of the hydrogen atom discussed in Section 7.2. As already shown, it is easy to see that $\hat{H}_0$, $\hat{L}^2$, $\hat{L}_3$, $\hat{S}^2$ and $\hat{S}_3$ form a complete set of observables as well as $\hat{H}_0$, $\hat{L}^2$, $\hat{S}^2$, $\hat{J}^2$, and $\hat{J}_3$, where the corresponding basis eigenvectors are denoted by $|n, l, m_3, s, s_3\rangle$ and $|n, l, s, j, j_3\rangle$, respectively. Since we are going to consider electrons only, hereafter we fix the quantum number of $\hat{S}^2$ to $\frac{1}{2}$ always. Note that both $|n, l, m_3, \frac{1}{2}, s_3\rangle$ and $|n, l, \frac{1}{2}, j, j_3\rangle$ are eigenvectors of $\hat{H}_0$ of eigenvalue $E_n^0 = -q_e^4 m_e / (2\hbar^2 n^2)$. By using both sets of eigenvectors we can study the two additional terms to $\hat{H}_0$ appearing in Eq. (11A.8.1), which in turn can be viewed as a perturbation.

Let us consider the case in which the spin–orbit term can be taken to be small with respect to the others; in this case it is convenient to adopt the basis $|n, l, m_3, s, s_3\rangle$, which exactly diagonalizes the Zeeman Hamiltonian

$$
\hat{H}_Z = \hat{H}_0 + \mu_B B \left(\frac{\hat{L}_3 + 2\hat{S}_3}{\hbar} \right),
\tag{11A.8.2}
$$

and in fact

$$\hat{H}_Z \left| n, l, m_3, \frac{1}{2}, s_3 \right\rangle = E^Z_{m_3, s_3} \left| n, l, m_3, \frac{1}{2}, s_3 \right\rangle, \tag{11A.8.3}$$

where

$$E^Z_{m_3, s_3} = \left[E^0_n + \mu_B (m_3 + 2 s_3) \right]. \tag{11A.8.4}$$

In this case the presence of the magnetic field implies a splitting of energy levels related to the third component of both orbital and spin angular momenta. In this basis the spin–orbit term has to be treated as a perturbation. Since the energy levels $E^Z_{m_3, s_3}$ depend on the combination $m_3 + 2 s_3$ only, they are in general degenerate. This can be checked by simply varying compatibly l, and/or by observing that, denoting with $p \equiv m_3 + 2 s_3$, there are two pairs of values (m_3, s_3) producing the same p, i.e. $(p - 1, \frac{1}{2})$ and $(p + 1, -\frac{1}{2})$. For this reason, in order to compute the first-order correction resulting from the spin–orbit term we should consider all matrix elements of

$$\hat{H}_{so} = 2 \mu_B^2 \frac{1}{r^3} \left(\frac{\hat{\vec{S}} \cdot \hat{\vec{L}}}{\hbar^2} \right) \tag{11A.8.5}$$

in the unperturbed energy eigenspaces. Fortunately, in this case such a calculation reduces to the diagonal terms only. This can be simply understood observing that, from the commutation relation $[\hat{L}^2, \hat{\vec{S}} \cdot \hat{\vec{L}}] = 0$,

$$\left\langle n, l', m_3', \frac{1}{2}, s_3' \left| \hat{\vec{S}} \cdot \hat{\vec{L}} \right| n, l, m_3, \frac{1}{2}, s_3 \right\rangle = 0 \quad \text{if} \quad l \neq l'. \tag{11A.8.6}$$

Moreover, assuming the same l, it can be shown that the only remaining off-diagonal term vanishes as well, i.e.

$$\left\langle n, l, p - 1, \frac{1}{2}, \frac{1}{2} \left| \hat{\vec{S}} \cdot \hat{\vec{L}} \right| n, l, p + 1, \frac{1}{2}, -\frac{1}{2} \right\rangle = 0. \tag{11A.8.7}$$

This result is trivial by just observing that $\hat{\vec{S}} \cdot \hat{\vec{L}} = \hat{L}_3 \hat{S}_3 + (\hat{L}_+ \hat{S}_- + \hat{L}_- \hat{S}_+)/2$ can only connect bra and kets differing for one unit in m_3, which is not the case for Eq. (11A.8.7). From the previous results, the first-order correction to $E^Z_{m_3, s_3}$ reads as

$$\begin{aligned}
\Delta E^Z_{n, l, m_3, s_3} &= 2 \mu_B^2 \left\langle n, l, m_3, \frac{1}{2}, s_3 \left| \frac{\hat{\vec{S}} \cdot \hat{\vec{L}}}{\hbar^2 r^3} \right| n, l, m_3, \frac{1}{2}, s_3 \right\rangle \\
&= \frac{q_e^8 m_e}{2 \hbar^4 c^2} \frac{m_3 s_3}{n^3 l (l + 1) (l + \frac{1}{2})} \\
&= -E^0_n \frac{\alpha^2}{n} \frac{m_3 s_3}{l (l + 1) (l + \frac{1}{2})},
\end{aligned} \tag{11A.8.8}$$

where we have used the result

$$\left\langle n, l, m_3, \frac{1}{2}, s_3 \left| r^{-3} \right| n, l, m_3, \frac{1}{2}, s_3 \right\rangle = \frac{m_e^3 q_e^6}{n^3 \hbar^6} \frac{1}{l (l + 1)(l + \frac{1}{2})}. \tag{11A.8.9}$$

Note that the correction (11A.8.8) is of second order in the fine-structure constant $\alpha \equiv q_{\mathrm{e}}^2/(\hbar c) \approx \frac{1}{137}$. Moreover, $\Delta E_{n,l,m_3,s_3}^Z$ is defined on s-waves as well, since in this case the limit $l, m_3 \to 0$ has to be taken at the same time.

11A.9 Relativistic corrections (α^2) to the hydrogen atom

As we have discussed in previous chapters, we use for the kinetic term of the hydrogen atom the non-relativistic approximation

$$T = \frac{\vec{p}^{\,2}}{2m_{\mathrm{e}}}. \tag{11A.9.1}$$

Such an expression can be corrected by considering a first-order expansion of its relativistic expression

$$T = \sqrt{c^2\vec{p}^{\,2} + m_{\mathrm{e}}^2 c^4} - m_{\mathrm{e}}c^2 \approx \frac{\vec{p}^{\,2}}{2m_{\mathrm{e}}} - \frac{1}{2m_{\mathrm{e}}c^2}\left(\frac{\vec{p}^{\,2}}{2m_{\mathrm{e}}}\right)^2. \tag{11A.9.2}$$

By recalling that for the hydrogen atom

$$\hat{H}_0 = \frac{\hat{\vec{p}}^{\,2}}{2m_{\mathrm{e}}} - \frac{q_{\mathrm{e}}^2}{r}, \tag{11A.9.3}$$

the contribution of relativistic corrections can be written as

$$\hat{H}' = -\frac{1}{2m_{\mathrm{e}}c^2}\left(\frac{\hat{\vec{p}}^{\,2}}{2m_{\mathrm{e}}}\right)^2 = -\frac{1}{2m_{\mathrm{e}}c^2}\left(\hat{H}_0 + \frac{q_{\mathrm{e}}^2}{r}\right)^2. \tag{11A.9.4}$$

The perturbation term $\hat{H}'$ commutes with $\hat{L}^2$, $\hat{S}^2$, $\hat{J}^2$, and $\hat{J}_3$, hence we can choose the basis $|n,l,s,j,j_3\rangle$ to evaluate the first-order corrections to E_n^0. In particular,

$$\begin{aligned}
\delta_{nl}^{\mathrm{rel}} &= -\frac{1}{2m_{\mathrm{e}}c^2}\left\langle n,l,\tfrac{1}{2},j,j_3\right|\left(E_n^0 + \frac{q_{\mathrm{e}}^2}{r}\right)^2\left|n,l,\tfrac{1}{2},j,j_3\right\rangle \\
&= -\frac{1}{2m_{\mathrm{e}}c^2}\left((E_n^0)^2 + 2q_{\mathrm{e}}^2 E_n^0\left\langle n,l,\tfrac{1}{2},j,j_3\right|r^{-1}\left|n,l,\tfrac{1}{2},j,j_3\right\rangle \right. \\
&\quad \left. + q_{\mathrm{e}}^4\left\langle n,l,\tfrac{1}{2},j,j_3\right|r^{-2}\left|n,l,\tfrac{1}{2},j,j_3\right\rangle\right) \\
&= -\frac{(E_n^0)^2}{2m_{\mathrm{e}}c^2}\left(\frac{8n}{(2l+1)} - 3\right) = E_n^0\frac{\alpha^2}{4n^2}\left(\frac{8n}{(2l+1)} - 3\right). \tag{11A.9.5}
\end{aligned}$$

As shown in the previous expression, the relativistic correction due to the kinematical term is of the same order of magnitude of the spin–orbit contribution (11A.8.8). For this reason, by re-evaluating in the $|n,l,\frac{1}{2},j,j_3\rangle$ basis the spin–orbit correction and adding the result to (11A.9.5) the complete fine-structure constant splitting of hydrogen energy levels is obtained. Bearing in mind that $\hat{\vec{L}} \cdot \hat{\vec{S}} = \frac{(\hat{J}^2 - \hat{L}^2 - \hat{S}^2)}{2}$, we can write

$$+ 2\mu_B^2 \left\langle n, l, \frac{1}{2}, j, j_3 \left| \frac{\hat{\vec{S}} \cdot \hat{\vec{L}}}{\hbar^2 r^3} \right| n, l, \frac{1}{2}, j, j_3 \right\rangle$$

$$= \frac{q_e^2 \hbar^2}{4 m_e^2 c^2} \left(j(j+1) - l(l+1) - \frac{3}{4} \right) \left\langle n, l, \frac{1}{2}, j, j_3 \left| r^{-3} \right| n, l, \frac{1}{2}, j, j_3 \right\rangle$$

$$= \frac{q_e^2 \hbar^2}{4 m_e^2 c^2} \left(j(j+1) - l(l+1) - \frac{3}{4} \right) \frac{m_e^3 q_e^6}{n^3 \hbar^6} \frac{1}{l(l+1)(l+1/2)}$$

$$= -E_n^0 \frac{\alpha^2}{2n} \frac{\left(j(j+1) - l(l+1) - \frac{3}{4} \right)}{l(l+1)(l+\frac{1}{2})}. \tag{11A.9.6}$$

Note that also in this case, since $\hat{H}_{so}$ commutes with the set of operators $\hat{L}^2$, $\hat{S}^2$, $\hat{J}^2$, and $\hat{J}_3$, we can restrict our analysis for the first-order energy correction in the $|n, l, \frac{1}{2}, j, j_3\rangle$ basis to the diagonal terms only.

Hence, by summing Eqs. (11A.9.6) and (11A.9.5), we get the complete α^2 correction to E_n^0

$$\Delta E_{nj}^0(\alpha^2) = E_n^0 \frac{\alpha^2}{n^2} \left(\frac{n}{j + \frac{1}{2}} - \frac{3}{4} \right), \tag{11A.9.7}$$

which turns out to depend on the quantum numbers n and j only.

In the basis $|n, l, \frac{1}{2}, j, j_3\rangle$ it is easy to evaluate, in addition to the spin–orbit, the anomalous Zeeman effect as well, if one recalls the relation connecting such a basis with $|n, l, m_3, \frac{1}{2}, s_3\rangle$. In the previous section we reported the interaction term between a magnetic field oriented along the third axis and the orbital and spin angular momenta, i.e.

$$\hat{H}_z = \mu_B B \left(\frac{\hat{L}_3 + 2\hat{S}_3}{\hbar} \right) = \mu_B B \left(\frac{\hat{J}_3 + \hat{S}_3}{\hbar} \right). \tag{11A.9.8}$$

Of course, we can focus for the moment on the term proportional to $\hat{S}_3$ only, the other being diagonal in the chosen basis. Since $\hat{S}_3$ commutes with $\hat{L}^2$, $\hat{S}^2$ and $\hat{J}_3$ its first-order contribution should be evaluated in the two-dimensional eigenspace of E_n^0 for given l and $s = \frac{1}{2}$ and j_3. For this purpose, it is worth recalling the simple relations

$$\left| n, l, \frac{1}{2}, l + \frac{1}{2}, m + \frac{1}{2} \right\rangle = \sqrt{\frac{(l+m+1)}{(2l+1)}} \left| n, l, m, \frac{1}{2}, \frac{1}{2} \right\rangle$$

$$+ \sqrt{\frac{(l-m)}{(2l+1)}} \left| n, l, m+1, \frac{1}{2}, -\frac{1}{2} \right\rangle, \tag{11A.9.9}$$

$$\left| n, l, \frac{1}{2}, l - \frac{1}{2}, m + \frac{1}{2} \right\rangle = -\sqrt{\frac{(l-m)}{(2l+1)}} \left| n, l, m, \frac{1}{2}, \frac{1}{2} \right\rangle$$

$$+ \sqrt{\frac{(l+m+1)}{(2l+1)}} \left| n, l, m+1, \frac{1}{2}, -\frac{1}{2} \right\rangle, \tag{11A.9.10}$$

where of course the terms involving $(l - m)$ are written only for $m \neq l$. Nevertheless, since the spin–orbit term (11A.9.7) has removed the degeneracy in j, restrict the computation must be restricted to the diagonal terms only, i.e.

$$\left\langle n,l,\frac{1}{2},l+\frac{1}{2},m+\frac{1}{2}\middle|\hat{S}_3\middle|n,l,\frac{1}{2},l+\frac{1}{2},m+\frac{1}{2}\right\rangle = \frac{\hbar}{2}\frac{(2m+1)}{(2l+1)} = \frac{\hbar j_3}{(2l+1)},$$

$$\left\langle n,l,\frac{1}{2},l-\frac{1}{2},m+\frac{1}{2}\middle|\hat{S}_3\middle|n,l,\frac{1}{2},l-\frac{1}{2},m+\frac{1}{2}\right\rangle = -\frac{\hbar}{2}\frac{(2m+1)}{(2l+1)} = -\frac{\hbar j_3}{(2l+1)}.$$

$$(11A.9.11)$$

By using the above results we obtain the anomalous Zeeman contribution to be added to the spin–orbit term $\Delta E_{nj}^0(\alpha^2)$

$$\Delta E_{nljj_3}^0(B) = \mu_B B\left\langle n,l,\frac{1}{2},j,j_3\middle|\left(\frac{\hat{J}_3+\hat{S}_3}{\hbar}\right)\middle|n,l,\frac{1}{2},j,j_3\right\rangle$$

$$= \mu_B B j_3\left(1 \pm \frac{1}{(2l+1)}\right), \qquad (11A.9.12)$$

where the sign corresponds to $j = l \pm \frac{1}{2}$.

11A.10 Variational method

The variational method relies on the observation that with any operator (Hermitian or self-adjoint) we may associate the expectation-value function

$$e_A(\psi) = \frac{\langle\psi|A|\psi\rangle}{\langle\psi|\psi\rangle}, \qquad (11A.10.1)$$

which is real-valued. The important result is that critical points of e_A, say those for which $de_A = 0$, are given by eigenvectors and the values of e_A at these critical points are the corresponding eigenvalues. By taking the differential of the expectation-value function,

$$de_A(\psi) = \frac{\langle d\psi|A|\psi\rangle + \langle\psi|A|d\psi\rangle}{\langle\psi|\psi\rangle} - \frac{\langle\psi|A|\psi\rangle\big(\langle d\psi|\psi\rangle + \langle\psi|d\psi\rangle\big)}{\langle\psi|\psi\rangle^2}$$

$$= \langle d\psi|\left(A|\psi\rangle - \frac{\langle\psi|A|\psi\rangle}{\langle\psi|\psi\rangle}|\psi\rangle\right)\frac{1}{\langle\psi|\psi\rangle}$$

$$+ \left(\langle\psi|A - \frac{\langle\psi|A|\psi\rangle}{\langle\psi|\psi\rangle}\langle\psi|\right)|d\psi\rangle\frac{1}{\langle\psi|\psi\rangle}, \qquad (11A.10.2)$$

which vanishes at critical points. By using the independence of $\langle d\psi|$ and $|d\psi\rangle$ we find

$$A|\psi\rangle = \frac{\langle\psi|A|\psi\rangle}{\langle\psi|\psi\rangle}|\psi\rangle, \qquad (11A.10.3)$$

$$\langle\psi|A = \frac{\langle\psi|A|\psi\rangle}{\langle\psi|\psi\rangle}\langle\psi|. \qquad (11A.10.4)$$

Thus, the search for critical points of the expectation-value function is equivalent to the eigenvalue equation. A practical use of this method consists of considering the expectation-value functions restricted to a sub-manifold of states parametrized by parameters or coordinates $a_1, a_2, \ldots, a_m$. We have encountered an example when dealing with coherent states; therein $a_1, a_2, \ldots, a_m$ were represented by $z_1, z_2, \ldots, z_m$. We find in this way

$$e_A(\psi(a_1, \ldots, a_m)).$$

The critical points of e_A are then evaluated in the manifold M with coordinates $(a_1, \ldots, a_m)$ as

$$\frac{\partial e_A}{\partial a_j} = 0, \ j = 1, \ldots, m. \tag{11A.10.5}$$

The solutions of these equations correspond to values of $e_A(a_1, \ldots, a_m)$ and provide approximate values for eigenvalues. In particular, the lowest of the values gives the best approximation for the ground-state energy E_0 when $A = H$, within the considered trial vectors we have chosen in the Hilbert space. In general, the result depends on the trial vectors (wave functions we have chosen). In the course of choosing the trial manifold of states we may be guided by symmetry properties and by the possibility of computing explicitly the expectation-value functions on the selected states.

11A.11 Time-dependent formalism

The time-dependent formalism in perturbation theory is appropriate for the description of quantum systems that are weakly interacting with other physical systems, so that the full Hamiltonian reads

$$H = H_0 + \varepsilon V(t), \tag{11A.11.1}$$

where H_0 is an essentially self-adjoint comparison Hamiltonian independent of time, V is a given function of time and ε is a small dimensionless parameter. For example, this may happen for an atom in an electromagnetic field. One is then mainly interested in the probability that the system, initially in an eigenstate of the unperturbed Hamiltonian H_0, performs a transition to any other eigenstate. Typically, these are the probabilities that an atom can receive or emit energy by virtue of the interaction with electromagnetic radiation. The mathematical description of such processes is as follows.

Let $\psi(t)$ be the state of the system at time t. In quantum mechanics, this solves the equation

$$i\hbar \frac{\mathrm{d}}{\mathrm{d}t} \psi(t) = H(t)\psi(t), \tag{11A.11.2}$$

for a given initial condition $\psi(t_0)$. At the risk of repeating ourselves, we recall that the wave function $\psi(\vec{x}, t)$ is recovered from the state vector $\psi(t)$ in a way that becomes clear from using the Dirac notation. On denoting by $|x\rangle$ a generalized solution of the eigenvalue equation for the position operator $\hat{x}$, with associated bra written as $\langle x|$, and writing $|\psi\rangle_t$ for the state vector $\psi(t)$ at time t,

$$\psi(\vec{x}, t) = \langle x|\psi\rangle_t = \int_{\mathbf{R}^3} \delta^3(\vec{x} - \vec{\xi})\psi(\vec{\xi}, t) \, \mathrm{d}^3\xi. \tag{11A.11.3}$$

The propagator $U(t, t_0)$ maps, by definition, $\psi(t_0)$ into $\psi(t)$:

$$\psi(t) = U(t, t_0)\psi(t_0), \tag{11A.11.4}$$

and hence satisfies the equation

$$i\hbar\frac{\mathrm{d}}{\mathrm{d}t}U(t, t_0) = H(t)U(t, t_0), \tag{11A.11.5}$$

with the initial condition $U(t_0, t_0) = 1$. Since the time-dependent part $\varepsilon V(t)$ in the Hamiltonian is supposed to be a 'small' perturbation of H_0 (see below), the idea is that U should differ by a small amount from the unperturbed propagator:

$$U_0(t, t_0) = U_0(t - t_0) = \exp\left[-i(t - t_0)\frac{H_0}{\hbar}\right]. \tag{11A.11.6}$$

we are thus led to write the perturbation in a multiplicative way to deal with unitary operators

$$U = U_0 \, W, \tag{11A.11.7}$$

where W is another unitary operator, which encodes all the effects resulting from the interaction. In other words, since the full Hamiltonian depends on time, the propagator $U(t, t_0)$ cannot depend on the difference $t - t_0$, but its evaluation is reduced to a series of corrections of the unperturbed propagator $U_0(t, t_0) = U_0(t - t_0)$, thanks to the introduction of the unitary operator W. The insertion of Eq. (11A.11.7) into Eq. (11A.11.5) yields

$$i\hbar\frac{\mathrm{d}}{\mathrm{d}t}W(t, t_0) = \varepsilon U_0^\dagger(t - t_0)V(t)U_0(t - t_0)W(t, t_0). \tag{11A.11.8}$$

We now look for a solution in the form

$$W(t, t_0) \sim \sum_{n=0}^\infty \varepsilon^n W_n(t, t_0). \tag{11A.11.9}$$

Once more, we should stress that we are dealing with formal series. The right-hand side may or may not converge for a given form of $V(t)$, and we are assuming that $W(t, t_0)$ has an asymptotic expansion. With this understanding, and defining

$$V_{\mathrm{int}}(t, t_0) \equiv U_0^\dagger(t, t_0)V(t)U_0(t, t_0), \tag{11A.11.10}$$

we find, $\forall n \geq 0$, a recursive set of differential equations

$$i\hbar\frac{\mathrm{d}}{\mathrm{d}t}W_{n+1}(t, t_0) = V_{\mathrm{int}}(t, t_0)W_n(t, t_0). \tag{11A.11.11}$$

The initial condition $W(t_0, t_0) = 1$ leads to the initial data

$$W_0(t_0, t_0) = 1, \tag{11A.11.12}$$

$$W_n(t_0, t_0) = 0, \quad \forall n > 0, \tag{11A.11.13}$$

and hence the various terms are given by

$$W_1(t, t_0) = -\frac{i}{\hbar} \int_{t_0}^{t} V_{int}(\tau, t_0) d\tau, \tag{11A.11.14}$$

$$W_2(t, t_0) = -\frac{i}{\hbar} \int_{t_0}^{t} V_{int}(\tau, t_0) W_1(\tau, t_0) d\tau, \tag{11A.11.15}$$

$$W_{n+1}(t, t_0) = -\frac{i}{\hbar} \int_{t_0}^{t} V_{int}(\tau, t_0) W_n(\tau, t_0) d\tau. \tag{11A.11.16}$$

We can now insert into each integral the form of W_n given by the previous equation. The resulting series is the Dyson series (Dyson 1949), with

$$W_2(t, t_0) = \left(\frac{-i}{\hbar}\right)^2 \int_{t_0}^{t} dt_1 \int_{t_0}^{t_1} dt_2 V_{int}(t_1, t_0) V_{int}(t_2, t_0), \tag{11A.11.17}$$

$$W_3(t, t_0) = \left(\frac{-i}{\hbar}\right)^3 \int_{t_0}^{t} dt_1 \int_{t_0}^{t_1} dt_2 \int_{t_0}^{t_2} dt_3 V_{int}(t_1, t_0) V_{int}(t_2, t_0) V_{int}(t_3, t_0), \tag{11A.11.18}$$

and so on. Note that, in the multiple integral defining $W_n(t, t_0)$, the operators $V_{int}(t, t_0)$ are ordered chronologically in that, if $t > t'$, $V_{int}(t, t_0)$ is to the left of $V_{int}(t', t_0)$. Thus, on using the symbol of chronological ordering for which

$$T H(t_1) H(t_2) \cdots H(t_n) = \theta(t_1, t_2, \ldots, t_n) H(t_1) H(t_2) \cdots H(t_n), \tag{11A.11.19}$$

with

$$\theta(t_1, t_2, \ldots, t_n) = 1 \quad \text{if } t_1 > t_2 > \cdots > t_n, \ 0 \ \text{otherwise}, \tag{11A.11.20}$$

we find

$$W(t, t_0) = T \exp\left[-\frac{i\varepsilon}{\hbar} \int_{t_0}^{t} V_{int}(\tau, t_0) d\tau\right]. \tag{11A.11.21}$$

Needless to say, the calculation of the right-hand side of Eq. (11A.11.21) is, in general, quite cumbersome, despite the elegance of the formula.

11A.12 Harmonic perturbations

We can now evaluate, approximately, the transition amplitude from an eigenstate φ_n of H_0 to another eigenstate φ_m under the action of the perturbation $\varepsilon V(t)$. To first order in ε,

$$A_{n \to m}(t, t_0) = \left(\varphi_m, U(t, t_0) \varphi_n\right)$$

$$= \exp\left[-\frac{i}{\hbar} E_m(t - t_0)\right] \left(\varphi_m, W(t, t_0) \varphi_n\right), \tag{11A.12.1}$$

where

$$\left(\varphi_m, W(t, t_0)\varphi_n\right) \sim \delta_{nm} - \frac{i\varepsilon}{\hbar} \int_{t_0}^{t} \left(\varphi_m, V_{\text{int}}(\tau, t_0)\varphi_n\right) d\tau + O(\varepsilon^2)$$

$$= \delta_{nm} - \frac{i\varepsilon}{\hbar} \int_{t_0}^{t} e^{[(i/\hbar)(E_m - E_n)(\tau - t_0)]} \left(\varphi_m, V(\tau)\varphi_n\right) d\tau$$

$$+ O(\varepsilon^2). \tag{11A.12.2}$$

It is quite interesting to consider the case of *harmonic perturbations*, for which $V(t)$ takes the form

$$V(t) = B\, e^{-i\omega t} + B^\dagger\, e^{i\omega t}, \tag{11A.12.3}$$

where B is a given operator in the Hilbert space of the problem. Note that, if such a perturbation is studied in a finite interval $(0, T)$, for $m \neq n$,

$$A_{n \to m}(T) = \left(\varphi_m, \varepsilon B \varphi_n\right) \frac{1 - e^{i(\omega_{mn} - \omega)T}}{\hbar(\omega_{mn} - \omega)}$$

$$+ \left(\varphi_m, \varepsilon B^\dagger \varphi_n\right) \frac{1 - e^{i(\omega_{mn} + \omega)T}}{\hbar(\omega_{mn} + \omega)}, \tag{11A.12.4}$$

having defined $\omega_{mn} \equiv \frac{E_m - E_n}{\hbar}$. Interestingly, this shows that the perturbation may lead to transitions even between states with Bohr frequencies ω_{mn} that differ from the frequency ω in Eq. (11A.12.3). However, the transition probability $|A_{n \to m}(T)|^2$ receives the dominant contribution either from the region $\omega_{mn} \cong \omega$, or from the region $\omega_{mn} \cong -\omega$. The former corresponds to *resonant absorption* (the quantum system receives energy from the external perturbation, and the final state has energy $E_m > E_n$), while the latter corresponds to *resonant emission*. It should be stressed that, when first-order effects provide a good approximation,

$$\left(\varphi_m, \varepsilon B \varphi_n\right) \ll \hbar\omega \text{ for resonant absorption}, \tag{11A.12.5}$$

$$\left(\varphi_m, \varepsilon B^\dagger \varphi_n\right) \ll \hbar\omega \text{ for resonant emission}. \tag{11A.12.6}$$

In the case of resonant absorption only the first term on the right-hand side of Eq. (11A.12.4) can be kept, and this leads to the following formula for the transition probability:

$$|A_{n \to m}(T)|^2 \sim \frac{|(\varphi_m, \varepsilon B \varphi_n)|^2}{\hbar^2} \left\{ \frac{\sin\left[\frac{(\omega - \omega_{mn})T}{2}\right]}{\frac{(\omega - \omega_{mn})}{2}} \right\}^2. \tag{11A.12.7}$$

Bearing in mind that

$$\lim_{\tau \to \infty} \frac{\sin^2 \alpha\tau}{\alpha^2\tau} = \pi\delta(\alpha),$$

we can then define the *transition probability per unit time* (see Figure 11A.1):

$$R \equiv \lim_{T \to \infty} \frac{1}{T} |A_{n \to m}(T)|^2 \sim \frac{2\pi}{\hbar} |(\varphi_m, \varepsilon B \varphi_n)|^2 \delta(\hbar\omega - E_m + E_n). \tag{11A.12.8}$$

Note that this expression for R is symmetric in T, i.e. $R(-T) = R(T)$.

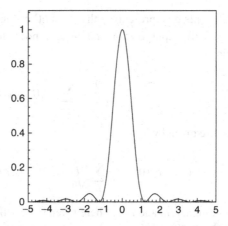

The curve shows the behaviour of the transition probability P_{nm} as a function of ω at a fixed instant of time t in the presence of harmonic perturbations $V_0 e^{i\omega t}$. Such a transition probability is substantially different from zero when the energy E_n of the final state lies in a narrow neighbourhood of $(E_n \pm \hbar\omega)$. When a resonance occurs, P_{nm} attains a maximum equal to $\frac{|(\varphi_m, V_0\varphi_n)|^2}{\hbar^2}$. When the difference $|\omega - \omega_{mn}|$ increases, P_{nm} oscillates in between 0 and the value $\frac{4|(\varphi_m, V_0\varphi_n)|^2}{\hbar^2(\omega - \omega_{mn})^2}$.

11A.13 Fermi golden rule

So far, we have assumed that the system has an entirely discrete energy spectrum. However, if we allow for transitions from a discrete level E_i to a final level E_f belonging to the continuous spectrum, we have to integrate the expression (11A.12.8), taking into account the number of energy eigenstates in between E and $E + dE$. This number is denoted by $n(E)dE$, and the formula for R becomes

$$R \sim \frac{2\pi}{\hbar}|(\varphi_f, B\varphi_i)|^2 n(E_f), \tag{11A.13.1}$$

where $E_f = E_i + \hbar\omega$. The formula (11A.13.1) is said to describe the Fermi golden rule (cf. Fermi 1932).

11A.14 Towards limiting cases of time-dependent theory

If the eigenvalue problem for the unperturbed Hamiltonian H_0 is completely solved, with eigenvectors φ_n and eigenvalues $E_n^{(0)}$, the previous calculations are equivalent to expanding the state vector $\psi(t)$ for the full Hamiltonian H in the form (assuming, for simplicity, a purely discrete spectrum of H_0)

$$\psi(t) = \sum_{n=1}^{\infty} c_n(t) e^{-iE_n^{(0)}\frac{t}{\hbar}} \varphi_n, \tag{11A.14.1}$$

with coefficients c_n representing the 'weight' at time t with which the stationary state φ_n contributes to the superposition that makes it possible to build $\psi(t)$. Such coefficients solve the system

$$i\hbar\frac{dc_n(t)}{dt} = \sum_{m=1}^{\infty} c_m(t)\Big(\varphi_n, \varepsilon V_{\text{int}}(t)\varphi_m\Big), \tag{11A.14.2}$$

and hence are given by

$$c_n(t) = c_n(0) - \frac{i}{\hbar}\sum_{m=1}^{\infty}\int_0^t c_m(t')\Big(\varphi_n, \varepsilon V_{\text{int}}(t')\varphi_m\Big)dt'. \tag{11A.14.3}$$

The Hamiltonians H_0 and H are here taken to be self-adjoint on the same domain, with the perturbation modifying the expansion of ψ valid for the stationary theory and breaking the invariance under time translations. The recurrence formulae for the various terms in the operator W lead to the following perturbative scheme for the evaluation of $c_n(t)$:

$$c_n(t) = c_n^{(0)} + c_n^{(1)}(t) + c_n^{(2)}(t) + \cdots, \tag{11A.14.4}$$

where, for all $r = 0, 1, 2, \ldots,$

$$c_n^{(r+1)}(t) = -\frac{i}{\hbar}\sum_{m=1}^{\infty}\int_0^t c_m^{(r)}(t')\Big(\varphi_n, \varepsilon V_{\text{int}}(t')\varphi_m\Big)dt'. \tag{11A.14.5}$$

According to the scheme outlined in the previous section, if the system is initially in the state φ_i, the quantity

$$P_{\text{if}}(t) = |c_f(t)|^2 \tag{11A.14.6}$$

can be interpreted as the probability that the system performs the transition from φ_i to the state φ_f as a result of the perturbation switched on in between the instants of time 0 and t. A picture of the transition can be obtained from the state φ_i to the state φ_f by considering the probability amplitude at the various orders of the perturbation expansion. The diagram in Figure 11A.2 represents the amplitude $c_f(t)\exp[-iE_f^{(0)}t/\hbar]$ relative to the state φ_f in the superposition (11A.14.1), reached at time t and to first order according to the formula

$$c_f^{(1)}(t) = -\frac{i}{\hbar}\int_0^t\Big(\varphi_f, \varepsilon V_{\text{int}}(t')\varphi_i\Big)dt'. \tag{11A.14.7}$$

The segments in Figure 11A.2 describe the temporal evolution of the system according to the unperturbed Hamiltonian H_0: from time 0 to time t', with t' in between 0 and t, the system remains in state φ_i, and its state vector is simply multiplied by the phase factor $e^{\frac{-iE_i^{(0)}t}{\hbar}}$. At time t' the perturbation $\varepsilon V(t')$ leads to the transition from the state φ_i to the state φ_f, hence the matrix element $\big(\varphi_f, \varepsilon V_{\text{int}}\varphi_i\big)$ and the factor $-\frac{i}{\hbar}$. Lastly, the system evolves towards the final state φ_f from the instant t' to the instant t according to the unperturbed Hamiltonian H_0. This yields an evolution factor $e^{\frac{-iE_f^{(0)}(t-t')}{\hbar}}$. Since the instant

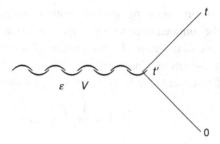

Fig. 11A.2   Transition amplitude to first order in perturbation theory.

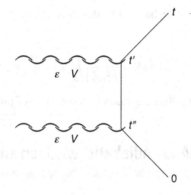

Fig. 11A.3   Transition amplitude to second order in perturbation theory.

t' is a generic instant of time between 0 and t, we have to 'sum' over all possible values of t'. We therefore consider

$$c_f^{(1)}(t)e^{-\frac{iE_f^{(0)}t}{\hbar}} = -\frac{i}{\hbar}\int_0^t e^{-\frac{iE_f^{(0)}(t-t')}{\hbar}}\left(\varphi_f, \varepsilon V(t')\varphi_i\right)e^{-\frac{iE_i^{(0)}t'}{\hbar}}\,dt', \tag{11A.14.8}$$

which is equal to

$$-\frac{i}{\hbar}\int_0^t \left(\varphi_f, \varepsilon V_{int}(t')\varphi_i\right)e^{\frac{-iE_f^{(0)}t}{\hbar}}\,dt'.$$

In the same way, the diagram in Figure 11A.3 represents the same amplitude evaluated at second order by using the formula

$$c_f^{(2)}(t) = \frac{1}{2!}\left(\frac{-i}{\hbar}\right)^2\sum_{m=1}^{\infty}\int_0^t dt'\int_0^t dt''$$

$$T\left(\varphi_f, \varepsilon V_{int}(t')\varphi_m\right)\left(\varphi_m, \varepsilon V_{int}(t'')\varphi_i\right). \tag{11A.14.9}$$

In this case, two interactions with the perturbation occur, at the instant t'' and at the instant $t' \geq t''$. The first interaction is responsible for the transition of the system from the state φ_i to the intermediate state φ_m, while the second leads to the transition from the state φ_m to the final state φ_f. In the intermediate state the system evolves between t'' and t' according to the unperturbed Hamiltonian. The instant t' is always subsequent to t'' by virtue of the action of the time-ordering operator. Both the intermediate state φ_m and the instants t' and t'' should be summed over in all possible ways. The intermediate state is therefore a *virtual state*, i.e. one of the infinitely many intermediate states through which the system can pass on its way towards the final state φ_f.

In the first-order formula for the transition probability

$$P_{if}(t) = \frac{1}{\hbar^2} \left| \int_0^t \left(\varphi_f, \varepsilon V_{int}(t')\varphi_i \right) \mathrm{d}t' \right|^2$$

$$= \frac{1}{\hbar^2} \left| \int_0^t \left(\varphi_f, \varepsilon V(t')\varphi_i \right) e^{i\omega_{fi}t'} \mathrm{d}t' \right|^2, \tag{11A.14.10}$$

if the perturbation has a finite duration, i.e. $V(t) \neq 0$ only if $t \in]0, \tau[$, it is possible integrate by parts to find

$$P_{if}(\tau) = \frac{1}{\hbar^2 \omega_{fi}^2} \left| \int_0^\tau e^{i\omega_{fi}t} \frac{\mathrm{d}}{\mathrm{d}t} \left(\varphi_f, \varepsilon V(t)\varphi_i \right) \mathrm{d}t \right|^2. \tag{11A.14.11}$$

The following limiting cases turn out to be of physical interest.

11A.15 Adiabatic switch on and off of the perturbation

The variation of interaction energy during an oscillation period of the system is small with respect to the energy jump in between the initial and final state, i.e.

$$\left| \frac{\mathrm{d}}{\mathrm{d}t} \left(\varphi_f, \varepsilon V(t)\varphi_i \right) \right| \ll \hbar \omega_{fi}^2. \tag{11A.15.1}$$

Since the time derivative of the matrix element of the perturbation remains basically constant during the interval $]0, \tau[$, the factor $e^{i\omega_{fi}t}$ in the integrand is the only important one, and hence

$$P_{if}(\tau) = \frac{4}{\hbar^2 \omega_{fi}^4} \left| \frac{\mathrm{d}}{\mathrm{d}t} \left(\varphi_f, \varepsilon V(t)\varphi_i \right) \right|^2 \sin^2 \left(\frac{1}{2}\omega_{fi}\tau \right). \tag{11A.15.2}$$

Thus, if the condition (11A.15.1) holds, the transition probability is much smaller than 1, i.e. the initial state φ_i is not abandoned after a time τ. This expression is also symmetric in τ, but the physical interpretation of it as a transition rate is possible only for $\tau > 0$.

11A.16 Perturbation suddenly switched on

Under such conditions, we have instead (cf. Eq. (11A.15.1))

$$\left| \frac{\mathrm{d}}{\mathrm{d}t} \left(\varphi_f, \varepsilon V(t)\varphi_i \right) \right| \gg \hbar \omega_{fi}^2. \tag{11A.16.1}$$

The dominant contribution to the integral occurring in Eq. (11A.14.11) is obtained from the integrand when the perturbation is switched on. If V_{fi} is the peak value of the matrix element of the perturbation, then

$$P_{\mathrm{if}}(\tau) = \frac{1}{\hbar^2 \omega_{\mathrm{fi}}^2} |V_{\mathrm{fi}}|^2. \tag{11A.16.2}$$

It should be stressed that we are working under the assumption that first-order perturbation theory can be successfully applied. For example, for a harmonic perturbation

$$V(t) \equiv V_0 e^{\mathrm{i}\omega t}, \tag{11A.16.3}$$

this means that (cf. Eq. (11A.12.5))

$$\left| \left(\varphi_{\mathrm{f}}, V_0 \varphi_{\mathrm{i}} \right) \right| \ll \hbar \omega_{\mathrm{fi}}. \tag{11A.16.4}$$

11A.17 Two-level system

For a two-level system, however, the perturbation

$$V(t) \equiv V_0(x) \left[e^{\mathrm{i}\omega t} \varphi_1 (\varphi_2, \cdot) + e^{-\mathrm{i}\omega t} \varphi_2 (\varphi_1, \cdot) \right] \tag{11A.17.1}$$

can be treated exactly. In such a case, only two eigenvectors of the unperturbed Hamiltonian exist, for which

$$H_0 \varphi_1 = E_1^{(0)} \varphi_1, \tag{11A.17.2}$$

$$H_0 \varphi_2 = E_2^{(0)} \varphi_2, \tag{11A.17.3}$$

and the expansion (11A.14.1) reduces to

$$\psi(t) = c_1(t) e^{-\mathrm{i}\frac{E_1^{(0)} t}{\hbar}} \varphi_1 + c_2(t) e^{-\mathrm{i}\frac{E_2^{(0)} t}{\hbar}} \varphi_2. \tag{11A.17.4}$$

Its insertion into the Schrödinger equation leads to the system

$$\mathrm{i}\hbar \frac{dc_1(t)}{dt} = \varepsilon \gamma_{11} c_2(t) e^{\mathrm{i}(\delta\omega)t} + \varepsilon \gamma_{12} c_1(t) e^{-\mathrm{i}\omega t}, \tag{11A.17.5}$$

$$\mathrm{i}\hbar \frac{dc_2(t)}{dt} = \varepsilon \gamma_{12}^* c_2(t) e^{\mathrm{i}\omega t} + \varepsilon \gamma_{22} c_1(t) e^{-\mathrm{i}(\delta\omega)t}, \tag{11A.17.6}$$

where $\gamma_{\alpha\beta} \equiv \left(\varphi_\alpha, V_0 \varphi_\beta \right)$ (which reduces to $\gamma \delta_{\alpha\beta}$ if $V_0 = \text{constant } \gamma$), $(\delta\omega) \equiv \omega - \omega_0$ and ω_0 is defined by $\hbar\omega_0 \equiv E_2^{(0)} - E_1^{(0)}$. The solution of such a system of equations becomes easier if the frequency of the perturbation is close to the proper frequency of the system, i.e. if

$$|(\delta\omega)| \ll \omega_0. \tag{11A.17.7}$$

In such a case the terms oscillating with frequency ω have vanishing average over time intervals comparable with the oscillation period $2\tau = \frac{2\pi}{\omega}$, and hence can be neglected with respect to the terms oscillating slowly with frequency $(\delta\omega)$. Thus, on defining the variables

$$a_{1,2}(t) \equiv \frac{1}{2\tau} \int_{t-\tau}^{t+\tau} c_{1,2}(t') \, dt', \tag{11A.17.8}$$

we find the differential equations

$$i\hbar \frac{da_1(t)}{dt} = \varepsilon \gamma_{11} a_2(t) e^{i(\delta\omega)t}, \tag{11A.17.9}$$

$$i\hbar \frac{da_2(t)}{dt} = \varepsilon \gamma_{22} a_1(t) e^{-i(\delta\omega)t}, \tag{11A.17.10}$$

which imply

$$\left[\frac{d^2}{dt^2} - i(\delta\omega)\frac{d}{dt} + \Omega^2 \right] a_1(t) = 0, \tag{11A.17.11}$$

$$\left[\frac{d^2}{dt^2} + i(\delta\omega)\frac{d}{dt} + \Omega^2 \right] a_2(t) = 0, \tag{11A.17.12}$$

having defined

$$\hbar^2 \Omega^2 \equiv \varepsilon^2 \gamma_{11} \gamma_{22}. \tag{11A.17.13}$$

The solutions of Eqs. (11A.17.11) and (11A.17.12) are therefore of the type

$$a_1(t) = e^{i\frac{(\delta\omega)t}{2}} \left(A \cos \frac{\alpha t}{2} + B \sin \frac{\alpha t}{2} \right), \tag{11A.17.14}$$

$$a_2(t) = e^{-i\frac{(\delta\omega)t}{2}} \left(C \cos \frac{\alpha t}{2} + D \sin \frac{\alpha t}{2} \right), \tag{11A.17.15}$$

where

$$\alpha \equiv \sqrt{(\delta\omega)^2 + 4\Omega^2}. \tag{11A.17.16}$$

If V_0 reduces to a constant γ as we mentioned earlier, so that $\gamma_{11} = \gamma_{22} = \gamma$, and if the system is initially in its ground state, for which

$$a_1(0) = 1, \quad a_2(0) = 0, \tag{11A.17.17}$$

then

$$A = 1, \quad C = 0. \tag{11A.17.18}$$

On inserting these solutions into the first-order system we obtain

$$B = -i\frac{(\delta\omega)}{\alpha}, \quad D = -2i\frac{\gamma\varepsilon}{\hbar\alpha}. \tag{11A.17.19}$$

The probability of finding the system in its excited state is therefore

$$|a_2(t)|^2 = \frac{4\Omega^2}{[(\delta\omega)^2 + 4\Omega^2]} \sin^2 \left(\frac{\alpha t}{2} \right), \tag{11A.17.20}$$

while the probability of again finding the system in the ground state is given by

$$|a_1(t)|^2 = \cos^2\left(\frac{\alpha t}{2}\right) + \frac{(\delta\omega)^2}{[(\delta\omega)^2 + 4\Omega^2]} \sin^2\left(\frac{\alpha t}{2}\right).$$

(11A.17.21)

The evolution in time of $|a_2(t)|^2$ exhibits a clearly visible maximum in the limiting case for which $(\delta\omega) \to 0$. This is therefore a typical resonating behaviour. As time passes, the system oscillates between the ground state and the excited state with frequency $\frac{\alpha}{2\pi}$.

11A.18 The quantum K^0–$\overline{K^0}$ system

Particles known as neutral kaons provide an interesting example of a two-level open quantum system. In the following we will sketch such an example, and avoid providing many concepts about particle physics that is beyond the scope of our book (the language will therefore be less technical). Let us consider the neutral component of particles known as K mesons, i.e. K^0 and its companion particle $\overline{K^0}$. Neutral kaons are unstable particles, in fact by virtue of weak interactions[1] they can decay into two particles called pions, i.e. $K^0, \overline{K^0} \to \pi\pi$. Moreover, since the reverse process $\pi\pi \to K^0, \overline{K^0}$ is also possible, there can be mixing via virtual intermediate π states, i.e. $K^0 \leftrightarrow \overline{K^0}$.

In quantum mechanics, the weak interaction connecting K^0 and $\overline{K^0}$ can be parametrized by a non-Hermitian Hamiltonian H_{eff}, which is described by the combination of two Hermitian 2×2 matrices M and Γ. In the basis $K^0, \overline{K^0}$ it reads

$$H_{\text{eff}} = Mc^2 - \frac{i}{2}\Gamma = \begin{pmatrix} M_{11}c^2 - \frac{i\Gamma_{11}}{2} & M_{12}c^2 - \frac{i\Gamma_{12}}{2} \\ M_{12}^*c^2 - \frac{i\Gamma_{12}^*}{2} & M_{22}c^2 - \frac{i\Gamma_{22}}{2} \end{pmatrix}.$$

(11A.18.1)

Note that the Γ matrix takes into account the neutral kaons, decay rate, hence it is related to their lifetimes, and M is the mass matrix. The diagonal elements M_{11} and M_{22} are the masses of K^0 and $\overline{K^0}$ respectively, whereas Γ_{11} and Γ_{22} are their lifetimes. Since the CPT symmetry is so far considered as an exact symmetry of nature, particles and antiparticles must share the same mass and lifetime; this implies that $M_{11} = M_{22}$ and $\Gamma_{11} = \Gamma_{22}$.

The mass and lifetime eigenvalues and eigenvectors are obtained by diagonalizing H_{eff}. This yields for the eigenvalues

$$M_S c^2 - \frac{i}{2}\Gamma_S = M_{11}c^2 - \frac{i}{2}\Gamma_{11} - \sqrt{\left(M_{12}c^2 - \frac{i}{2}\Gamma_{12}\right)\left(M_{12}^*c^2 - \frac{i}{2}\Gamma_{12}^*\right)}$$

$$M_L c^2 - \frac{i}{2}\Gamma_L = M_{11}c^2 - \frac{i}{2}\Gamma_{11} + \sqrt{\left(M_{12}c^2 - \frac{i}{2}\Gamma_{12}\right)\left(M_{12}^*c^2 - \frac{i}{2}\Gamma_{12}^*\right)},$$

(11A.18.2)

---

[1] This is the mechanism responsible for the weak nuclear force. The force is said to be weak because its field strength over a given distance is typically several orders of magnitude less than that of the so-called strong nuclear force and electromagnetism.

where the subscript 'S' or 'L' stands for short-lived K_S^0 and long-lived K_L^0 eigenvectors, respectively. Their expression in terms of a mixture of K^0 and $\overline{K^0}$ can be simply reported by defining the states

$$
|K_1^0\rangle = \frac{1}{\sqrt{2}} \left(|K^0\rangle - |\overline{K^0}\rangle \right),
$$
$$
|K_2^0\rangle = \frac{1}{\sqrt{2}} \left(|K^0\rangle + |\overline{K^0}\rangle \right). \tag{11A.18.3}
$$

By using $|K_1^0\rangle$ and $|K_2^0\rangle$ we can express the eigenvectors as

$$
|K_S^0\rangle = \frac{1}{\sqrt{1 + |\varepsilon|^2}} \left(|K_1^0\rangle + \varepsilon |K_2^0\rangle \right),
$$
$$
|K_L^0\rangle = \frac{1}{\sqrt{1 + |\varepsilon|^2}} \left(|K_2^0\rangle + \varepsilon |K_1^0\rangle \right), \tag{11A.18.4}
$$

with

$$
\varepsilon = \frac{\sqrt{\left(M_{12} c^2 - \mathrm{i}\frac{\Gamma_{12}}{2} \right)} - \sqrt{\left(M_{12}^* c^2 - \mathrm{i}\frac{\Gamma_{12}^*}{2} \right)}}{\sqrt{\left(M_{12} c^2 - \mathrm{i}\frac{\Gamma_{12}}{2} \right)} + \sqrt{\left(M_{12}^* c^2 - \mathrm{i}\frac{\Gamma_{12}^*}{2} \right)}}. \tag{11A.18.5}
$$

It can be easily shown that the CP-violation effects in the neutral kaon system is encoded in the imaginary parts of M_{12} and Γ_{12}. From the experimental point of view, the CP-violating terms are much smaller than the CP-preserving ones; hence, for simplicity the former can be neglected, and therefore $\varepsilon = 0$, $|K_S^0\rangle = |K_1^0\rangle$ and $|K_L^0\rangle = |K_2^0\rangle$. Under such a simplifying ansatz, the strangeness oscillations can be studied.

Let us consider a generic state in the $K^0 - \overline{K^0}$ Hilbert space, hereafter denoted by $|\psi(t)\rangle$. Its time evolution is ruled by the effective Hamiltonian H_{eff}, i.e.

$$
|\psi(t)\rangle = \exp\left\{ -\frac{\mathrm{i}}{\hbar} H_{\mathrm{eff}} t \right\} |\psi(0)\rangle. \tag{11A.18.6}
$$

By using the basis $|K_S^0\rangle$, $|K_L^0\rangle$ that diagonalized H_{eff}, we can rewrite the previous expression as

$$
|\psi(t)\rangle = \exp\left\{ -\frac{(\mathrm{i} M_S c^2 + \frac{\Gamma_S}{2})t}{\hbar} \right\} \langle K_S^0 | \psi(0)\rangle | K_S^0\rangle
$$
$$
+ \exp\left\{ -\frac{(\mathrm{i} M_L c^2 + \frac{\Gamma_L}{2})t}{\hbar} \right\} \langle K_L^0 | \psi(0)\rangle | K_L^0\rangle. \tag{11A.18.7}
$$

Let us assume that we have produced at $t = 0$ a K^0 from the $\pi^- p$ scattering. This means that $|\psi(0)\rangle = |K^0\rangle$; hence we get

$$
|\psi(t)\rangle = \exp\left\{ -\frac{(\mathrm{i} M_S c^2 + \frac{\Gamma_S}{2})t}{\hbar} \right\} \langle K_S^0 | K^0\rangle | K_S^0\rangle
$$
$$
+ \exp\left\{ -\frac{(\mathrm{i} M_L c^2 + \frac{\Gamma_L}{2})t}{\hbar} \right\} \langle K_L^0 | K^0\rangle | K_L^0\rangle
$$

$$= \frac{1}{\sqrt{2}} \exp \left\{ -\frac{(iM_S c^2 + \frac{\Gamma_S}{2})t}{\hbar} \right\} |K_S^0\rangle$$

$$+ \frac{1}{\sqrt{2}} \exp \left\{ -\frac{(iM_L c^2 + \frac{\Gamma_L}{2})t}{\hbar} \right\} |K_L^0\rangle. \qquad (11A.18.8)$$

From this expression we can deduce the probability to measure a K^0 or a $\overline{K^0}$ after an interval of time t, if the system consisted of a K^0 at $t = 0$. These probabilities read as

$$P\left(K^0(t=0) \to K^0(t) \right) = \left| \frac{1}{\sqrt{2}} \exp \left\{ -(iM_S c^2 + \Gamma_S/2)\frac{t}{\hbar} \right\} \langle K^0 | K_S^0\rangle \right.$$

$$\left. + \frac{1}{\sqrt{2}} \exp \left\{ -\frac{(iM_L c^2 + \frac{\Gamma_L}{2})t}{\hbar} \right\} \langle K^0 | K_L^0\rangle \right|^2$$

$$= \frac{1}{4} \left[e^{-\frac{\Gamma_S t}{\hbar}} + e^{-\frac{\Gamma_L t}{\hbar}} \right.$$

$$\left. + 2\cos\left(\frac{(M_L - M_S)c^2}{\hbar} t \right) e^{-\frac{t}{2\hbar}(\Gamma_S + \Gamma_L)} \right], \qquad (11A.18.9)$$

and analogously

$$P(K^0(t=0) \to \overline{K^0}(t)) = \left| \frac{1}{\sqrt{2}} \exp \left\{ -(iM_S c^2 + \Gamma_S/2)\frac{t}{\hbar} \right\} \langle \overline{K^0} | K_S^0\rangle \right.$$

$$\left. + \frac{1}{\sqrt{2}} \exp \left\{ -\frac{(iM_L c^2 + \frac{\Gamma_L}{2})t}{\hbar} \right\} \langle \overline{K^0} | K_L^0\rangle \right|^2$$

$$= \frac{1}{4} \left[e^{-\frac{\Gamma_S t}{\hbar}} + e^{-\frac{\Gamma_L t}{\hbar}} \right.$$

$$\left. - 2\cos\left(\frac{(M_L - M_S)c^2}{\hbar} t \right) e^{-\frac{t}{2\hbar}(\Gamma_S + \Gamma_L)} \right]. \qquad (11A.18.10)$$

11A.19 The quantum system of three active neutrinos

Neutrino oscillations were first proposed by Bruno Pontecorvo in the late 1950s inspired by the K^0–$\overline{K^0}$ system. At that time the electron neutrino ν_e was the only known, and Pontecorvo assumed the existence of an extra sterile neutrino. When, in the early 1960s, the muon neutrino was discovered, Pontecorvo also suggested that ν_e might transform (oscillate) in the Sun in a muon neutrino ν_μ ($\nu_e \to \nu_\mu$). This intuition was later confirmed by the Homestake experiment, and it is at the basis of the solution of the solar neutrino problem. At present we know that there are only three active neutrinos, and even though the search for extra sterile degrees of freedom is still ongoing, for the aim of this section we can safely restrict the analysis to active degrees of freedom only.

Neutrino oscillations are a quantum mechanical phenomenon, due to the fact that neutrinos are produced by weak interactions in the so-called flavour eigenstates, which are a linear superposition of mass eigenstates. If the energy resolution of energy and momentum measurements of all particles involved in the process were so high as to allow for the inference of the mass of the produced neutrino, no oscillations would be detected. On the other hand, if differences among neutrino masses are exceedingly small, energy and momentum resolutions are typically not good enough to distinguish among neutrino mass eigenstates. In this case the outgoing wave packet is a linear superposition of different mass states and the effect of interference (oscillations) can be observed. For simplicity, in the following we will consider neutrinos as plane waves and will restrict our analysis to the relativistic regime, following Lesgourgues *et al.* (2013).

A particular flavour ket eigenstate can be written as

$$|\nu_\alpha\rangle = \sum_{k=1}^{3} U^*_{\alpha k} |\nu_k\rangle \quad \text{with} \quad \alpha = e, \mu, \tau, \tag{11A.19.1}$$

where the unitary matrix U is the 3×3 *square* mixing matrix, which transforms a state from one basis into the other. We assume the normalization conditions $\langle \nu_\alpha | \nu_\beta \rangle = \delta_{\alpha\beta}$ and $\langle \nu_i | \nu_j \rangle = \delta_{ij}$.

Since massive neutrinos are Hamiltonian eigenstates, a generic $|\nu_\alpha\rangle$ produced at $t = 0$ evolves in time as

$$|\nu_\alpha(t)\rangle = \sum_{k=1}^{3} U^*_{\alpha k} \, e^{-i\frac{E_k t}{\hbar}} |\nu_k\rangle. \tag{11A.19.2}$$

From the unitarity of U we get $|\nu_k\rangle = \sum_\beta U_{\beta k} |\nu_\beta\rangle$, and we can rewrite Eq. (11A.19.2) in the following way:

$$|\nu_\alpha(t)\rangle = \sum_{\beta=e,\mu,\tau} \left(\sum_{k=1}^{3} U^*_{\alpha k} \, e^{-i\frac{E_k t}{\hbar}} U_{\beta k} \right) |\nu_\beta\rangle. \tag{11A.19.3}$$

In general, we are interested in the evolution of neutrino flavour eigenstates (denoted by the Greek indexes). This is easily understood by bearing in mind that any detection technique for neutrinos is based on weak interactions. In the combination (11A.19.3) the coefficient proportional to $|\nu_\beta\rangle$ is given by

$$\langle \nu_\beta | \nu_\alpha(t) \rangle = \sum_{k=1}^{3} U^*_{\alpha k} \, e^{-i\frac{E_k t}{\hbar}} U_{\beta k}. \tag{11A.19.4}$$

From this expression we obtain the probability that, by performing a measurement of neutrino flavour at time t, a flavour β is detected:

$$P_{\nu_\alpha \to \nu_\beta}(t) = \left| \langle \nu_\beta | \nu_\alpha(t) \rangle \right|^2 = \sum_{k,j=1}^{3} U^*_{\alpha k} \, U_{\beta k} U_{\alpha j} \, U^*_{\beta j} \, e^{-i\frac{(E_k - E_j)t}{\hbar}}. \tag{11A.19.5}$$

For relativistic neutrinos[2]

$$\frac{E_i}{c} = \sqrt{|\vec{p}|^2 + m_i^2} \approx |\vec{p}| + \frac{m_i^2}{2|\vec{p}|}, \tag{11A.19.6}$$

---

[2] This is typically the case for all neutrino sources considered in flavour oscillation experiments.

and by substituting the previous expression in (11A.19.5) eventually we get

$$P_{\nu_\alpha \to \nu_\beta}(t) = \left| \langle \nu_\beta | \nu_\alpha(t) \rangle \right|^2 = \sum_{k,j=1}^{3} U_{\alpha k}^* \, U_{\beta k} U_{\alpha j} \, U_{\beta j}^* \, e^{-i \frac{\Delta m_{kj}^2}{2|\vec{p}|} \frac{ct}{\hbar}}, \tag{11A.19.7}$$

where $\Delta m_{kj}^2 \equiv m_k^2 - m_j^2$.

In a neutrino experiment the time dependence of the flavour transition probability cannot be followed, but rather we can measure how the probabilities depend on L, the distance of the detection point from the origin of the neutrino beam. It is therefore customary to replace the time dependence by the variable L/c, hence Eq. (11A.19.7) reads as

$$P_{\nu_\alpha \to \nu_\beta}(L) = \sum_{k,j=1}^{3} U_{\alpha k}^* \, U_{\beta k} U_{\alpha j} \, U_{\beta j}^* \, e^{-i \frac{\Delta m_{kj}^2}{2|\vec{p}|} \frac{L}{\hbar}}. \tag{11A.19.8}$$

Note that for $L = 0$ Eq. (11A.19.8) becomes $P_{\nu_\alpha \to \nu_\beta}(0) = \sum_{k,j=1}^{3} U_{\alpha k}^* \, U_{\beta k} U_{\alpha j} \, U_{\beta j}^* = \delta_{\alpha \beta}$.

By observing that $\sum_{k,j=1}^{3} = \sum_{k>j} + \sum_{j>k} + \sum_{k=j}$, we can split the expression (11A.19.8) into two parts

$$P_{\nu_\alpha \to \nu_\beta}(L) = \sum_{k=1}^{3} |U_{\alpha k}|^2 |U_{\beta k}|^2 + 2 \operatorname{Re} \left[\sum_{k>j} U_{\alpha k}^* \, U_{\beta k} U_{\alpha j} \, U_{\beta j}^* \, e^{-i \frac{\Delta m_{kj}^2}{2|\vec{p}|} \frac{L}{\hbar}} \right], \tag{11A.19.9}$$

or analogously

$$P_{\nu_\alpha \to \nu_\beta}(L) = \delta_{\alpha \beta} - 4 \sum_{k>j} \operatorname{Re} \left[U_{\alpha k}^* \, U_{\beta k} U_{\alpha j} \, U_{\beta j}^* \right] \sin^2 \left(\frac{\Delta m_{kj}^2}{4|\vec{p}|} \frac{L}{\hbar} \right)$$

$$+ 2 \sum_{k>j} \operatorname{Im} \left[U_{\alpha k}^* \, U_{\beta k} U_{\alpha j} \, U_{\beta j}^* \right] \sin \left(\frac{\Delta m_{kj}^2}{2|\vec{p}|} \frac{L}{\hbar} \right). \tag{11A.19.10}$$

The probability that starting with a ν_α the same flavour is still detected at a distance L is called *survival probability* and it is obtained from the previous expression for $\alpha = \beta$

$$P_{\nu_\alpha \to \nu_\alpha}(L) = 1 - 4 \sum_{k>j} |U_{\alpha k}|^2 |U_{\alpha j}|^2 \sin^2 \left(\frac{\Delta m_{kj}^2}{2|\vec{p}|} \frac{L}{\hbar} \right). \tag{11A.19.11}$$

The total transition probability, i.e. the probability of observing a change in flavour is given by $1 - P_{\nu_\alpha \to \nu_\alpha}$, by virtue of unitarity.

If the neutrino production or detection occurs over a distance whose uncertainty is much larger than the oscillation lengths $L_{kj}^{\text{osc}} = \frac{4\pi |\vec{p}| \hbar}{\Delta m_{kj}^2}$, the expression (11A.19.9) should be averaged over distances larger than a L_{kj}^{osc} and thus, the second term on the right-hand side of Eq. (11A.19.9) cancels out. In this case the probability for flavour oscillation becomes

$$\langle P_{\nu_\alpha \to \nu_\beta} \rangle = \sum_{k=1}^{3} |U_{\alpha k}|^2 |U_{\beta k}|^2. \tag{11A.19.12}$$

The same expression is valid in case of incoherent detection or production as well.

Jeffreys–Wentzel–Kramers–Brillouin method

11B.1 The JWKB method

There are at least three milestones in the historical routes to this subject, and they are as follows:

(i) Lord Rayleigh obtains approximate solutions for wave equations (Rayleigh 1912).
(ii) The work of Sir Harold Jeffreys (Jeffreys 1923).
(iii) The work of Wentzel, Kramers and Brillouin (Brillouin 1926a,b; Kramers 1926, Wentzel 1926).

This method consists of a series expansion in $\hbar$ to find approximate solutions of the Schrödinger equation. As quantum mechanics 'passes over' into classical mechanics when $\hbar \to 0$, it should be expected that these approximations are pretty good for high quantum numbers, where classical theory is nearly correct. However, this expansion is only semiconvergent and cannot yield an exact solution of the problem. For this reason we shall limit the discussion to the approximation obtained by using the first two terms of an expansion in terms of $\hbar$. In general, is considered the Schrödinger equation for stationary states and looks for solutions of the kind

$$\psi(x) = A(x)e^{i\frac{S(x)}{\hbar}}, \tag{11B.1.1}$$

where A and S are real-valued functions. Often it is also convenient to re-absorb the amplitude A through the formula

$$\psi(x) = e^{i\frac{W(x)}{\hbar}}, \tag{11B.1.2}$$

with $W \equiv S - i\hbar \log(A)$, where S and $\log(A)$ are even functions of $\hbar$ (A and S are the amplitude and the phase, respectively). By substituting either ansatz in the Schrödinger equation we obtain

$$\hbar^2 A'' + \left[2m(E - V) - (S')^2\right]A = 0, \tag{11B.1.3}$$

$$2A'S' + S''A = 0. \tag{11B.1.4}$$

This latter equation has the immediate solution

$$A = \text{const.}(S')^{-\frac{1}{2}}. \tag{11B.1.5}$$

For the other equation we set

$$A = A_0 + \hbar^2 A_2 + \cdots , \tag{11B.1.6}$$

$$S = S_0 + \hbar^2 S_2 + \cdots . \tag{11B.1.7}$$

Again, by introducing these expansions in the equation we have derived we find

$$A_0\left[2m(E - V) - (S_0')^2\right] = 0, \tag{11B.1.8}$$

so that

$$S_0' = \pm[2m(E - V)]^{\frac{1}{2}}, \quad A_0 = \text{const.}(S_0')^{-\frac{1}{2}}. \tag{11B.1.9}$$

By equating coefficients of $\hbar^2$,

$$A_0'' - 2S_0'S_2'A_0 = 0, \tag{11B.1.10}$$

and

$$S_2' = \frac{\left[(S_0')^{-\frac{1}{2}}\right]''}{2(S_0')^{\frac{1}{2}}}. \tag{11B.1.11}$$

Let us discuss the equation

$$(S')^2 \cong (S_0')^2 = 2m[E - V]. \tag{11B.1.12}$$

First we consider:

(1) $E > V(x)$, and define

$$\lambda(x) = \frac{\hbar}{\sqrt{2m(E - V)}}, \tag{11B.1.13}$$

with the corresponding equation for S

$$S' \cong \pm\frac{\hbar}{\lambda}. \tag{11B.1.14}$$

A general solution is a linear superposition of oscillating functions

$$\psi(x) = \alpha\sqrt{\lambda}\cos\left(\int \frac{dx}{\lambda} + \varphi\right), \tag{11B.1.15}$$

with α, φ arbitrary constants.

(2) $E < V(x)$ (classically forbidden region). The assumption suggests defining

$$l(x) = \frac{\hbar}{\sqrt{2m(V - E)}}, \tag{11B.1.16}$$

while S solves the equation

$$S' \cong \pm i\frac{\hbar}{l}. \tag{11B.1.17}$$

The resulting solution is a superposition of exponential functions

$$\psi(x) = \sqrt{l}\left[\gamma \exp\int \frac{dx}{l} + \delta \exp\left(-\int \frac{dx}{l}\right)\right], \tag{11B.1.18}$$

with γ, δ arbitrary constants.

Clearly, the previous approximate solutions are valid if $\hbar^2$ corrections are negligible. Such a correction amounts to multiplying the previous solutions by $\exp(i\hbar S_2)$, thus we need $|\hbar S_2| << 1$. If we insert $S = S_0 + \hbar^2 S_2$ in the equation

$$(S')^2 = 2m(E - V) + \hbar^2 \left[\frac{3}{4} \left(\frac{S''}{S'} \right)^2 - \frac{1}{2} \frac{S'''}{S'} \right], \qquad (11B.1.19)$$

and equate coefficients in $\hbar^2$ we get

$$2S_0' S_1' = \frac{\left[(S_0')^{-\frac{1}{2}} \right]''}{[S_0']^{-\frac{1}{2}}} = \frac{3}{4} \left(\frac{S_0''}{S_0'} \right)^2 - \frac{1}{2} \frac{S_0'''}{S_0'}. \qquad (11B.1.20)$$

We distinguish again two cases.

(1) $E > V$, $S_0' = \pm \frac{\hbar}{\lambda}$, which implies that

$$\hbar S_2' = \pm \left(\frac{1}{4} \lambda' - \frac{1}{8} \int \frac{(\lambda')^2}{\lambda} dx \right). \qquad (11B.1.21)$$

The condition $|\lambda| << 1$ implies that

$$\lambda << \frac{2m(E - V)}{mV'}. \qquad (11B.1.22)$$

In simple terms, the de Broglie wavelength must be small with respect to the distances over which the potential changes appreciably.

(2) $E < V$. In a similar manner we find

$$|l'| << 1. \qquad (11B.1.23)$$

In most of the cases where we apply this method the various requirements are satisfied, except at points where $E = V(x)$ (turning points for classical motion). Among turning points connection formulae are needed such that sines and cosines can smoothly join on to exponential functions. One method of approximation involves solving exactly an approximated equation obtained by replacing $V(x)$ with a linear approximation around $x = a$. Thus, we have to solve

$$\psi'' - \frac{2m}{\hbar^2} \frac{dV}{dx} \bigg|_{x=a} (x - a)\psi = 0. \qquad (11B.1.24)$$

Solutions are obtained in terms of Bessel functions of order $\pm \frac{1}{3}$. These solutions are used to connect JWKB solutions on one side with those on the other side. Connection formulae may be given by assuming $E > V$ (respectively $E < V$) for $x > a$ (respectively $x < a$), if the barrier is on the left.

A general solution will be a superposition of ψ_1 and ψ_2 which, in asymptotic terms, are

$$\psi_1 \sim \sqrt{l} \exp \left(\int_x^a \frac{dx}{l} \right) \quad \text{if } x << a, \qquad (11B.1.25)$$

$$\psi_1 \sim \sqrt{\lambda} \sin \left(\int_a^x \frac{dx}{\lambda} - \frac{\pi}{4} \right) \quad \text{if } x >> a, \qquad (11B.1.26)$$

$$\psi_2 \sim \sqrt{l} \exp\left(-\int_x^a \frac{dx}{l}\right) \text{ if } x << a, \tag{11B.1.27}$$

$$\psi_2 \sim \sqrt{\lambda} \cos\left(\int_a^x \frac{dx}{\lambda} - \frac{\pi}{4}\right) \text{ if } x >> a. \tag{11B.1.28}$$

The number of wavelengths contained in the interval (x_1, x_2) is given by

$$\frac{1}{2\pi} \int_{x_1}^{x_2} \frac{dx}{\lambda} \text{ or } \frac{i}{2\pi} \int_{x_1}^{x_2} \frac{dx}{l}.$$

Previous connection formulae are valid if:

(a) at turning points $(E - V)$ goes to zero as $(x - a)$ does, and keeps the same behaviour, i.e. proportional to $(x - a)$, in a region of a few wavelengths;
(b) each *turning region* is connected on both sides with an asymptotic region (which extends over various wavelengths) over which the JWKB approximation is valid.

11B.2 Potential barrier

Let us consider in one dimension a potential $V(x) = -V_0 < 0$ if $x \in]0, a[$, while it is positive and monotonically decreasing in the open intervals $]a, b[$ and $]b, \infty[$, tending to 0 at ∞. The three intervals are denoted by I, II and III, respectively, and we set, for $x >> b$ (up to an arbitrary constant)

$$\psi_{\text{III}} = \sqrt{\lambda}\left[\cos\left(\int_b^x \frac{dx}{\lambda} - \frac{\pi}{4}\right) + i \sin\left(\int_b^x \frac{dx}{\lambda} - \frac{\pi}{4}\right)\right]. \tag{11B.2.1}$$

This may be extended to region II as

$$\sqrt{\lambda}\, e^{-i\frac{\pi}{4}} e^{i\int_b^x \frac{dx}{\lambda}} \rightarrow \frac{1}{\sqrt{i}} e^{-i\frac{\pi}{4}} \sqrt{l}\, e^{-\int_b^x \frac{dx}{l}}, \tag{11B.2.2}$$

if $a << x << b$. Then

$$\psi_{\text{II}} = -i\sqrt{l}\, e^{\left(\int_x^b \frac{dx}{l}\right)} = -i\sqrt{l}\, e^\tau e^{\left(-\int_a^x \frac{dx}{l}\right)}, \tag{11B.2.3}$$

where $\tau \equiv \int_a^b \frac{dx}{l}$.

   In region I

$$\psi_{\text{I}} = A \sin[k(x - a) + \delta], \tag{11B.2.4}$$

where $k \equiv \frac{\sqrt{2m(E - V_0)}}{\hbar}$. The constants A and δ may be computed by imposing continuity conditions for ψ and its derivatives at a, i.e.

$$k \cot \delta = -\frac{1}{l(a)}, \tag{11B.2.5}$$

$$A \sin \delta = -i\sqrt{l(a)}\, e^\tau. \tag{11B.2.6}$$

The stationary state ψ_I is a superposition of the incident wave and of the reflected wave. The incident wave is

$$-\frac{1}{2}iA\,e^{i(k(x-a)+\delta)},$$

with flux $\frac{1}{4}|A|^2\frac{\hbar k}{m}$, where

$$\frac{k|A|^2}{4} = \frac{kl(a)}{4}\frac{e^{2\tau}}{\sin^2\delta} = \frac{kl(a)}{4}e^{2\tau}(1+\cot^2\delta)$$

$$= e^{2\tau}\frac{(1+k^2l^2(a))}{4kl(a)}. \tag{11B.2.7}$$

As for the flux of the transmitted wave

$$T = \frac{4kl(a)}{(1+k^2l^2(a))}e^{-2\tau} = 4\frac{\sqrt{(V_a-E)(E-V_0)}}{(V_a-V_0)}e^{-2\tau}. \tag{11B.2.8}$$

This result is acceptable if V changes slowly in regions II and III. Moreover, V should be approximable by a linear function of x in a region several wavelengths distant from the turning point. This means that the barrier has a width of several wavelengths, which implies that $\tau >> 2\pi$ and $T << 1$.

11B.3   Energy levels in a potential well

Let us assume that the potential V depends on x according to a parabolic law in the first quadrant, and that the regions I, II and III correspond to the three open intervals $]0,a[$, $]a,b[$ and $]b,\infty[$. The line $V=E=$ constant, parallel to the x-axis, intersects the parabola at points P_1 and P_2 from which the parallels to the V-axis intersect the x-axis at a and b, respectively. The stationary states may be approximated in the form

$$\psi_I = \frac{C}{2}\sqrt{l}\exp\left(-\int_x^a\frac{dx}{l}\right), \text{ if } x << a, \tag{11B.3.1}$$

$$\psi_{III} = \frac{C'}{2}\sqrt{l}\exp\left(-l\int_b^x\frac{dx}{l}\right), \text{ if } x >> b. \tag{11B.3.2}$$

They may be continued in region II with (here $a << x << b$)

$$\psi_a(x) = C\sqrt{\lambda}\cos\left(\int_a^x\frac{dx}{\lambda}-\frac{\pi}{4}\right), \tag{11B.3.3}$$

$$\psi_b(x) = C'\sqrt{\lambda}\cos\left(\int_x^b\frac{dx}{\lambda}-\frac{\pi}{4}\right). \tag{11B.3.4}$$

But $\psi_a(x) = \psi_b(x) = \psi_{II}$ if

$$I \equiv \int_a^b\frac{dx}{\lambda} = \int_a^b\frac{\sqrt{2m(E-V)}}{\hbar}dx = \left(n+\frac{1}{2}\right)\pi, \; n\in Z_+. \tag{11B.3.5}$$

This rule determines the energy levels (discrete spectrum). The validity condition of the JWKB (linearity of V for several wavelengths around turning points) becomes $n >> 1$.

In the particular case of an harmonic oscillator, the JWKB approximation provides exact results for the energy levels, including the ground state.

The condition $I = \left(n + \frac{1}{2}\right)\pi$ may be seen as a condition for stationary states: the interval (a, b) should contain $n + \frac{1}{2}$ wavelengths. It differs from the Bohr–Sommerfeld quantization conditions by a term $\frac{1}{2}$. The integral $\tilde{I} \equiv 2\hbar \oint p \, dq$ may be extended, via Stokes' theorem, to

$$\omega(E) = \int\int_{H \leq E} dp \, dq. \tag{11B.3.6}$$

Therefore $\omega(E) = \left(n + \frac{1}{2}\right)\hbar$, so that the phase-space area $\omega(E)$ increases by $\hbar$ in the transition from one level to the next.

The extension of the energy band $(E, E + \delta E)$ when measured in units of $\hbar$ is equal to the number of bound states contained in such a band (of course, we are assuming $n \gg 1$).

11B.4 α-decay

Two kinds of spontaneous emission are known to exist in nature: α- and β-decay. Experiments in which a sufficiently large number of α-particles enter a chamber with a thin mica window and are collected show that α-rays correspond to positively charged particles with a $\frac{q}{m}$ ratio that is the same as that of a doubly ionized helium atom: He^{++}. Their identification with He^{++} is made possible because, when a gas of α-particles produces light, it indeed reveals the spectroscopic lines of He^{++}. An important quantitative law of α-decay was discovered, at the experimental level, by Geiger and Nuttall in 1911. It states that the velocity v of the α-particle, and the mean life T of the emitting nucleus, are related by the empirical equation

$$\log(T) = A + B\log(v), \tag{11B.4.1}$$

where the constants A and B can be determined in the laboratory. We are now aiming to develop an elementary theory of α-decay, following Born (1969). For further details, we refer the reader to the work in Gamow (1928) and Fermi (1950).

Let $V(r)$ be the potential for a particle emitted by a nucleus of atomic number Z. By virtue of the shielding effect, it is as if the particle were 'feeling' the effect of a field with residual atomic number $(Z - 2)$. At large distances, i.e. for $r > r_0$, the resulting potential is Coulombian, while for $r < r_0$ the precise form of $V(r)$ is not known. However, phenomenological reasons we expect it should have the shape of a *crater* (the α-particle being 'trapped' in the field of the nucleus). In other words, we consider a potential such that

$$V(r) = -V_0 < 0 \quad \text{if } r \in]0, r_0[, \tag{11B.4.2}$$

$$V(r) = \frac{2(Z-2)q_e^2}{r} \quad \text{if } r > r_0, \tag{11B.4.3}$$

and a positive value of the energy E such that $E < V$ if $r \in]r_0, r_1[$, $E = V$ at $r = r_1$ and $E > V$ if $r > r_1$. The emission frequency, ν, can be expressed as the product of the number n of times that the α-particle hits against the crater with probability p that it can 'tunnel' through the barrier given by the crater: $\nu = np$. To find p, the stationary Schrödinger equation must be solved with the lowest possible value of angular momentum. The part of the stationary state depending on spatial variables is written as $\varphi(r) = \frac{\psi(r)}{r}$, where $\psi(r)$ satisfies the equation

$$\left\{ \frac{d^2}{dr^2} + \frac{2m}{\hbar^2}[E - V(r)] \right\} \psi = 0. \tag{11B.4.4}$$

In between r_0 and r_1, the wave function is exponentially damped, and hence, approximately,

$$p = \frac{|\psi(r_1)|^2}{\psi(r_0)|}. \tag{11B.4.5}$$

In such an intermediate region the potential is well approximated by the Coulomb part (11B.4.3). For large values of the atomic number Z, we thus look for solutions of Eq. (11B.4.4) in the form

$$\psi = \exp\left[\frac{y(r)}{\hbar} \right], \tag{11B.4.6}$$

which leads to a non-linear equation for y:

$$\hbar y'' + y'^2 - F(r) = 0, \tag{11B.4.7}$$

having defined

$$F(r) \equiv 2m\left[-E + \frac{2(Z-2)q_e^2}{r} \right]. \tag{11B.4.8}$$

If the term $\hbar y''$ can be neglected in Eq. (11B.4.7),

$$y(r) = \int_a^r \sqrt{F(x)} \, dx, \tag{11B.4.9}$$

$$\frac{\psi(r_1)}{\psi(r_0)} = \exp\left[\frac{y(r_1) - y(r_0)}{\hbar} \right] = \exp\left[\frac{1}{\hbar} \int_{r_0}^r \sqrt{F(x)} \, dx \right]. \tag{11B.4.10}$$

The integral occurring in Eq. (11B.4.10) is non-trivial but only involves standard techniques (e.g. Gradshteyn and Ryzhik (1965)), so that the probability can eventually be expressed as

$$p = \exp\left[-(2n_0 - \sin 2n_0) \frac{8\pi q_e^2}{h} \frac{(Z-2)}{\nu} \right], \tag{11B.4.11}$$

where n_0 is a root of the equation

$$\cos^2 n_0 = \frac{r_0 E}{2(Z-2)q_e^2}. \tag{11B.4.12}$$

For small values of the energy E of α-particles, and for a deep crater, the probability can be approximated as follows:

$$p = \exp\left[-8\pi^2 q_e^2 \frac{(Z-2)}{hv} + \frac{16\pi q_e}{h}\sqrt{m(Z-2)r_0}\right]. \tag{11B.4.13}$$

The decay constant, v_D, is thus given by

$$\log(v_D) = \log\left(\frac{h}{4mr_0^2}\right) - 8\pi^2 q_e^2 \frac{(Z-2)}{hv} + \frac{16\pi q_e}{h}\sqrt{m(Z-2)r_0}. \tag{11B.4.14}$$

Indeed, Eq. (11B.4.14) differs from the empirical law (11B.4.1), because it predicts that $\log(v_D)$ depends linearly on $\frac{1}{v}$. Moreover, it introduces a dependence on Z and r_0. However, the discrepancy is not very severe from the point of view of the actual values taken by the physical parameters: the velocity varies between 1.4×10^9 and 2×10^9 cm s^{-1}, and the variation of Z and r_0 is negligible for the elements of a radioactive series. Note also that the range of values of the decay constant is quite large, since $\frac{8\pi^2 q_e^2}{h}$ is very large.

Scattering theory

11C.1 Aims and problems of quantum scattering theory

Scattering theory is the branch of physics concerned with interacting systems on a scale of time and/or distance that is large compared with the scale of the interaction itself, and it provides the most powerful tool for studying the microscopic nature of the world. In quantum mechanics, a typical scattering process involves a beam of particles prepared with a given energy and with a more or less defined linear momentum. One then studies either the collision with a similar wave packet, or the collision of the given wave packet against a fixed target. In particular, for a two-body elastic scattering problem with conservative forces, we work in the centre of mass frame, hence reducing the original problem to the analysis of a particle in an external field of forces. Indeed, in Section 7.2 we already used such a technique in the investigation of the hydrogen atom. The crucial difference with respect to Section 7.2 is that, in scattering problems, the continuous spectrum and its perturbations are studied, whereas the Balmer formula has to do with bound states. In the following section we outline the basic aspects of what is called time-dependent scattering.

11C.2 Time-dependent scattering

The physical situation we would like to study is as follows: a beam of particles is prepared in a state that is approximately free and allowed to evolve towards the scatterer. After the collision we detect the scattered beam far away from the interaction region. In mathematical terms, the Schrödinger equation describes the moving beam with Hamiltonian operator $\widehat{H} = \widehat{H}_0 + \widehat{V}$, where $\widehat{H}_0$ is the comparison Hamiltonian, usually called the *free Hamiltonian* and represented by $-\frac{\hbar^2}{2m}\Delta$ in the coordinate system associated with the position operators. In the same coordinate system the interaction operator $\widehat{V}$ is represented by multiplication by $V(x)$, a potential function assumed to be of *short range*, i.e. suitably vanishing at infinity. We often say that $V(x)$ decays at infinity sufficiently rapidly. Since we would like to describe the situation by means of the 1-parameter group of unitary transformations $\widehat{U}(t) = e^{-i\frac{\widehat{H}t}{\hbar}}$, we require that $\widehat{H}$ is an essentially self-adjoint operator. We want to compare $\widehat{U}(t)$ when $t \to \pm\infty$ with the asymptotic dynamics described by $\widehat{U}_0(t) = e^{-i\frac{\widehat{H}_0 t}{\hbar}}$. We consider a coordinate system for the Hilbert space so as to obtain $\mathcal{L}^2(\mathbf{R}^3)$.

Asymptotic states $\widehat{U}(t)|\psi\rangle$, called $|\varphi_\pm\rangle$, are characterized by the relation

$$\widehat{U}(t)|\psi\rangle \approx \widehat{U}_0(t)|\varphi_\pm\rangle,$$

for $t \to \pm\infty$. More precisely, we require that for a given $|\psi\rangle$ there is a state $|\varphi_+\rangle$ and a state $|\varphi_-\rangle$ such that

$$\lim_{t\to\pm\infty} \left\| e^{-i\frac{\widehat{H}t}{\hbar}}|\psi\rangle - e^{-i\frac{\widehat{H}_0 t}{\hbar}}|\varphi_\pm\rangle \right\| = 0. \tag{11C.2.1}$$

This relation motivates the introduction of the wave operator, or Möller operator $\widehat{W}_\pm$ as it is usually called:

$$\lim_{t\to\pm\infty} \left\| |\psi\rangle - e^{i\frac{\widehat{H}t}{\hbar}} e^{-i\frac{\widehat{H}_0 t}{\hbar}}|\varphi_\pm\rangle \right\| = \lim_{t\to\pm\infty} \left\| |\psi\rangle - \widehat{W}_\pm|\varphi_\pm\rangle \right\| = 0. \tag{11C.2.2}$$

Thus, we may define $|\psi\rangle$ to be a scattering state if it is possible to find two states (depending on $|\psi\rangle$), denoted by $|\varphi_+\rangle$ and $|\varphi_-\rangle$ such that the previous limit exists, and then we say that the wave operator exists. It should be stressed that there might be states for which $\widehat{W}_+$ exists but there is no $\widehat{W}_-$, and vice versa. When the wave operators exist for $|\varphi\rangle \in \mathcal{H}$, given $|\varphi\rangle$ we can define

$$|\psi\rangle = \widehat{W}_\pm|\varphi\rangle, \tag{11C.2.3}$$

and it satisfies the relation

$$\lim_{t\to\pm\infty} \left\| e^{-i\frac{\widehat{H}t}{\hbar}}|\psi\rangle - e^{-i\frac{\widehat{H}_0 t}{\hbar}}|\varphi_\pm\rangle \right\| = 0. \tag{11C.2.4}$$

Note also that

$$\left\| \widehat{W}_\pm|\varphi_\pm\rangle \right\| = \||\varphi_\pm\rangle\|, \tag{11C.2.5}$$

and hence $\widehat{W}_\pm$ is a partial isometry. The *partial-isometry* character may be necessary to take into account the 'portion' of Hilbert space that should be projected out to be able to define $\widehat{W}_\pm$. It is convenient to introduce the range of $\widehat{W}_\pm$, say $R_\pm$. If $R_+ = R_-$, we can introduce the *scattering operator*

$$\widehat{S}(\widehat{H},\widehat{H}_0) = \widehat{W}_+^\dagger(\widehat{H},\widehat{H}_0)\widehat{W}_-(\widehat{H},\widehat{H}_0), \tag{11C.2.6}$$

and show that it is unitary.

The relation $R_+ = R_-$, or a stronger requirement that $R_\pm$ are both equal to the orthogonal complement of the span of proper eigenspaces of $\mathcal{H}$, is called the completeness of the wave operator. The completeness in the stronger sense implies that $\widehat{H}$ has no singular continuous spectrum and that the absolutely continuous part of $\widehat{H}$ is unitarily equivalent to that of $\widehat{H}_0$. The transition from the solution $|\varphi_-(t)\rangle$ to $|\varphi_+(t)\rangle$ is the result of the interaction with the scatterer. The scattering operator $\widehat{S}$ maps $|\varphi_-\rangle$ into $|\varphi_+\rangle$. Under the assumptions made so far it is not difficult to show that

$$\widehat{S}\widehat{H}_0 = \widehat{H}_0\widehat{S} \tag{11C.2.7}$$

and

$$\widehat{S}^\dagger\widehat{S} = \widehat{S}\widehat{S}^\dagger = I. \tag{11C.2.8}$$

In concrete problems, asymptotic completeness is the most difficult part to prove, and there are various theorems that, by making proper requirements on the behaviour of $V(x)$, show asymptotic completeness. For instance, it is necessary to remove those states, if any, for which $\widehat{H}|\psi\rangle = E|\psi\rangle$ for proper eigenvectors $|\psi\rangle$ not in the generalized sense; indeed, in their evolution $|\psi(t)\rangle = \mathrm{e}^{-\mathrm{i}\frac{Et}{\hbar}}|\psi_0\rangle$, we would clearly violate the asymptotic condition with respect to a possible state of the comparison Hamiltonian. We have to refer to the large existing literature for further details.

To provide an idea of the origin of pathologies we have mentioned, we consider a classical system with one degree of freedom and a potential satisfying the required decay properties at infinity.

11C.3 An example: classical scattering

In this section we would like to pinpoint the possible origin of various problems we may encounter in the quantum regime in the identification of scattering states. We start our considerations by dealing with a classical Hamiltonian system described by the Hamiltonian function

$$H = \frac{p^2}{2m} + V(x), \; p, x \in \mathbf{R}^3. \tag{11C.3.1}$$

We assume that the system admits three functionally independent first integrals f_1, f_2, f_3 pairwise in involution

$$\{f_j, f_k\} = 0, \; \forall j, k = 1, 2, 3. \tag{11C.3.2}$$

By using collective coordinates which embody both x and p values, i.e. $\xi_j = x_j, \xi_{j+n} = p_j$, we assume that the matrix of partial derivatives $\left(\frac{\partial f_j}{\partial \xi_k}\right)$ has rank 3. The projectability onto the configuration space may be better expressed by requiring

$$\det\left(\frac{\partial f_j}{\partial p_k}\right) \neq 0.$$

It is then possible to find phase-space coordinates such that the equations of motion acquire the form

$$\frac{\mathrm{d}}{\mathrm{d}t} g_j = v_j(f), \tag{11C.3.3}$$

$$\frac{\mathrm{d}}{\mathrm{d}t} f_j = 0, \tag{11C.3.4}$$

where

$$\{g_j, f_k\} = \delta_{jk}. \tag{11C.3.5}$$

It follows that the equations of motion are easily integrated and give rise to the flow

$$g_j(t) = g_j(0) + t\, v_j(f), \; f_j(t) = f_j(0). \tag{11C.3.6}$$

It is quite evident that, whenever the invariant surfaces defined by f_1, f_2, f_3 are non-compact and the motion is unbounded, a prototypical example is provided by the free Hamiltonian

$$H_0 = \frac{p^2}{2m}. \tag{11C.3.7}$$

In this case, we have the flow associated with the equations of motion given by

$$p_i(t) = p_i(0), \tag{11C.3.8}$$

$$q^i(t) = q^i(0) + \delta^{ij} p_j \frac{t}{m}. \tag{11C.3.9}$$

We notice that, for any choice of $\{p_j\}$, the evolution determines a flow on the configuration space whose infinitesimal generator is $\Gamma = \frac{p_j}{m} \frac{\partial}{\partial q_j}$. Each different choice will determine a different flow on the configuration space. The previous considerations mean that, in those regions of phase space where the conditions required by the formulated theorem hold true, our system is diffeomorphic with the free system (in particular, in the identified region, there is a complete integral of the associated Hamilton–Jacobi equation).

In general, we start in a way very similar to the one that we have considered for perturbation theory for systems with a discrete spectrum, i.e. we assume that the free Hamiltonian, i.e. 'comparison Hamiltonian', and the Hamiltonian we are considering give rise to a 1-parameter group of canonical transformations on relevant parts of the phase space, which identify what will be called the *space of scattering states*.

To identify and visualize the variety of problems which may arise, we shall consider a simple one-dimensional problem described by the following Hamiltonian function in dimensionless units:

$$H = \frac{p^2}{2m} + \frac{k}{(x^4 - x^2 + a^2)}. \tag{11C.3.10}$$

The shape of the potential function is given in Figure 11C.1, and the resulting phase portrait in phase space is shown in Figure 11C.2.

It is easy to see that phase space is divided into several regions, and in some of them clearly the orbits are not behaving like those of a free particle. For many of them, the behaviour at $t \to -\infty$ and at $t \to +\infty$ resembles that of the free particle; for this reason we usually considers orbits in the far past and in the far future. The so-called asymptotic condition is the requirement that, at very large (positive and negative) values of t the system

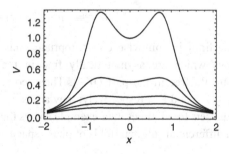

Fig. 11C.1 Shape of the potential function $V(x)$. The upper curve is obtained when $a = 1$, whereas the lower curves correspond to other values of a.

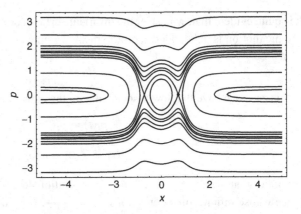

Fig. 11C.2   Phase portrait for the potential function $V(x)$ in Figure 11C.1.

we are considering should behave like a free system. In some sense, the prepared initial states and the detected final states can be considered as states of a free system. The simplest way of expressing the asymptotic condition is to require that, for each trajectory of our system, there are two other states (free trajectories) which asymptotically, one in the past and one in the future, will approximate our trajectory.

The particular example we are considering shows that there are regions of phase space where trajectories are bounded, therefore cannot be asymptotically free. There are trajectories which are asymptotically free either in the past or in the future but not both; they correspond in the picture to that trajectories terminate at those points where we have maxima of the potential function. We denote by T_0 the 1-parameter group

$$T_0(t) : (\vec{x}, \vec{p}) \to \left(\vec{x} + \frac{\vec{p}}{m}t, \vec{p}\right),$$

and we replace points (x, p) with $p = 0$ because the resulting trajectories would be points instead of lines. We denote by T the 1-parameter group associated with our system and defined by the flow of the differential equation

$$\left(\dot{\vec{x}}, \dot{\vec{p}}\right) = \left(\frac{\vec{p}}{m}, -\vec{\nabla}V\right).$$

We define the *wave operators*

$$W^{\pm} \equiv \lim_{t \to \pm\infty} T(-t) \cdot T_0(t). \tag{11C.3.11}$$

When the limit is pointwise on appropriate ranges, these ranges are identified by the trajectories which are asymptotically free in the future (W^+) and those which are asymptotically free in the past (W^-). Those corresponding regions of phase space are denoted by Γ^+ and Γ^-, respectively; they are complements of the trajectories that remain in compact sets for all times. Our example shows that Γ^+ need not coincide with Γ^-. They are open differential sub-manifolds of phase space and satisfy

$$T(t)W^{\pm} = W^{\pm}T_0(t). \tag{11C.3.12}$$

It follows that the restriction of the equations of motion to the sub-manifolds where $W^{\pm}$ are defined can be canonically transformed into the equations of motion of the free system in the region of phase space where $p \neq 0$. We define

$$\Gamma \equiv \Gamma^+ \cap \Gamma^- \tag{11C.3.13}$$

as the region of phase space where the trajectories of our system are asymptotically free, both in the past and in the future. On this region of phase space we define the scattering transformation by setting

$$S \equiv \left(W^+\right)^{-1} \cdot W^-. \tag{11C.3.14}$$

It is defined on $\left(W^-\right)^{-1} \Gamma$ and it maps this region canonically onto $\left(W^+\right)^{-1} \Gamma$; both sets are open and invariant under the action of $T_0(t)$, the free comparison system. These concepts have a counterpart in the Hilbert space language when dealing with quantum systems.

In the specialized literature it is possible to find conditions on the potential V which allow complete control on the regions Γ, Γ^+ and Γ^-. Properties of W^+, W^- and S are then derived with respect to the symplectic structure and the Poisson brackets. In particular, in dealing with states as trajectories, it is more convenient to consider the properties of the scattering transformation on the algebra of functions on phase space, and to introduce the notion of asymptotic constants of motion. The corresponding quantum version would describe scattering in the Heisenberg picture. For useful background material we refer the reader to the work in Hunziker (1974), Thirring (1997).

11C.4 Time-independent scattering

This description of scattering theory was historically developed first, and it provided Born with the appropriate setting to introduce the probabilistic interpretation of the wave function. In this approach the wave operator $\widehat{W}_{\pm}$ and thereafter the scattering operator $\widehat{S}$ are constructed as integral operators in the coordinate system for $\mathcal{H}$ associated with position operators. This approach will be discussed with more details in the following sections for the realistic treatment of scattering theory in three dimensions. Before doing that, we would like to summarize very briefly some of the material covered in Sections 6.2 and 6.3, in particular Section 6.3.1, and present it in a form that is a preparation for the three-dimensional situation.

11C.4.1 One-dimensional stationary description of scattering

The main question we would like to answer is as follows. For a system described by the Hamiltonian operator

$$\widehat{H} = -\frac{\hbar^2}{2m}\frac{d^2}{dx^2} + V(x),$$

if a beam of nearly monoenergetic particles with mean momenta $\langle p \rangle = \hbar k_0$ comes from the far left ($x \to -\infty$) onto a potential of compact support $V(x)$, what fraction T will get transmitted and what fraction R will get reflected? It turns out that, if the individual momentum-space wave functions are sharply peaked at $\hbar k_0$, the reflection and the transmission probabilities depend only on k_0 and not on the detailed shapes of the wave functions. Thus, it is possible to evaluate $R(k_0)$ and $T(k_0)$ that apply to every particle in the beam. With the assumptions on the potential, comparing eigensolutions of $\widehat{H}$ with those of $\widehat{H}_0$ we may say that the solutions $\psi_1(x, k)$, $\psi_2(x, k)$ of $\widehat{H}|\psi\rangle = E|\psi\rangle$, when written in the position representation, will coincide with those of $\widehat{H}_0|\psi\rangle = E|\psi\rangle$ for $x \to \infty$ and $x \to -\infty$. More formally, we can write that

$$\lim_{x \to -\infty} e^{ikx} \psi_2(x, k) = 1, \quad \lim_{x \to -\infty} e^{ikx} \psi_2'(x, k) = -ik, \tag{11C.4.1}$$

while

$$\lim_{x \to \infty} e^{-ikx} \psi_1(x, k) = 1, \quad \lim_{x \to \infty} e^{ikx} \psi_1'(x, k) = ik, \tag{11C.4.2}$$

where $k = \frac{p}{\hbar}$. In conclusion, we can say that each solution $\psi(x, k)$ of the interacting Hamiltonian will be uniquely determined by means of its behaviour on the left of the potential barrier and the behaviour on the right of the potential barrier. For example,

$$\psi_1(x, k) \to e^{ikx} + B_1 e^{-ikx} \text{ on the far left}$$

$$\psi_1(x, k) \to A_1 e^{ikx} \text{ on the far right.}$$

The physical interpretation, which also motivates the specific choice we have made, is that on the left we have an incident plus a reflected plane wave while, on the right, we only have an outgoing plane wave. Similarly, another solution, independent of the one we have described, would be given by

$$\psi_2(x, k) \to e^{-ikx} + B_2 e^{ikx} \text{ on the far right}$$

$$\psi_2(x, k) \to A_2 e^{-ikx} \text{ on the far left.}$$

By using the conservation of probability expressed in terms of the probability current associated with

$$j_1 = \frac{\hbar}{2mi}\left(\psi_1^* \frac{d}{dx}\psi_1 - \psi_1 \frac{d}{dx}\psi_1^*\right) \tag{11C.4.3}$$

we find

$$1 = |A_1|^2 + |B_1|^2. \tag{11C.4.4}$$

This relations shows that the transmitted part plus the reflected one is equal to the incident probability. The *scattering matrix* with the above notation would be

$$S = \begin{pmatrix} A_1 & B_1 \\ B_2 & A_2 \end{pmatrix}, \tag{11C.4.5}$$

where other considerations require that

$$A_1 = A_2, \quad B_2 = -\frac{B_1^* A_1}{A_1^*}. \tag{11C.4.6}$$

This matrix is unitary. The scattering matrix transforms incoming particles into outgoing ones. Usually $T = |A_1 A_1^*| = |A_2 A_2^*|$ is called the transmission coefficient while $R = |B_1 B_1^*| = |B_2 B_2^*|$ is the reflection coefficient. If $\varphi_1(k)$ and $\varphi_2(k)$ are wave packets, i.e. functions with compact support concentrated around $k_0 > 0$, we can construct solutions of the evolutionary Schrödinger equation

$$i\hbar \frac{\partial}{\partial t} \tilde{\psi} = H \tilde{\psi}$$

by means of the superpositions

$$\tilde{\psi}_{1,2}(x,t) = \int_0^\infty \varphi_{1,2}(k) \psi_{1,2}(x,k) e^{-i\frac{\hbar k^2 t}{2m}} \, dk. \tag{11C.4.7}$$

In the regions $x \to +\infty$ and $x \to -\infty$ the expressions can be simplified to get

$$\tilde{\psi}_1(x,t) = \int_0^k \varphi_1(k) A_1(k) e^{i\left(kx - \frac{\hbar k^2 t}{2m}\right)} \, dk, \quad x \to +\infty, \tag{11C.4.8}$$

$$\tilde{\psi}_1(x,t) = \int_0^k \varphi_1(k) \left[e^{i\left(kx - \frac{\hbar k^2 t}{2m}\right)} + B_1(k) e^{-i\left(kx + \frac{\hbar k^2 t}{2m}\right)} \right] \, dk, \quad x \to -\infty, \tag{11C.4.9}$$

and similarly for $\tilde{\psi}_2(x,t)$.

11C.5 Integral equation for scattering problems

What we have outlined may be expressed in abstract terms and usually goes under the name of the Lippmann–Schwinger equation. Let us assume again that the Hamiltonian may be written in the form $\widehat{H} = \widehat{H}_0 + \widehat{V}$, with $\widehat{H}_0$ the comparison Hamiltonian. If $|\varphi\rangle$ is an (improper) eigenket of the momentum, say

$$\vec{P}|\varphi\rangle = \hbar \vec{k}|\varphi\rangle, \tag{11C.5.1}$$

it is clear that $|\varphi\rangle$ is also an eigenket of the comparison Hamiltonian $\widehat{H}_0$, with eigenvalue $E = \frac{\hbar^2 k^2}{2m}$.

We would like to determine the (improper) eigenstate, say $|\psi\rangle$, which is an eigenstate of $\widehat{H}$ with the same eigenvalue E of the free Hamiltonian $\widehat{H}_0$. We would have

$$(\widehat{H}_0 + \widehat{V})|\psi\rangle = E|\psi\rangle, \tag{11C.5.2}$$

and it makes sense to set

$$|\psi\rangle = |\varphi\rangle + |\psi'\rangle, \tag{11C.5.3}$$

so that

$$(\widehat{H}_0 - EI)|\psi\rangle = 0 + (\widehat{H}_0 - EI)|\psi'\rangle, \tag{11C.5.4}$$

which in turn implies

$$-\widehat{V}|\psi\rangle = (\widehat{H}_0 - EI)|\psi'\rangle \tag{11C.5.5}$$

and therefore, formally,

$$|\psi'\rangle = \frac{1}{(EI - \widehat{H}_0)}\widehat{V}|\psi\rangle. \tag{11C.5.6}$$

In conclusion, we would end up with

$$|\psi\rangle = |\varphi\rangle + \frac{1}{(EI - \widehat{H}_0)}\widehat{V}|\psi\rangle. \tag{11C.5.7}$$

This equation only determines $|\psi\rangle$ implicitly. However, $EI - \widehat{H}_0$ need not be invertible, as it admits the null eigenvalue when evaluated on $|\varphi\rangle$. To remedy this situation the operator is usually deformed very slightly by replacing $EI - \widehat{H}_0$ with $EI - \widehat{H}_0 \pm i\varepsilon$; then we would have

$$|\psi_\pm\rangle = |\varphi\rangle + \frac{1}{(EI - \widehat{H}_0 \pm i\varepsilon)}\widehat{V}|\psi_\pm\rangle, \tag{11C.5.8}$$

with the understanding that, eventually, we should let ε tend to zero. It is quite evident that, for any $\varepsilon > 0$, the spectrum of $EI - \widehat{H}_0 \pm i\varepsilon$ does not contain the null eigenvalue.

It is now possible to consider the previous abstract setting in the position representation, where we would get

$$\psi_\pm(\vec{x}) = \langle\vec{x}|\psi_\pm\rangle = \langle\vec{x}|\varphi\rangle + \langle\vec{x}|\frac{1}{(EI - \widehat{H}_0 \pm i\varepsilon)}\widehat{V}|\psi_\pm\rangle$$

$$= \varphi(\vec{x}) + \int d^3x' \langle\vec{x}|\frac{1}{(EI - \widehat{H}_0 \pm i\varepsilon)}\widehat{V}(\vec{x}')|\vec{x}'\rangle\psi_\pm(\vec{x}'). \tag{11C.5.9}$$

By using the fact that $|\varphi\rangle$ is an eigenvector of the momentum operator, we get

$$\varphi(\vec{x}) = \frac{e^{i\vec{k}\cdot\vec{x}}}{(2\pi\hbar)^{\frac{3}{2}}}. \tag{11C.5.10}$$

Since the general reader might not be familiar with Green functions in quantum (or even classical) scattering, we now perform a few explicit calculations. Thus, after setting

$$k \equiv |\vec{k}| \equiv \frac{\sqrt{2mE}}{\hbar}, \tag{11C.5.11}$$

$$\mathcal{U}(\vec{x}) \equiv \frac{2m}{\hbar^2}V(\vec{x}), \tag{11C.5.12}$$

$$\phi_k(\vec{x}) \equiv \mathcal{U}(\vec{x})\psi_k(\vec{x}), \tag{11C.5.13}$$

the Schrödinger equation for stationary states reads as[3]

$$\left(\triangle + k^2\right)\psi_k(\vec{x}) = \phi_k(\vec{x}). \tag{11C.5.14}$$

The full integral of (11C.5.14) consists of the general integral of the homogeneous equation

$$\left(\triangle + k^2\right)\psi_k(\vec{x}) = 0,$$

---

[3] Strictly speaking we should be writing $\psi_{\vec{k}}(\vec{x})$, but hereafter we omit the vector symbol upon the subscript k to avoid writing too many mathematical symbols.

i.e. $\psi_k(\vec{x}) = e^{\pm i\vec{k}\cdot\vec{x}}$, plus a particular solution of (11C.5.14). For this purpose, we need to invert the operator $\left(\triangle + k^2\right)$, so that $\psi_k(\vec{x})$ may be expressed by an integral operator acting on $\phi_k(\vec{x})$. This is achieved by first finding the Green function G_k of (11C.5.14). For this purpose, we may introduce polar coordinates, assuming that G_k depends only on the modulus $r = |\vec{x}|$. In other words, we first consider the problem of finding a $G_k(r)$ such that

$$\left(\frac{d^2}{dr^2} + \frac{2}{r}\frac{d}{dr} + k^2\right) G_k(r) = 0 \quad \forall r > 0, \tag{11C.5.15}$$

with the understanding that, in the integral equation we are going to derive, we shall replace r by the modulus

$$|\vec{x} - \vec{x}'| \equiv \sqrt{x^2 + x'^2 - 2xx'\cos\alpha},$$

where α is the angle between the vectors $\vec{x}$ and $\vec{x}'$. The solution of (11C.5.15) is found to be

$$G_k(r) = \frac{e^{i\varepsilon kr}}{\beta r}, \tag{11C.5.16}$$

where $\varepsilon = \pm 1$, and β takes the value -4π, which is fixed by the well-known property of the Laplace operator

$$\triangle(r^{-1}) = 0 \quad \forall r > 0, \tag{11C.5.17}$$

jointly with the condition

$$\int_{\mathcal{A}} \triangle(r^{-1}) d^3x = -4\pi, \tag{11C.5.18}$$

if the origin lies in $\mathcal{A}$. The formula (11C.5.16) is obtained by re-expressing the Helmholtz equation (11C.5.15) for the Green function in the form

$$\left(\frac{d^2}{dr^2} + k^2\right)(rG) = 0 \quad \forall r > 0, \tag{11C.5.19}$$

which is solved by

$$G = A\frac{e^{ikr}}{r} + B\frac{e^{-ikr}}{r},$$

for some constants A and B.

Hence we obtain the integral equation for the continuous spectrum

$$\psi_k^{\pm}(\vec{x}) = e^{\pm i\vec{k}\cdot\vec{x}} - \frac{1}{4\pi} \int \frac{e^{\pm ik|\vec{x}-\vec{x}'|}}{|\vec{x} - \vec{x}'|} \mathcal{U}(\vec{x}')\psi_k^{\pm}(\vec{x}')d^3x', \tag{11C.5.20}$$

which is the explicit expression of (11C.5.9) up to the inessential multiplicative factor $(2\pi\hbar)^{-\frac{3}{2}}$. The structure of this integral equation is quite general, and can be used in all cases where, to the free Hamiltonian H_0, an interaction term V is added. The advantages of the integral formulation are twofold:

(i) the boundary conditions on the scattering solution are automatically encoded;
(ii) a series solution can be found.

An equivalent way of expressing Eq. (11C.5.20) is

$$\psi_E^{\pm} = \psi_E + G_E^{\pm} V \psi_E^{\pm}, \tag{11C.5.21}$$

which is called the Lippmann–Schwinger equation (Lippmann and Schwinger 1950).

If $|\vec{x}| >> |\vec{x}'|$, we can approximate, writing x for $|\vec{x}|$ and x' for $|\vec{x}'|$ (α being the angle formed by the vectors $\vec{x}$ and $\vec{x}'$)

$$|\vec{x} - \vec{x}'| = x\left[1 - (\cos\alpha)\frac{x'}{x} + \frac{1}{2}(\sin\alpha)^2\left(\frac{x'}{x}\right)^2 + O\left(\frac{x'}{x}\right)^3\right]$$

$$= x - (\cos\alpha)x' + xO\left(\frac{x'}{x}\right)^2$$

$$\sim r - \vec{e}_x \cdot \vec{x}', \tag{11C.5.22}$$

where $\vec{e}_x$ is the unit vector $\frac{\vec{x}}{|\vec{x}|}$, and hence the integral on the right-hand side of Eq. (11C.5.20) can be approximated, for $\psi_k^+(\vec{x})$, by

$$\frac{e^{ikr}}{r}\int e^{-i\vec{\kappa}\cdot\vec{x}'}\mathcal{U}(\vec{x}')\psi_k^+(\vec{x}')d^3x',$$

where we have defined $\vec{\kappa} \equiv k\vec{e}_x$. We then find for the first set of solutions, at large distances from the scattering centre,

$$\psi_k^+(\vec{x}) \sim e^{i\vec{k}\cdot\vec{x}} + f_{k\kappa}\frac{e^{ikr}}{r}, \tag{11C.5.23}$$

where

$$f_{k\kappa} \equiv -\frac{1}{4\pi}\int e^{-i\vec{\kappa}\cdot\vec{x}'}\mathcal{U}(\vec{x}')\psi_k^+(\vec{x}')d^3x'. \tag{11C.5.24}$$

This describes the joint effect of a plane wave, and of a spherical wave modulated by the factor $f_{k\kappa}$. To recover the time-dependent formulation, we have to build the wave packet

$$\psi(\vec{x}, t) = \int C(\vec{k})\psi_k^+(\vec{x})e^{-i\frac{\hbar k^2 t}{2m}}d^3k, \tag{11C.5.25}$$

where C is a function with compact support. Thus, on applying the stationary phase method we find that, when $t < 0$, the spherical wave term does not contribute, whereas, for $t > 0$, it plays a crucial role, and represents a spherical wave propagating from the centre with amplitude depending on $f_{k\kappa}$. The term $e^{i\vec{k}\cdot\vec{x}}$ represents instead, upon integration, the motion of a free wave packet.

It should be stressed that the solutions ψ_k^- can be legitimately considered, from a purely mathematical point of view. However, on applying the stationary-phase method, they give rise to a wave packet which, for $t \to -\infty$, consists of a plane wave and a wave directed towards the centre, whereas, for $t \to \infty$, they give rise to a plane wave propagating in the direction $-\vec{k}$. It remains unclear how to realize such a situation in the actual experiments, since the coherence properties of the wave function are required to hold on macroscopic length scales. Thus, on physical grounds, the stationary states ψ_k^- are ruled out, and only the states ψ_k^+ are worked with (they are in an extended space which spans the Hilbert space) to build a scattered wave packet. Of course, the full Hilbert space of the problem might include bound states as well.

In conclusion, the stationary wave function may be written as a superposition of an incoming plane wave (associated with a particle of momentum $\hbar\vec{k}$) and a second contribution, resulting from the scatterer, which represents an outgoing spherical wave

$|\psi_+\rangle$. If $\hbar\vec{k}'$ is the momentum after collision, conservation of momentum implies $|\vec{k}| = |\vec{k}'|$. The direction of $\vec{k}'$ after collision, at the detector, is essentially the same as $\vec{r}$.

As far as the scattering cross-section is concerned, we exploit again conservation of probability stated in terms of the continuity equation for the probability current. By introducing the flow of incoming particles and that of the scattered particles we have

$$\vec{j}_{\text{inc}} \cdot \mathrm{d}\vec{\sigma} = r^2 \vec{j}_{\text{sc}} \cdot \vec{n}\mathrm{d}\Omega, \tag{11C.5.26}$$

where $\frac{\mathrm{d}\sigma}{\mathrm{d}\Omega}$ is called the differential cross-section. In the approximation we have considered, the incoming current reads as

$$\vec{j}_{\text{inc}} = \frac{\hbar}{2mi}\left(e^{-i\vec{k}\cdot\vec{x}}\vec{\nabla}e^{i\vec{k}\cdot\vec{x}} - e^{i\vec{k}\cdot\vec{x}}\vec{\nabla}e^{-i\vec{k}\cdot\vec{x}}\right) = \frac{\hbar\vec{k}}{m}, \tag{11C.5.27}$$

while the scattering part is given by

$$\vec{j}_{\text{sc}} = \frac{\hbar}{2mi}\left(\psi_{\text{sc}}^* \vec{\nabla}\psi_{\text{sc}} - \psi_{\text{sc}}\vec{\nabla}\psi_{\text{sc}}^*\right). \tag{11C.5.28}$$

It is convenient to write the gradient operator in polar coordinates, i.e.

$$\vec{\nabla} = \widehat{e}_r\frac{\partial}{\partial r} + \widehat{e}_\theta\frac{1}{r}\frac{\partial}{\partial\theta} + \widehat{e}_\phi\frac{1}{r\sin\theta}\frac{\partial}{\partial\phi}, \tag{11C.5.29}$$

with $\widehat{e}_r, \widehat{e}_\theta, \widehat{e}_\phi$ the unit vectors in the corresponding directions. When $r \to \infty$ it is clear that the first term will dominate with respect to the other two, and by using

$$\psi_{\text{sc}} \approx \frac{e^{ikr}}{r}f_k(\theta,\varphi) \tag{11C.5.30}$$

we get

$$\frac{\partial}{\partial r}f_k(\theta,\varphi)\frac{e^{ikr}}{r} = f_k(\theta,\varphi)ik\frac{e^{ikr}}{r} + \mathrm{O}(r^{-2}), \tag{11C.5.31}$$

and hence the scattered current is expressed through the scattering amplitude by

$$\vec{j}_{\text{sc}} = \frac{\widehat{e}_r}{r^2}|f_k|^2\frac{\hbar k}{m}. \tag{11C.5.32}$$

From the previous formulae obtain eventually

$$\frac{\mathrm{d}\sigma}{\mathrm{d}\Omega} = |f_k(\theta,\varphi)|^2, \tag{11C.5.33}$$

showing that, in the time-independent picture, the calculation of $\frac{\mathrm{d}\sigma}{\mathrm{d}\Omega}$ reduces to the evaluation of the scattering amplitude $f_k(\theta,\varphi)$.

11C.6 The Born series

Having decided that we only consider scattering with mutual interaction in a two-particle system, we focus on the stationary states ψ_k^+, for which the Lippmann–Schwinger equation (11C.5.21) can be written in the form

$$\psi_E^+ = \psi_E + G_E^+ V\psi_E^+, \tag{11C.6.1}$$

which implies that

$$\left(I - G_E^+ V\right)\psi_E^+ = \psi_E, \tag{11C.6.2}$$

and hence, upon using the inverse operator $\left(I - G_E^+ V\right)^{-1}$,

$$\psi_E^+ = \left(I - G_E^+ V\right)^{-1} \psi_E. \tag{11C.6.3}$$

This applies only to the continuous spectrum (more generally for states labelled by the same labels as for the free system). To lowest order, $\psi_E^+ \sim \psi_E$, and the resulting scattering amplitude is that of the Born approximation, i.e. (cf. Eq. (11C.5.24))

$$f_{k\kappa}^{\text{Born}} = -\frac{m}{2\pi\hbar^2} \int e^{i(\vec{k}-\vec{\kappa})\cdot\vec{x}} V(\vec{x}) d^3 x, \tag{11C.6.4}$$

which is proportional to the Fourier transform of the potential, evaluated for $\vec{q} = \vec{\kappa} - \vec{k}$.

In general, we can try to express the inverse operator $\left(I - G_E^+ V\right)^{-1}$ as a (formal) series

$$\left(I - G_E^+ V\right)^{-1} \sim \sum_{n=0}^{\infty} (G_E^+ V)^n, \tag{11C.6.5}$$

where the right-hand side is called the Born series. There are, in fact, some rigorous results which provide sufficient conditions for the convergence of the Born series. They are as follows (Reed and Simon 1979).

Theorem 11.1 Let V be Lebesgue-summable on $\mathbf{R}^3$:

$$\int_{\mathbf{R}^3} |V(\vec{x})| d^3 x < \infty \tag{11C.6.6}$$

and in the Rollnik class, i.e. such that (Rollnik 1956)

$$\|V\|_{\text{Rollnik}}^2 \equiv \int\int_{\mathbf{R}^6} \frac{|V(\vec{x})||V(\vec{y})|}{|\vec{x}-\vec{y}|^2} d^3 x d^3 y < \infty. \tag{11C.6.7}$$

Under such conditions, there is an energy E_0 such that the Born series is convergent $\forall E > E_0$.

Theorem 11.2 If the assumptions of Theorem 11.1 hold, and the Rollnik norm of V satisfies the inequality

$$\|V\|_{\text{Rollnik}} < 4\pi, \tag{11C.6.8}$$

the Born series converges $\forall E > 0$.

For the detailed proof of such theorems we refer the reader to Reed and Simon (1979). However, we remark that, if both V and V^{-1} are bounded functions, the integral operator $G_E^+ V$ is related by the following transformation to

$$\widetilde{G} \equiv \sqrt{V} \, G_E^+ \, \sqrt{V}. \tag{11C.6.9}$$

We then use the property

$$\|\widetilde{G}\|^2 = (4\pi)^{-2} \|V\|_{\text{Rollnik}}^2 \tag{11C.6.10}$$

to prove Theorem 11.2, because the maximal eigenvalue of $G_E^+ V$ is smaller than 1 if (11C.6.8) is satisfied, from (11C.6.9) and (11C.6.10).

Potentials V that belong to $\mathcal{L}^1(\mathbf{R}^3)$ and are in the Rollnik class play an important role in proving the uniqueness of the solution of the Lippmann–Schwinger equation. More precisely, if $H_0 = -\Delta$ on $\mathcal{L}^2(\mathbf{R}^3)$, and if $H = H_0 + V$, with V satisfying Eqs. (11C.6.6) and (11C.6.7), we can prove that there is a subset $\mathcal{E}$ of $\mathbf{R}_+$, closed and with zero Lebesgue measure such that, if $k^2 \notin \mathcal{E}$, there is a unique solution of Eq. (11C.5.21) (Reed and Simon 1979).

11C.7 Partial wave expansion

As we know from Chapter 7, in a central potential the self-adjoint operators $\hat{H}, \hat{L}^2, \hat{L}_z$ have common eigenfunctions $u_{n,l,m}(r, \theta, \varphi)$ (the effects of spin are neglected here). Moreover, the z-axis can be arranged along the direction of the incoming beam. Such a beam has, in general, cylindrical symmetry about the propagation direction, which implies that we are eventually dealing with eigenfunctions of $\hat{L}_z$ belonging to the eigenvalue $m = 0$. Thus, bearing in mind that the spherical harmonic Y_l^0 is proportional to the Legendre polynomial $P_l(\cos \theta)$, the stationary states can be written in the form

$$\psi_k^+(\vec{x}) = \sum_{l=0}^{\infty} C_l R_{kl}(r) P_l(\cos \theta). \tag{11C.7.1}$$

If the potential has compact support:

$$V(r) = 0 \quad \forall r > a, \tag{11C.7.2}$$

we find

$$R_{kl}(r) = A_l j_l(kr) + B_l n_l(kr) \quad \forall r > a, \tag{11C.7.3}$$

in terms of the spherical Bessel functions. Thus, by virtue of the well-known asymptotic formulae valid as $r \to \infty$,

$$j_l(kr) \sim \frac{1}{kr} \sin\left(kr - \frac{l\pi}{2}\right), \tag{11C.7.4}$$

$$n_l(kr) \sim -\frac{1}{kr} \cos\left(kr - \frac{l\pi}{2}\right), \tag{11C.7.5}$$

and defining

$$\frac{B_l}{A_l} \equiv -\tan(\delta_l), \tag{11C.7.6}$$

$$\frac{A_l}{\cos(\delta_l)} \equiv a_l, \tag{11C.7.7}$$

we find

$$R_{kl}(r) = a_l\left[\cos(\delta_l)j_l(kr) - \sin(\delta_l)n_l(kr)\right]$$

$$\sim \frac{a_l}{kr}\sin\left(kr - \frac{l\pi}{2} + \delta_l\right). \tag{11C.7.8}$$

The insertion of (11C.7.8) into the expansion (11C.7.1) therefore yields the asymptotic formula

$$\psi_k^+(r) \sim \sum_{l=0}^{\infty} \frac{a_l(k)}{kr}P_l(\cos\theta)\sin\left(kr - \frac{l\pi}{2} + \delta_l\right). \tag{11C.7.9}$$

On the other hand, from Eq. (11C.5.23) we find another useful formula for $\psi_k^+(r)$, i.e. (with $z \equiv r\cos\theta$)

$$\psi_k^+(r) \sim e^{ikz} + f_k(\theta)\frac{e^{ikr}}{r}. \tag{11C.7.10}$$

Further progress is made by using the expansion of a plane wave, in spherical coordinates, in terms of Legendre polynomials, i.e.

$$e^{ikr\cos\theta} = \sum_{l=0}^{\infty}(2l+1)i^l\frac{\sin\left(kr - \frac{l\pi}{2}\right)}{kr}P_l(\cos\theta). \tag{11C.7.11}$$

The next step is now to compare the asymptotic formulae (11C.7.9) and (11C.7.10), and re-express the sin e functions with the help of the identity $e^{ix} = \cos(x) + i\sin(x)$. This leads to the equation

$$f_k(\theta)\frac{e^{ikr}}{r} + \sum_{l=0}^{\infty}(2l+1)\frac{i^l}{2ikr}e^{-i\frac{l\pi}{2}}P_l(\cos\theta)e^{ikr}$$

$$+ \sum_{l=0}^{\infty}(2l+1)\frac{i^l}{2kr}ie^{i\frac{l\pi}{2}}P_l(\cos\theta)e^{-ikr}$$

$$= \sum_{l=0}^{\infty}a_l(k)\frac{1}{2ikr}e^{-i\frac{l\pi}{2}}e^{i\delta_l}P_l(\cos\theta)e^{ikr}$$

$$+ \sum_{l=0}^{\infty}a_l(k)\frac{i}{2kr}e^{i\frac{l\pi}{2}}e^{-i\delta_l}P_l(\cos\theta)e^{-ikr}. \tag{11C.7.12}$$

This can be re-expressed in the form

$$T_{1,k}e^{-ikr} + T_{2,k}e^{ikr} = 0, \tag{11C.7.13}$$

which is satisfied $\forall r$ if and only if

$$T_{1,k} = 0, \quad T_{2,k} = 0. \tag{11C.7.14}$$

Now, from (11C.7.12) and (11C.7.14),

$$a_l(k) = (2l + 1)i^l e^{i\delta_l}, \tag{11C.7.15}$$

$$f_k(\theta) = \sum_{l=0}^{\infty} \frac{(2l + 1)}{k} e^{i\delta_l} \sin(\delta_l) P_l(\cos\theta). \tag{11C.7.16}$$

Thus, the *phase shifts* $\delta_l(k)$ determine the scattering amplitude completely, and we find, for the total cross-section,

$$\sigma = \int |f_k(\theta)|^2 d\Omega = \frac{4\pi}{k^2} \sum_{l=0}^{\infty} (2l + 1) \sin^2 \delta_l(k). \tag{11C.7.17}$$

A number of rigorous results exist on the partial wave expansion. In particular, we would like to emphasize the following (Reed and Simon 1979).

Theorem 11.3 Let V be a central potential with $e^{\alpha|\vec{x}|}V$ in the Rollnik class, for some $\alpha > 0$. Given $E \in \mathbf{R}_+/\mathcal{E}$, the partial wave expansion

$$F(E, \cos\theta) = \sum_{l=0}^{\infty} (2l + 1) f_l(E) P_l(\cos\theta) \tag{11C.7.18}$$

converges uniformly for $\theta \in [0, 2\pi]$.

Theorem 11.4 Let $V(r)$ be a central potential, piecewise continuous on $[0, \infty[$. Suppose that

$$\int_0^1 r|V(r)|dr < \infty, \qquad \int_1^\infty |V(r)|dr < \infty. \tag{11C.7.19}$$

Let E be positive and $l \geq 0$. If these conditions hold, there is a unique function $\varphi_{l,E}(r)$ on $(0, \infty)$, which is of class C^1, piecewise C^2, and which satisfies the equation

$$\left[-\frac{d^2}{dr^2} + V_l(r) \right] \varphi_{l,E}(r) = E\varphi_{l,E}(r), \tag{11C.7.20}$$

where

$$V_l(r) \equiv V(r) + \frac{l(l + 1)}{r^2}, \tag{11C.7.21}$$

jointly with the boundary conditions

$$\lim_{r \to 0} \varphi_{l,E}(r) = 0, \tag{11C.7.22}$$

$$\lim_{r \to 0} r^{-l-1} \varphi_{l,E}(r) = 1. \tag{11C.7.23}$$

Moreover, there is a constant b such that

$$\lim_{r \to \infty} \left[b\varphi_{l,E}(r) - \sin\left(kr - \frac{l\pi}{2} + \delta_l(E) \right) \right] = 0. \tag{11C.7.24}$$

Note that we are concerned with $\varphi_{l,E}(r)$, which is obtained from the radial wave function by the relation $R_{l,E}(r) = \frac{\varphi_{l,E}(r)}{r}$. We should also acknowledge that the condition (11C.7.22) is implied by the limit (11C.7.23), whereas the converse does not hold. They are both written explicitly to make it easier to perform a comparison with other kinds of boundary-value problems occurring in theoretical physics.

11C.8 *s*-Wave scattering states in the square-well potential

As a simple application of the formulae of the previous section, let us assume that only low-energy particles interact with the square-well potential (7.3.1), so that only *s*-waves are scattered. If $r < a$, the stationary states solve the equation

$$\left(\frac{d^2}{dr^2} + k^2 + k_0^2\right) y = 0,$$

(11C.8.1)

having defined

$$k^2 \equiv \frac{2mE}{\hbar^2}, \quad k_0^2 \equiv \frac{2mV_0}{\hbar^2}.$$

(11C.8.2)

Regularity at the origin means that $y(0) = 0$, and this leads to the interior solution

$$y(r) = A \sin(k_1 r),$$

(11C.8.3)

where

$$k_1^2 = k^2 + k_0^2.$$

(11C.8.4)

If $r > a$, the stationary states instead solve the equation

$$\left(\frac{d^2}{dr^2} + k^2\right) y = 0,$$

(11C.8.5)

which yields the exterior solution

$$y(r) = B \sin(kr + \delta_0),$$

(11C.8.6)

with the *s*-wave scattering phase shift denoted by δ_0. Since the square-well potential (7.3.1) has a finite discontinuity at $r = a$, we can impose the familiar matching conditions for stationary states and their first derivative,

$$A \sin(k_1 a) = B \sin(ka + \delta_0),$$

(11C.8.7)

$$k_1 A \cos(k_1 a) = kB \cos(ka + \delta_0).$$

(11C.8.8)

On taking the ratio of these equations we find

$$\frac{1}{k_1} \tan(k_1 a) = \frac{1}{k} \frac{(\tan(ka) + \tan \delta_0)}{(1 - \tan(ka) \tan \delta_0)}.$$

(11C.8.9)

Multiplication of (11C.8.9) by $kk_1(1 - \tan(ka) \tan \delta_0)$ leads eventually to

$$k \tan(k_1 a) - k_1 \tan(ka) = (\tan \delta_0)(k_1 + k \tan(ka) \tan(k_1 a)).$$

(11C.8.10)

Now the approximations come into play, because the assumption of low-energy particles leads to $ka << 1$, while a shallow potential implies that $k_1 a \cong k_0 a$. Thus, Eq. (11C.8.10) can be solved approximately for δ_0 in the form

$$\delta_0 \cong \tan \delta_0 \cong ka \left(\frac{\tan(k_1 a) - k_1 a}{k_1 a} \right). \tag{11C.8.11}$$

If only s-waves are scattered, the cross-section is isotropic and hence

$$\sigma_T = \frac{4\pi}{k^2} \sin^2 \delta_0 \cong \frac{4\pi}{k^2} \delta_0^2$$

$$\cong 4\pi a^2 \left(\frac{\tan(k_0 a) - k_0 a}{k_0 a} \right)^2. \tag{11C.8.12}$$

In this particular case the cross-section is therefore independent of the energy of the particles.

11C.9 Problems

11.P1. A particle trapped in an infinite one-dimensional potential well is in the ground state, and is subject for a time δ to a constant force. Find, to the first non-vanishing order in perturbation theory, the transition probability to the second excited level after a time $t > \delta$. Find the dominant contribution when $\delta = \frac{\hbar}{8E_0}$, where E_0 is the ground-state energy.

11.P2. A one-dimensional harmonic oscillator of electric charge q is subject to a uniform electric field, which depends on time according to the law

$$\vec{E} = \vec{E}_0 e^{-\left(\frac{t}{\tau}\right)^2}. \tag{11C.9.1}$$

Assuming that the harmonic oscillator was in the ground state at $t = -\infty$, find to second order in perturbation theory the probability of finding it in the first excited level as $t \to \infty$.

11.P3. In suitable units, and setting $\hbar = 1$, a particle is described by the Hamiltonian

$$H = A^\dagger A, \ A \equiv F(x) + ip, \ F(x) \equiv x(1 - gx). \tag{11C.9.2}$$

Find the sign of the ground-state energy, and compute such an energy up to second order in g.

11.P4. By using perturbation theory, compute the correction to the first non-vanishing order in g to the energy of the Hamiltonian H when the quantum number $l = 1$, having set

$$H = H_0 + V, \ H_0 = AL^2 + BL_z, \ V = gBL_x, \tag{11C.9.3}$$

where A, B, g are constants and $\vec{L}$ is the orbital angular momentum. Find the eigenvalues of H from an exact calculation, and compare it with the perturbative result.

11.P5. Referring to Section 11C.5 on the partial-wave expansion, prove the optical theorem, according to which the total cross-section $\sigma(k)$ and the scattering amplitude $f_k(\theta)$ are related by

$$\sigma(k) = \frac{4\pi}{k} \mathrm{Im} f_k(0). \tag{11C.9.4}$$

11.P6. Prove that, for a central potential $V(r)$, the scattering amplitude in the Born approximation is given by

$$f(\theta, \varphi) = -\frac{2m}{\kappa \hbar^2} \int_0^\infty r V(r) \sin(\kappa r)\, \mathrm{d}r, \tag{11C.9.5}$$

where $\kappa \equiv \vec{k} - \vec{k}'$, $\vec{k}$ and $\vec{k}'$ being the wave vectors of incoming and scattered particles, respectively, while θ is the angle between these wave vectors.

11.P7. Find the differential and total scattering cross-section for a particle interacting with the Fermi pseudo-potential

$$V(\vec{r}) = \frac{2\pi k^2}{m} a \delta^{(3)}(\vec{r}). \tag{11C.9.6}$$

11.P8. Consider the stationary Schrödinger equation

$$-\frac{\hbar^2}{2m} \triangle \psi(\vec{r}) + \int_{\mathbf{R}^3} V(\vec{r}, \vec{r}')\psi(\vec{r}')\mathrm{d}^3 r' = E\psi(\vec{r}), \tag{11C.9.7}$$

where the kernel takes the form

$$V(\vec{r}, \vec{r}') = V(r, r') = \frac{\hbar^2}{2m}\lambda u(r)u(r'), \quad r = |\vec{r}|, \; r' = |\vec{r}'|. \tag{11C.9.8}$$

(i) Prove that only the s-wave is affected by the interaction.

(ii) Derive the integral equation corresponding to the stationary Schrödinger equation, with the associated boundary conditions.

(iii) In light of the particular form of the kernel $V(r, r')$, such an integral equation can be reduced to an algebraic equation. Exploit this property to show that, on defining

$$v(k) \equiv \int_0^\infty \frac{\sin(kr)}{kr} u(r) r^2 \mathrm{d}r, \tag{11C.9.9}$$

the scattering amplitude reads as

$$f(k) = \frac{4\pi\lambda |v(k)|^2}{\left[1 + \frac{2\lambda}{\pi} \int \mathrm{d}^3 q \frac{|v(q)|^2}{(k^2 - q^2 + \mathrm{i}\epsilon)}\right]}. \tag{11C.9.10}$$

(iv) Compare this formula with the Born series.

12 Modern pictures of quantum mechanics

The main body of this chapter introduces the reader to quantum mechanics on phase space and the tomographic picture, and the various pictures of quantum mechanics are then applied to the study of a two-level system. Finally, we introduce composite systems, and this prepares the ground for quantum statistics. The topic of identical particles is eventually studied.

12.1 Quantum mechanics on phase space

In previous chapters, we have seen that quite often, to perform concrete computations, it is very convenient to work in the position representation; in this way the operator associated with the potential will act as a multiplication function. In some cases we have found it is also convenient to work in the momentum representation. In dealing with coherent states we have seen that, in such a coordinate system for the Hilbert space, it is natural to consider wave functions and operators defined directly on phase space. In specific terms, the basic observables $\vec{Q}$ and $\vec{P}$, when acting on functions of the positions, will be

$$(\vec{Q}\psi)(\vec{x}) = \vec{x}\psi(\vec{x}), \quad (\vec{P}\psi)(\vec{x}) = \frac{\hbar}{i}\vec{\nabla}\psi(\vec{x}), \tag{12.1.1}$$

while in the momentum representation

$$(\vec{Q}\psi)(\vec{p}) = i\hbar\vec{\nabla}\psi(\vec{p}), \quad (\vec{P}\psi)(\vec{p}) = \vec{p}\psi(\vec{p}). \tag{12.1.2}$$

Similarly, when expressed in terms of creation and annihilation operators in these various realizations we find

$$\vec{Q}\vec{P} - \vec{P}\vec{Q} = i\hbar I. \tag{12.1.3}$$

These realizations are usually called the quantum canonical commutation relations (CCR). These realizations are unitarily equivalent, i.e. there is a unitary transformation mapping $\psi(x)$ onto $\psi(p)$ such that operators in Eq. (12.1.1) are mapped into operators in Eq. (12.1.2). This unitary transformation is provided by the Fourier transform.

On, closer scrutiny Eq. (12.1.3) shows that these operator equations cannot be valid in a strict sense on the full space of square-integrable functions, because the right-hand side, being a multiple of the identity, is defined on all possible elements of the Hilbert space, while the left-hand side is defined only on differentiable functions, because it contains differential operators. More generally, it has been proved by Wintner (1947) that $\widehat{P}$ and $\widehat{Q}$ cannot be both bounded operators.

In his book on quantum mechanics, Weyl (1931) introduced an exponentiated form of the canonical commutation relations, thus avoiding dealing with unbounded operators. If we consider the simple case of phase space, $\mathbf{R}^2 = T^*\mathbf{R}$, with coordinates $(x, \alpha) \in T^*\mathbf{R}$, a Weyl map is defined to be a correspondence between pairs (x, α) and unitary operators on some Hilbert space $\mathcal{H}$ satisfying the following properties:

$$W(x, \alpha) \equiv e^{-i(\alpha \widehat{Q} - x \widehat{P})}, \tag{12.1.4}$$

$$W(x_1, \alpha_1) \cdot W(x_2, \alpha_2) = W(x_1 + x_2, \alpha_1 + \alpha_2) e^{\frac{i}{2}(\alpha_1(x_2) - \alpha_2(x_1))}, \tag{12.1.5}$$

where we stress that α is a covector and x a vector, so that α can be evaluated on x, as a linear map. Moreover, the operators $W(x, \alpha)$ are unitary with

$$W^{-1}(x, \alpha) = W^\dagger(x, \alpha) = W(-x, -\alpha), \tag{12.1.6}$$

and $W(0, 0) = I$ is the identity operator.

The operators $W(x, \alpha)$ are known as Weyl operators and, from the mathematical point of view, constitute a projective (i.e. determined up to a phase factor) unitary representation of the Weyl form of canonical commutation relations. We have already encountered these operators when dealing with coherent states. In particular, displacement operators $D(z)$ when written in terms of real variables become the Weyl operators. Indeed, if we consider $\mathbf{R}^2$ as the complex numbers $\mathbf{C}$, by introducing complex numbers $z = \frac{1}{\sqrt{2}}(x + i\alpha)$ we find that $W(x, \alpha)$ is nothing but the displacement operator $D(z)$ we have already encountered when dealing with coherent states. In the complex form,

$$W(z_1)W(z_2) = e^{i\mathrm{Im}(z_1^* z_2)} W(z_1 + z_2), \quad z_1, z_2 \in \mathbf{C}. \tag{12.1.7}$$

A specific action of the Weyl operators can be given on the Hilbert space of square-integrable functions on $\mathbf{R}$, i.e.

$$(W(x, \alpha)\psi)(y) = e^{i\alpha(y)} \psi(y + x). \tag{12.1.8}$$

This particular realization corresponds to the Schrödinger representation. In a similar way, it is possible to write the momentum representation. It should be stressed that the action on $L^2(\mathbf{R})$ is irreducible, i.e. the only operators A such that

$$AW(x, \alpha) = W(x, \alpha)A \tag{12.1.9}$$

for any value of x and α are multiples of the identity, $A = aI$. By using

$$(W(x, 0)\psi)(y) = \psi(y + x), \tag{12.1.10}$$

we see that $e^{ix\widehat{P}} = W(x, 0)$ implies that $\widehat{P}$ is the generator of translations. By differentiating with respect to x we find also $i\widehat{P}\psi = \frac{\partial \psi}{\partial x}$. As for $W(0, \alpha)$ we see that it is simply the operator of multiplication by $e^{i\alpha(y)}$. More specifically, we see that (y being the point at which we perform all evaluations)

$$\left(e^{a\frac{d}{dx}}\psi\right)(y) = \psi(y) + \left(a\frac{d}{dx}\psi\right)(y) + \left(\frac{a^2}{2!}\frac{d^2}{dx^2}\psi\right)(y) + \cdots, \tag{12.1.11}$$

where we have used the Taylor expansion for the exponential operator. The right-hand side is recognized to be the Taylor expansion of ψ about y, with parameter a, i.e.

$$\psi(y+a) = \psi(y) + \left(a\frac{\mathrm{d}}{\mathrm{d}x}\psi\right)(y) + \cdots . \qquad (12.1.12)$$

Thus, if we interpret $\mathrm{e}^{a\frac{\mathrm{d}}{\mathrm{d}x}}$ as the displacement operator, say we define

$$\left(\mathrm{e}^{a\frac{\mathrm{d}}{\mathrm{d}x}}\psi\right)(y) = \psi(y+a), \qquad (12.1.13)$$

it is immediately seen that it makes sense on any function, and in particular it represents a unitary operator when we restrict it to square-integrable functions. The expression in terms of derivatives would require ψ to be real-analytic.

By using the Weyl association

$$\mathrm{e}^{-\mathrm{i}(\alpha q-xp)} \longrightarrow \mathrm{e}^{-\mathrm{i}(\alpha\widehat{Q}-x\widehat{P})} \qquad (12.1.14)$$

it is possible to associate an operator with every function which can be expanded via a Fourier transform. For any function f we define its Fourier transform $\mathcal{F}(f)$

$$\mathcal{F}(f)(x,\alpha) \equiv (2\pi)^{-n} \int_{\mathbf{R}^{2n}} \mathrm{d}q\,\mathrm{d}p f(q,p)\mathrm{e}^{\mathrm{i}(\alpha q-xp)}. \qquad (12.1.15)$$

With the previous replacement, out of $\mathcal{F}(f)$ we construct an operator

$$\widehat{f} \equiv (2\pi)^{-n} \int_{\mathbf{R}^{2n}} \mathrm{d}x\,\mathrm{d}\alpha\,\mathcal{F}^{-1}(f)(x,\alpha)W(x,\alpha), \qquad (12.1.16)$$

which is the integral Schrödinger representation of $\mathcal{F}^{-1}(f)$. The map which associates an operator with a function on phase space

$$f \to \widehat{f}$$

is bijective and makes it possible to induce a product structure on functions on phase space which are images of operators according to the previous correspondence.

Remark   The bijectivity holds true if some requirements on functions and operators are made. For these technical aspects we refer to the specialized literature. For instance, if f is smooth and rapidly decreasing, i.e. f is in the Schwarz space, then $\widehat{f}$ is an integral operator of trace class.

We define a $*$-product on functions by means of the operator product

$$\widehat{f*g} = \widehat{f}\cdot\widehat{g}, \qquad (12.1.17)$$

i.e. the product structure on the space of operators induces a $*$-product structure on functions on phase space. This product may be written in some explicit form rather than in the implicit form following from the previous equation. We proceed at the same informal level we have been considering so far. For any operator A we define a function f_A on phase space by setting

$$f_A(x,\alpha) \equiv \mathrm{Tr}\left(AW^\dagger(x,\alpha)\right). \qquad (12.1.18)$$

We define

$$(f_A * f_B)(x, \alpha) \equiv \mathrm{Tr}\Big(ABW^\dagger(x, \alpha)\Big). \tag{12.1.19}$$

To reconstruct the operator we consider the association

$$A = \int f_A(x, \alpha) W(x, \alpha) \mathrm{d}x \, \mathrm{d}\alpha, \tag{12.1.20}$$

and use

$$\mathrm{Tr}\Big(W(x, \alpha) W^\dagger(x', \alpha')\Big) = \delta(x - x')\delta(\alpha - \alpha'). \tag{12.1.21}$$

By establishing this one-to-one correspondence between operators and functions on phase space we find that *both classical and quantum mechanics may be described on the same carrier space*; what makes the difference is the multiplication rule we have to use for observables, i.e. the pointwise product for classical description and the $*$-product for the quantum description.

If we use the representation of states by means of convex combinations of rank-1 projectors, say

$$\rho = \sum_j p_j \frac{|\psi_j\rangle\langle\psi_j|}{\langle\psi_j|\psi_j\rangle}, \tag{12.1.22}$$

with the probability vector $p = (p_1, p_2, \ldots, p_n, \ldots)$, the p_j being such that

$$\sum_j p_j = 1, \; p_j \geq 0,$$

we find

$$f_\rho(x, \alpha) \equiv \mathrm{Tr}(\rho W^\dagger(x, \alpha)), \tag{12.1.23}$$

which is called Wigner function. By using the representation of states by means of Hermitian operators it is possible to write eigenvalue problems using the $*$-product and functions on phase space, i.e.

$$f_H * f_{\rho E} = E f_{\rho E}. \tag{12.1.24}$$

Similarly, for the equations of motion,

$$\frac{\mathrm{d}}{\mathrm{d}t} f_A = \frac{1}{i\hbar}(f_H * f_A - f_A * f_H). \tag{12.1.25}$$

In conclusion, it is possible to reformulate quantum mechanics entirely on phase space, once the appropriate $*$-product has been defined. We have to write in some explicit manner the $*$-product and identify autonomously functions which are suitable to represent observables. A specific way is provided by the machinery of the group-algebra representation.

12.2 Representations of the group algebra

If we analyze our construction of operators, we find that the unitary Weyl operators behave like a 'basis' $W(x, \alpha)$, which we multiply by 'coefficients' $f(x, \alpha)$ and 'sum over', integrate, along the points (x, α) of the vector space. This construction is known in the mathematical literature as a representation of the algebra of a group. To describe the situation better and to avoid technicalities we restrict ourselves for the moment to finite groups, while we refer to the book we have already quoted by H. Weyl (1931) and to the more recent book by Balachandran *et al.* (2010) for details.

Consider a finite group G and a representation $U : G \rightarrow \mathcal{U}(\mathcal{H})$ from G to unitary operators on some Hilbert space $\mathcal{H}$. For any function $f : G \rightarrow \mathbf{C}$ we construct the operator

$$F = \sum_g f(g)U(g). \tag{12.2.1}$$

We have defined a map from functions on G to operators on $\mathcal{H}$. This map is clearly a linear map from the vector space of functions on G to the vector space of operators on $\mathcal{H}$. The vector space of functions has dimensions equal to the order of the group, which is the number of elements of the finite-dimensional group.

Given two operators F and H, i.e.

$$\sum_g f(g)U(g), \quad \sum_g h(g)U(g),$$

it is possible to consider the product

$$\sum_{g,g'} f(g)h(g')U(g \cdot g') = \sum_{g''} K(g'')U(g''). \tag{12.2.2}$$

Thus, we are defining a new product, which is different from the pointwise product, non-local on the vector space of functions on G,

$$K(g'') \equiv \sum_{g \cdot g' = g''} f(g)h(g'). \tag{12.2.3}$$

This sum is extended over all pairs of elements g, g' whose product is g''. This new product is usually called the *convolution product* on $\mathcal{F}(G)$.

In this particular case, where the product on functions is induced from operators when $U : G \rightarrow \mathcal{U}(\mathcal{H})$ is a representation, the previously introduced $*$-product coincides with the convolution product on G. The particular function, which is 1 when evaluated at the identity of the group and 0 otherwise, is associated with the identity in the group algebra. Thus, a representation of the group defines a representation of the group algebra, and vice versa. We recall that a function on the group such that $f(g_1g_2) = f(g_2g_1)$ is called a *class function* operators corresponding to class functions constitute the centre of the algebra, i.e. they commute with any other element in the algebra. For any $F = \sum_g f(g)U(g)$, when $U(g)$ is a unitary representation, we have some other properties, i.e.

$$F^\dagger = \sum_g f^*(g)U^\dagger(g) = \sum_g f^*(g)U(g^{-1}) = \sum_g f^*(g^{-1})U(g). \tag{12.2.4}$$

Thus, the Hermitian conjugate of the operator F makes it possible to define the Hermitian conjugate on $\mathcal{F}(G)$, the map $f \rightarrow f^\dagger$ being defined by

$$f^\dagger(g) \equiv f^*(g^{-1}). \tag{12.2.5}$$

Hermitian conjugate matrices are associated with conjugate elements in a unitary representation, and this property characterizes unitary representations. An element of the algebra will be said to be real if it coincides with its conjugate. If $U : G \to \mathcal{U}(\mathcal{H})$ is an n-dimensional irreducible representation of our finite group, and $A = \| a_{jk} \|$ is a given $n \times n$ matrix, the element in $\mathcal{F}(G)$ associated with A is

$$a(g) = \frac{n}{h}\mathrm{Tr}(AU^{\dagger}(g)), \tag{12.2.6}$$

where h is the order of the group and n is the dimension of the representation. The correspondence associating matrices with functions on G generates all complex $n \times n$ matrices. Thus, we have defined a one-to-one correspondence between $n \times n$ matrices and functions on G.

Remark The association operators $\to$ functions on G, by means of the trace, may be defined even if $U : G \to \mathcal{U}(\mathcal{H})$ is not a representation. In this circumstance, the induced $*$-product on functions corresponding to operators need not coincide with the convolution product. If the association $U : G \to \mathcal{U}(\mathcal{H})$ is paired with another association $D : G \to \mathcal{U}(\mathcal{H})$ such that

$$\mathrm{Tr}\Big(D(g_1)U^{\dagger}(g_2)\Big) = \delta\Big(g_1 g_2^{-1}\Big), \tag{12.2.7}$$

it is then possible to reconstruct the operator matrix A from

$$f_A(g) = \mathrm{Tr}\Big(AU^{\dagger}(g)\Big) \tag{12.2.8}$$

by using

$$A = \sum_g f_A(g)D(g). \tag{12.2.9}$$

When the group G is not finite but is replaced by a Lie group, the sums must be replaced by integrals, and the measure on the group is usually taken to be the Haar measure (Balachandran *et al.* 2010). Then,

$$A_f = \int f(g)U(g)\mathrm{d}\mu_g, \tag{12.2.10}$$

$$(f * h)(g) = \int f(gs^{-1})h(s)\mathrm{d}\mu_s = \int f(s)h(s^{-1}g)\mathrm{d}\mu_s. \tag{12.2.11}$$

This time, the role of the identity in the algebra is played by a distribution. We conclude that the general ideas developed for the finite-dimensional groups can be extended to the case of Lie groups when sums are replaced by integrals and we accept as coefficients also distributions.

Let us give a simple example of the previous construction, to help the reader become familiar with the group-algebra concept (while a broader-perspective treatment will be presented in Section 12.6). We consider the following Abelian group of four elements, say g_0, g_1, g_2, g_3, with the multiplication rule

$$g_1 g_2 = g_3 = g_2 g_1, \ g_2 g_3 = g_1 = g_3 g_2, \ g_3 g_1 = g_2 = g_1 g_3, \tag{12.2.12}$$

$$g_0^2 = g_0 = g_1 g_1 = g_2 g_2 = g_3 g_3, \tag{12.2.13}$$

and

$$g_0 g_j = g_j, \tag{12.2.14}$$

i.e. g_0 is the identity in the group. We construct a Weyl system by realizing the elements as unitary matrices on $\mathbf{C}^2$. It is easy to see that (here $U(g_\mu) = \sigma_\mu$)

$$U(g_j) U(g_k) U^{-1}(g_j) U^{-1}(g_k) = -U(g_0), \tag{12.2.15}$$

for $j \neq k \neq 0$. With any 2×2 matrix A we associate a function f_A on the group such that

$$f_A(k) = \text{Tr}(A \sigma_k), \ \forall k = 0, 1, 2, 3. \tag{12.2.16}$$

Clearly, complex-valued functions on $G = \{g_0, g_1, g_2, g_3\}$ can be identified with $\mathbf{C}^4$. The product on elements of $\mathbf{C}^4$ is clearly non-local and non-commutative. If $A = \sigma_1, B = \sigma_2$ the product has support on g_3 while σ_1 and σ_2 have support on g_1 and g_2, respectively.

We are now reverting to Weyl operators $W(x, \alpha)$, and try to read our construction in terms of convolution products. We have commented that the association

$$(x, \alpha) \rightarrow e^{-i(\alpha \widehat{Q} - x \widehat{P})}$$

is not a unitary representation but rather a projective unitary representation; this means that the induced product, on going from operators to functions, is not a convolution but a *twisted convolution*, i.e. we have to take into account the contribution resulting from $e^{\frac{i}{2}(\alpha_1(x_2) - \alpha_2(x_1))}$.

One way to consider the induced product as again being a convolution product instead of a twisted convolution is to deal with Weyl operators as arising from unitary representations of an extended group, known as the Heisenberg–Weyl group. In the simple case $\mathbf{R}^2 \times \mathbf{R}$, with parameters (x, α, s), the Heisenberg–Weyl group has the following composition rule:

$$(x, \alpha, s)(y, \eta, t) = \left(x + y, \alpha + \eta, s + t + \frac{1}{2}(\alpha(y) - \eta(x)) \right). \tag{12.2.17}$$

By using unitary representations of this group, and by repeating the construction of the representation of the group algebra, when restricted to $\mathbf{R}^2$, we obtain the twisted convolution on $\mathbf{R}^2$ by means of the convolution on $\mathbf{R}^2 \times \mathbf{R}$.

Once we understand $\mathbf{R}^2$ as the classical phase space, by using the $*$-product we have defined, it is possible to deal with quantum mechanics on phase space. The Heisenberg picture is implemented therein by means of a non-local product on functions, which is also non-commutative and encodes all properties required to properly describe quantum systems. It is interesting that on the same manifold, phase space, *the automorphism group of the $*$-product reflects the quantum geometry while the diffeomorphism group represents classical geometry.*

12.3 Moyal brackets

On phase space, by using the $*$-product, it is possible to write the equations of motion in the form

$$i\hbar \frac{d}{dt}f = [f,H]_{M},$$
(12.3.1)

where

$$[f,H]_{M} = f * H - H * f,$$
(12.3.2)

and we have used the letter M for Moyal. In this formalism, states which are functions associated with rank-1 projectors satisfy the conditions

$$\rho * \rho = \rho, \quad 1 = \frac{1}{2\pi h}\int \rho^{2}(q,p)dq\,dp.$$
(12.3.3)

The eigenvalue problem for the Hamiltonian H reads as

$$H * \rho_{E_{j}} = E_{j}\rho_{E_{j}},$$
(12.3.4)

and the function H may be decomposed into the sum

$$H = \sum_{j} E_{j}\rho_{E_{j}},$$
(12.3.5)

where

$$E_{j} = \frac{1}{2\pi\hbar}\int (H * \rho_{E_{j}})(q,p)dq\,dp = \frac{1}{2\pi\hbar}\int H(q,p)\rho_{E_{j}}(q,p)dq\,dp.$$
(12.3.6)

Time evolution is ruled by the first-order differential equation

$$i\hbar\frac{d}{dt}e^{-i\frac{Ht}{\hbar}} = H * e^{-i\frac{Ht}{\hbar}},$$
(12.3.7)

where

$$e^{-i\frac{Ht}{\hbar}} = \sum_{n=0}^{\infty}\frac{1}{n!}\left(-\frac{it}{\hbar}\right)^{n}(H*)^{n},$$
(12.3.8)

having defined

$$(H*)^{n} \equiv H * H * \ldots * H \text{ (n factors)}.$$
(12.3.9)

The Moyal picture is a complete alternative picture of quantum mechanics on phase space. In this picture, the quantum-to-classical transition may be dealt with directly at the level of the Heisenberg picture, instead of the Schrödinger picture in terms of wave functions.

In conclusion, it is possible to replace the classical phase space by an Abelian vector group with a symplectic structure ω. We consider the algebra of complex-valued functions on the Abelian group (say separable and locally compact). We restrict our attention to the sub-algebra of functions which are the Fourier transform of integrable functions. They will mimic the algebra of bounded operators on some Hilbert space. The complex conjugation on functions replaces the Hermitian adjoint. The symplectic structure is used to build a multiplier

$$m(x, \alpha; x', \alpha') \equiv e^{\frac{i}{2}\hbar\omega[(x,\alpha),(x',\alpha')]}. \tag{12.3.10}$$

The explicit form of the product is then given by

$$\left(f_1 *_m f_2\right)(\xi) \equiv \int d\mu f_1(\eta) f_2(\xi - \eta) m(\eta, \xi - \eta), \tag{12.3.11}$$

for any $\xi = (x, \alpha)$ in the group. To explore the dependence of the algebra on $\hbar$, the multiplier may be expanded in powers of $\hbar$. It is convenient to restrict the algebra to Schwarz functions in (x, α). The terms in the series can then be explicitly evaluated. For instance, the constant term is just the usual pointwise, associative and commutative product

$$(f_1 \cdot f_2)(\xi) = f_1(\xi) f_2(\xi), \tag{12.3.12}$$

while the next term becomes, after some calculations, $\frac{i\hbar}{2} \{f_1, f_2\}$, i.e. the Poisson bracket associated with the symplectic structure.

We conclude by stressing that it is quite remarkable that both classical and quantum mechanics may be described on the same carrier space phase space, what makes the difference is the multiplication rule we have to use for the 'observables'. *The evolution is quantum or classical depending on whether it is an automorphism for one product or the other.*

12.4 Tomographic picture: preliminaries

We have seen in Chapter 4 and elsewhere that wave functions do not have an immediate experimental interpretation; what may be deduced from experiments is rather the associated probability distribution. We have also seen that a probability distribution associated with a vector in the Hilbert space may be defined whenever we have an orthonormal basis therein, i.e. if $\left\{|e_j\rangle\right\}_{j\in\mathbf{N}}$ satisfies

$$\langle e_j | e_k \rangle = \delta_{jk}, \quad \sum_{j=1}^{\infty} |e_j\rangle\langle e_j| = I. \tag{12.4.1}$$

We associate a probability distribution with a vector $|\psi\rangle$ by setting

$$p_j = \frac{\langle \psi | e_j \rangle \langle e_j | \psi \rangle}{\langle \psi | \psi \rangle} \geq 0, \tag{12.4.2}$$

and noticing that $\sum_{j=1}^{\infty} p_j = 1$. The vector $(p_1, p_2, \ldots, p_n, \ldots)$ of non-negative numbers, with the additional requirement that their sum equals 1, is called a probability vector. When a basis for $\mathcal{H}$ is given by improper functions, as in the case of the eigenvectors of the position or the momentum operator on the real line, we have

$$\int_{\mathbf{R}} |x\rangle dx \langle x| = I, \tag{12.4.3}$$

and we find the probability distribution density given by

$$\frac{\langle \psi | x \rangle \, dx \, \langle x | \psi \rangle}{\langle \psi | \psi \rangle}.$$

By introducing rank-1 projectors to represent physical states, say

$$\rho_\psi \equiv \frac{| \psi \rangle \langle \psi |}{\langle \psi | \psi \rangle}, \tag{12.4.4}$$

we have

$$p_j = \text{Tr} \rho_\psi \rho_j \tag{12.4.5}$$

in the discrete case, and

$$p_x dx = \text{Tr} \rho_\psi \rho_x dx \tag{12.4.6}$$

in the continuum case.

In his book, *General Principles of Quantum Mechanics*, Wolfgang Pauli, in a footnote in Chapter II, says: 'The mathematical problem as to whether, for given probability densities $W(p)$ and $W(x)$, the wave function ψ, if such a function exists, is always uniquely determined, has still not been investigated in all its generality.' It is a natural question to ask if the wave function can be determined from knowing the associated probability distribution in the momentum and the position space. The answer is negative as the following counterexample shows. Let us consider the two wave functions

$$\psi_1(q) = N \exp(-\alpha q^2 + i\beta q), \tag{12.4.7}$$

and

$$\psi_2(q) = N \exp(-\alpha^* q^2 + i\beta q), \tag{12.4.8}$$

with $\text{Re}\,\alpha \geq 0, \beta = \beta^*, N = \left(\frac{\alpha + \alpha^*}{\pi} \right)^{\frac{1}{4}}$. These stationary states have the following expressions in the momentum representation:

$$\psi_1(p) = \frac{N}{\sqrt{2\alpha}} \exp\left[-\frac{(\beta + p)^2}{4\alpha} \right], \tag{12.4.9}$$

$$\psi_2(p) = \frac{N}{\sqrt{2\alpha^*}} \exp\left[-\frac{(\beta + p)^2}{4\alpha^*} \right]. \tag{12.4.10}$$

The associated probability distributions are

$$\psi_1^*(q)\psi_1(q) = \psi_2^*(q)\psi_2(q) = |N|^2 \exp(-(\alpha + \alpha^*)q^2), \tag{12.4.11}$$

$$\psi_1^*(p)\psi_1(p) = \psi_2^*(p)\psi_2(p) = \frac{|N|^2}{2|\alpha|} \exp\left(-(\beta + p)^2 \frac{\alpha + \alpha^*}{4\alpha\alpha^*} \right); \tag{12.4.12}$$

nevertheless,

$$|\langle \psi_1 | \psi_2 \rangle|^2 = \frac{\alpha + \alpha^*}{2\sqrt{\alpha\alpha^*}} = \frac{\text{Re}\,\alpha}{|\alpha|} \neq 1, \tag{12.4.13}$$

which implies that the two states are different.

In general, to be able to reconstruct the state from probability distributions, it is necessary to give not just two of them, but a whole family. The general problem of reconstruction of states in this quantum setting has been studied in a wider context and goes under the name of tomography. The reconstruction problem amounts to the determination of the internal structure, or some property of the internal structure of an object without having to cut, crack, or otherwise damage the object. With such generality the problem appears in a wide variety of situations; for example, X-rays, γ-rays, visible light, microwaves, protons, neutrons, sound waves and nuclear magnetic resonance signals have been used to study various objects. The method has found applications in medicine, astronomy, nuclear magnetic resonance, optics, geophysics and stress analysis of materials. Recently, it has been presented also as an alternative picture of quantum mechanics (Ibort *et al.* 2009, 2010, 2011). In the following sections we shall present this novel approach in its essential parts, while referring to the literature for further details.

12.5 Tomographic picture

The relation between quantum and classical mechanics, the quantum–classical transition, is rather subtle and we cannot claim it is fully understood. In previous chapters we have tried to address this issue whenever it has been possible. Because of this desire to understand, it is very useful to pursue and develop approaches where classical and quantum descriptions may be more easily compared. One such approach is the tomographic one. At the quantum level we might trace back an attempt in the Wigner proposal to formulate quantum mechanics in such a way that it resembles classical mechanics. Wigner searched for a distribution function $\rho(q,p)$ associated with a state on phase space, such that for any observable A, represented say by f_A, we would have

$$\langle A \rangle_\psi = \int f_A(q,p)\rho_\psi(q,p)\mathrm{d}q\,\mathrm{d}p. \qquad (12.5.1)$$

In this way we would obtain a description of quantum mechanics very similar to the description of classical statistical mechanics, where $\rho(q,p)$ would be the classical state, i.e. a probability distribution, and f the classical observable whose average we are interested in. As mentioned in Section 12.1, the function $\rho_\psi(q,p)$ is usually called the Wigner function associated with the rank-1 projector ρ_ψ when viewed as an operator, and we follow the procedure presented in Section 12.1. Unfortunately, by virtue of the quantum indetermination relations, the Wigner function associated with a quantum state cannot be non-negative over the whole phase space, therefore it cannot be a fair probability distribution and has been named quasi-distribution.

What tomography achieves is to replace both fair classical distribution functions and quantum quasi-distributions, defined on phase space, by functions defined on a common space in such a way that both classical and quantum states are represented by fair probability distributions. Thus, the emerging description makes it possible then to compare classical and quantum systems described on the same space by means of functions which are probability distributions.

Let us try to assess the idea behind such a description, it goes back to Radon who formulated it on the basis of X-ray radiography. As considered in Radon (1917), the problem is as follows: if we integrate a function of two variables, say a function f defined on the plane $\mathbf{R}^2$, which satisfies suitable regularity conditions along an arbitrary straight line λ, is it possible to recover f by the knowledge of $F(\lambda)$ for all straight lines?

The obvious way we may adapt this problem to our description of classical and quantum systems is to consider the plane as representing a phase space, and the straight lines as possible identifications of configuration spaces. In higher-dimensional phase spaces, say $\mathbf{R}^{2n}$, these lines are replaced by n-dimensional planes made of vectors whose associated Weyl operators commute pairwise. It could be said that they represent the joint continuous spectrum of a maximal set of commuting operators, all of which have a continuous spectrum.

12.5.1 Classical tomography

Classical statistical mechanics describes states by means of non-negative smooth functions $\rho(q,p) \geq 0$, which are normalized:

$$\int_{\mathbf{R}^2} \rho(q,p)\mathrm{d}q \; \mathrm{d}p = 1. \tag{12.5.2}$$

It is possible to consider straight lines defined by $X - \mu q - \nu p = 0$ and to define the transformation, denoted by $\mathcal{R}\rho$, by means of the following integral written using a Dirac delta:

$$(\mathcal{R}\rho)(X, \mu, \nu) = \int_{\mathbf{R}^2} \rho(q,p)\delta(X - \mu q - \nu p)\mathrm{d}q \; \mathrm{d}p. \tag{12.5.3}$$

This Radon transform of $\rho(q,p)$ has the interesting property of mapping probability distributions on phase space into probability distributions on the space of real lines. To adhere to the common notation, we write $W(X, \mu, \nu)$ for $(\mathcal{R}\rho)(X, \mu, \nu)$, and this $W(X, \mu, \nu)$ is called a *tomogram*.

Note that the tomogram depends on the random variable X and two real parameters μ, ν. The random variable is a linear combination of position and momentum $X = \mu q + \nu p$ where μ and ν can be expressed, in turn, as

$$\mu = s \cos\theta, \; \nu = s^{-1} \sin\theta. \tag{12.5.4}$$

In other words, X may be interpreted as the particle position in a configuration space as identified in a new reference frame for phase space, with a θ-rotated and a s-scaled axis. To introduce dynamical evolution equations, we may consider states with a time dependence; then, for each fixed value of t,

$$W(X, \mu, \nu, t) \geq 0, \; \int W(X, \mu, \nu, t)\mathrm{d}X = 1. \tag{12.5.5}$$

The Radon transform that defines the tomogram may be inverted to express the probability density in the form

$$f(q,p,t) = \frac{1}{4\pi^2} \int W(X, \mu, \nu, t)\mathrm{e}^{\mathrm{i}(X - \mu q - \nu p)}\mathrm{d}X \; \mathrm{d}\mu \; \mathrm{d}\nu. \tag{12.5.6}$$

These formulae establish a one-to-one correspondence between probability densities on phase space and tomographic probability densities, so that the classical tomographic representation is equivalent to the classical phase-space representation. The classical tomogram obeys an evolution equation, which is obtained by Radon transform of the Liouville equation

$$\left[\frac{\partial}{\partial t} + p\frac{\partial}{\partial q} - \frac{\partial V}{\partial q}\frac{\partial}{\partial p}\right]f(q,p,t) = 0. \tag{12.5.7}$$

Thus, it reads as

$$\left[\frac{\partial}{\partial t} - \mu\frac{\partial}{\partial \nu} - \frac{\partial V}{\partial q}\bigg|_{q=q^*} \nu\frac{\partial}{\partial X}\right]W(X,\mu,\nu,t) = 0, \tag{12.5.8}$$

where $q^* \equiv -\frac{\partial}{\partial \mu}\left(\frac{\partial}{\partial X}\right)^{-1}$, the inverse of the differential operator $\frac{\partial}{\partial X}$ being defined using the Fourier transform as

$$\left(\frac{\partial}{\partial X}\right)^{-1}\int f(k)e^{ikX}\mathrm{d}k = \int \frac{f(k)}{ik}e^{ikX}\mathrm{d}k. \tag{12.5.9}$$

The evolution tomographic equation is therefore integro-differential.

The tomograms $W(X,1,0,t)$ and $W(X,0,1,t)$ are the marginal probability distributions of position and momentum, respectively, and the statistics can be written concisely as

$$\langle X^n\rangle_{\mu,\nu,t} = \int X^n W(X,\mu,\nu,t)\mathrm{d}X. \tag{12.5.10}$$

For a free particle described by the phase-space density

$$f(q,p,t) = f_0(q - pt, p), \tag{12.5.11}$$

the tomogram reads as

$$W(X,\mu,\nu,t) = W_0(X,\mu,\nu,t) = \int f_0(q - pt, p)\delta(X - \mu q - \nu p)\mathrm{d}q\,\mathrm{d}p. \tag{12.5.12}$$

On defining the tomographic classical free motion propagator

$$K(X,\mu,\nu,X',\mu',\nu',t) \equiv \delta(X - X')\delta(\mu - \mu')\delta(\nu + \mu t - \nu'), \tag{12.5.13}$$

the classical tomogram can be re-expressed in the propagating form

$$W(X,\mu,\nu,t) = \int K(X,\mu,\nu,X',\mu',\nu',t)W_0(X',\mu',\nu',0)\mathrm{d}X'\,\mathrm{d}\mu'\,\mathrm{d}\nu'. \tag{12.5.14}$$

12.5.2 Quantum tomography

First, note that the classical tomogram previously defined can be written as a mean value

$$W(X,\mu,\nu,t) = \int f(q,p,t)\mathrm{d}q\,\mathrm{d}p\int e^{ik(X-\mu q-\nu p)}\frac{\mathrm{d}k}{2\pi}$$

$$= \left\langle\int e^{ik(X-\mu q-\nu p)}\frac{\mathrm{d}k}{2\pi}\right\rangle_f. \tag{12.5.15}$$

This formula suggests defining the quantum tomogram as the operator-valued version of such a mean value, i.e.

$$W(X, \mu, \nu, t) \equiv \left\langle \int e^{ik(\hat{X} - \mu\hat{q} - \nu\hat{p})} \frac{dk}{2\pi} \right\rangle_\rho = \mathrm{Tr}\left[\hat{\rho}(t) \int e^{ik(\hat{X} - \mu\hat{q} - \nu\hat{p})} \frac{dk}{2\pi} \right]. \quad (12.5.16)$$

This formula tells us that a quantum tomogram may be viewed as a quantum Radon transform of a quantum density state. It is physically relevant because it is possible to measure in quantum optical experiments (Smithey *et al.* 1993) the tomographic probability density.

In complete analogy with the classical case, the quantum tomogram is the marginal probability density of the random variable X, corresponding to the spectral variable of the Hermitian operator (Man'ko *et al.* 2013)

$$\hat{X}_{\mu,\nu} \equiv \mu\hat{q} + \nu\hat{p}. \quad (12.5.17)$$

Moreover, the quantum tomograms $W(X, 1, 0, t)$ and $W(X, 0, 1, t)$ are the marginal probability distributions of the position and the momentum operator, respectively, and the quantum Radon transform defining the tomogram can be inverted to reconstruct the density state according to

$$\hat{\rho}(t) = \frac{1}{2\pi} \int W(X, \mu, \nu, t) e^{i(\hat{X} - \mu\hat{q} - \nu\hat{p})} dX \, d\mu \, d\nu. \quad (12.5.18)$$

We can say that there are two main steps to define quantum tomography. First, we use Weyl operators to associate functions on phase space with an operator acting on a Hilbert space. Second, the resulting function is Radon-transformed to get a tomogram. Therefore, for both classical and quantum tomography we start with functions on phase space and consider their Radon transform. The resulting functions will be classical and quantum tomograms, respectively. Properties of the starting functions on phase space are different and they will correspond to different properties for their associated marginal distributions.

On continuing the comparison with classical theory, we can say that the evolution equation of a quantum tomogram is obtained by a quantum Radon transform of the von Neumann equation for the density operator $\hat{\rho}(t)$. It therefore reads as

$$\left[\frac{\partial}{\partial t} - \mu \frac{\partial}{\partial \nu} + i(V(q)|_{q=\tilde{q}} - \text{c.c.}) \right] W(X, \mu, \nu, t) = 0, \quad (12.5.19)$$

where

$$\tilde{q} \equiv -\frac{\partial}{\partial \mu} \left(\frac{\partial}{\partial X} \right)^{-1} + \frac{i}{2} \nu \frac{\partial}{\partial X}. \quad (12.5.20)$$

In particular, for a free particle, the above evolution equation reduces to the classical one, i.e.

$$\left[\frac{\partial}{\partial t} - \mu \frac{\partial}{\partial \nu} \right] W(X, \mu, \nu, t) = 0. \quad (12.5.21)$$

We can therefore exploit the same classical propagator the previous subsection.

12.6 Pictures of quantum mechanics for a two-level system

To compare the various pictures of quantum mechanics we have considered so far we shall analyze a two-level system. We should stress that this is not a simple academic example but it is actually used to describe the ammonia molecule in its lowest energy state, where we disregard vibrational and rotational aspects. Another two-state quantum system is that of neutral K mesons (see Section 11A.18). This consists of two states distinguished from one another by having opposite strangeness quantum numbers. Weak interactions do not preserve the strangeness quantum number and therefore provide a coupling between the two states. If we also take into account the possibility for the K^0 mesons to decay, the operator describing the evolution would not be Hermitian. In the following, we shall concentrate on the ammonia molecule.

The ammonia molecule, denoted by NH_3, consists of one nitrogen atom out of the plane formed by three hydrogen atoms lying at the vertices of a triangle. The nitrogen atom can be found above or below the plane identified by the three hydrogen atoms. These two states are associated with two isoenergetic configurations. There is a finite probability that we have a tunnelling from one configuration to another the nitrogen atom tunnels through the configuration where it would be in the same plane of the hydrogen atoms to which there is a corresponding a strong repulsion. The various aspects are described by the following energy matrix:

$$H = \begin{pmatrix} E_0 & -A \\ -A & E_0 \end{pmatrix}. \tag{12.6.1}$$

If the two basis states are denoted by $|e_1\rangle$ and $|e_2\rangle$, we have

$$\langle e_1|H|e_1\rangle = E_0 = \langle e_2|H|e_2\rangle, \tag{12.6.2}$$

$$\langle e_1|H|e_2\rangle = \langle e_2|H|e_1\rangle = -A. \tag{12.6.3}$$

In the Schrödinger picture, the equation of motion associated with this Hamiltonian H is

$$i\hbar \frac{d}{dt}|\psi\rangle = H|\psi\rangle. \tag{12.6.4}$$

By setting

$$\langle e_1|\psi\rangle = z_1, \ \langle e_2|\psi\rangle = z_2, \tag{12.6.5}$$

and

$$\langle \psi|e_1\rangle = z_1^*, \ \langle \psi|e_2\rangle = z_2^*, \tag{12.6.6}$$

we find

$$i\hbar \frac{d}{dt}\begin{pmatrix} z_1 \\ z_2 \end{pmatrix} = \begin{pmatrix} E_0 & -A \\ -A & E_0 \end{pmatrix}\begin{pmatrix} z_1 \\ z_2 \end{pmatrix} \tag{12.6.7}$$

and its Hermitian conjugate. The solution of this equation may be described by exponentiating the Hamiltonian matrix. We notice that $H = E_0 \sigma_0 - A\sigma_1$, where σ_0 and σ_1 are the commuting matrices

$$\sigma_0 = \begin{pmatrix} 1 & 0 \\ 0 & 1 \end{pmatrix}, \ \sigma_1 = \begin{pmatrix} 0 & 1 \\ 1 & 0 \end{pmatrix};$$

therefore, the motion is found by the composition of two independent motions, given by

$$e^{-i\frac{E_0 t}{\hbar}\sigma_0}, \ e^{i\frac{At}{\hbar}\sigma_1}.$$

To compute the second exponential we notice that even powers are proportional to σ_0 with coefficients having alternating signs. On combining, we find

$$\phi(t) = e^{-i\frac{E_0 t}{\hbar}} \left(\cos \frac{At}{\hbar}\sigma_0 + i \sin \frac{At}{\hbar}\sigma_1 \right). \tag{12.6.8}$$

By application to some initial state $|e_1\rangle$,

$$|\psi(t)\rangle = e^{-i\frac{E_0 t}{\hbar}} \left(\cos \frac{At}{\hbar}|e_1\rangle + i \sin \frac{At}{\hbar}|e_2\rangle \right). \tag{12.6.9}$$

In components, this state vector reads as

$$z_1(t) = e^{-i\frac{E_0 t}{\hbar}} \left(\cos \frac{At}{\hbar}z_1(0) + i \sin \frac{At}{\hbar}z_2(0) \right), \tag{12.6.10}$$

$$z_2(t) = e^{-i\frac{E_0 t}{\hbar}} \left(\cos \frac{At}{\hbar}z_2(0) + i \sin \frac{At}{\hbar}z_1(0) \right). \tag{12.6.11}$$

Another direct way to solve these simple linear equations,

$$i\hbar\frac{d}{dt}z_1 = E_0 z_1 - Az_2, \tag{12.6.12}$$

$$i\hbar\frac{d}{dt}z_2 = E_0 z_2 - Az_1, \tag{12.6.13}$$

is by decoupling the two equations; for instance, by adding and subtracting them we get

$$i\hbar\frac{d}{dt}(z_1 + z_2) = E_0(z_1 + z_2) - A(z_1 + z_2), \tag{12.6.14}$$

$$i\hbar\frac{d}{dt}(z_1 - z_2) = E_0(z_1 - z_2) + A(z_1 - z_2). \tag{12.6.15}$$

The solution may easily be written in the form

$$(z_1 + z_2)(t) = e^{-\frac{i}{\hbar}(E_0 - A)t}(z_1(0) + z_2(0)), \tag{12.6.16}$$

$$(z_1 - z_2)(t) = e^{-\frac{i}{\hbar}(E_0 + A)t}(z_1(0) - z_2(0)). \tag{12.6.17}$$

Yet another way to solve the equations is to diagonalize the energy matrix by finding the invariant eigenspaces. By setting the secular determinant to zero,

$$\det \begin{pmatrix} E_0 - \lambda & -A \\ -A & E_0 - \lambda \end{pmatrix} = 0, \tag{12.6.18}$$

i.e. $(E_0 - \lambda)^2 - A^2 = 0$, which is solved by $\lambda_1 = E_0 - A$, $\lambda_2 = E_0 + A$. The corresponding eigenvectors are $\begin{pmatrix} 1 \\ 1 \end{pmatrix}$ and $\begin{pmatrix} 1 \\ -1 \end{pmatrix}$. The unitary matrix connecting the initial energy matrix $\begin{pmatrix} E_0 & -A \\ -A & E_0 \end{pmatrix}$ with the diagonalized one $\begin{pmatrix} E_0 - A & 0 \\ 0 & E_0 + A \end{pmatrix}$ is

$$T = \frac{1}{\sqrt{2}} \begin{pmatrix} 1 & 1 \\ 1 & -1 \end{pmatrix}. \tag{12.6.19}$$

The probability distribution associated with the state $|\psi(t)\rangle$ is given by

$$\langle e_1 | \psi(t) \rangle \langle \psi(t) | e_1 \rangle = \cos^2 \frac{At}{\hbar}, \tag{12.6.20}$$

$$\langle e_2 | \psi(t) \rangle \langle \psi(t) | e_2 \rangle = \sin^2 \frac{At}{\hbar}. \tag{12.6.21}$$

The term $\cos^2 \frac{At}{\hbar}$ gives the probability of finding the molecule in the state $|e_1\rangle$ (the nitrogen atom above the plane formed by the hydrogen atoms) at time t, having assumed that the initial state was $|e_1\rangle$.

12.6.1 von Neumann picture

We now consider the description of a two-level system in the von Neumann picture, which provides a Schrödinger equation on pure states represented by rank-1 projectors. Let us start with the equations of motion on the space of states. We have seen that, using the probabilistic interpretation, quantum states are not described by vectors in the Hilbert space but rather they are associated with rank-1 projectors

$$\rho_\psi = \frac{|\psi\rangle\langle\psi|}{\langle\psi|\psi\rangle}. \tag{12.6.22}$$

It is easily seen that states, in this representation, are invariant under the transformation

$$|\psi\rangle \to \lambda e^{i\alpha} |\psi\rangle, \tag{12.6.23}$$

with $\lambda > 0$. The Schrödinger equation gives rise to the equation

$$i\hbar \frac{d}{dt} \rho_\psi = [H, \rho_\psi]. \tag{12.6.24}$$

The state in the von Neumann representation is given by

$$\rho_\psi = |e_1\rangle\langle e_1|\psi\rangle\langle\psi|e_1\rangle\langle e_1| + |e_1\rangle\langle e_1|\psi\rangle\langle\psi|e_2\rangle\langle e_2|$$
$$+ |e_2\rangle\langle e_2|\psi\rangle\langle\psi|e_2\rangle\langle e_2| + |e_2\rangle\langle e_2|\psi\rangle\langle\psi|e_1\rangle\langle e_1|. \tag{12.6.25}$$

In more familiar terms, the corresponding matrix is obtained by taking the row-by-column product of the representatives, i.e.

$$\rho_\psi = \begin{pmatrix} z_1 z_1^* & z_1 z_2^* \\ z_2 z_1^* & z_2 z_2^* \end{pmatrix}, \tag{12.6.26}$$

with $z_1 z_1^* + z_2 z_2^* = \langle \psi | \psi \rangle = 1$. We verify immediately that

$$\rho_\psi = \rho_\psi^\dagger, \ \rho_\psi \cdot \rho_\psi = \rho_\psi, \ \mathrm{Tr}\rho_\psi = 1. \tag{12.6.27}$$

Since a pure state is an Hermitian operator, we can expand it in a basis of the Hermitian matrices provided by Pauli matrices to get

$$\rho_\psi = Y_0 \sigma_0 + \vec{Y} \cdot \vec{\sigma}, \tag{12.6.28}$$

where $Y_\mu = \frac{1}{2}\mathrm{Tr}(\sigma_\mu \rho_\psi)$. We are using here the scalar product on complex-valued matrices defined by $\langle A | B \rangle \equiv \mathrm{Tr}(A^\dagger B)$. For the Pauli matrices to be orthonormal we should use

$$\langle \sigma_\mu | \sigma_\nu \rangle = \frac{1}{2}\mathrm{Tr}(\sigma_\mu \sigma_\nu). \tag{12.6.29}$$

Thus, from the condition of unit trace for ρ_ψ we find $Y_0 = \frac{1}{2}$. Moreover, the projector condition on ρ_ψ implies $Y_0^2 + \vec{Y} \cdot \vec{Y} = \frac{1}{2}$, which requires $\vec{Y} \cdot \vec{Y} = \frac{1}{4}$. In conclusion, the space of pure states ρ_ψ is in one-to-one correspondence with the points of the sphere $S^2 \subset \mathbf{R}^3$ of radius $\frac{1}{2}$. This sphere is usually called the Bloch sphere. The projection from S^3, space of normalized states, to the space of pure states, S^2, is usually called the Hopf projection.

Remark If we use a parametrization of the sphere in terms of homogeneous coordinates, say $\xi = \frac{z_1}{z_2}$, the equation corresponding to the Schrödinger equation becomes

$$i\hbar \frac{\mathrm{d}}{\mathrm{d}t} \begin{pmatrix} z_1 \\ z_2 \end{pmatrix} = \begin{pmatrix} H_{11} & H_{12} \\ H_{21} & H_{22} \end{pmatrix} \begin{pmatrix} z_1 \\ z_2 \end{pmatrix}, \tag{12.6.30}$$

which, when written in homogeneous coordinates, becomes

$$i\hbar \frac{\mathrm{d}}{\mathrm{d}t} \xi = \frac{(H_{11}z_1 + H_{12}z_2)}{z_2} - \frac{z_1}{z_2} \frac{(H_{21}z_1 + H_{22}z_2)}{z_2}$$
$$= H_{12} + (H_{11} - H_{22})\xi - \xi^2 H_{21}. \tag{12.6.31}$$

The resulting equation is non-linear and it is known in the literature as the Riccati equation.

A few considerations are now in order.

(1) It is clear that on C^2 and therefore on S^2 the equation of motion integrates to a 1-parameter group of unitary transformations because H is self-adjoint. Therefore the singularity of the equation, when written in homogeneous coordinates, is simply an artifact of the chosen coordinates, which has no physical origin. Note that this phenomenon may appear very often in the description of physical situations; for instance care must be taken in describing solutions of the Einstein equations, to allow for discrimination of true singularities from fake singularities. Further details on the non-linear equation associated with the Schrödinger equation may be found in Chaturvedi *et al.* (2007).

(2) As the equation is non-linear, it seems to result is problems with the superposition rule of solutions, which is necessary to describe interference. Again, it is clear that the non-linear equation inherits a superposition rule for solutions resulting from the description on the Hilbert space (Ercolessi *et al.* 2010). Let us now consider the Heisenberg picture.

12.6.2 Heisenberg picture

We take, as a starting point, the operators on $\mathbf{C}^2$. We recall that they can be decomposed into real and imaginary parts according to

$$A = \frac{1}{2}(A + A^\dagger) + \frac{1}{2}(A - A^\dagger) = A_{\mathrm{re}} + iA_{\mathrm{im}}, \tag{12.6.32}$$

where A_{re} and A_{im} are both Hermitian. By introducing an orthonormal basis on $\mathbf{C}^2$, say

$$|e_1\rangle = \begin{pmatrix} 1 \\ 0 \end{pmatrix}, \quad |e_2\rangle = \begin{pmatrix} 0 \\ 1 \end{pmatrix},$$

it is possible to associate with every Hermitian operator on $\mathbf{C}^2$ a 2×2 Hermitian matrix. A basis of matrices of this type is given by the identity and Pauli matrices, i.e.

$$\sigma_0 = \begin{pmatrix} 1 & 0 \\ 0 & 1 \end{pmatrix} = |e_1\rangle\langle e_1| + |e_2\rangle\langle e_2|, \tag{12.6.33}$$

$$\sigma_1 = \begin{pmatrix} 0 & 1 \\ 1 & 0 \end{pmatrix} = |e_1\rangle\langle e_2| + |e_2\rangle\langle e_1|, \tag{12.6.34}$$

$$\sigma_2 = \begin{pmatrix} 0 & -i \\ i & 0 \end{pmatrix} = i\Big(|e_2\rangle\langle e_1| - |e_1\rangle\langle e_2|\Big), \tag{12.6.35}$$

$$\sigma_3 = \begin{pmatrix} 1 & 0 \\ 0 & -1 \end{pmatrix} = |e_1\rangle\langle e_1| - |e_2\rangle\langle e_2|. \tag{12.6.36}$$

A generic Hermitian matrix can be written in the form

$$A = \frac{1}{2}\sum_{k=0}^{3}\sigma_k\mathrm{Tr}(A\sigma_k) = \begin{pmatrix} Y_A^0 + Y_A^3 & Y_A^1 - iY_A^2 \\ Y_A^1 + iY_A^2 & Y_A^0 - Y_A^3 \end{pmatrix}. \tag{12.6.37}$$

A scalar product on the space of Hermitian matrices is given by

$$\langle A|B\rangle \equiv \mathrm{Tr}(AB). \tag{12.6.38}$$

We can consider the unitary operator associated with every Hermitian matrix through the exponentiation, i.e.

$$U_t(A) \equiv e^{iAt}, \tag{12.6.39}$$

where

$$A = Y_A^0\sigma_0 + \vec{Y}_A \cdot \vec{\sigma} = Y_A^0\sigma_0 + |\vec{Y}_A|\vec{n}_A \cdot \vec{\sigma}, \tag{12.6.40}$$

and we point out that, for all $k = 1, 2, 3$,

$$U(\sigma_k) = \sigma_0\cos(t) + i\sigma_k\sin(t), \tag{12.6.41}$$

to find

$$U_t(A) = e^{itY_A^0\sigma_0}\Big[\sigma_0\cos(t|\vec{Y}_A|) + i\sin(t|\vec{Y}_A|)\vec{n}_A \cdot \vec{\sigma}\Big]. \tag{12.6.42}$$

Of course, the linear combinations of Pauli matrices with complex coefficients generate all matrices $gl(2, C)$, i.e. 2×2 matrices with complex elements. For these matrices, the scalar product reads as

$$(A|B) \equiv \mathrm{Tr}(A^\dagger B). \tag{12.6.43}$$

12.6.3 Unitary group $U(2)$

The Heisenberg differential equations associated with Hermitian operators,

$$i\hbar \frac{\mathrm{d}}{\mathrm{d}t} A = [A, H], \tag{12.6.44}$$

when integrated, give rise to 1-parameter groups of unitary operators. These constitute the unitary group $U(2)$. More precisely, the group $U(2)$ of 2×2 unitary matrices consists of the matrices

$$\begin{pmatrix} a & b \\ -b^* & a^* \end{pmatrix} \begin{pmatrix} e^{i\varphi} & 0 \\ 0 & e^{i\varphi} \end{pmatrix}, \tag{12.6.45}$$

with the complex numbers a, b such that $aa^* + bb^* = 1$. Note that the decomposition is not unique, indeed we may consider

$$\begin{pmatrix} a & b \\ -b^* & a^* \end{pmatrix} \begin{pmatrix} -1 & 0 \\ 0 & -1 \end{pmatrix},$$

and

$$\begin{pmatrix} -1 & 0 \\ 0 & -1 \end{pmatrix} \begin{pmatrix} e^{i\varphi} & 0 \\ 0 & e^{i\varphi} \end{pmatrix},$$

providing an alternative decomposition. For this reason we say that

$$U(2) = SU(2) \times \frac{U(1)}{Z_2}. \tag{12.6.46}$$

Note also that $SU(2) = S^3 \subset \mathbf{R}^4$, because the unit determinant condition becomes

$$aa^* + bb^* = \sum_{i=1}^{4} x_i^2 = 1, \tag{12.6.47}$$

having set $a = x_1 + ix_2, b = x_3 + ix_4$. A different parametrization of $U(2)$ is provided by

$$\begin{pmatrix} e^{-\frac{i}{2}\alpha} e^{-\frac{i}{2}\gamma} \cos\frac{\beta}{2} & -e^{-\frac{i}{2}\alpha} e^{\frac{i}{2}\gamma} \sin\frac{\beta}{2} \\ e^{\frac{i}{2}\alpha} e^{-\frac{i}{2}\gamma} \sin\frac{\beta}{2} & e^{\frac{i}{2}\alpha} e^{\frac{i}{2}\gamma} \cos\frac{\beta}{2} \end{pmatrix}. \tag{12.6.48}$$

The formula (12.6.48) makes it clear, as already stated in Chapter 3, that this matrix corresponds to a rotation in the space $\mathbf{R}^3$ associated with the angles (α, β, γ). The single rotations are

$$R(\alpha) = \begin{pmatrix} \cos\alpha & \sin\alpha & 0 \\ -\sin\alpha & \cos\alpha & 0 \\ 0 & 0 & 1 \end{pmatrix}, \tag{12.6.49}$$

$$R(\beta) = \begin{pmatrix} \cos\beta & 0 & -\sin\beta \\ 0 & 1 & 0 \\ \sin\beta & 0 & \cos\beta \end{pmatrix}. \tag{12.6.50}$$

This correspondence makes it clear, as we already said in Section 3.4.1, that a two-to-one homomorphism exists between $SU(2)$, the group of 2×2 unitary matrices with unit determinant, and $SO(3)$, the group of 3×3 orthogonal matrices with unit determinant.

12.6.4 A closer look at states in the Heisenberg picture

As we have seen, for the case of 2×2 matrices, pure states are elements of the Bloch sphere S^2. By taking convex combinations of pure states all possible states are obtained, as it follows easily from the spectral decomposition of an Hermitian operator which is positive and normalized, i.e.

$$\rho_\omega = \sum_n |\psi_n\rangle p_n \langle\psi_n|, \tag{12.6.51}$$

with $p_n \geq 0$, $\sum_n p_n = 1$. A sequence

$$\vec{p} = (p_1, p_2, \ldots, p_n, \ldots)$$

with $p_k \geq 0$ and $\sum_k p_k = 1$ is called a *probability vector*.

Since, for pure states, $\rho_\omega \cdot \rho_\omega = \rho_\omega$ and $\mathrm{Tr}\rho_\omega = 1$, it follows that $\mathrm{Tr}\rho_\omega^2 < 1$ if ω is not pure, and usually a trace of this kind is called the *purity parameter*. In a pure state, ρ_ψ,

$$\mathrm{Tr}\rho_\psi A = \frac{\langle\psi|A|\psi\rangle}{\langle\psi|\psi\rangle}. \tag{12.6.52}$$

Thus, we define expectation values of an observable A in a state ρ as

$$e_A(\rho) \equiv \mathrm{Tr}(\rho A). \tag{12.6.53}$$

In our running example of 2×2 matrices, the space of states will be the ball whose boundary is given by the Bloch sphere. They are characterized by

$$\rho = \frac{1}{2}(\sigma_0 + p_i \sigma_i), \tag{12.6.54}$$

with the requirement $\vec{p} \cdot \vec{p} = 1$ for purity. Indeed, from $\rho \cdot \rho = \rho$, i.e.

$$\frac{1}{4}\left(\sigma_0 + \vec{p} \cdot \vec{p}\sigma_0 + 2p_i\sigma_i\right),$$

we derive $\vec{p} \cdot \vec{p} = 1$ if it must be equal to

$$\frac{1}{2}(\sigma_0 + p_i\sigma_i).$$

On the other hand, if the state is not pure,

$$\mathrm{Tr}\rho^2 = \frac{1}{2}(1 + \vec{p} \cdot \vec{p}) < 1,$$

i.e. $\vec{p} \cdot \vec{p} < 1$.

By using the expression (12.6.51), we find that p_n is the probability to find the system in the pure state associated with $|\psi_n\rangle$. As should be clear by now, the formalism of density states is a generalization of the Schrödinger picture.

12.6.5 Weyl picture

To describe our two-level system in the Weyl picture, we consider an Abelian group G with four elements, say g_0, g_1, g_2, g_3. The multiplication table for this group is (as already seen in Section 12.2)

$$g_0 g_j = g_j g_0 = g_j, \tag{12.6.55}$$

i.e. g_0 is the identity. We further have

$$g_j g_k = |\varepsilon_{jkl}| g_l = g_k g_j, \tag{12.6.56}$$

and

$$g_j g_j = g_0, \tag{12.6.57}$$

for all $j, k \in \{1, 2, 3\}$. A Weyl map from G to unitary operators acting on $\mathcal{H} = \mathbf{C}^2$ is provided by

$$W(g_0) = \sigma_0, \quad W(g_j) = \sigma_j, \tag{12.6.58}$$

with σ_0, σ_j the Pauli matrices. It is easily seen that

$$W(g_j) W(g_k) = m_{jk} W(g_j g_k), \tag{12.6.59}$$

where m_{jk} is a complex number of unit modulus.

Complex-valued functions on the group G are a vector space, denoted by $\mathcal{F}(G, \mathbf{C})$, isomorphic with $\mathbf{C}^4$. With every complex linear operator $A : \mathbf{C}^2 \to \mathbf{C}^2$ we associate a function on G by simply setting

$$f_A(g) \equiv \frac{1}{2} \mathrm{Tr}\left(W^\dagger(g) A\right). \tag{12.6.60}$$

Conversely, with any function f on G we associate an operator A_f on $\mathbf{C}^2$ by setting

$$A_f = \sum_j f(g_j) W(g_j). \tag{12.6.61}$$

It is easy to see that the group algebra on G, when represented by the unitary operators $W(g)$, becomes a *twisted convolution algebra*, with 'twist' provided by m_{jk}. Clearly, the twisted convolution product coincides with the matrix product on $gl(2, \mathbf{C})$.

The function

$$f_A(g_j) \equiv \frac{1}{2} \mathrm{Tr}(\sigma_j A) \tag{12.6.62}$$

may be represented by the four-component vector $(a_0, \vec{a})$, where $a_0 \in \mathbf{C}, \vec{a} \in \mathbf{C}^3$. If A is Hermitian, the four-component vector is real. The product induced on these functions is given by

$$\left(f_A * f_B\right)(g_\mu) \equiv \frac{1}{2}\mathrm{Tr}(\sigma_\mu AB), \tag{12.6.63}$$

i.e.

$$(a_0, \vec{a}) * (b_0, \vec{b}) = \left[a_0 b_0 + \vec{a} \cdot \vec{b}, a_0\vec{b} + b_0\vec{a} + \mathrm{i}(\vec{a} \wedge \vec{b})\right]. \tag{12.6.64}$$

Interestingly, this product is non-local, i.e. the support of $f_A * f_B$ is not included in the intersection of the support of f_A with the support of f_B; moreover, it is non-Abelian. This product captures the essential feature of quantum mechanics, non-locality and non-commutativity. The skew product $f_A * f_B - f_B * f_A$ defines a Lie algebra and it is easily seen, in our example, that it is given by the vector product $\mathrm{i}(\vec{a} \wedge \vec{b})$ and is therefore isomorphic with the rotation Lie algebra.

There is another way to implement a Weyl-type picture, where we do not require the carrier space to be an Abelian group with the Weyl map providing a projective representation. We consider $\mathbf{R}^4$ as a set of real numbers v_0, v_1, v_2, v_3. We construct the unitary operator

$$U(v_0, v_1, v_2, v_3) = e^{(\mathrm{i}v_0\sigma_0)}\, e^{(\mathrm{i}\vec{v}\cdot\vec{\sigma})}. \tag{12.6.65}$$

By setting

$$v^j\sigma_j = vn^j\sigma_j, \quad \hat{n}v = \vec{v}, \tag{12.6.66}$$

we find

$$U(v_0, \vec{v}) = e^{\mathrm{i}v_0}\left(\cos(v)\sigma_0 + \mathrm{i}\sin(v)n^j\sigma_j\right). \tag{12.6.67}$$

Remark We are passing from the space $\mathbf{R}^4$, which may be thought of as parametrizing the Lie algebra of $U(2)$, to the association of unitary operators with points, but we do not consider neither the vector-space structure nor the Lie-algebra structure.

We can construct the association of complex-valued functions on $\mathbf{R}^4$ with operators on $\mathbf{C}^2$ by setting

$$f_A(v_0, \vec{v}) \equiv \frac{1}{2}\mathrm{Tr}\left(U^\dagger(v_0, \vec{v})A\right) = e^{-\mathrm{i}v_0}\left(a_0 \cos(v) - \mathrm{i}(\vec{a} \cdot \vec{n}) \sin(v)\right), \tag{12.6.68}$$

and it is clear that the association $A \to f_A$ is one-to-one. Indeed, by fixing different values for v_0, v_1, v_2, v_3 it is possible to find enough linear operations whose solutions provide us with a_0, a_1, a_2, a_3. By using this one-to-one correspondence we may define a non-local product on these functions by taking the function which corresponds to the product of the operators. We have

$$(f_A * f_B)(v_0, \vec{v}) = f_{AB}(v_0, \vec{v}) = \frac{1}{2}\mathrm{Tr}(ABU^\dagger(v_0, \vec{v})), \tag{12.6.69}$$

and we find

$$(f_A * f_B)(v_0, \vec{v}) = e^{-\mathrm{i}v_0}\left\{\cos(v)(a_0 b_0 + \vec{a} \cdot \vec{b})\right.$$
$$\left. - \mathrm{i}\sin(v)\left[a_0(\vec{b} \cdot \vec{n}) + b_0(\vec{a} \cdot \vec{n}) + \mathrm{i}\vec{n} \cdot (\vec{a} \wedge \vec{b})\right]\right\}. \tag{12.6.70}$$

Since any operator may be decomposed in terms of Pauli matrices, it is useful to consider the functions associated with them, i.e.

$$f_{\sigma_0}(v_0, \vec{v}) = e^{-iv_0} \cos(v), \qquad (12.6.71)$$

$$f_{\sigma_1}(v_0, \vec{v}) = e^{-iv_0} n_1 \sin(v), \qquad (12.6.72)$$

$$f_{\sigma_2}(v_0, \vec{v}) = ie^{-iv_0} n_2 \sin(v), \qquad (12.6.73)$$

$$f_{\sigma_3}(v_0, \vec{v}) = ie^{-iv_0} n_3 \sin(v). \qquad (12.6.74)$$

The commutator bracket, i.e. the Moyal bracket necessary to write the equations of motion in this formalism, is given by

$$\{f_A, f_B\}(v) = -\sin(v)e^{-iv_0}\left(\vec{n} \cdot \vec{a} \wedge \vec{b}\right). \qquad (12.6.75)$$

We stress again that this classical looking bracket should not be confused with a classical picture. Indeed, here the relevant associative product is non-local.

12.6.6 Probability distributions and states

Let us now discuss the way we reconstruct states out of fair probability distributions. We have seen that, given a resolution of the identity and a state ρ_ψ, it is possible to construct a probability distribution. Given a resolution of the identity, say

$$I = |1\rangle\langle 1| + |2\rangle\langle 2|, \qquad (12.6.76)$$

we find

$$p_\psi(1) = \langle \psi|1\rangle\langle 1|\psi\rangle = \mathrm{Tr}\left(\rho_\psi |1\rangle\langle 1|\right), \qquad (12.6.77)$$

$$p_\psi(2) = \langle \psi|2\rangle\langle 2|\psi\rangle = \mathrm{Tr}\left(\rho_\psi |2\rangle\langle 2|\right), \qquad (12.6.78)$$

where we are using $\langle \psi|\psi\rangle = 1$.

How many different probability distributions do we need to be able to reconstruct ρ_ψ? In our simple case it is clear that, if we have a family of orthonormal eigenvectors, say associated with a matrix

$$A(\theta, \phi) = \begin{pmatrix} \cos\theta & e^{-i\phi}\sin\theta \\ e^{i\phi}\sin\theta & -\cos\theta \end{pmatrix}, \qquad (12.6.79)$$

where $A \equiv \vec{n} \cdot \vec{\sigma}$, $\vec{n}$ is the vector

$$\vec{n} = (\sin\theta\cos\phi, \sin\theta\sin\phi, \cos\theta), \qquad (12.6.80)$$

we have

$$|+, \theta, \phi\rangle = e^{-i\frac{\phi}{2}}\cos\frac{\theta}{2}|1\rangle + e^{i\frac{\phi}{2}}\sin\frac{\theta}{2}|2\rangle, \qquad (12.6.81)$$

$$|-, \theta, \phi\rangle = e^{-i\frac{\phi}{2}} \sin\frac{\theta}{2}|1\rangle - e^{i\frac{\phi}{2}} \sin\frac{\theta}{2}|2\rangle, \tag{12.6.82}$$

where $+$ and $-$ denote the eigenvalues $+1$ and -1, respectively. The vectors $|1\rangle$ and $|2\rangle$ are the vectors we use to represent the Pauli operators as matrices. The corresponding projectors are

$$|+, \theta, \phi\rangle\langle\theta, \phi, +| = \begin{pmatrix} \cos^2\frac{\theta}{2} & e^{-i\phi}\cos\frac{\theta}{2}\sin\frac{\theta}{2} \\ e^{i\phi}\sin\frac{\theta}{2}\cos\frac{\theta}{2} & \sin^2\frac{\theta}{2} \end{pmatrix}, \tag{12.6.83}$$

$$|-, \theta, \phi\rangle\langle\theta, \phi, -| = \begin{pmatrix} \sin^2\frac{\theta}{2} & -e^{-i\phi}\cos\frac{\theta}{2}\sin\frac{\theta}{2} \\ -e^{i\phi}\sin\frac{\theta}{2}\cos\frac{\theta}{2} & \cos^2\frac{\theta}{2} \end{pmatrix}. \tag{12.6.84}$$

We notice that each projector may be written in the form

$$P(\theta, \phi) = \frac{1}{2}(\sigma_0 + \vec{n}\cdot\vec{\sigma}) = \frac{1}{2}\begin{pmatrix} 1+\cos\theta & e^{-i\phi}\sin\theta \\ e^{i\phi}\sin\theta & 1-\cos\theta \end{pmatrix}. \tag{12.6.85}$$

The reconstruction of the state out of given probability distributions requires that the probability distributions should form a 'kind of basis'. The space of Hermitian 2×2 matrices is four-dimensional; if we write a basis in terms of rank-1 projectors we need four of them; thus, the associated probability distributions will be independent. Since a state has to satisfy the condition $\mathrm{Tr}\rho_\psi = 1$, it follows that the number of parameters required to reconstruct it will be only three. Thus, by fixing different values of θ and ϕ so that they define at least three independent Hermitian operators, we will be able, with the corresponding probability distributions, to reconstruct the state. We remark that probability distributions may be composed using convex combinations, and they form a convex body. Extremal points correspond to pure states while convex combinations correspond to mixed states.

12.6.7 Ehrenfest picture

Let us now discuss the picture in terms of expectation values. With every operator A, if Hermitian, we associate the decomposition

$$A = a_0\sigma_0 + a_j\sigma_j, \tag{12.6.86}$$

with $a_0, a_1, a_2, a_3 \in \mathbf{R}$. The corresponding expectation value

$$e_A(\psi) = \frac{\langle\psi|A|\psi\rangle}{\langle\psi|\psi\rangle} \tag{12.6.87}$$

can be written for $\psi \in \mathcal{H} \equiv \mathbf{C}^2$ by setting

$$\langle e_1|\psi\rangle = z_1 = x_1 + iy_1, \tag{12.6.88}$$

$$\langle e_2|\psi\rangle = z_2 = x_2 + iy_2. \tag{12.6.89}$$

Once more, it is enough to consider the expectation values associated with Pauli matrices, so that

$$e_{\sigma_0}(\psi) = 1, \tag{12.6.90}$$

$$e_{\sigma_1}(\psi) = \frac{2(x_1 x_2 + y_1 y_2)}{(x_1^2 + y_1^2 + x_2^2 + y_2^2)}, \tag{12.6.91}$$

$$e_{\sigma_2}(\psi) = \frac{2(y_2 x_1 - y_1 x_2)}{(x_1^2 + y_1^2 + x_2^2 + y_2^2)}, \tag{12.6.92}$$

$$e_{\sigma_3}(\psi) = \frac{x_1^2 + y_1^2 - (x_2^2 + y_2^2)}{(x_1^2 + y_1^2 + x_2^2 + y_2^2)}. \tag{12.6.93}$$

For a generic A,

$$e_A(\psi) = a_0 + \frac{2}{(x_1^2 + y_1^2 + x_2^2 + y_2^2)} \Big[a_1(x_1 x_2 + y_1 y_2) + a_2(y_2 x_1 - y_1 x_2)$$
$$+ a_3(x_1^2 + y_1^2 - (x_2^2 + y_2^2)) \Big]. \tag{12.6.94}$$

With these expectation-value functions we construct the Ehrenfest picture, and it is possible to describe the dynamics using Poisson brackets. Before discussing the equations of motion, notice that the knowledge of the expectation values for $\sigma_0, \sigma_1, \sigma_2, \sigma_3$ makes it possible for us to reconstruct the state, i.e. x_1, y_1, x_2, y_2, with the constraint

$$x_1^2 + y_1^2 + x_2^2 + y_2^2 = 1, \tag{12.6.95}$$

and that corresponding to our previous condition $y_0 = \frac{1}{2}$. We stress that, even though the expectation-value functions are defined on the 2-sphere, the space of pure states, we are describing them on $\mathbf{R}^4$ with the constraint (12.6.95). The description on the vector space $\mathbf{R}^4$ is more convenient from the point of view of differential calculus, because we can consider the usual differential calculus on $\mathbf{R}^4$ without using constraints or, equivalently, different charts to perform our computations. Thus, the natural Poisson bracket on $\mathbf{R}^4$, which would be

$$\{x_1, y_1\} = 1 = \{x_2, y_2\}, \tag{12.6.96}$$

$$\{x_1, x_2\} = \{y_1, y_2\} = \{x_1, y_2\} = \{x_2, y_1\} = 0, \tag{12.6.97}$$

should be amended to ensure that, when applied to two expectation-value functions, it again provides a function of the same nature. To achieve this we have to define

$$\{x_1, y_1\} = \{x_2, y_2\} \equiv x_1^2 + x_2^2 + y_1^2 + y_2^2, \tag{12.6.98}$$

and use the property according to which, for any Hermitian operator A,

$$\left\{ \frac{\langle \psi | A | \psi \rangle}{\langle \psi | \psi \rangle}, \langle \psi | I | \psi \rangle \right\} = 0. \tag{12.6.99}$$

In this way, even though the previous Poisson bracket does not satisfy the Jacobi identity on generic functions on $\mathbf{R}^4$, it does indeed satisfy the Jacobi identity *when it is restricted to expectation-value functions of Hermitian operators*. In general it is found that the Poisson bracket of two expectation values is given by

$$\{e_A, e_B\} = ie_{[A,B]}. \tag{12.6.100}$$

This result can be derived by starting with the Schrödinger equation applied to $e_A(\psi)$. Then,

$$\frac{\mathrm{d}}{\mathrm{d}t}e_A(\psi) = [\langle\psi|\psi\rangle]^{-1}\left[\langle\psi| - \frac{H}{i\hbar}A|\psi\rangle + \langle\psi|A\frac{H}{i\hbar}|\psi\rangle\right]$$
$$= ie_{[A,H]}(\psi). \tag{12.6.101}$$

Clearly, the denominator $\langle\psi|\psi\rangle$ behaves like a numerical quantity with respect to derivations associated with Hermitian operators. From the stated Poisson brackets we find the equations of motion in the form

$$\frac{\mathrm{d}}{\mathrm{d}t}e_A = \{e_H, e_A\}. \tag{12.6.102}$$

In this formulation, equations of motion, when written in terms of expectation values, can be written by using only a classical-type object like the Poisson bracket. We should bear in mind, however, that in light of our definition it provides derivations for the $*$-product, which is non-local and non-commutative as we have remarked. We may wonder how and if this picture is equivalent to the standard Heisenberg picture. We first point out that, by giving the expectation values of A as a function on the space of states, it is possible to recover A exactly in the same way that we recovered states out of probability distributions. Indeed, starting from

$$e_A(\psi) = \mathrm{Tr}(A\rho_\psi), \tag{12.6.103}$$

we realize immediately that, when we consider sufficiently many states so as to obtain a basis of Hermitian operators, the 'components' $e_A(\psi)$ will make it possible to reconstruct the 'vector' within the vector space of operators. The Poisson bracket allows the reconstruction of the skew-symmetric bracket operator; therefore, to reconstruct the full algebra structure of the space of operators we also need the symmetrized product $AB + BA$ for any two Hermitian operators A, B. This suggests that in addition to the skew-symmetric product given by the Poisson bracket, we also introduce a symmetrized product

$$(e_A, e_B) \equiv e_{AB+BA}. \tag{12.6.104}$$

It is possible to write the symmetrized product by means of a symmetric bidifferential operator, counterpart of the skew-symmetric bidifferential operator describing the Poisson bracket. If we denote it by G, instead of Λ as we did for the Poisson bracket, we have

$$G(\mathrm{d}e_A, \mathrm{d}e_B) \equiv (e_A, e_B) - e_A \cdot e_B. \tag{12.6.105}$$

This relation determines G completely, as is easily seen by considering the symmetrized product of expectation-value functions associated with rank-1 projectors. The two bidifferential operators, when considered as tensor fields in contravariant form, define an Hermitian tensor field $G + i\Lambda$, which is the tensorial version of the Hermitian product on the space of states. These considerations form the basis of the geometrical formulation of quantum mechanics on the space of states (Ercolessi *et al.* 2010 and references therein, Marmo and Volkert 2010). If we carry on the various steps, we find

$$G = \left(x_1^2 + y_1^2 + x_2^2 + y_2^2\right) \sum_{j=1}^{2} \left(\frac{\partial}{\partial x_j} \otimes \frac{\partial}{\partial x_j} + \frac{\partial}{\partial y_j} \otimes \frac{\partial}{\partial y_j}\right)$$
$$- \sum_{l,m} \left(x^l \frac{\partial}{\partial x^l} \otimes x^m \frac{\partial}{\partial x^m} + y^l \frac{\partial}{\partial y^l} \otimes y^m \frac{\partial}{\partial y^m}\right)$$
$$- \sum_{l,m} \left(y^l \frac{\partial}{\partial x^l} - x^l \frac{\partial}{\partial y^l}\right) \otimes \left(y^m \frac{\partial}{\partial x^m} - x^m \frac{\partial}{\partial y^m}\right). \tag{12.6.106}$$

Similarly, for Λ one finds

$$\Lambda = \left(x_1^2 + y_1^2 + x_2^2 + y_2^2\right) \sum_{j=1}^{2} \left(\frac{\partial}{\partial y_j} \otimes \frac{\partial}{\partial x_j} - \frac{\partial}{\partial x_j} \otimes \frac{\partial}{\partial y_j}\right)$$
$$- \sum_{l,m} \left[\left(y^l \frac{\partial}{\partial y^l} + x^l \frac{\partial}{\partial x^l}\right) \otimes \left(y^m \frac{\partial}{\partial x^m} - x^m \frac{\partial}{\partial y^m}\right)\right.$$
$$\left. - \left(y^m \frac{\partial}{\partial x^m} - x^m \frac{\partial}{\partial y^m}\right) \otimes \left(y^l \frac{\partial}{\partial y^l} + x^l \frac{\partial}{\partial x^l}\right)\right]. \tag{12.6.107}$$

The symmetric tensor defines a new algebra structure on the space of expectation values, which is called a Jordan algebra and was considered by the founding fathers of quantum mechanics (Jordan *et al.* 1934).

This product plays an important role in formulating uncertainty relations in a covariant form (Robertson 1930). We consider the inequality

$$\langle \psi | F^\dagger F | \psi \rangle \geq 0, \tag{12.6.108}$$

which holds for an arbitrary operator F. Let us consider the particular operator

$$F = C_1 Q + C_2 P, \tag{12.6.109}$$

where C_1, C_2 are complex numbers and Q, P are canonical operators representing position and momentum operators. The expectation value for $F^\dagger F$ may be re-written in the form

$$C_1^* C_2 \langle QP \rangle_\psi + C_1 C_2^* \langle PQ \rangle_\psi + C_1 C_1^* \langle Q \rangle_\psi + C_2 C_2^* \langle P \rangle_\psi \geq 0, \tag{12.6.110}$$

for all complex values of C_1, C_2. We use the identity

$$\langle PQ \rangle_\psi = \frac{1}{2} \langle PQ + QP \rangle_\psi + \frac{1}{2} \langle PQ - QP \rangle_\psi, \tag{12.6.111}$$

and the positivity of the quadratic form to find

$$\det \begin{pmatrix} \langle Q^2 \rangle_\psi & \frac{1}{2} \langle QP + PQ \rangle_\psi + \frac{i}{2} \hbar \\ \frac{1}{2} \langle QP + PQ \rangle_\psi - \frac{i}{2} \hbar & \langle P^2 \rangle_\psi \end{pmatrix} \geq 0, \tag{12.6.112}$$

i.e.

$$\langle Q^2 \rangle_\psi \langle P^2 \rangle_\psi - \frac{1}{4} \langle QP + PQ \rangle_\psi \geq \frac{1}{4} \hbar. \tag{12.6.113}$$

If we further replace Q with $Q - \langle Q \rangle_\psi$ and P with $P - \langle P \rangle_\psi$, the previous relation becomes the Robertson–Schrödinger uncertainty relation

$$\sigma_{QQ}\,\sigma_{PP} - \sigma_{PQ}^2 \geq \frac{1}{4}\hbar, \tag{12.6.114}$$

where the variances of position and momentum are

$$\sigma_{QQ} \equiv \langle Q^2 \rangle_\psi - \langle Q \rangle_\psi^2, \tag{12.6.115}$$

$$\sigma_{PP} \equiv \langle P^2 \rangle_\psi - \langle P \rangle_\psi^2, \tag{12.6.116}$$

$$\sigma_{QP} \equiv \frac{1}{2}\langle QP + PQ \rangle_\psi - \langle Q \rangle_\psi \langle P \rangle_\psi, \tag{12.6.117}$$

or, equivalently,

$$\sigma_{QQ} \equiv G(\mathrm{d}e_Q, \mathrm{d}e_Q), \tag{12.6.118}$$

$$\sigma_{PP} \equiv G(\mathrm{d}e_P, \mathrm{d}e_P), \tag{12.6.119}$$

$$\sigma_{QP} \equiv G(\mathrm{d}e_P, \mathrm{d}e_Q). \tag{12.6.120}$$

We stress that *the space of pure states, being a manifold rather than a vector space, requires the formulation of quantum mechanics in geometrical terms*, rather than by means of vectors and operators.

12.7 Composite systems

If we consider two systems $\mathcal{S}_1$ and $\mathcal{S}_2$, with states described by the complex Hilbert spaces $\mathcal{H}_1$ and $\mathcal{H}_2$, respectively, the composite system $(\mathcal{S}_1, \mathcal{S}_2)$ is described by the tensor product $\mathcal{H}_1 \otimes \mathcal{H}_2$. We recall that the tensor product may be described in terms of an orthonormal basis for $\mathcal{H}_1$, say $|e_j\rangle$, and an orthonormal basis for $\mathcal{H}_2$, say $|f_k\rangle$, by means of a basis given by all possible pairs

$$|e_j\rangle |f_k\rangle.$$

Such a basis is also denoted by $|e_j, f_k\rangle$ or $|e_j\rangle \otimes |f_k\rangle$. A generic state will be associated with a matrix $\|a_{jk}\|$ according to

$$\rho = \sum_{j,k} a_{jk} |e_j\rangle \otimes |f_k\rangle. \tag{12.7.1}$$

This condition is fulfilled linearity of the tensor product is required with respect to both arguments.

A state

$$|\psi\rangle \otimes |\varphi\rangle = \sum_{j,k} (a_j|e_j\rangle) \otimes (b_k|f_k\rangle) = \sum_{j,k} a_j b_k |e_j, f_k\rangle, \tag{12.7.2}$$

i.e. where $a_{jk} = a_j b_k$, is called a separable state. When a_{jk} cannot be factorized in this way the state ρ is said to be *entangled*. An entangled state of a composite system is a state for which a single state vector cannot be ascribed to either of the components. Entanglement is one of the main features of quantum mechanical systems to be used in quantum computation. As stressed by Schrödinger, entanglement is not *one* but *the* main feature of quantum mechanics.

12.7.1 Inner product in tensor spaces

In the tensor product $\mathcal{H}_1 \otimes \mathcal{H}_2$ it is possible to define a scalar product out of the products available in $\mathcal{H}_1$ and $\mathcal{H}_2$. For separable states, e.g. $|\psi_1\rangle \otimes |\varphi_1\rangle$, $|\psi_2\rangle \otimes |\varphi_2\rangle$,

$$h\Big(|\psi_1\rangle \otimes |\varphi_1\rangle, |\psi_2\rangle \otimes |\varphi_2\rangle\Big) = \langle\psi_1|\psi_2\rangle\langle\varphi_1|\varphi_2\rangle, \qquad (12.7.3)$$

and then we extend by (sesqui)linearity to any combination. If the tensor products in Eq. (12.7.3) can be written as

$$|\psi_1\rangle \otimes |\varphi_1\rangle = \sum_{j,k} a_{j,k} |e_j\rangle \otimes |f_k\rangle, \qquad (12.7.4)$$

$$|\psi_2\rangle \otimes |\varphi_2\rangle = \sum_{j,k} b_{j,k} |e_j\rangle \otimes |f_k\rangle, \qquad (12.7.5)$$

the left-hand side of Eq. (12.7.3) becomes equal to

$$\sum_{j,k,l,m} a_{j,k}^* b_{l,m} \langle e_j|e_l\rangle \langle f_k|f_m\rangle = \sum_{j,k} a_{j,k}^* b_{j,k}.$$

12.7.2 Complex linear operators in tensor spaces

If we consider linear maps

$$A : \mathcal{H}_1 \rightarrow \mathcal{F}_1, \ B : \mathcal{H}_2 \rightarrow \mathcal{F}_2,$$

we may build the linear map

$$A \otimes B : \mathcal{H}_1 \otimes \mathcal{H}_2 \rightarrow \mathcal{F}_1 \otimes \mathcal{F}_2$$

by setting

$$(A \otimes B)|\psi_1\rangle \otimes |\varphi_1\rangle \equiv (A|\psi_1\rangle) \otimes (B|\varphi_1\rangle), \qquad (12.7.6)$$

and then extending the action on general states by linearity. A product can be constructed on these operators when their domain and codomain coincide: $\mathcal{H}_1 = \mathcal{F}_1, \mathcal{H}_2 = \mathcal{F}_2$, so that

$$(A_1 \otimes B_1)(A_2 \otimes B_2) = (A_1 \cdot A_2) \otimes (B_1 \cdot B_2). \qquad (12.7.7)$$

By using again our running example, we construct the tensor product of $\mathbf{C}^2$ with itself by using in both of them a basis

$$|e_1\rangle = |f_1\rangle = \begin{pmatrix} 1 \\ 0 \end{pmatrix}, \ |e_2\rangle = |f_2\rangle = \begin{pmatrix} 0 \\ 1 \end{pmatrix}. \qquad (12.7.8)$$

If we consider

$$|\psi_1\rangle = \frac{1}{\sqrt{2}}(|e_1\rangle + |e_2\rangle), \qquad (12.7.9)$$

and

$$|\varphi_1\rangle = i|f_1\rangle, \qquad (12.7.10)$$

we find

$$|\psi_1, \varphi_1\rangle = |\psi_1\rangle \otimes |\varphi_1\rangle = \frac{i}{\sqrt{2}}(|e_1\rangle \otimes |f_1\rangle + |e_2\rangle \otimes |f_1\rangle). \qquad (12.7.11)$$

The state

$$|\Sigma\rangle = \frac{1}{\sqrt{2}}(|e_1\rangle \otimes |f_1\rangle + |e_2\rangle \otimes |f_2\rangle) \qquad (12.7.12)$$

is an entangled state and cannot be expressed as a product state.

Remark   A product state $|\psi_1\rangle \otimes |\varphi_1\rangle$ may be considered as a linear map from the dual of $\mathcal{H}_1$ into $\mathcal{H}_2$. Indeed, the map

$$(|\psi_1\rangle \otimes |\varphi_1\rangle) : \mathcal{H}_1^* \to \mathcal{H}_2 \qquad (12.7.13)$$

is given by

$$\langle\psi| \to \langle\psi|\psi_1\rangle|\varphi_1\rangle. \qquad (12.7.14)$$

Thus, a state of a bipartite system (i.e. a system built out of two subsystems) may be represented by a matrix once an orthonormal basis has been chosen. For the previous example,

$$|\psi_1, \varphi_1\rangle = \frac{i}{\sqrt{2}} \begin{pmatrix} 1 & 0 \\ 1 & 0 \end{pmatrix}, \qquad (12.7.15)$$

$$|\Sigma\rangle = \frac{1}{\sqrt{2}} \begin{pmatrix} 1 & 0 \\ 0 & 1 \end{pmatrix}. \qquad (12.7.16)$$

12.7.3 Composite systems and Kronecker products

Our previous remark concerning tensor products and bilinear (or multilinear) maps shows that, out of a representation of vectors by means of columns, say, we may associate a matrix with a composite system. If

$$|\psi_1\rangle = \begin{pmatrix} a_1 \\ a_2 \end{pmatrix}, \ |\varphi\rangle = \begin{pmatrix} b_1 \\ b_2 \end{pmatrix}, \qquad (12.7.17)$$

we find

$$|\psi_1\rangle \otimes |\varphi_1\rangle = \begin{pmatrix} a_1 b_1 & a_1 b_2 \\ a_2 b_1 & a_2 b_2 \end{pmatrix}. \qquad (12.7.18)$$

More generally, for two matrices A and B the Kronecker product is defined by

$$A \otimes B = \begin{pmatrix} a_{11}B & a_{12}B & \dots & a_{1n}B \\ a_{21}B & a_{22}B & \dots & a_{2n}B \\ \dots & \dots & \dots & \dots \\ a_{m1}B & a_{m2}B & \dots & a_{mn}B \end{pmatrix}. \tag{12.7.19}$$

Some relevant formulae are

$$(A \otimes B) \cdot (C \otimes D) = (AC \otimes BD), \tag{12.7.20}$$

$$(A \otimes B)^{\dagger} = A^{\dagger} \otimes B^{\dagger}, \tag{12.7.21}$$

$$\text{Tr}(A \otimes B) = (\text{Tr}\,A)(\text{Tr}\,B), \tag{12.7.22}$$

$$e^{A \otimes I + I \otimes B} = e^{A} \otimes e^{B}. \tag{12.7.23}$$

We may wonder why in quantum mechanics composite systems are described by tensor products $\mathcal{H}_1 \otimes \mathcal{H}_2$ while in classical mechanics we would considered the Cartesian or direct product, i.e. the sum of vector spaces. An intuitive argument is as follows. If we considered each system described by the same Hamiltonian operator with non-degenerate eigenvalues, e.g. with n_1 eigenvalues for H_1 and n_2 eigenvalues for H_2, respectively, we would have for the composite system a number of possible eigenvalues given by $n_1 n_2$. Thus, the total space would have dimensions $n_1 n_2$ and not $(n_1 + n_2)$.

12.8 Identical particles

12.8.1 Product basis

Let us consider a system consisting of different parts, e.g. a multiparticle system, where we denote for simplicity the different components with A, B, C etc. If the Hamiltonian $\widehat{H}$ of the system can be decomposed as a sum of terms corresponding to A, B, C etc., respectively

$$\widehat{H} = \widehat{h}_A \otimes I \otimes I \otimes \cdots \otimes I + I \otimes \widehat{h}_B \otimes I \otimes \cdots \otimes I + I \otimes I \otimes \widehat{h}_C \otimes \cdots \otimes I + \cdots, \tag{12.8.1}$$

this Hamiltonian describes independent components (hereafter considered for simplicity as particles), because it lacks coupling terms $\widehat{h}_{AB}$ that depend on A and B simultaneously. In this case the eigenstates of this *independent-particle* Hamiltonian can always be chosen to be product states, for which

$$\widehat{H}|\Psi\rangle = (E_A + E_B + E_C + \cdots)|\psi_A\rangle \otimes |\psi_B\rangle \otimes |\psi_C\rangle \cdots. \tag{12.8.2}$$

Such *product states* describe a system of particles that are independent of, or uncorrelated with, one another. As a consequence of this, a measurement on A will not affect the state of B, and vice versa. Therefore, product states are also called *independent particle* states, which is appropriate since they describe the state of each particle independently. Nevertheless, even though a product state describes a multiparticle system, it is not mandatory that every multiparticle wavefunction should be written in this form. Let us consider for example a multiparticle state consisting of the linear combination of two product states, i.e.

$$|\Psi\rangle = c_1 |\psi_A^1\rangle \otimes |\psi_B^1\rangle \otimes |\psi_C^1\rangle \cdots + c_2 |\psi_A^2\rangle \otimes |\psi_B^2\rangle \otimes |\psi_C^2\rangle \cdots . \tag{12.8.3}$$

For this state, a measurement on A affects the outcome of measurements on B, C and D: if we find that A is in state 1, then B, C and D must collapse to state 1, while A in state 2 implies B, C and D will collapse to state 2. This correlation between different particles is called entanglement and plays a very important role in multiparticle quantum mechanics.

12.8.2 Exchange symmetry

Two particles are called identical when there is no way to distinguish one from the another, even in principle. In practice, systems of identical particles can consist of elementary particles, and, in some particular cases, of composite states such as atoms and molecules. There are two main categories of identical particles: bosons, which can share quantum states, and fermions, which do not share quantum states by virtue of the Pauli exclusion principle.

Two particles can be distinguished by using the possible difference in their intrinsic physical characteristics, such as mass, electric charge, and spin. Nevertheless, in the microscopic world we may have to consider systems consisting of particles with identical intrinsic properties; in this case there remains a second method for distinguishing between particles, which is to track the trajectory of each particle. In classical mechanics, since each particle can be precisely followed through its trajectory, once their identity is fixed at a starting time it is kept unchanged during the whole motion. This does not occur in quantum mechanics when the distance between particles is short enough that their wave functions overlap in a substantial way. In this case the single identity is lost, since we have a common region in space where both particles can be detected. For this reason we expect that in quantum mechanics, if a system of two particles is represented by the full wave function $\psi(x_1, x_2)$, the operation of exchanging the particles does not have to modify the physical observables on the system.

In order to state clearly the previous concept, let us consider a system of N identical particles where each of them can be found in a state among the set $|\psi_1\rangle, \ldots, |\psi_M\rangle$ with $M \neq N$ in general. If the N particles were distinguishable, one might associate to the global system the quantum state

$$|\psi_{p_1}\rangle \ldots |\psi_{p_N}\rangle, \tag{12.8.4}$$

belonging to the Hilbert space which is the n-fold tensor product of the single-particle Hilbert space, i.e. $\mathcal{H} \otimes \cdots \otimes \mathcal{H}$. On this state one can consider the action of an operator P_ξ, which acts by exchanging the state of each single particle in (12.8.4) according to a permutation ξ of integers $1, \ldots, N$, i.e.

$$P_\xi |\psi_{p_1}\rangle \ldots |\psi_{p_N}\rangle = |\psi_{p_{\xi(1)}}\rangle \ldots |\psi_{p_{\xi(N)}}\rangle. \tag{12.8.5}$$

Note that $P_\xi^2 = I$ for any ξ, hence such operators can have eigenvalues $\lambda_P = \pm 1$ only. We define an eigenstate of P_ξ as symmetric or antisymmetric under the permutation ξ if it corresponds to the eigenvalue $+1$ or -1, respectively.

According to the previous considerations, the action of P_ξ cannot modify the state of a system in a physically relevant way, i.e. it represents a symmetry for it. In this case the energy eigenvectors appear as P_ξ eigenvectors too, which can be split into symmetric (S) or antisymmetric (A) states, i.e.

$$|\psi_{p_1}, \ldots, \psi_{p_N}; S\rangle = \sqrt{\frac{N_1! \ldots N_m!}{N!}} \sum_{\xi'} |\psi_{p_{\xi(1)}}\rangle \ldots |\psi_{p_{\xi(N)}}\rangle, \qquad (12.8.6)$$

where the sum for ξ' is taken over all permutations producing different states, and $N_1, \ldots, N_m$ denote the numbers of times in which in the sequence $p_1, \ldots, p_N$ a state appears ($m \leq N$ and $\sum_{i=1}^m N_i = N$). In the antisymmetric case we have instead

$$|\psi_{p_1}, \ldots, \psi_{p_N}; A\rangle = \sqrt{\frac{1}{N!}} \sum_{\xi} \mathrm{sgn}(\xi) |\psi_{p_{\xi(1)}}\rangle \ldots |\psi_{p_{\xi(N)}}\rangle, \qquad (12.8.7)$$

where the sum for ξ is taken over all permutations. Note that in this case, unlike the symmetric one, each particle must have a different quantum state, i.e. if two particles were sharing the same state this would imply a vanishing multiparticle wavefunction (12.8.7). This property, which is the consequence of the antisymmetric behaviour only under a generic permutation of particle labels in Eq. (12.8.7), it is also known as *Pauli's exclusion principle*. This feature of Eq. (12.8.7) is more evident by observing that a multiparticle wave function corresponding to this expression can also be written as the well-known Slater determinant

$$\Psi^A_{p_1 p_2 \ldots p_N}(q_1, q_2, \ldots, q_N) \equiv \langle q_1, q_2, \ldots, q_n | \psi_{p_1}, \ldots, \psi_{p_N}; A\rangle$$

$$= \sqrt{\frac{1}{N!}} \begin{vmatrix} \psi_{p_1}(q_1) & \psi_{p_1}(q_2) & \ldots & \psi_{p_1}(q_N) \\ \psi_{p_2}(q_1) & \psi_{p_2}(q_2) & \ldots & \psi_{p_2}(q_N) \\ \ldots & \ldots & \ldots & \ldots \\ \psi_{p_N}(q_1) & \psi_{p_N}(q_2) & \ldots & \psi_{p_N}(q_N) \end{vmatrix}. \qquad (12.8.8)$$

A rather profound result holds, which is related to the above considerations, called the *spin statistics theorem* that proves two properties:

(1) All particles fall into two classes; fermions, which are always antisymmetric under exchange, and bosons, which are always symmetric under exchange.

(2) Fermions always have half-odd spin (e.g. electrons, protons, He^3, etc.). Bosons always have integer spin (e.g. photons, He^4, etc.), showing that this theorem is definitely beyond the soope of this book; hence, showing that we will just take it as given.

12.8.3 Exchange interaction

In the previous section we parametrized the single-particle wave functions as depending on general coordinates $q_1, \ldots, q_N$. In a generic quantum system, by virtue of the presence of spin, we know that for a single particle we are generally forced to consider both spatial $\vec{r}_i$ and spin s_i coordinates. However, as long as we do not consider the presence of magnetic fields, and neglect spin–orbit-like terms, we can forget about the spin wave function of the system and just focus on the purely spatial part. This means that the solution of the Schrödinger equation can be found in the factorized form

$$\Psi(q_1, \ldots, q_N) = \chi(s_1, \ldots, s_N)\, \psi(\vec{r}_1, \ldots, \vec{r}_N), \qquad (12.8.9)$$

where χ denotes the spin wave function, whereas ψ is the orbital one. Even though the spin wave function does not play a role in determining directly the energy of the state via a corresponding term in the Hamiltonian, it does, however, affect the energy level via the indistinguishability principle of identical particles.

Let us discuss such a property by considering a simple system consisting of 2 identical particles. If the total Hamiltonian is $\widehat{H} = \widehat{h}_1 \otimes \widehat{I}_2 + \widehat{I}_1 \otimes \widehat{h}_2$ with the single-particle terms identical, the energy eigenvectors can be found by using global wave functions with a definite property of symmetry under particle exchange $(1 \leftrightarrow 2)$, i.e. symmetric or antisymmetric functions. If the two particles have vanishing spin, since they are bosons, their global wave function $\Psi(\vec{r}_1, \vec{r}_2)$ must be symmetric under the particle exchange $1 \leftrightarrow 2$. This means that all total energy eigenvalues, corresponding to antisymmetric wave functions, are forbidden to the system. In particular, in the centre of mass reference frame, the spatial coordinates of the two particles turn out to be $\vec{r}_1 = -\vec{r}_2 = \vec{r}$; this means that the particle exchange operation $1 \leftrightarrow 2$ is equivalent to *parity*. In this case, if we denote by l the quantum number of total orbital angular momentum of the system, since the orbital parity is $(-1)^l$, only even orbital angular-momentum states can be occupied by the two-particle system.

If the two particles are electrons, since they have spin $\frac{1}{2}$ they are fermions, hence their full wave function must be globally antisymmetric under particle exchange. This means that only the combination of an antisymmetric factor multiplied by a symmetric one is allowed, i.e. $\chi(s_1, s_2; A)\, \psi(\vec{r}_1, \vec{r}_2; S)$ or $\chi(s_1, s_2; S)\, \psi(\vec{r}_1, \vec{r}_2; A)$. Observing that an antisymmetric spin wave function for the two electrons means a total spin equal to 1, whereas a symmetric χ yields a vanishing total spin, we see that the property of symmetry of $\psi(\vec{r}_1, \vec{r}_2)$, and hence the energy levels accessible, are function of the total spin of the system. Such an effect, which implies the influence via the exchange symmetry of the total spin on the energy level of a system, even though the spin does not occur directly in the energy contribution, is called *exchange interaction*.

12.8.4 Two-electron atoms

To illustrate the above formalism let us consider as an example helium-like atoms (i.e. He, Li$^+$, etc.). Assuming that the nucleus is infinitely massive, the Hamiltonian for the two electrons is given by

$$\widehat{H} = \frac{\vec{p}_1^2}{2m} + \frac{\vec{p}_1^2}{2m} - \frac{Z q_e^2}{|\vec{r}_1|} - \frac{Z q_e^2}{|\vec{r}_2|} + \frac{q_e^2}{|\vec{r}_1 - \vec{r}_2|}. \qquad (12.8.10)$$

If we denote the single-particle terms by

$$\widehat{h}_1 = \frac{\vec{p}_1^2}{2m} - \frac{Zq_{\mathrm{e}}^2}{|\vec{r}_1|},$$

$$\widehat{h}_2 = \frac{\vec{p}_2^2}{2m} - \frac{Zq_{\mathrm{e}}^2}{|\vec{r}_2|}, \tag{12.8.11}$$

and the interaction term by

$$\widehat{V}_{12} = \frac{q_{\mathrm{e}}^2}{|\vec{r}_1 - \vec{r}_2|}, \tag{12.8.12}$$

the total Hamiltonian can be written in terms of the unperturbed term $\widehat{H}_0 \equiv \widehat{h}_1 + \widehat{h}_2$ as

$$\widehat{H} = \widehat{H}_0 + \widehat{V}_{12}. \tag{12.8.13}$$

It is worth observing that the electron–electron repulsion $\widehat{V}_{12}$ is typically of the same order of magnitude as the independent particle energy, and hence a perturbative approach will not generally be a good approximation. However, it is convenient because the eigenstates of $\widehat{H}_0$ are the only ones known.

If we denote with $| \uparrow \rangle$ and $| \downarrow \rangle$ the spin eigenstates for $\widehat{S}$ of eigenvalues $+\frac{1}{2}$ and $-\frac{1}{2}$, respectively, a single electron in the $1s$ orbital can be described by the two possible states: $|1s, \uparrow \rangle \equiv |1s\rangle | \uparrow \rangle$ and $|1s, \downarrow \rangle \equiv |1s\rangle | \downarrow \rangle$. In terms of these the ground state of $\widehat{H}_0$ can be easily found in the form

$$\begin{aligned}|\Psi^0\rangle &= \frac{1}{\sqrt{2}} \left(|1s, \uparrow\rangle |1s, \downarrow\rangle - |1s, \downarrow\rangle |1s, \uparrow\rangle \right) \\ &= |1s\rangle |1s\rangle \frac{1}{\sqrt{2}} \left(| \uparrow\rangle | \downarrow\rangle - | \downarrow\rangle | \uparrow\rangle \right), \end{aligned} \tag{12.8.14}$$

which is totally antisymmetric under the particle exchange, and consists of a symmetric orbital wave function times an antisymmetric spinorial one corresponding to the singlet ($S = 0$) state. This is reasonable because, in the absence of spin–orbit coupling, the electronic states can always be chosen to be spin eigenfunctions. Note that, under particle exchange, a total spin eigenfunction takes the symmetric factor $(-1)^{S+1}$; this means that spin singlets need symmetric spatial parts, whereas triplets require antisymmetric spatial parts. Unfortunately, zeroth-order ground-state energy computed by acting with $\widehat{H}_0$ on $|\Psi^0\rangle$ provides a value of -108.8 eV (eight times the ground level of hydrogen atom), which is very far from the experimental value of -78.86 eV.

The computation improves by considering first-order corrections in the perturbation $\widehat{V}_{12}$, which implies the evaluation of the matrix element

$$E^{(1)} = \langle \Psi^0 | \widehat{V}_{12} | \Psi^0 \rangle. \tag{12.8.15}$$

This is a fairly complicated six-dimensional integral, but it can be evaluated, and the resulting first-order correction is positive and equal to 34 eV (the sign is due to the electron repulsion). Thus, to first order, the total energy is $-108.8 + 34 = -74.8$ eV, which improves the naive zeroth-order approximation significantly. Eventually, the second-order correction provides an additional contribution of -3.8 eV. This yields an approximate ground-state energy of $-74.8 - 3.8 = -78.6$ eV, which is in fair agreement with the experimental result. Note that the perturbation approach, which works quite well for the ground state of helium-like systems, accounts for the fact that the electrons are not independent but rather they tend to avoid each other to minimize the Coulomb repulsion. For identical particles, the definition of entanglement needs further qualification (Grabowski *et al.* 2011).

12.9 Generalized paraFermi and paraBose oscillators

We have discussed two kinds of assemblies of identical particles. The Bose statistics applies to integer-spin particles, which are completely symmetric under the interchange of any two particles, and the Fermi statistics holds for cases in which the states are completely antisymmetric under the interchange of any two identical particles, and applies to half-integer-spin particles. Can we have a more general statistics in quantum theory?

This question was answered in the affirmative by H. S. Green, who gave the actual construction (Green 1953) of a more general statistics for integer-spin particles (now called paraBose statistics) and a family of statistics for half-integer-spin particles (now called paraFermi statistics). Green's recipe was to take the direct sum of mutually anti-commuting Bose oscillators to realize the paraBose statistics; they then satisfy trilinear commutation relations, similar to the Duffin–Kemmer algebra. For paraFermi statistics the construction was obtained by taking the direct sum of mutually commuting Fermi oscillators, which also satisfied similar trilinear commutation relations. In modern particle physics, the quarks, of which the hadrons are constituted, exhibit such paraFermi statistics referred to as *colour*, which is useful in classifying the various hadrons. In this context the individual commuting Fermi operators are said to possess colour, and observable hadrons are said to be colour singlets.

While Green's ansatz for paraFermi oscillators is the direct sum of mutually commuting Fermi operators, there are more general solutions, which correspond to paraFermi oscillators that cannot be constructed in this way; they behave as if they were paraFermi oscillators of a real non-integer order (Sudarshan *et al.* 1995).

12.10 Problems

12.P1. Let $\{f_n\}$ be an orthonormal basis in the Hilbert space $L^2[a, b]$ of square-integrable functions on the closed interval $[a, b]$, and denote by P_n the projection operators defined by

$$P_n g \equiv \langle f_n, g \rangle f_n. \tag{12.10.1}$$

(i) Prove that P_n is linear and self-adjoint.

(ii) Prove that

$$P_n P_m = \delta_{mn} P_n. \tag{12.10.2}$$

(iii) Find eigenvalues and eigenfunctions of P_n.

(iv) If the $\{f_n\}$ are a basis of eigenstates corresponding to a self-adjoint operator with non-degenerate spectrum, show that

$$A = \sum_n a_n P_n, \quad [A, P_n] = 0. \tag{12.10.3}$$

12.P2. Prove that, if the linear operator A is self-adjoint, all its eigenvalues are real. Second, provide an example of an operator having only real eigenvalues, which is not even symmetric. Third, provide an operator which is symmetric but not self-adjoint.

12.P3. The correlator of two linear operators A, B is defined as

$$\Gamma(A, B) \equiv \frac{1}{2}\langle AB + BA \rangle - \langle A \rangle \langle B \rangle. \tag{12.10.4}$$

(i) Prove that, if A and B are self-adjoint, then their correlator is real-valued.

(ii) Prove that, on defining

$$\widetilde{A} \equiv A - \langle A \rangle, \quad \widetilde{B} \equiv B - \langle B \rangle, \tag{12.10.5}$$

then

$$\Gamma(A, B) = \frac{1}{2}\langle \widetilde{A}\widetilde{B} + \widetilde{B}\widetilde{A} \rangle. \tag{12.10.6}$$

(iii) Finally, prove that

$$(\Delta A)^2 (\Delta B)^2 \geq \Gamma^2(A, B) + \frac{1}{4}\langle [A, B] \rangle. \tag{12.10.7}$$

Discuss the relevance of this inequality for the proof of the uncertainty relations.

12.P4. A particle of mass m is bound to move on a circle of radius r but is otherwise free. Find the energy eigenvalues and the corresponding eigenfunctions.

12.P5. The Wigner transform of a wave function ψ is defined by the relation

$$W_\psi(x, p) \equiv \frac{1}{\pi \hbar} \int_{-\infty}^{\infty} dy \, \exp\left(\frac{2ipy}{\hbar}\right) \psi^*(x+y)\psi(x-y). \tag{12.10.8}$$

Prove that, by virtue of this definition, W_ψ fulfills the conditions

$$\int_{-\infty}^{\infty} dp \, W_\psi(x, p) = |\psi(x)|^2, \tag{12.10.9}$$

$$\int_{-\infty}^{\infty} dx \, W_\psi(x, p) = |\widetilde{\psi}(p)|^2, \tag{12.10.10}$$

where $\widetilde{\psi}(p)$ is the Fourier transform of $\psi(x)$. Compute the Wigner transform of the first excited state of the harmonic oscillator and try to interpret the result.

13 Formulations of quantum mechanics and their physical implications

In this chapter, the physical content of previous chapters is summarized, and the reader is suided towards some key areas of modern physics: the Feynman formulation, the unificaton of quantum theory with special relativity, and dualities in quantum mechanics. This very short chapter is intended to be a bridge aimed at motivating readers to attend first course in quantum field theory.

13.1 Towards an overall view

At this stage, the student or the general reader might want to know what picture of the physical world emerges from the material covered so far and from the areas related to it but not covered in our book, so as to be introduced to what is in sight not only for the beginner but also for the scientific community. For this purpose, we summarize very briefly the previous pictures and we arrive at three key concepts: the Lagrangian in quantum mechanics via the Feynman sum over histories; the unification of quantum mechanics and special relativity; and new duality symmetries in quantum mechanics.

13.2 From Schrödinger to Feynman

Quantum mechanics provides a probabilistic description of the world on atomic or sub-atomic scale. It tells us that, on such scales, the world can be described by a Hilbert space structure, or suitable generalizations. Even in the relatively simple case of the hydrogen atom, the appropriate Hilbert space is infinite-dimensional, but finite-dimensional Hilbert spaces play a role as well. For example, the space of spin-states of a spin-s particle is $\mathbf{C}^{2s+1}$ and is therefore finite-dimensional. Various pictures or formulations of quantum mechanics have been developed over the years, and their key elements can be summarized as follows, at the risk of repeating ourselves:

(i) In the *Schrödinger picture*, we are dealing with wave functions evolving in time according to a first-order equation. More precisely, in an abstract Hilbert space $\mathcal{H}$, we study the Schrödinger equation

$$i\hbar\frac{d\psi}{dt} = \hat{H}\psi, \tag{13.2.1}$$

where the state vector ψ belongs to $\mathcal{H}$, while $\hat{H}$ is the Hamiltonian operator. In wave mechanics, the emphasis is more immediately put on partial differential equations, with the wave function viewed as a complex-valued map $\psi : (x, t) \rightarrow \mathbf{C}$ obeying the equation

$$i\hbar \frac{\partial \psi}{\partial t} = \left(-\frac{\hbar^2}{2m} \Delta + V \right) \psi, \tag{13.2.2}$$

where $-\Delta$ is the Laplacian in Cartesian coordinates on $\mathbf{R}^3$ (with this sign convention, its symbol is positive-definite).

(ii) In the *Heisenberg picture*, what evolves in time are instead the operators, according to the first-order equation

$$i\hbar \frac{d\hat{A}}{dt} = [\hat{A}, \hat{H}]. \tag{13.2.3}$$

Heisenberg performed a quantum mechanical re-interpretation of kinematic and mechanical relations (Heisenberg 1925) because he wanted to formulate quantum theory in terms of observables only.

(iii) In the *Dirac quantization*, from an assessment of the Heisenberg approach and of classical Poisson brackets (Dirac 1926), we discover that quantum mechanics can be made to rely upon the basic commutation relations involving position and momentum operators:

$$[\hat{q}^j, \hat{q}^k] = [\hat{p}_j, \hat{p}_k] = 0, \tag{13.2.4}$$

$$[\hat{q}^j, \hat{p}_k] = i\hbar \delta^j_{\ k}. \tag{13.2.5}$$

For generic operators depending on $\hat{q}, \hat{p}$ variables, their formal Taylor series, jointly with the application of Eqs. (13.2.4) and (13.2.5), should yield their commutator.

(iv) *Weyl quantization.* The operators satisfying the canonical commutation relations (13.2.5) cannot both be bounded (Esposito *et al.* 2004), whereas it would be good to have quantization rules not involving unbounded operators and domain problems. For this purpose, we can consider the strongly continuous 1-parameter unitary groups having position and momentum as their infinitesimal generators. These read as $V(t) \equiv e^{it\hat{q}}$, $U(s) \equiv e^{is\hat{p}}$, and satisfy the Weyl form of canonical commutation relations, which is given by

$$U(s)V(t) = e^{ist\hbar} V(t)U(s). \tag{13.2.6}$$

Here the emphasis was, for the first time, on group-theoretical methods, with a substantial departure from the historical development, that instead relied heavily on quantum commutators and their relation with classical Poisson brackets.

(v) *Feynman quantization* (i.e. Lagrangian approach). The Weyl approach is very elegant and far-sighted, with several modern applications (Esposito *et al.* 2004) but still has to do with a more rigorous way of carrying canonical quantization, which is not suitable for the inclusion of relativity. A space–time approach to ordinary quantum mechanics was instead devised by Feynman (Feynman 1948) (and, first, by Dirac himself (Dirac 1933)), who proposed that the Green kernel of the Schrödinger equation should be expressed in the form

$$G[x_f, t_f; x_i, t_i] = \int_{\text{all paths}} e^{i\frac{S}{\hbar}} d\mu, \tag{13.2.7}$$

where $d\mu$ is a suitable (putative) measure on the set of all space–time paths (including continuous, piecewise continuous, or even discontinuous paths) matching the initial and final conditions. This point of view has enormous potentialities in the quantization of theories with infinitely many degrees of freedom (see the following section), since it preserves manifest covariance and the full symmetry group, as it is derived from a Lagrangian.

It should be stressed that quantum mechanics regards wave functions only as a technical tool to study bound states (corresponding to the discrete spectrum of the Hamiltonian operator $\hat{H}$), scattering states (corresponding instead to the continuous spectrum of $\hat{H}$), and to evaluate probabilities (of finding the values taken by the observables of the theory). Although the geometry of Hilbert space was known since von Neumann published his famous book on quantum mechanics (von Neumann 1955), a fully geometric view, not relying upon Hilbert space as a primary object, has been developed only recently. Within this framework, the basic structures are probability distributions in the first place. As we stressed already in Section 12.3, since the space of pure states is a manifold in the first place, it requires formulating quantum mechanics in geometrical terms, rather than by means of vectors and (unbounded) operators. We have learned in Sections 12.6 and 12.7 that classical and quantum states can be described as tomographic probability densities, which provide an alternative description for both probability densities on phase space in classical theory and state vectors or density operators in quantum theory.

13.2.1 Remarks on the Feynman approach

In his doctoral dissertation 'The Principle of Least Action in Quantum Mechanics', Feynman tried to find a generalization of quantum mechanics applicable to a system whose classical analogue is described by a principle of least action, and not necessarily by equations of motion in Hamiltonian form. Feynman obtained a formal solution of the problem by writing the probability amplitude $(q_t'|q_T')$ for finding at time t in position q_t' a particle known to be at time T in position q_T' as follows (Cartier and DeWitt–Morette 2000):

$$\langle q_t'|q_T'\rangle = \int \int, \ldots, \int \langle q_t'|q_m'\rangle dq_m' \langle q_m'|q_{m-1}'\rangle dq_{m-1}', \ldots, \langle q_2'|q_1'\rangle dq_1' \langle q_1'|q_T'\rangle, \quad (13.2.8)$$

where the interval $[T, t]$ has been divided into a large number of small intervals

$$[T, t_1], \ldots, [t_m, t],$$

q_k' denotes $q_{t_k}' \equiv q(t_k)$, and, denoting by $L(\dot{q}, q)$ the Lagrangian for the classical system under consideration, the symbol $(q_{t+\delta t}'|q_t')$ is often equal to

$$\exp \frac{i}{\hbar} \left[L\left(\frac{q_{t+\delta t}' - q_t'}{\delta t}, q_{t+\delta t}' \right) \delta t \right]$$

up to a normalization constant, in the limit as $\delta t \to 0$. Feynman was vague on the actual value of this normalization constant, whose absolute value was obtained in Morette (1951) by requiring that $(q_{t+\delta t}'|q_t')$ should satisfy a unitarity condition. The phase of the complex value is still an open question. Each q'_k is integrated over its full domain, and the limit for large m is a sum over all continuous paths

$$q : [T, t] \to \mathbf{R}$$

with fixed end points, called a path integral.

The original goal, i.e. action at a distance in the quantum theory of electrodynamics, was not achieved by path integrals; ordinary quantum mechanics had been formulated in terms of the action functional

$$S(q) = \int_T^t L(\dot{q}(s), q(s)) \mathrm{d}s, \tag{13.2.9}$$

a program initiated by Dirac himself as we said before (see also Section 32 of his book on quantum mechanics (Dirac 1958)). A formulation in terms of an action functional is well suited for a relativistically invariant formulation of quantum mechanics, whereas ordinary quantum mechanics relies upon the Schrödinger equation, which is not invariant under Lorentz transformations of space–time coordinates. We may therefore expect that Feynman's approach has greater potentialities than merely reformulating the quantum mechanics of Schrödinger, Heisenberg, Dirac or Weyl. This is indeed the case, and the program has been systematically pursued in the 'space of histories (or global) approach to quantum theory' (DeWitt 2003).

Let us now rewrite the defining formula (13.2.8) by collecting the phase factors resulting from the approximate formula for $\langle q'_{t+\delta t} | q'_t \rangle$. This yields an approximation (Cartier and DeWitt–Morette 2000)

$$\langle q'_t | q'_T \rangle \approx \int_{\mathbf{R}^m} \Gamma_m \prod_{k=1}^m \mathrm{d}q'_k \, \mathrm{e}^{\frac{iS_m}{\hbar}}, \tag{13.2.10}$$

where Γ_m is the as-yet unspecified normalization constant, and

$$S_m = \sum_{k=1}^m L\left(\frac{q'_{k+1} - q'_k}{t_{k+1} - t_k}, q'_{k+1} \right) (t_{k+1} - t_k), \tag{13.2.11}$$

where $t'_{m+1} \equiv t$, $q'_{m+1} \equiv q'_t$. Interestingly, S_m is the Riemann sum approximation to the action (13.2.9), by taking a path

$$q : [T, t] \to \mathbf{R}^D$$

with end points $q(T) = q'_T$, $q(t) = q'_t$. Feynman dared to take the limit $m = \infty$ inside the integral (13.2.10) and hence he arrived at a suggestive formula for the transition amplitude, written as

$$\langle q'_t | q'_T \rangle = \int \mathcal{D}q \, \mathrm{e}^{\frac{iS(q)}{\hbar}}. \tag{13.2.12}$$

The symbol $\mathcal{D}q$ represents a putative measure

$$\mathcal{D}q = \lim_{m \to \infty} \Gamma_m \prod_{k=1}^m \mathrm{d}q'_k = \Gamma \prod_{T \le s \le t} \mathrm{d}q(s), \tag{13.2.13}$$

and the integral on the right-hand side of Eq. (13.2.12) is viewed as the 'sum over all possible paths as thoroughly as we can'. To say that Eq. (13.2.12) is the limit of (13.2.10) as $m \to \infty$ implies that the short-time amplitude $\langle q'_{t+\delta t} | q'_t \rangle$ is known, and that the limit exists. Even the experts recommend regarding $\mathcal{D}q$ as a mere symbol, while all that matters are the rules for using it (Cartier and DeWitt–Morette 2000). Nowadays, however, Feynman integrals find several applications that can be put on a firm mathematical ground (e.g. those in Glimme and Jaffe (1987)).

As far as ordinary quantum mechanics is concerned, we can, however, consider an apparently unrelated problem, which makes it possible to obtain a meaningful definition of the original Feynman path integral. For this purpose, we recall that Wiener had noticed in the 1920s that the integral of a functional $F(q)$ of a path q of the Brownian motion can be evaluated as a limit for $m \to \infty$ of an integral

$$\int_{\mathbf{R}^m} C_m \prod_{k=1}^m dq'_k e^{-S_m(q'_1 \cdots q'_m)} F(q),$$

where the term in the exponential factor is given by

$$S_m = \frac{1}{2C} \sum_{k=1}^m \frac{(q'_{k+1} - q'_k)^2}{(\tau_{k+1} - \tau_k)}. \tag{13.2.14}$$

Since the oscillating exponential $e^{i\frac{S_m}{\hbar}}$ has been replaced by the exponentially damped factor e^{-S_m}, a convergent integral is obtained as $m \to \infty$. Following the notation used for Feynman path integrals, the limiting form of the Wiener integral can be denoted by

$$\int \mathcal{D}q \, e^{-S_c(q)} F(q).$$

Note that in this case the action, derived from the kinetic energy alone, is given by

$$S_c(q) = \frac{1}{2C} \int_T^t \dot{q}(\tau)^2 d\tau = \frac{1}{2C} \int_T^t \left(\frac{dq}{d\tau}\right)^2 d\tau. \tag{13.2.15}$$

The transition amplitude $\langle q'_t | q'_T \rangle$ can be used in quantum mechanics to solve the Schrödinger equation. For example, for a particle under an exterior force field, ruled by the Lagrangian

$$L(\dot{q}, q) = \frac{m}{2}(\dot{q})^2 - V(q), \tag{13.2.16}$$

the Schrödinger equation takes the form familiar from Chapter 4, and the Feynman solution is given by

$$\psi(t, q'_t) = \int dq'_T \langle q'_t | q'_T \rangle \psi(T, q'_T), \tag{13.2.17}$$

with the Green kernel given formally by Eq. (13.2.12). If the so-called 'imaginary time' $\tau = it$ is introduced, the Schrödinger equation is turned into a diffusion equation

$$\frac{\partial \varphi}{\partial \tau} = \frac{C}{2} \Delta \varphi - W\varphi, \tag{13.2.18}$$

where $C \equiv \hbar/m$ and $W \equiv V/\hbar$. If the imaginary time variable is introduced directly into the Feynman path integral, $iS(q)/\hbar$ is turned into $-S^I(q)$, where

$$S^I(q) = \int_{iT}^{it} \left[\frac{1}{2C} \left(\frac{dq}{d\tau} \right)^2 + W(q) \right] d\tau, \tag{13.2.19}$$

while the amplitude $\langle q'_t | q'_T \rangle$ becomes the kernel

$$K \langle q'_{it} | q'_{iT} \rangle = \int \mathcal{D}q \, e^{-S^I(q)}. \tag{13.2.20}$$

Now from the notation

$$E[F(q)] \equiv \int \mathcal{D}q \, e^{-\frac{1}{2C} \int_T^t \left(\frac{dq}{d\tau} \right)^2 d\tau} F(q) \tag{13.2.21}$$

we can write

$$K \langle q'_t | q'_T \rangle = E \left[e^{-\int_T^t W(q(\tau)) d\tau} \right], \tag{13.2.22}$$

and we eventually obtain the fundamental formula

$$\mathcal{D}q \, e^{\left[\frac{i}{\hbar} S(q) \right]} = E \left[e^{-\frac{1}{\hbar} \int_{iT}^{it} V(q(\tau)) d\tau} \right]. \tag{13.2.23}$$

The imaginary limits of integration on the right-hand side mean that the two integrals therein, a path integral over q and an ordinary integral over τ, should be performed with real values of τ, followed by an analytic continuation in this variable. In other words, *one first obtains the Green function for a heat equation*, which is a well-posed problem, *and its analytic continuation back to real time helps in defining and evaluating the Green function of the Schrödinger equation.* From the point of view of the general formalism, however, Wiener's integral spoils the unified picture of kinetic and potential parts of the action functional.

13.3 Path integral for systems interacting with an electromagnetic field

If the system under consideration is a particle of charge q and mass m, subject to the action of an electromagnetic field, with vector potential $\vec{A}$ and scalar potential ϕ, the resulting Lagrangian in $\mathbf{R}^3$ may be written as in Eq. (3.2.34), and the Green kernel of ordinary quantum mechanics can be obtained from the limiting procedure described by the heuristic formula

$$G(\vec{z}, t; \vec{y}, 0)$$

$$= \lim_{N \to \infty} \int d^3 x_1 \ldots d^3 x_N \left(\frac{m}{2\pi i \hbar \varepsilon} \right)^{3 \frac{N}{2}} e^{\frac{i\varepsilon}{\hbar} \sum_{j=0}^{N-1} \left[\frac{m}{2} \left(\frac{|\vec{x}_{j+1} - \vec{x}_j|}{\varepsilon} \right)^2 - q\phi(\vec{x}_j) \right] + C_A}, \tag{13.3.1}$$

where we denote by C_A the contribution of the vector potential to the classical action, i.e.

$$C_A \equiv \frac{q}{c}\int_0^t \frac{dx^i}{ds}A_i(\vec{x})ds = \frac{q}{c}\int dx^i A_i(\vec{x}). \tag{13.3.2}$$

This term can be viewed as the limit, as $\varepsilon \to 0$, of the sum

$$C_A = \sum_{j=0}^{N-1} \frac{q}{c}A_i(\vec{x})\left(x^i(t_{j+1}) - x^i(t_j)\right). \tag{13.3.3}$$

In other words, we are evaluating integrals according to the prescription

$$\int_{t_0}^{t_N} f(t)dt = \lim_{\varepsilon\to 0}\sum_{j=0}^{N-1}(t_{j+1} - t_j)f(\hat{t}_j), \tag{13.3.4}$$

where we choose

$$\hat{t}_j = \frac{t_{j+1} + t_j}{2}. \tag{13.3.5}$$

For our system, the wave function at time $t_0 + \varepsilon$ is obtained from its value at time t_0 from the integral formula

$$\psi(\vec{x}, t_0 + \varepsilon) = \int d^3y \left(\frac{m}{2\pi i\hbar\varepsilon}\right)^{\frac{3}{2}} e^{\frac{i\varepsilon}{\hbar}\left[\frac{m}{2}\left(\frac{\vec{x}-\vec{y}}{\varepsilon}\right)^2 - q\phi(\vec{y}) + \frac{q}{c}(\vec{x}-\vec{y})\cdot\vec{A}\left(\frac{\vec{x}+\vec{y}}{2}\right)\right]} \psi(\vec{y}, t_0). \tag{13.3.6}$$

Now we define the vector $\vec{\xi} \equiv \vec{x} - \vec{y}$, we denote $q\phi$ by V, and expand $\psi(\vec{y}, t_0)$, $\vec{A}$ and V in the neighbourhood of $\vec{\xi} = 0$, i.e.

$$\psi(\vec{x}, t_0 + \varepsilon) = \int d^3\xi \left(\frac{m}{2\pi i\hbar\varepsilon}\right)^{\frac{3}{2}} \left[e^{\frac{im|\vec{\xi}|^2}{2\varepsilon\hbar}} e^{-\frac{i\varepsilon}{\hbar}(V(\vec{x}) + \vec{\nabla}V(\vec{x})\cdot\vec{\xi} + \cdots)}\right.$$

$$\left. e^{-\frac{iq}{\hbar c}\vec{\xi}\cdot\left[\vec{A}(\vec{x}) + \frac{(\vec{\xi}\cdot\vec{\nabla})\vec{A}(\vec{x})}{2} + \cdots\right]}\right]$$

$$\times \left(\psi(\vec{x}, t_0) + \vec{\xi}\cdot\vec{\nabla}\psi + \frac{1}{2}\sum_{k,l=1}^{3}\xi_k\xi_l\frac{\partial^2\psi}{\partial x_k\partial x_l} + \cdots\right). \tag{13.3.7}$$

We are only interested in first-order terms in the small parameter ε, and we exploit the basic formula

$$\int dy\, y^p e^{-ay^2} = \sqrt{\frac{\pi}{a}}\frac{f(p)}{a^{\frac{p}{2}}} \text{ if } p \text{ is even, } 0 \text{ if } p \text{ is odd,} \tag{13.3.8}$$

where f is a function of p. This implies that first order in ε is attained by retaining terms of second order in $\vec{\xi}$ when expanding the exponentials and $\psi(\vec{y}, t_0)$, i.e.

$$\psi(\vec{x}, t_0 + \varepsilon) = \left(1 - \frac{i\varepsilon V(\vec{x})}{\hbar}\right)\left(\frac{m}{2\pi i\hbar\varepsilon}\right)^{\frac{3}{2}}\int d^3\xi\, e^{\frac{im|\vec{\xi}|^2}{2\varepsilon\hbar}}$$

$$\times \left[1 - \frac{iq}{\hbar c}\xi_n A_n - \frac{iq}{2\hbar c}\xi_n\xi_r\vec{\nabla}_r A_n - \frac{1}{2}\frac{q^2}{\hbar^2 c^2}\xi_n A_n\xi_r A_r\right]$$

$$\times \left[\psi + \xi_l\vec{\nabla}_l\psi + \frac{1}{2}\xi_l\xi_k\vec{\nabla}_l\vec{\nabla}_k\psi\right]. \tag{13.3.9}$$

From the Gaussian-integral formula (13.3.8), all terms containing odd powers of ξ, and all terms containing products like $\xi_n \xi_l$ with $n \neq l$, give vanishing contribution to the integral. Thus, only integrals containing products like $\xi_n \xi_n$ give a non-vanishing contribution, and we find

$$\psi(\vec{x}, t_0 + \varepsilon) = \left[I - \frac{i\varepsilon}{\hbar} V + \frac{\varepsilon q}{2mc}(\vec{\nabla} \cdot \vec{A}) + \frac{i\varepsilon\hbar}{2m} \triangle - \frac{i\varepsilon q^2}{2\hbar mc}|\vec{A}|^2 + \frac{\varepsilon q}{mc}(\vec{A} \cdot \vec{\nabla}) \right] \psi.$$

(13.3.10)

Finally, we expand $\psi(\vec{x}, t_0 + \varepsilon)$ in the neighbourhood of $\varepsilon = 0$; we only keep terms up to first order in ε and multiply by $\hbar$ to obtain

$$i\hbar \frac{\partial \psi}{\partial t} = \left[\frac{1}{2m} \left(\hat{p} - \frac{q}{c}\hat{A} \right)^2 + \hat{V} \right] \psi,$$

(13.3.11)

which is the Schrödinger equation for a particle of charge q in the presence of an electromagnetic field. Note that, had we evaluated the vector potential at $\vec{x}_j$ or $\vec{x}_{j+1}$, rather than at the intermediate point as we have done, there would have been a 'missing' $\frac{1}{2}$ in the term $(\vec{\xi} \cdot \vec{\nabla})\vec{A}$ in the integral formula for $\psi(\vec{x}, t_0 + \varepsilon)$ and hence, upon integrating, we would have obtained $\frac{i\hbar q}{mc}\psi(\vec{\nabla} \cdot \vec{A})$ rather than $\frac{i\hbar q}{2mc}\psi(\vec{\nabla} \cdot \vec{A})$. Thus, we would not have obtained the correct form of the Schrödinger equation.

To sum up, the Green kernel for the Schrödinger equation in the presence of electromagnetic fields has been shown to be

$$G(\vec{z}, t; \vec{y}, 0) = \int_{y,0}^{z,t} d\vec{x}(s) e^{\frac{i}{\hbar} \int_0^t ds \left[\frac{m}{2}|\dot{\vec{x}}|^2 + \frac{q}{c}\dot{\vec{x}} \cdot \vec{A} - V(\vec{x}) \right]}.$$

(13.3.12)

It is also interesting to derive the gauge-transformation property of the Green kernel. Indeed, if the vector potential changes by the gradient of a freely specifiable smooth function φ, we find the gauge-transformation property

$$\int_y^z dx^i A_i(\vec{x}(s)) \rightarrow \int_y^z dx^i A_i(\vec{x}(s)) + \int_y^z dx^i \partial_i \varphi.$$

(13.3.13)

The second integral on the right-hand side is equal to $\varphi(\vec{z}) - \varphi(\vec{y})$ only if the variations of φ are evaluated at the intermediate point. This recipe leads to the gauge-transformation law

$$G(\vec{z}, t; \vec{y}, 0) \rightarrow e^{\frac{iq\varphi(\vec{z})}{\hbar c}} G(\vec{z}, t; \vec{y}, 0) e^{-\frac{iq\varphi(\vec{y})}{\hbar c}}.$$

(13.3.14)

13.4 Unification of quantum theory and special relativity

One of the first steps towards the unification of quantum theory and special relativity can be seen to consist of the Dirac derivation of a relativistically invariant wave equation for electrons, unlike the Pauli derivation of Chapter 9, which is non-relativistic. The steps of the Dirac argument are as follows (Dirac 1928).

It is indeed well known that the relativistic Hamiltonian of a particle of mass m is

$$H = c\sqrt{m^2 c^2 + p_1^2 + p_2^2 + p_3^2},$$

(13.4.1)

and this leads to the wave equation

$$\left[p_0 - \sqrt{m^2c^2 + p_1^2 + p_2^2 + p_3^2}\right]\psi = 0, \tag{13.4.2}$$

where the p_μ should be regarded as operators:

$$p_\mu \equiv i\hbar \frac{\partial}{\partial x^\mu}. \tag{13.4.3}$$

Thus, acting upon Eq. (13.4.2) on the left by the operator

$$p_0 + \sqrt{m^2c^2 + p_1^2 + p_2^2 + p_3^2}$$

leads to the equation

$$\left[p_0^2 - m^2c^2 - p_1^2 - p_2^2 - p_3^2\right]\psi = 0, \tag{13.4.4}$$

which is a more appropriate starting point for a relativistic theory. Of course, Eqs. (13.4.2) and (13.4.4) are not completely equivalent: every solution of Eq. (13.4.2) is also, by construction, a solution of Eq. (13.4.4), whereas the converse does not hold. Only the solutions of Eq. (13.4.4) corresponding to positive values of p_0 are also solutions of Eq. (13.4.2).

However, Eq. (13.4.4), being quadratic in p_0, is not of the form desirable in quantum mechanics, where, since the derivation of Schrödinger's equation, we are familiar with the need to obtain wave equations which are linear in p_0. In order to combine relativistic invariance with linearity in p_0, and to obtain equivalence with Eq. (13.4.4), we look for a wave equation which is rational and linear in p_0, p_1, p_2, p_3:

$$\left[p_0 - \alpha_1 p_1 - \alpha_2 p_2 - \alpha_3 p_3 - \beta\right]\psi = 0, \tag{13.4.5}$$

where α and β are independent of p. Since we are studying the (idealized) case when the electron moves in the absence of the electromagnetic field, all space-time points are equivalent, and hence the operator in square brackets in Eq. (13.4.5) is independent of x. This implies in turn that α and β are independent of x, and commute with the p and x operators. At a deeper level, α and β make it possible to obtain a relativistic description of the spin of the electron.

At this stage, we can act upon Eq. (13.4.5) on the left by the operator

$$p_0 + \alpha_1 p_1 + \alpha_2 p_2 + \alpha_3 p_3 + \beta.$$

This leads to the equation

$$\left[p_0^2 - \sum_{i=1}^{3} \alpha_i^2 p_i^2 - \sum_{i \neq j}(\alpha_i \alpha_j + \alpha_j \alpha_i)p_i p_j \right.$$
$$\left. - \sum_{i=1}^{3}(\alpha_i \beta + \beta \alpha_i)p_i - \beta^2\right]\psi = 0. \tag{13.4.6}$$

Equations (13.4.4) and (13.4.6) agree, for an electron of mass m_e, if α_i and β satisfy the conditions

$$\alpha_i^2 = I \quad \forall i = 1, 2, 3, \tag{13.4.7}$$

$$\alpha_i \alpha_j + \alpha_j \alpha_i = 0 \quad \text{if } i \neq j, \tag{13.4.8}$$

$$\beta^2 = m_e^2 c^2 I, \tag{13.4.9}$$

$$\alpha_i \beta + \beta \alpha_i = 0 \quad \forall i = 1, 2, 3. \tag{13.4.10}$$

Thus, on setting

$$\beta = \alpha_0 m_e c, \tag{13.4.11}$$

it is possible to re-express the properties (13.4.7)–(13.4.10) by the single equation

$$\alpha_\mu \alpha_\nu + \alpha_\nu \alpha_\mu = 2\delta_{\mu\nu} I \quad \forall \mu, \nu = 0, 1, 2, 3. \tag{13.4.12}$$

So far, we have found that, if Eq. (13.4.12) holds, the wave equation (13.4.5) is equivalent to Eq. (13.4.4). Thus, we can *assume* that Eq. (13.4.5) is the correct relativistic wave equation for the motion of an electron in the absence of a field. However, Eq. (13.4.5) is not entirely equivalent to Eq. (13.4.2), but, as Dirac first pointed out, it allows for solutions corresponding to negative as well as positive values of p_0. The former are relevant for the theory of positrons, and will not be discussed in this book.

To obtain a representation of four anti-commuting α, as in Eq. (13.4.12), we has to consider 4×4 matrices. Following Dirac, it is convenient to express the α in terms of generalized Pauli matrices (see below), denoted here by $\Sigma_1, \Sigma_2, \Sigma_3$, and of a second set of anti-commuting matrices, say ρ_1, ρ_2, ρ_3. Explicitly,

$$\alpha_1 \equiv \rho_1 \Sigma_1, \tag{13.4.13}$$

$$\alpha_2 \equiv \rho_1 \Sigma_2, \tag{13.4.14}$$

$$\alpha_3 \equiv \rho_1 \Sigma_3, \tag{13.4.15}$$

$$\alpha_0 \equiv \rho_3, \tag{13.4.16}$$

where (Dirac 1958)

$$\Sigma_1 \equiv \begin{pmatrix} 0 & 1 & 0 & 0 \\ 1 & 0 & 0 & 0 \\ 0 & 0 & 0 & 1 \\ 0 & 0 & 1 & 0 \end{pmatrix}, \tag{13.4.17}$$

$$\Sigma_2 \equiv \begin{pmatrix} 0 & -i & 0 & 0 \\ i & 0 & 0 & 0 \\ 0 & 0 & 0 & -i \\ 0 & 0 & i & 0 \end{pmatrix}, \tag{13.4.18}$$

$$\Sigma_3 \equiv \begin{pmatrix} 1 & 0 & 0 & 0 \\ 0 & -1 & 0 & 0 \\ 0 & 0 & 1 & 0 \\ 0 & 0 & 0 & -1 \end{pmatrix}, \tag{13.4.19}$$

$$\rho_1 \equiv \begin{pmatrix} 0 & 0 & 1 & 0 \\ 0 & 0 & 0 & 1 \\ 1 & 0 & 0 & 0 \\ 0 & 1 & 0 & 0 \end{pmatrix}, \tag{13.4.20}$$

$$\rho_2 \equiv \begin{pmatrix} 0 & 0 & -i & 0 \\ 0 & 0 & 0 & -i \\ i & 0 & 0 & 0 \\ 0 & i & 0 & 0 \end{pmatrix}, \tag{13.4.21}$$

$$\rho_3 \equiv \begin{pmatrix} 1 & 0 & 0 & 0 \\ 0 & 1 & 0 & 0 \\ 0 & 0 & -1 & 0 \\ 0 & 0 & 0 & -1 \end{pmatrix}. \tag{13.4.22}$$

In this formalism for the electron, the wave function has four components, and they all depend on the whole set of x only. Unlike the non-relativistic formalism with spin, three are two extra components that reflect the ability of Eq. (13.4.5) to describe negative-energy states.

By virtue of Eqs. (13.4.13)–(13.4.16), Eq. (13.4.5) may be re-expressed as

$$\left[p_0 - \rho_1 \sigma_j p_j - \rho_3 m_e c \right] \psi = 0. \tag{13.4.23}$$

The generalization to the case when an external electromagnetic field is present is not difficult. For this purpose, it is enough to recall that

$$p_\mu = m v_\mu + \frac{q_e}{c} A_\mu \quad \forall \mu = 0, 1, 2, 3, \tag{13.4.24}$$

where A_μ are such that $A_\mu dx^\mu$ is the connection 1-form (or potential) of the theory. By raising indices with the metric of the background space–time, we obtain the corresponding 4-vector

$$A^\mu = g^{\mu\nu} A_\nu. \tag{13.4.25}$$

The desired wave equation of the relativistic theory of the electron in an external electromagnetic field turns out to be

$$\left[p_0 + \frac{q_e}{c} A_0 - \rho_1 \sigma_j \left(p_j + \frac{q_e}{c} A_j \right) - \rho_3 m_e c \right] \psi = 0. \tag{13.4.26}$$

With this notation, the wave function should be regarded as a column vector with four rows, while its *conjugate imaginary*, say $\overline{\psi}^\dagger$, is a row vector, i.e. a 1×4 matrix, and obeys the equation

$$\overline{\psi}^\dagger \left[p_0 + \frac{q_e}{c} A_0 - \rho_1 \sigma_j \left(p_j + \frac{q_e}{c} A_j \right) - \rho_3 m_e c \right] = 0, \tag{13.4.27}$$

where the momentum operators operate to the right.

In the literature, bearing in mind Eqs. (13.4.12), it is standard practice to define the γ-matrices by means of the anti-commutation property

$$\gamma^{\mu}\gamma^{\nu} + \gamma^{\nu}\gamma^{\mu} = 2Ig^{\mu\nu}. \tag{13.4.28}$$

This holds *in any* space–time dimension. In the *Dirac representation*, in four dimensions,

$$\gamma^{0} = \rho_{3}, \tag{13.4.29}$$

$$\gamma^{i} = \gamma^{0}\,\alpha^{i}\ \forall i = 1, 2, 3, \tag{13.4.30}$$

where the matrices α^{i} coincide with those defined in Eqs. (13.4.13)–(13.4.15) (the position of the indices should not confuse the reader, at this stage).

Denoting by σ_{0} the 2×2 identity matrix

$$\sigma_{0} \equiv \begin{pmatrix} 1 & 0 \\ 0 & 1 \end{pmatrix},$$

and by σ_{2} the second Pauli matrix

$$\sigma_{2} \equiv \begin{pmatrix} 0 & -i \\ i & 0 \end{pmatrix},$$

we can define two additional representations for the γ-matrices that are quite useful. In the *Majorana representation*,

$$\gamma^{\mu}_{\text{Majorana}} = U\,\gamma^{\mu}_{\text{Dirac}}\,U^{\dagger}, \tag{13.4.31}$$

where

$$U \equiv \frac{1}{\sqrt{2}} \begin{pmatrix} \sigma_{0} & \sigma_{2} \\ \sigma_{2} & -\sigma_{0} \end{pmatrix} = U^{\dagger}. \tag{13.4.32}$$

In the *chiral representation*, instead,

$$\gamma^{\mu}_{\text{chiral}} = V\,\gamma^{\mu}_{\text{Dirac}}\,V^{\dagger}, \tag{13.4.33}$$

with

$$V \equiv \frac{1}{\sqrt{2}} \begin{pmatrix} \sigma_{0} & -\sigma_{0} \\ \sigma_{0} & \sigma_{0} \end{pmatrix}. \tag{13.4.34}$$

Note that, strictly speaking, so far we have studied the γ-matrices in four-dimensional Minkowski space–time, which is a flat Lorentzian 4-manifold. When the Riemann curvature does not vanish, the property (13.4.28) remains a good starting point in the formalism of γ-matrices, but the details are more involved.

The unification of quantum theory with special relativity was rather successful after Dirac derived his famous wave equation. Several new features were then found to arise, e.g. creation and annihilation of particles are frequent phenomena, and the classical concept of electron has become completely superseded. Moreover, when we try to build the quantum theory of systems with infinitely many degrees of freedom, i.e. 'fields', it is precisely Einstein's special relativity that is responsible for the fact that the field operator at a space–time point with coordinates x^{μ} is too singular for the concept to be meaningful. Rather, such an operator is multiplied by a suitably smooth function to obtain a linear functional whose behaviour is under control after integration over (all) space. The relativistic quantum fields that the student will meet after reading this book were therefore found to be operator-valued distributions (Wightman 1996).

13.5 Dualities: quantum mechanics leads to new fundamental symmetries

The vacuum Maxwell equations in the absence of charges and currents

$$\varepsilon^{ij}{}_k \frac{\partial}{\partial x^j} B^k = \frac{1}{c} \frac{\partial E^i}{\partial t}, \ \varepsilon^{ij}{}_k \frac{\partial}{\partial x^j} E^k = -\frac{1}{c} \frac{\partial B^i}{\partial t}, \ \partial_k B^k = 0, \ \partial_k E^k = 0, \tag{13.5.1}$$

are symmetric under *duality*, i.e. a map replacing electric field with magnetic field, and magnetic field with minus electric field, i.e.

$$\vec{E} \to \vec{B}, \ \vec{B} \to -\vec{E}. \tag{13.5.2}$$

If charges and currents are non-vanishing, the duality symmetry is preserved if both electric and magnetic charges and currents are added. So far, however, no magnetic charges have been observed, which spoils duality. At a deeper level, duality seems to be violated when the magnetic field is derived from a vector potential and the electric field is obtained from a scalar potential (think of electrostatics). The vector potential, however, is not an optional ingredient, because it is needed in the following circumstances (Witten 1997):

(i) To write the Schrödinger equation for an electron in a magnetic field.

(ii) To derive the Maxwell equations from a Lagrangian, which leads to the wave equation for the full 4-potential in the Lorenz gauge.

(iii) To build the unified theory of electromagnetic, weak and strong interactions in Minkowski space–time.

Modern physical theories indeed provide examples where both electric and magnetic charges occur. In the case of weak coupling, the region where understands the behaviour of quantum field theories is better understood, electric and magnetic charges behave in completely different ways. For electrodynamics, weak coupling means that the fine structure constant

$$\alpha = \frac{q_e^2}{4\pi \hbar c} \tag{13.5.3}$$

is small, where the electron charge q_e plays the role of a coupling constant of the quantum theory of the electromagnetic field.

In general, for weak coupling, electric charges appear as elementary quanta. The electric charge Q_e of any particle is an integer multiple of q_e: $Q_e = n q_e$. On the other hand, magnetic monopoles arise for weak coupling as collective excitations of elementary particles. Such collective excitations appear, in the weak-coupling limit, as extended solutions of non-linear equations, called *solitons*. The magnetic charge q_m of any particle that is obtained by quantizing such a soliton is an integer multiple of a fundamental quantum of magnetic charge, according to the Dirac formula

$$q_{\mathrm{m}} = N \frac{2\pi \hbar c}{q_{\mathrm{e}}},$$ (13.5.4)

where N is an integer.

While there seem to be very profound reasons for a lack of symmetry between electricity and magnetism, the work in Montonen and Olive (1977) noted instead a surprising symmetry between electricity and magnetism in the classical limit of a certain four-dimensional field theory. The authors therein realized that in their model the mass M of any particle of electric charge q_{e} and magnetic charge q_{m} was given by the formula

$$M = \langle \phi \rangle \sqrt{q_{\mathrm{e}}^2 + q_{\mathrm{m}}^2},$$ (13.5.5)

where $\langle \phi \rangle$ is a constant. These authors conjectured that the theory displays an exact symmetry that exchanges q_{e} and q_{m}. But such a symmetry should exchange the quantum of electric charge with a multiple of the quantum of magnetic charge. For example, in the Montonen–Olive case the transformation is

$$q_{\mathrm{e}} \leftrightarrow \frac{4\pi \hbar c}{q_{\mathrm{e}}},$$ (13.5.6)

which implies

$$\alpha \leftrightarrow \frac{1}{\alpha}.$$ (13.5.7)

This symmetry exchanges electric and magnetic fields, and hence, to a classical observer, it resembler the duality of Maxwell's equations. Finally, such a symmetry should exchange elementary quanta with collective excitations because, for weak coupling, electric charges arise as elementary quanta and magnetic charges arise as collective excitations.

As stressed in Witten (1997), the above picture suggests a change in the logical role of quantum mechanics. The whole subject has evolved and is currently being taught at undergraduate level with emphasis on quantum systems obtained by quantizing classical systems. However, in the search for a unified theory of all fundamental interactions, theoretical physics is currently considering a quantum theory that has the previously known theories as different classical limits. *The competing classical limits are equally significant, and none of them is distinguished* (Witten 1997). The different classical limits are related by dualities that generalize the Montonen–Olive duality, and from the occurrence of $\hbar$ on the right-hand side we see that such dualities are symmetries that exist only in the quantum world (Witten 1997).

One might therefore argue that, on the path towards unification of all interactions and all guiding principles, the modern role of quantum mechanics lies in making it possible to obtain new fundamental symmetries.

Exam problems

14.1 End-of-year written exams

14.1. A spin-$\frac{1}{2}$ particle on the line is described, in the basis where σ_3 is diagonal, by the following wave function:

$$\psi = N \begin{pmatrix} 6e^{-\frac{x^2}{a^2}} \\ 7i\frac{x}{a}e^{-\frac{(x-7a)^2}{a^2}} \end{pmatrix}, \tag{14.1.1}$$

where N is a normalization constant and the constant a has the dimension of length. Find

(i) the possible values of a measurement of spin, with the associated probabilities;

(ii) the probability of finding the particle on the positive half-line.

14.2. A particle that is forced to move on a circle of radius R is described by the following Hamiltonian:

$$H = \frac{p^2}{2m} + \varepsilon \cos kx. \tag{14.1.2}$$

Compute the spectrum of H:

(i) exactly in the free case ($\varepsilon = 0$);

(ii) perturbatively when ε does not vanish.

(iii) Under which conditions is the perturbative calculation meaningful?

14.3. At the initial time $t = 0$, a particle is in the state described by the following wave function:

$$\Psi_0 = Y_{1,0}(\theta, \phi)f(r), \tag{14.1.3}$$

where $f(r)$ is chosen so as to achieve normalization of Ψ_0, and the coordinates chosen have the z-axis as the polar axis. Assuming that the Hamiltonian ruling the system is

$$H = \alpha(L_+ + L_-), \tag{14.1.4}$$

with α a real constant, compute

(i) the possible values of a measurement of L_z;

(ii) their probabilities as functions of time.

Moreover, show that the sum of all probabilities satisfies the unitarity condition.

14.4. Consider a particle described by the Hamiltonian (in dimensionless variables)

$$H = \frac{1}{2}\left(p_x^2 + p_y^2 + x^2 + y^2\right) + \varepsilon x^4. \tag{14.1.5}$$

Compute how the ground-state energy is corrected to first order in ε.

14.5. Consider a particle that can only move on the negative half-line. This means that an infinite potential barrier is taken to be located at $x = 0$, so that the particle is prevented from entering the positive half-line.

(i) Study the spectrum of the Hamiltonian operator and analyze both proper and improper eigenfunctions.

(ii) Use the improper eigenfunctions to describe a wave packet which, being localized at $x \to -\infty$ as $t \to -\infty$, hits the barrier as time evolves and is eventually reflected. Discuss the properties of the reflection process and the limiting behaviour of the wave packet as $t \to \infty$. (Suggestion: put together the plane-wave elementary solutions by convolution with a Gaussian in the momentum variable, centred about a generic k_0 with variance $\triangle k_0 << k_0$.)

14.6. Consider the one-dimensional harmonic oscillator Hamiltonian

$$H = \frac{p^2}{2m} + \frac{1}{2}m\omega^2 x^2. \tag{14.1.6}$$

(i) Consider the mean value of both kinetic and potential energy, as a function of time, in a generic energy eigenstate denoted by ψ_n.

(ii) Compare the obtained relation between the mean values of kinetic and potential energy in the quantum case, and their mean values, over the same period, in the classical case.

14.7. Consider the state of a quantum system in one dimension represented by the wave function $\psi(x)$, where x is the coordinate on the real line in a reference system S. If such a system is translated by an amount a to get a new reference system S' such that $x' = x - a$, a new wave function is defined in S' so that

$$\Psi'(x') = \Psi(x) = \Psi(x' + a). \tag{14.1.7}$$

Find the translation operator U such that

$$U\Psi(x) = \Psi(x + a). \tag{14.1.8}$$

14.8. At the initial time $t = 0$, a particle is in an eigenstate of L^2 and L_x, with quantum number $l = 1$. Assuming that the Hamiltonian ruling the system reads as

$$H = gBL_z, \tag{14.1.9}$$

with g a real constant, compute

(i) the possible values of a measurement of L_z;

(ii) their probabilities as functions of time;

knowing that, on the initial state, $\Delta L_z^2 = \hbar^2$.

14.9. Consider a particle described by the Hamiltonian

$$H = \frac{1}{2m}(p_x^2 + p_y^2) + \frac{1}{2}m\omega^2(x^2 + y^2) + \varepsilon L_z^2. \tag{14.1.10}$$

Compute the correction to the ground-state energy to first order in ε. Is it possible to solve the eigenvalue problem for H in exact form? If so, describe the method that you have used to diagonalize H.

14.10. A one-dimensional harmonic oscillator with Hamiltonian

$$H = \frac{p^2}{2m} + \frac{1}{2}m\omega^2 x^2,$$

is, at the initial time, in the quantum state

$$\psi_0(x) = N\exp\left\{-\frac{1}{2}\left(\sqrt{\frac{m\omega}{\hbar}}x - 1 - i\right)^2\right\}, \tag{14.1.11}$$

where N is a normalization constant. Compute

(i) the possible values of a measurement of H;

(ii) their probabilities as functions of time.

14.11. Consider a particle described by the Hamiltonian

$$H = \frac{1}{2m}(p_x^2 + p_y^2) + \frac{1}{2}m\left(\omega_x^2 x^2 + \omega_y^2 y^2\right) + \varepsilon L_z \cos(\Omega t), \tag{14.1.12}$$

with $\omega_x \neq \omega_y$, where the time-dependent term can be viewed as a perturbation. Compute, to second order in perturbation theory, the transition probability at time t from the ground state to the state corresponding to the energy level $\frac{3\hbar(\omega_x + \omega_y)}{2}$.

14.12. A particle is in a one-dimensional infinite potential well in the interval $x \in [0, a]$. Find

(i) eigenvalues and eigenvectors of the Hamiltonian;

(ii) the momentum distribution of the particle when it is in the quantum state with energy $E = \frac{2\pi^2\hbar^2}{(ma^2)}$.

14.13. Find the energy levels and the wave functions of a spin-1 particle with the Hamiltonian

$$H = A\widehat{S}_x^2 + B\widehat{S}_y^2 + C\widehat{S}_z^2, \tag{14.1.13}$$

where A, B, C are real constants.

14.14. A one-dimensional harmonic oscillator with Hamiltonian

$$H = \frac{p^2}{2} + \frac{x^2}{2} \tag{14.1.14}$$

is, at the initial time, in the quantum state

$$\psi_0(x) = N \exp\left\{-\frac{1}{2}(x - \alpha)^2\right\}, \tag{14.1.15}$$

where N is a normalization constant and α is a real parameter. Compute

(i) the possible values of a measurement of H;

(ii) the mean value $\langle x \rangle$ as a function of time.

14.15. Consider a system described by the Hamiltonian

$$H = \frac{\vec{L}^2}{2I} + \alpha(L_+ + L_-), \tag{14.1.16}$$

where $\vec{L}$ is the angular momentum and L_+, L_- are the raising and lowering operators for the L_z component. The parameters I and α are positive constants.

Find the eigenvalues of the Hamiltonian corresponding to $l = 1$. Moreover, find the mean value of H in the state $|l = 1, m_z = 1\rangle$.

14.16. A system with two possible quantum states, denoted by $|1\rangle$ and $|2\rangle$, is described by the Hamiltonian

$$H = \begin{pmatrix} 0 & m \\ m & M \end{pmatrix}, \tag{14.1.17}$$

with $m \ll M$. Compute the probability that, by preparing the system in state $|1\rangle$ at an initial time, it is found in the same state at a later time t. To simplify the calculation, approximate the formulae that you find to the first non-vanishing order in the parameter $\frac{m}{M}$.

14.17. Add to the Hamiltonian of the previous problem an energy term V which, in the $|1\rangle$, $|2\rangle$ basis, reads as

$$V = \begin{pmatrix} \delta & 0 \\ 0 & 0 \end{pmatrix}. \tag{14.1.18}$$

Repeat the calculations of the previous problem by treating V as a constant perturbation.

14.18. A particle of mass m is moving on the real line, subject to the Hamiltonian

$$H = -\frac{\hbar^2}{2m}\frac{d^2}{dx^2} + V_1\delta(x + a) + V_2\delta(x - a), \tag{14.1.19}$$

with $V_1, V_2 < 0$. Find whether bound states exist. If so, compute their energy.

14.19. Consider a one-dimensional harmonic oscillator whose Hamiltonian is corrected by relativistic effects up to quartic order in momentum. We then get the expression

$$\hat{H} = \frac{\hat{p}^2}{2m} + \frac{1}{2}m\omega^2\hat{x}^2 - \frac{1}{8m^3c^2}\hat{p}^4. \tag{14.1.20}$$

(i) Show how the additional term is obtained.

(ii) By regarding such a term as a perturbation, find the first-order correction to the ground-state energy, as well as the first-order-corrected ground state.

14.20. A spin-$\frac{1}{2}$ particle that moves on a line while being subject to an elastic force, is in the following quantum state:

$$\Psi = N \left(|\tfrac{1}{2}\rangle \otimes e^{i\hat{a}^\dagger} |0\rangle + |-\tfrac{1}{2}\rangle \otimes e^{-i\hat{a}^\dagger} |0\rangle \right), \tag{14.1.21}$$

where N is a normalization constant, $|\tfrac{1}{2}\rangle$ and $|-\tfrac{1}{2}\rangle$ are the eigenfunctions of $\hat{s}_3 = \frac{\hbar\sigma_3}{2}$, and $|0\rangle$ is the harmonic oscillator ground state. Compute

(i) the possible values of a measurement of the third component of the spin operator, and their respective probabilities;

(ii) the probability of finding the particle in the first excited state of the harmonic oscillator.

14.21. Consider a particle described by an harmonic oscillator Hamiltonian (in dimensionless variables) plus a perturbative term

$$H = \frac{1}{2}(P^2 + Q^2) + \varepsilon Q^4 \cos(\omega t). \tag{14.1.22}$$

Compute the probability, to the first non-vanishing order in ε, that the system, which is in the ground state at $t = 0$, ends up in the state $|2\rangle$ after a time t.

14.22. A one-dimensional harmonic oscillator of mass m and frequency ω is, at the initial time $t = 0$, in the quantum state described by the wave function

$$\Psi_0 = N(1 + Q)\exp\left(-\frac{Q^2}{2}\right), \tag{14.1.23}$$

where Q is the dimensionless position variable, and N is a normalization constant. Compute the time evolution of the state and the mean value of the position operator in such a state.

14.23. Add to the Hamiltonian of the previous problem the perturbation $\varepsilon Q^4 \cos(2\omega t)$, and compute the transition probability from the ground state $|0\rangle$ to the state $|2\rangle$ to first order in ε, as a function of t. What is the transition rate (defined as the time derivative of the transition probability) at large time intervals?

14.24. Consider the wave function defined on the whole real line by

$$\Psi(x) = N\exp(-\lambda|x|), \tag{14.1.24}$$

where N is a normalization constant and λ is a positive parameter. Compute the value of linear momentum for which the probability distribution achieves its maximum in momentum space.

14.25. A system of two distinguishable particles having spin $\frac{1}{2}$ is described by the Hamiltonian

$$H = \alpha \vec{S}_1^2 + \beta \vec{S}_2^2 + \gamma \vec{S}_1 \cdot \vec{S}_2. \tag{14.1.25}$$

Assume that, at the initial time $t = 0$, the quantum state of the system reads as

$$\psi_0 = |\frac{1}{2}, \frac{1}{2}\rangle_1 \, |\frac{1}{2}, -\frac{1}{2}\rangle_2, \tag{14.1.26}$$

in the basis $\vec{S}_1^2, S_{1z}, \vec{S}_2^2, S_{2z}$.

(i) Compute the probability that, at time t, the measurements of S_{1z} and S_{2z} yield $-\frac{\hbar}{2}$ and $+\frac{\hbar}{2}$, respectively.

(ii) Compute the same probability, to first order in perturbation theory, when the following term is added to the Hamiltonian:

$$V = \delta (S_{1x} + S_{2x}). \tag{14.1.27}$$

14.26. A particle moves in the closed interval $x \in [0, L]$, and its initial state is given by

$$\Psi(x) = \sqrt{\frac{30}{L^5}} \, x(x - L). \tag{14.1.28}$$

(i) Find the probability that, after a measurement of energy, the system is in the state

$$\psi_n(x) = \sqrt{\frac{2}{L}} \sin\left(\frac{n\pi x}{L}\right), \tag{14.1.29}$$

for which the energy takes the value $E_n = \frac{\hbar^2 \pi^2 n^2}{2mL^2}$.

(ii) If a measurement of energy yields the result $\frac{9\hbar^2 \pi^2 n^2}{2mL^2}$ what is the probability that, in a subsequent measurement of position, the particle is found in an interval of width dx centred about $x = \frac{L}{2}$?

(iii) If, in the previous question, no energy measurement is performed, what is the resulting probability of a position measurement?

14.27. Compute the time dependence of the probability that a spin-$\frac{1}{2}$ particle in a constant magnetic field along the z-axis, with magnitude B, is found in an eigenstate of $\hat{S}_y$, if the initial state is an eigenstate of $\hat{S}_x$ belonging to the eigenvalue $\frac{\hbar}{2}$.

14.28. Compute the mean values of energy and momentum of a particle whose state is represented by the wave function

$$\Psi(\vec{r}) = C \exp(-\alpha(\vec{r} - \vec{r}_0)^2), \tag{14.1.30}$$

where $\alpha^2 \equiv \frac{m\omega}{2\hbar}$, and the Hamiltonian is given by

$$\hat{H} = \frac{\hat{p}^2}{2m} + \frac{m\omega^2}{2}\hat{r}^2. \tag{14.1.31}$$

Find also the mean value and the mean quadratic deviation of the position operator.

14.29. A spin-$\frac{1}{2}$ nucleus is subject to a stationary magnetic field $\vec{B}_0$ directed along the z-axis, supplemented by a rotating component in the x–y plane of amplitude B_1 and frequency ω. If the nucleus points initially (at $t = 0$) towards the positive-z direction, what is the probability that the spin is again aligned along the same axis at a generic time t?

Suggestion: write the spin wave function in the form

$$\Psi(t) = \begin{pmatrix} \alpha(t) \\ \beta(t) \end{pmatrix},$$ (14.1.32)

and consider the Schrödinger-type equation

$$i\frac{\partial}{\partial t}\Psi(t) = -\mu_N\vec{\sigma} \cdot \vec{B}(t).$$ (14.1.33)

14.30. Compute Δx and Δp of the following wave function:

$$\Psi(x) = \left(\frac{1}{2\pi\sigma^2}\right)^{\frac{1}{4}} \exp\left\{-\frac{(x - x_0)^2}{4\sigma^2}\right\} \exp(ik_0x),$$ (14.1.34)

and check that the Heisenberg uncertainty relation is fulfilled.

14.31. Consider a particle of mass m and charge q subject to a magnetic field $\vec{B} = B\vec{j}$ and an electric field $\vec{E} = E\vec{k}$, both of them being uniform. Compute the energy eigenvalues and eigenfunctions. (Suggestion: the calculation can be carried out by analogy with that for Landau levels.)

14.32. Consider a particle of mass m moving in one dimension and subject to the potential

$$V(x) = 0 \text{ if } x < 0, \ -V_0 \text{ if } x > 0.$$ (14.1.35)

Consider an energy eigenstate corresponding to a beam of incoming particles, with momentum p, coming from $x \to \infty$. Find the amplitude of reflected and transmitted particles, with the associated reflection and transmission coefficients, and check that their sum equals 1.

15 Definitions of geometric concepts

15.1 Outline

Geometry is an appropriate mathematical language for the description of classical mechanics, special relativity, quantum mechanics, as well as all fundamental interactions known in nature. For example, when Einstein unified space and time in his formulation of special relativity, the appropriate mathematical concept was the space–time manifold studied by Minkowski, jointly with the theory of the Lorentz and Poincaré groups. Hereafter we present an incomplete glossary of geometry, aimed at introducing the concepts more frequently used in the main body of the text. The advanced reader may skip it, but we hope that third-year undegraduate students (we only assume familiarity with the concepts of vector space and its dual, and with differentiable manifolds and elementary definition of tensors) will find it useful. The key concepts of the geometric formulation we are interested in are elementary group theory, vector fields, 1-forms, Lie algebras and maps of manifolds.

15.2 Groups

A group G is a set of elements endowed with a binary internal composition law, say $B : G \times G \to G$, often denoted by multiplication, i.e. $B(a, b) = ab$, and having the following properties:

(i) B is associative, i.e. $(ab)c = a(bc)$ for all $a, b, c \in G$.
(ii) There is an element of G, denoted by I, called the neutral or unit element, such that $Ia = aI = a$ for all $a \in G$.
(iii) For all $a \in G$, there exists $a^{-1} \in G$, called the inverse element, such that $a^{-1}a = aa^{-1} = I$.

   In a group the composition law is not, in general, commutative. This would mean that

(iv) $ab = ba$ for all $a, b \in G$.

The groups in which (iv) holds are said to be commutative or Abelian. In the Abelian groups the composition law is generally denoted by addition. In this case the neutral element is called *zero* and denoted by 0, and the inverse of an element γ is said to be its opposite and denoted by $-\gamma$. The sets $\mathbf{Z}, \mathbf{Q}, \mathbf{R}, \mathbf{C}$ are Abelian groups with respect to ordinary addition, but they are not even groups with respect to ordinary multiplication.

15.3 Lie groups

A very important class of groups consists of Lie groups. They are groups which are also a manifold endowed with a C^∞ structure such that the maps

$$(g, h) \to g \cdot h,$$

$$g \to g^{-1}$$

are C^∞ functions. It was a non-trivial result of Lie-group theory to arrive at a C^∞ structure, since the original definition demanded that the product and inverse maps should be analytic (Montgomery and Zippin 1940, Chevalley 1946, Montgomery and Zippin 1955). Examples of Lie groups, among many, are as follows:

(i) The vector space $\mathbf{R}^n$ endowed with the addition operation.

(ii) The circle S^1 viewed as the quotient space $\frac{\mathbf{R}}{\mathbf{Z}}$.

(iii) If G and H are Lie groups, their product $G \times H$ is also a Lie group.

(iv) By virtue of (iii), the torus $S^1 \times S^1$ is a Lie group.

(v) The general linear group $GL(n, \mathbf{R})$ of all $n \times n$ real non-singular matrices.

(vi) The orthogonal group $O(n)$, i.e. the proper subset of $GL(n, \mathbf{R})$ consisting of all matrices A such that

$$A \in GL(n, \mathbf{R}) : AA^t = I. \tag{15.3.1}$$

With respect to the usual base in $\mathbf{R}^n$, the matrix A represents a linear map which is an isometry, i.e. the Euclidean inner product and the associated norm.

(vii) If $H \subset G$ is both a subgroup of a Lie group G and a closed sub-manifold of G, then H is a Lie group. Thus, $S^1 \subset \mathbf{R}^2$ emerges once more as a Lie group, being the sub-manifold of $\mathbf{R}^2$ consisting of

$$z \in \mathbf{C} : z\bar{z} = x^2 + y^2 = 1. \tag{15.3.2}$$

(viii) Among all n-spheres S^n, only S^1 and S^3 are Lie groups. The 3-sphere S^3 can be viewed as the Lie group of quaternions of unit norm. The quaternions x are built from three imaginary unit symbols i, j, k according to the rules

$$i^2 = j^2 = k^2 = -1, \tag{15.3.3}$$

$$ij = -ji = k, \ jk = -kj = i, \ ki = -ik = j, \tag{15.3.4}$$

$$x \equiv x_1 + x_2 i + x_3 j + x_4 k. \tag{15.3.5}$$

The conjugate quaternion $\bar{x}$ is then given by

$$\bar{x} \equiv x_1 - x_2 i - x_3 j - x_4 k, \tag{15.3.6}$$

and we find the unit norm condition

$$x\bar{x} = \sum_{i=1}^{4} x_i^2 = 1,$$ (15.3.7)

which indeed is also the defining equation for the 3-sphere.

(ix) The group $SO(n)$ of orthogonal $n \times n$ matrices with unit determinant. A matrix realization of this group is provided by means of Pauli matrices, i.e. $\sigma_0, i\sigma_1, i\sigma_2, i\sigma_3$, σ_0 being the 2×2 identity matrix.

(x) The group $E(n)$ of all isometries of the Euclidean space $\mathbf{R}^n$. Each element of $E(n)$ can be written uniquely in the form $A \cdot \tau$, where A is an orthogonal $n \times n$ matrix, and τ is a translation. The group $E(n)$ may be realized from $(n+1) \times (n+1)$ matrices.

15.4 Symmetry

Symmetry pervades all our descriptions, modelling or understanding of natural phenomena. From microscopic physics to cosmology, from chemistry to biology, symmetry properties or considerations are ubiquitous in scientific research.

In ancient times, symmetry meant mostly '*harmony in the proportions*', and in modern natural science it emerges mostly as the '*invariance with respect to a transformation group*'. In abstract terms, any set $\mathcal{S}$ is associated with a group Aut $\mathcal{S}$, the family of all invertible maps $\varphi : \mathcal{S} \to \mathcal{S}$, which may be composed:

$$(\varphi_1 \cdot \varphi_2)(s) = \varphi_1(\varphi_2(s)),$$ (15.4.1)

and the composition is associative, i.e.

$$(\varphi_1 \cdot \varphi_2) \cdot \varphi_3 = \varphi_1 \cdot (\varphi_2 \cdot \varphi_3),$$ (15.4.2)

it allows for the identity transformation

$$I : s \to s, \ \forall s \in \mathcal{S},$$ (15.4.3)

and an inverse φ^{-1} such that

$$\varphi(\varphi^{-1}(s)) = \varphi^{-1}(\varphi(s)) = I \cdot s = s.$$ (15.4.4)

Any subset, pattern, or configuration $\mathcal{P}$ in $\mathcal{S}$ defines a subgroup of Aut $\mathcal{S}$; by selecting only those transformations which preserve or keep invariant $\mathcal{P}$, for any $p \in \mathcal{P} \subset \mathcal{S}$ we consider only those transformations $\varphi \in$ Aut $\mathcal{S}$ such that $\varphi(p) \in \mathcal{P}$, $\forall p \in \mathcal{P}$. This subgroup is never empty because it contains at least the identity transformation that constitutes the smallest subgroup in Aut $\mathcal{S}$. In addition to subsets, we may select subgroups of Aut $\mathcal{S}$, by requiring that properties or relations among pairs or more points are preserved. This strict connection between configurations and transformations was explicitly introduced by F. Klein and S. Lie. In the Erlangen program (1872) of F. Klein, a geometric theory is defined as the study of those properties of the space, $\mathcal{S}$, and of its subsets (figures) which are preserved with respect to a selected subgroup of Aut $\mathcal{S}$. Similarly, in physics, starting with Galileo and culminating with Einstein, every physical theory carries with it its *covariance* group.

It was P. Curie (1894) who (translated and) incorporated into physics the role of symmetry as a working tool in our formalization of the external world, by stating that 'symmetries in the cause should be reflected in the effects, and the lack of symmetries in the effects should be searched for in the causes'. Of course, the occurrence of *natural symmetry breaking* would require a better formulation of Curie's principle. Symmetries as invariance with respect to a selected group of transformations are dealt with in physics as invariance principles. Explicitly we find the requirement of invariance in special relativity with respect to the Poincaré group (see below); this brings in the notion of the reference system, their equivalence or lack of.

When $\mathcal{S}$ is discrete, Aut $\mathcal{S}$ is the permutation group. When $\mathcal{S}$ is vector space, Aut $\mathcal{S}$ is the general linear group. When $\mathcal{S}$ is a differentiable manifold, Aut $\mathcal{S}$ is the diffeomorphism group. When we endow $\mathcal{S}$ with a Euclidean inner product, the general linear group reduces to the group of isometries, $E(n)$, if $n = \dim \mathcal{S}$.

The use of transformations for classification purposes has introduced in physics the transformation method, i.e. a given system is transformed into an equivalent one considered as a model system or a *normal form* of it. Being connected by the action of an element of G is an equivalence relation, because it is reflexive, symmetric and transitive. In heuristic terms we can say that to similar problems there correspond similar solutions (similar means *connected by symmetries*). In analytic mechanics, this approach has brought in the use of canonical transformations, Hamilton and Jacobi. In this approach, evolution itself, the dynamics, is presented as a 1-parameter group of transformations. In modern physics, the formulation of special relativity and general relativity on the one hand, and the extension of canonical transformations in the quantum setting due to P. Dirac on the other hand, has intimately connected physics, geometry and transformation groups. In mathematics, this research line has culminated in the study of Lie groups, Lie algebras, representation theory and, in particular, unitary representations.

By definition, a *symmetry* of a dynamical equation is any transformation which maps bijectively the set of solutions onto itself. Although not strictly needed, we can add the further requirement that symmetries should preserve the parametrization of solutions. In the case of a dynamical system obeying Newton-type equations of motion, i.e.

$$\frac{dq^i}{dt} = u^i, \quad \frac{du^i}{dt} = f^i, \tag{15.4.5}$$

we may or may not require the relation between positions and velocities to be preserved. In the former case, symmetry transformations are further qualified as *point symmetries*.

15.5 Various definitions of vector fields

When we first learn the intrinsic definition of tangent vectors, we think of them as an equivalence class of curves at a point p of some manifold M. The equivalence relation is that two curves are tangent at the point p. The equivalence class at p of a particular curve σ is denoted by $[\sigma]_p$. The collection of all equivalence classes of curves at p is a vector space denoted by $T_p(M)$. The union of all these vector spaces, i.e.

$$\cup_{p \in M} \{p\} \times T_p(M) \equiv TM, \tag{15.5.1}$$

is called the tangent bundle of M.

A *vector field* X on a C^∞ manifold M is a *smooth* assignment of a tangent vector $X_p \in T_p(M)$ for each point $p \in M$. By smooth we mean that, for all $f \in C^\infty(M)$, the function

$$p \in M \to (Xf)(p) \equiv X_p(f)$$

is C^∞. Thus, a vector field may be viewed as a map

$$X : C^\infty(M) \to C^\infty(M)$$

in which $f \to X(f)$ is defined as above. The function $X(f)$ is called the Lie derivative (see also Section 15.8) of the function f along the vector field X, and is usually denoted by $L_X f$.

The map $f \to X(f)$ has the following properties:

$$X(f + h) = X(f) + X(h) \quad \forall f, h \in C^\infty(M), \tag{15.5.2}$$

$$X(rf) = rX(f) \quad \forall f \in C^\infty(M) \text{ and } r \in \mathbf{R}, \tag{15.5.3}$$

$$X(fh) = fX(h) + hX(f) \quad \forall f, h \in C^\infty(M). \tag{15.5.4}$$

These formulae should be compared with the ones defining a *derivation* at a point $p \in M$. These are maps $v : C^\infty(M) \to \mathbf{R}$ such that

$$v(f + h) = v(f) + v(h) \quad \forall f, h \in C^\infty(M), \tag{15.5.5}$$

$$v(rf) = rv(f) \quad \forall f \in C^\infty(M) \text{ and } r \in \mathbf{R}, \tag{15.5.6}$$

$$v(fh) = f(p)v(h) + h(p)v(f) \quad \forall f, h \in C^\infty(M). \tag{15.5.7}$$

The set of all derivations at $p \in M$ is denoted by $D_p M$.

Equations (15.5.2) and (15.5.3) show that X is a linear map from the vector space $C^\infty(M)$ into itself, and Eq. (15.5.4) shows that X is a derivation of the set $C^\infty(M)$. It is therefore possible to *define* a vector field as a derivation of the set of functions $C^\infty(M)$ satisfying Eqs. (15.5.2)–(15.5.4). Such a map assigns a derivation, in the sense of Eqs. (15.5.5)–(15.5.7), to each point $p \in M$, denoted X_p and defined by

$$X_p(f) \equiv [X(f)](p), \tag{15.5.8}$$

for each $f \in C^\infty(M)$. By construction, X_p satisfies Eqs. (15.5.5)–(15.5.7). But then the map $p \to X_p$ assigns a field of tangent vectors, and this assignment is smooth. Thus, a completely equivalent way of defining a vector field in the first place is to regard it as a derivation on the set $C^\infty(M)$. This alternative approach, where we go from a vector field to a derivation at a point $p \in M$, is useful in many applications. If (U, ϕ) is a local coordinate chart on the manifold M, the derivations X_p associated with a vector field X defined on U make it possible to express $(Xf)(p)$ as

$$(Xf)(p) = X_p(f) = \sum_{\mu=1}^{m} X_p(x^\mu) \left(\frac{\partial}{\partial x^\mu} \right)_p f, \tag{15.5.9}$$

which implies that

$$X = \sum_{\mu=1}^{m} X(x^\mu) \frac{\partial}{\partial x^\mu}. \tag{15.5.10}$$

This formula shows the precise sense in which a vector field may be viewed as a homogeneous first-order differential operator on the functions on a manifold. The functions $X(x^\mu)$ are defined on the open set U that defines the coordinate chart with the coordinate functions x^μ, and are the components of the vector field X with respect to this coordinate system. It is a theorem by Willmore that all derivations of $C^\infty(M)$ are given by vector fields.

A crucial question is whether or not one can compose two vector fields X and Y to get a third field. Indeed, we know that X and Y can be viewed as linear maps of $C^\infty(M)$ into itself, and hence we can define the composite map $X \cdot Y : C^\infty(M) \to C^\infty(M)$ as

$$X \cdot Y(f) \equiv X(Yf). \tag{15.5.11}$$

By construction, this is a linear map, in that it satisfies Eqs. (15.5.2) and (15.5.3). However, Eq. (15.5.11) is not a vector field since it fails to satisfy Eq. (15.5.4). By contrast, we find

$$X \cdot Y(fh) = X(Y(fh)) = X(hY(f) + fY(h))$$
$$= X(h)Y(f) + hX(Y(f))$$
$$+ X(f)Y(h) + fX(Y(h)), \tag{15.5.12}$$

which does not equal $hX(Y(f)) + fX(Y(h))$. However, in addition,

$$Y \cdot X(fh) = Y(h)X(f) + hY(X(f)) + Y(f)X(h) + fY(X(h)). \tag{15.5.13}$$

We can now subtract Eqs. (15.5.12) and (15.5.13) to find

$$(X \cdot Y - Y \cdot X)(fh) = h(X \cdot Y - Y \cdot X)(f) + f(X \cdot Y - Y \cdot X)(h). \tag{15.5.14}$$

We have thus found that $X \cdot Y - Y \cdot X$ is a vector field, even though the individual pieces $X \cdot Y$ and $Y \cdot X$ are not. The new vector field just obtained is called the commutator of X and Y, and is denoted by $[X, Y]$. If X, Y and Z are any three vector fields on M, their commutators satisfy the Jacobi identity:

$$[X, [Y, Z]] + [Y, [Z, X]] + [Z, [X, Y]] = 0. \tag{15.5.15}$$

The consideration of such structures becomes natural within the framework of Lie algebras, as shown in Section 15.7.

In this book we are interested in the interpretation of vector fields suggested by systems of first-order differential equations. More precisely, we know that, given any differential equation

$$\dot{x} = F(x, y), \tag{15.5.16}$$

$$\dot{y} = G(x, y), \tag{15.5.17}$$

the derivative of functions are defined along solutions by setting

$$\frac{d}{dt}f = \frac{\partial f}{\partial x}\frac{dx}{dt} + \frac{\partial f}{\partial y}\frac{dy}{dt} = \left(F\frac{\partial}{\partial x} + G\frac{\partial}{\partial y}\right)f. \tag{15.5.18}$$

The homogeneous first-order differential operator

$$X = F\frac{\partial}{\partial x} + G\frac{\partial}{\partial y} \tag{15.5.19}$$

is called a vector field. If our differential equation admits solutions for any initial condition from $-\infty$ to $+\infty$ in time, it defines the action of a 1-parameter group of transformations. The corresponding vector field is said to be complete and is called the *infinitesimal generator* of the 1-parameter group ϕ_t.

15.6 Covariant vectors and 1-form fields

A covariant vector ω at a point p of the manifold M is a real-valued linear function on the tangent space $T_p(M)$. If u is a vector at p, the number into which ω maps u is written $\langle \omega, u \rangle$. Then linearity of the map implies that

$$\langle \omega, \alpha u + \beta v \rangle = \alpha \langle \omega, u \rangle + \beta \langle \omega, v \rangle \tag{15.6.1}$$

holds for all $\alpha, \beta \in \mathbf{R}$ and for all $u, v \in T_p(M)$.

Given a basis $\{E_a\}$ of vectors at p, we can define a unique set of n covectors $\{E^a\}$ by the condition that E^i should map any vector u to the number u^i, which is the ith component of u with respect to the basis $\{E_a\}$. In particular, this yields

$$\langle E^a, E_b \rangle = \delta^a{}_b. \tag{15.6.2}$$

For example, if $(x^1, \ldots, x^n)$ are local coordinates on the manifold M, the set of differentials $(dx^1, \ldots, dx^n)$ at p form the basis of covectors dual to the basis $\left(\frac{\partial}{\partial x^1}, \ldots, \frac{\partial}{\partial x^n}\right)$ of vectors at p, since

$$\left\langle dx^i, \frac{\partial}{\partial x^j} \right\rangle = \frac{\partial x^i}{\partial x^j} = \delta^i{}_j. \tag{15.6.3}$$

The differential of a function f, written as $df = \frac{\partial f}{\partial x^j}dx^j$, does not depend on the chosen coordinate system which we use, i.e.

$$df = \frac{\partial f}{\partial x^j}dx^j = \frac{\partial f}{\partial y^k}dy^k. \tag{15.6.4}$$

This means that we can identify

$$\mathrm{d} = \mathrm{d}x^j \frac{\partial}{\partial x^j} = \mathrm{d}y^k \frac{\partial}{\partial y^k}, \tag{15.6.5}$$

which acts as a derivation (on sums and products) and behaves like a *scalar operator* under any change of coordinates. In particular, this means that an expression like $g\,\mathrm{d}f$, with g and f functions, behaves like a scalar because it is made out of scalar objects. This property of 1-forms is the main motivation to formulate the Einstein–de Broglie relation (1.1.2) in terms of 1-forms instead of 4-vectors.

By defining linear combinations of covectors by the rules

$$\langle \alpha\omega + \beta\eta, u \rangle = \alpha\langle \omega, u \rangle + \beta\langle \eta, u \rangle \tag{15.6.6}$$

for any covector ω, η and any $\alpha, \beta \in \mathbf{R}, u \in T_p, \{E^a\}$ can be regarded as a basis of covectors, since any covector ω at p can be expressed as

$$\omega = \omega_i E^i, \quad \text{where } \omega_i = \langle \omega, E_i \rangle. \tag{15.6.7}$$

Thus, the set of all covectors at p forms an n-dimensional vector space at p, the dual space $T_p^*(M)$ of the tangent space $T_p(M)$, called the cotangent space at $p \in M$. The basis $\{E^a\}$ of covectors is therefore the dual basis to the basis $\{E_a\}$ of vectors. For any $\omega \in T_p^*(M)$ and $u \in T_p(M)$, the number $\langle \omega, u \rangle$ can be expressed in terms of the components ω_i, u^i of ω, u with respect to dual bases $\{E^a\}, \{E_a\}$ by the relation

$$\langle \omega, u \rangle = \langle \omega_i E^i, u^j E_j \rangle = \omega_i u^i. \tag{15.6.8}$$

We can formalize the previous construction and build the following space, frequently encountered in classical mechanics and in the theory of all fundamental interactions:

$$T^*M \equiv \cup_{p \in M} \{p\} \times T_p^*(M). \tag{15.6.9}$$

The space T^*M is called the cotangent bundle of M, and is dual to the concept of the tangent bundle defined in Section 15.5.

Recall now that, in the case of vectors, when a smooth assignment of the tangent vector X_p is made as p varies on M, we get a vector field. Similarly, when a smooth assignment of covector ω_p is made as p varies on M, we get a 1-form field (or covector field). Without the qualifying term *field*, we may therefore say that a covector is a 1-form. In old-fashioned language, we talk instead of contravariant vectors (i.e. our vectors) or covariant vectors, and their smooth assignments as p varies on M were then called contravariant vector fields or covariant vector fields. In component language, we may pass from components of a vector to components of the corresponding covector using a metric or any other $(0, 2)$ tensor field, i.e.

$$u_\mu = g_{\mu\nu}u^\nu, \quad u^\mu = g^{\mu\nu}u_\nu. \tag{15.6.10}$$

These equations express the isomorphism between the tangent and the cotangent space at $p \in M$: $T_p(M) \cong T_p^*(M)$. Such an isomorphism is said to be *not natural* since it involves an additional structure, i.e. the assignment of a metric g.

Note that no notation is completely satisfactory. For example, even though nowadays X is a standard notation for vector fields, and Ω is a standard notation for differential forms (the 1-forms defined here, or higher-order forms, which contain all the information about anti-symmetric tensors), if we were to re-invent geometry from scratch, or to describe our standard of knowledge to another (imaginary) civilization, we could hardly expect that any letter whatsoever contains enough information about the nature of the object we are considering. For this reason, instead of using the index-free intrinsic notation, or the component notation (which is inappropriate if we want to emphasize the geometric nature of physical laws), some authors have introduced the abstract-index notation, according to which a letter and a family of labels are needed. For example, with abstract-index notation, X^a are not the components X^μ of the vector X, but represent the vector X itself. The aim is therefore to exploit both the virtues of index-free notation, and the power of tensor calculus. Despite such a property, this notation, too, runs into trouble at certain stages, and its use of lower-case Latin indices is dual to their role in the theory of fundamental interactions, where we are dealing, for example, with 1-forms taking values in a suitable vector space (the Lie algebra defined in the following section). In that case, the a upstairs-index is used for 1-forms, unlike its use in abstract-index notation, where it denotes vectors. At a deeper level, we needs a purely pictorial, diagrammatic version for geometric objects and operations upon them, not too different from what humankind has invented for music. This is indeed available thanks to the work of Penrose (see Penrose and Rindler (1984)).

15.7 Lie algebras

The notion of Lie algebra makes it possible to deal with several ordinary differential equations at the same time. Their integration gives rise to the notion of the action of a Lie group, instead of a 1-parameter group of transformations.

A vector space E over the real or complex numbers is said to be a Lie algebra if a bilinear map $E \times E \to E$ exists, often denoted by [,], such that

$$[u, v] = -[v, u], \tag{15.7.1}$$

$$[u, [v, w]] = [[u, v], w] + [v, [u, w]]. \tag{15.7.2}$$

Equation (15.7.2) describes the fundamental Jacobi identity, and is written in a way that emphasizes the link with the Leibniz rule for derivations.

Examples of Lie algebras are provided by $n \times n$ matrices with real or complex coefficients, whose bilinear map is defined by

$$[A, B] \equiv A \cdot B - B \cdot A, \tag{15.7.3}$$

where the *dot* denotes the row by column product. Yet another familiar example is provided by the vector product on three-dimensional vector spaces:

$$[\vec{u}, \vec{v}] \equiv \vec{u} \times \vec{v}. \tag{15.7.4}$$

The reader may check that Eqs. (15.7.1) and (15.7.2) are then satisfied. If we associate with any vector $\vec{u}$ the linear transformation defined by

$$T_{\vec{u}}(\vec{v}) \equiv \vec{u} \times \vec{v}, \tag{15.7.5}$$

in a given orthonormal basis we can associate a 3×3 matrix with $T_{\vec{u}}$. We thus find, associated with the unit vectors $\vec{i}, \vec{j}, \vec{k}$,

$$T_{\vec{i}} \begin{pmatrix} 0 \\ 1 \\ 0 \end{pmatrix} = \begin{pmatrix} 0 \\ 0 \\ 1 \end{pmatrix}, \tag{15.7.6}$$

$$T_{\vec{i}} \begin{pmatrix} 1 \\ 0 \\ 0 \end{pmatrix} = \begin{pmatrix} 0 \\ 0 \\ 0 \end{pmatrix}, \tag{15.7.7}$$

$$T_{\vec{i}} \begin{pmatrix} 0 \\ 0 \\ 1 \end{pmatrix} = \begin{pmatrix} 0 \\ -1 \\ 0 \end{pmatrix}, \tag{15.7.8}$$

which implies

$$T_{\vec{i}} = \begin{pmatrix} 0 & 0 & 0 \\ 0 & 0 & -1 \\ 0 & 1 & 0 \end{pmatrix}. \tag{15.7.9}$$

A calculation along the same lines yields

$$T_{\vec{j}} = \begin{pmatrix} 0 & 0 & 1 \\ 0 & 0 & 0 \\ -1 & 0 & 0 \end{pmatrix}, \quad T_{\vec{k}} = \begin{pmatrix} 0 & -1 & 0 \\ 1 & 0 & 0 \\ 0 & 0 & 0 \end{pmatrix}. \tag{15.7.10}$$

The matrices $T_{\vec{i}}, T_{\vec{j}}$ and $T_{\vec{k}}$ define a Lie algebra, and the correspondence just derived is an example of Lie algebra *homomorphism*. We are actually dealing with the Lie algebra of the rotation group in three dimensions. In Chapter 3 we represented the same algebra by first-order differential operators.

By integrating simultaneously a set of vector fields, we find the action of a *multi-parameter* group. This is the basic concept of a Lie group, first encountered in Chapter 3. Since the integration may give rise to various problems, it is common to introduce Lie groups autonomously and then show that the infinitesimal generators of a Lie group define a Lie algebra.

15.8 Lie derivatives

In the intrinsic formulation of geometric properties, Lie derivatives belong to the set of operations that can be defined on any manifold without introducing additional structures.

For our purposes, we can say that the action of a homogeneous first-order differential operator on functions is called a Lie derivative. As we said in Section 15.1, given the vector field (summation over repeated indices is understood)

$$X = \alpha^i \frac{\partial}{\partial x^i}, \tag{15.8.1}$$

we define

$$L_X f \equiv X f \equiv \alpha^i \frac{\partial f}{\partial x^i}. \tag{15.8.2}$$

Given any matrix A, it is possible to associate with it a vector field

$$X_A = (x_1, x_2, \ldots, x_n) \begin{pmatrix} a_{11} & \cdots & a_{1n} \\ a_{n1} & \cdots & a_{nn} \end{pmatrix} \begin{pmatrix} \partial_1 \\ \partial_2 \\ \cdots \\ \partial_n \end{pmatrix}, \tag{15.8.3}$$

where $\partial_j \equiv \frac{\partial}{\partial x^j}$. The advantage of the vector field with respect to the matrix A is that, if we use quadratic polynomials instead of linear ones, we find another matrix representation on some higher dimensional vector space; similarly, if we use homogeneous polynomials of higher degree. In some sense, the *vector-field* representation is a kind of *universal* representation able to reproduce all higher tensor representations.

15.9 Symplectic vector spaces

A symplectic vector space is a pair (V, ω), where V is a vector space (e.g. over $\mathbf{R}$) and $\omega : V \times V \to \mathbf{R}$ is a bilinear form such that

$$\omega(u, v) = -\omega(v, u) \ \forall u, v \in V, \tag{15.9.1}$$

$$\omega(u, u) = 0 \ \forall u \in V, \tag{15.9.2}$$

$$\omega(u, v) = 0 \ \forall v \in V \Longrightarrow u = 0. \tag{15.9.3}$$

The bilinear form ω is said to be a symplectic form in this case, and the three properties above state that ω is skew-symmetric, totally isotropic and non-degenerate, respectively.

The standard symplectic space is $\mathbf{R}^{2n}$ with the symplectic form given by a non-singular, skew-symmetric matrix. Typically, ω is chosen to be the block matrix

$$\omega = \begin{pmatrix} 0 & I_n \\ -I_n & 0 \end{pmatrix}, \tag{15.9.4}$$

where I_n is the $n \times n$ identity matrix. In terms of basis vectors, consisting of the n-tuple $(e_1, \ldots, e_n)$ and of the n-tuple $(u_1, \ldots, u_n)$

$$\omega(e_i, u_j) = -\omega(u_j, e_i) = \delta_{ij}, \tag{15.9.5}$$

$$\omega(e_i, e_j) = \omega(u_i, u_j) = 0. \tag{15.9.6}$$

Our procedure to associate tensor fields with algebraic objects would give

$$\Omega = (dx_1, dx_2, \ldots, dx_n) \begin{pmatrix} 0 & I_n \\ -I_n & 0 \end{pmatrix} \begin{pmatrix} dx_1 \\ dx_2 \\ \cdots \\ dx_n \end{pmatrix}$$

$$= dx_j \otimes dx_{j+n} - dx_{j+n} \otimes dx_j. \tag{15.9.7}$$

By using the inverse matrix $\lambda \equiv \begin{pmatrix} 0 & -I_n \\ I_n & 0 \end{pmatrix}$ and contravariant vectors we would have

$$\Lambda = (\partial_1, \partial_2, \ldots, \partial_n) \begin{pmatrix} 0 & -I_n \\ I_n & 0 \end{pmatrix} \begin{pmatrix} \partial_1 \\ \partial_2 \\ \cdots \\ \partial_n \end{pmatrix}$$

$$= \frac{\partial}{\partial x_j} \otimes \frac{\partial}{\partial x_{j+n}} - \frac{\partial}{\partial x_{j+n}} \otimes \frac{\partial}{\partial x_j}. \tag{15.9.8}$$

The remarkable property of the object we define is that it also represents a bidifferential operator, so that we can write

$$\Lambda(df, dg) = \frac{\partial f}{\partial x_j} \frac{\partial g}{\partial x_{j+n}} - \frac{\partial f}{\partial x_{j+n}} \frac{\partial g}{\partial x_j} \equiv \{f, g\}. \tag{15.9.9}$$

Thus, tensor fields in contravariant form also allow for the interpretation in terms of multi-differential operators, independently of their being symmetric or skew-symmetric tensors.

15.10 Homotopy maps and simply connected spaces

Suppose that, in the open set Ω of $\mathbf{R}^2$, two continuous closed curves φ and ψ are assigned. We can assume that the parametric interval is the same for both curves and is given by the closed interval $[0, 1]$. The idea underlying the homotopy relation that we are going to introduce is as follows: one of the curves can be transformed into the other through a continuous deformation, while remaining always in Ω.

Definition. The continuous map

$$G : [0, 1] \times [0, 1] \to \Omega \tag{15.10.1}$$

is a *homotopy map* between φ and ψ if it has the properties

$$G(t, 0) = \varphi(t), \ G(t, 1) = \psi(t) \ \forall t \in [0, 1], \tag{15.10.2}$$
$$G(0, \lambda) = G(1, \lambda) \ \forall \lambda \in [0, 1]. \tag{15.10.3}$$

As the parameter λ varies in between 0 and 1 we therefore obtain a family of curves varying continuously from φ to ψ. The homotopy relation between closed curves is reflexive, symmetric and transitive, and is therefore an equivalence relation.

Definition. A connected open set Ω of $\mathbf{R}^n$ is said to be simply connected if every closed curve therein is homotopic to a constant (a constant curve is a curve whose image is a point).

15.10.1 Examples of spaces which are or are not simply connected

(i) Recall from elementary topology that an open set Ω of $\mathbf{R}^2$ is starred with respect to one of its points, say P, if every point of Ω can be joined with P by means of a segment completely contained in Ω (thus, convex sets are starred with respect to all their points). In general, in $\mathbf{R}^n$, an open set Ω, which is starred with respect to one of its points, is simply connected. To prove it, suppose for simplicity that Ω contains the origin O of $\mathbf{R}^n$ and is starred with respect to O. Given any closed curve φ in Ω, with a parametric interval $[0, 1]$, the map $[0, 1] \times [0, 1] \to \Omega$ that associates $\lambda \varphi(t)$ to the pair (t, λ) is a homotopy map which turns every curve into a constant curve having as image the origin of $\mathbf{R}^n$.

(ii) If P is an arbitrary point of $\mathbf{R}^2$, the set $\mathbf{R}^2 - \{P\}$ is not simply connected because, if we consider a circle γ having P as an internal point, it is not possible to reduce γ in a continuous way to a point without passing over P.

(iii) The subset of $\mathbf{R}^3$ given by the triplets (x_1, x_2, x_3) of real numbers such that

$$1 \le (x_1^2 + x_2^2 + x_3^2) \le 4, \tag{15.10.4}$$

is simply connected.

The property of being simply connected has several applications in mathematics and physics. For example, if the linear differential form

$$\omega = A(x, y)\mathrm{d}x + B(x, y)\mathrm{d}y \tag{15.10.5}$$

is of class C^1, is defined on a simply connected open set of $\mathbf{R}^2$, and fulfills the condition

$$\frac{\partial A}{\partial y} = \frac{\partial B}{\partial x}, \tag{15.10.6}$$

it is then an exact form, i.e. ω is the differential of a function.

15.11 Diffeomorphisms of manifolds

Given two manifolds M and N, a differentiable map $f : M \to N$ is called a *diffeomorphism* if it is a bijection and its inverse $f^{-1} : N \to M$ is differentiable as well. If these functions are r times continuously differentiable, f is called a C^r-diffeomorphism.

Two manifolds M and N are diffeomorphic if there is a diffeomorphism f from M to N. They are C^r diffeomorphic if there is an r times continuously differentiable bijective map between them whose inverse is also r times continuously differentiable.

If U, V are connected open subsets of $\mathbf{R}^n$ such that V is simply connected, a differentiable map $f : U \to V$ is a diffeomorphism if it is proper and if the differential $\mathrm{d}f_x : \mathbf{R}^n \to \mathbf{R}^n$ is bijective at each point x in U. Note that it is essential for U to be simply connected for the function f to be globally invertible. For example, consider the map

$$f : \mathbf{R}^2 - \{(0,0)\} \to \mathbf{R}^2 - \{(0,0)\}, \ (x,y) \to (x^2 - y^2, 2xy). \tag{15.11.1}$$

Then f is surjective and satisfies

$$\det \mathrm{d}f_x = 4(x^2 + y^2) \neq 0. \tag{15.11.2}$$

Thus, the differential is bijective at each point yet f is not invertible because it fails to be injective, e.g. one has $f(-1,0) = f(1,0) = (1,0)$.

Diffeomorphisms are necessarily between manifolds of the same dimension. Imagine that f were going from dimension n to dimension k. If $n < k$ then the differential could never be surjective, whereas if $n > k$ then the differential could never be injective. Hence in both cases the differential would fail to be a bijection, which contradicts what is expected of it.

15.12 Foliations of manifolds

A p-dimensional foliation F of an n-dimensional manifold M is a covering by charts U_i together with maps

$$\phi_i : U_i \to \mathbf{R}^n \tag{15.12.1}$$

such that, for overlapping pairs U_i, U_j the transition functions $\varphi_{ij} : \mathbf{R}^n \to \mathbf{R}^n$ defined by

$$\varphi_{ij} \equiv \phi_j \phi_i^{-1} \tag{15.12.2}$$

take the form

$$\varphi_{ij}(x,y) = \left(\varphi_{ij}^1(x,y), \varphi_{ij}^2(x,y) \right), \tag{15.12.3}$$

where x denotes the first $n - p$ coordinates, and y denotes the remaining p coordinates. In other words, we are dealing with maps

$$\varphi_{ij}^1 : \mathbf{R}^{n-p} \to \mathbf{R}^{n-p}, \tag{15.12.4}$$

and

$$\varphi_{ij}^2 : \mathbf{R}^p \to \mathbf{R}^p. \tag{15.12.5}$$

In the chart U_i, the stripes $x = $ constant match up with the stripes on other charts U_j. These stripes are called *leaves* of the foliation. In each chart, the leaves are p-dimensional sub-manifolds.

For example, consider an n-dimensional space, foliated as a product by subspaces consisting of points whose first $n - p$ coordinates are constant. This can be covered by a single chart. The existence of the foliation means that

$$\mathbf{R}^n = \mathbf{R}^{n-p} \times \mathbf{R}^p, \tag{15.12.6}$$

with the leaves being enumerated by $\mathbf{R}^{n-p}$. When $n = 3$ and $p = 2$ this just tells us that the two-dimensional leaves of a book are enumerated by a page number.

References

Asorey, M., Ibort, A. and Marmo, G. (2005) Global theory of quantum boundary conditions and topology change, *Int. J. Mod. Phys.* **A20**, 1001–26.

Balachandra, A. P., Marmo, G., Skagerstam, B. S. and Stern, A. (1983) *Gauge Symmetries and Fibre Bundles. Applications to Particle Dynamics* (Berlin, Springer).

Balachandran, A. P., Jo, S. G. and Marmo, G. (2010) *Group Theory and Hopf Algebras. Lectures for Physicists* (Singapore, World Scientific).

Balmer, J. J. (1885) Notiz über die spectrallinien des wasserstoffs, *Verh. d. Naturf. Ges.* **7**, 548–60.

Bergmann, P. G. (1949) Non-linear field theories, *Phys. Rev.* **75**, 680–5.

Bergmann, P. G. and Brunings, J. H. M. (1949) Non-linear field theories II. Canonical equations and quantization, *Rev. Mod. Phys.* **21**, 480–7.

Bimonte, G. and Musto, R. (2003a) Comment on quantitative wave-particle duality in multibeam interferometers, *Phys. Rev.* **A67**, 066101.

Bimonte, G. and Musto, R. (2003b) Interferometric duality in multibeam experiments, *J. Phys.* **A36**, 11481–502.

Bohr, N. (1913) On the constitution of atoms and molecules, in *The Old Quantum Theory*, ed. D. Ter Haar (Glasgow, Pergamon Press).

Born, M. and Wolf, E. (1959) *Principles of Optics* (London, Pergamon Press).

Born, M. (1969) *Atomic Physics* (London, Blackie & Son).

Bose, S. N. (1924) Plancks gesetz und lichtquantenhypothese, *Z. Phys.* **26**, 178–81.

Brillouin, L. (1926a) La mécanique ondulatoire de Schrödinger; une méthode générale de resolution par approximations successives, *Comptes Rendus* **183**, 24–6.

Brillouin, L. (1926b) Remarques sur la mécanique ondulatoire, *J. Phys. Radium* **7**, 353–68.

Cartan, E. (1938) *La Théorie des Spineurs I & II* (Paris, Hermann).

Cartier, P. and DeWitt–Morette, C. (2000) Functional integration, *J. Math. Phys.* **41**, 4154–87.

Chaturvedi, S., Ercolessi, E., Marmo, G., *et al.* (2007) Ray space 'Riccati' evolution and geometric phases for N-level quantum systems, *Pramana* **69**, 317–27.

Chevalley, C. (1946) *Theory of Lie Groups* (Princeton, Princeton University Press).

Cohen-Tannoudji, C., Diu, B. and Laloe, F. (1977a) *Quantum Mechanics. I* (New York, Wiley).

Cohen-Tannoudji, C., Diu, B. and Laloe, F. (1977b) *Quantum Mechanics. II* (New York, Wiley).

Compton, A. H. (1923a) A quantum theory of the scattering of X-rays by light elements, *Phys. Rev.* **21**, 483–502.

Compton, A.E. (1923b) The spectrum of scattered X-rays, *Phys. Rev.* **22**, 409–13.

Curie, P. (1894) Sur la symmetrie dans les phènomènes physiques. Symmetrie d'un champ electrique et d'un champ magnetique. *J. Phys.* (Paris) **3**.

Davisson, C. and Germer, L. H. (1927) Diffraction of electrons by a crystal of nickel, *Phys. Rev.* **30**, 705–40.

de Broglie, L. (1923) Waves and quanta, *Nature* **112**, 540.

de Ritis, R., Marmo, G. and Preziosi, B. (1999) A new look at relativity transformations, *Gen. Rel. Grav.* **31**, 1501–17.

DeWitt, B. S. (2003) *The Global Approach to Quantum Field Theory*, International Series of Monographs on Physics **114** (Oxford, Clarendon Press).

Dirac, P. A. M. (1926) The fundamental equations of quantum mechanics, *Proc. R. Soc. Lond.* **A109**, 642–53.

Dirac, P. A. M. (1928) The quantum theory of the electron, *Proc. R. Soc. Lond.* **A117**, 610–24.

Dirac, P. A. M. (1933) The Lagrangian in quantum mechanics, *Phys. Z. USSR* **3**, 64–72.

Dirac, P. A. M. (1958) *The Principles of Quantum Mechanics* (Oxford, Clarendon Press).

Dirac, P. A. M. (1964) *Lectures on Quantum Mechanics* (Belfer Graduate School of Science, New York, Yeshiva University).

DuBridge, L. A. (1933) Theory of the energy distribution of photoelectrons, *Phys. Rev.* **43**, 727–41.

Dyson, F. J. (1949) The radiation theories of Tomonaga, Schwinger and Feynman, *Phys. Rev.* **75**, 486–502.

Ehrenfest, P. (1927) Bemerkung über die angenäherte gültigkeit der klassischen mechanik innerhalb der quantenmechanik, *Z. Phys.* **45**, 455–7.

Einstein, A. (1905) On a heuristic point of view about the creation and conversion of light, in *The Old Quantum Theory*, ed. D. Ter Haar (Glasgow, Pergamon Press).

Einstein, A. (1917) On the quantum theory of radiation, in *Sources of Quantum Mechanics*, ed. B. L. van der Waerden (New York, Dover).

Elsasser, W. (1925) Bemerkungen zur quantenmechanik freier elektrone, *Naturwiss enschafter* **13**, 711.

Ercolessi, E., Marmo, G. and Morandi, G. (2010) From the equations of motion to the canonical commutation relations, *Riv. Nuovo Cim.* **33**, 401–590.

Esposito, G., Marmo, G. and Sudarshan, E. C. G. (2004) *From Classical to Quantum Mechanics* (Cambridge, Cambridge University Press).

Esteve, J. G., Falceto, F. and Garcia-Canal, C. (2010) Generalization of the Hellmann–Feynman theorem, *Phys. Lett.* **A374**, 819–22.

Fermi, E. (1932) Quantum theory of radiation, *Rev. Mod. Phys.* **4**, 87–132.

Fermi, E. (1950) *Nuclear Physics* (Chicago, The University of Chicago Press).

Feynman, R. P. (1939) Forces in molecules, *Phys. Rev.* **56**, 340–3.

Feynman, R. P. (1948) Space–time approach to non-relativistic quantum mechanics, *Rev. Mod. Phys.* **20**, 367–87.

Franck, J. and Hertz, G. (1914) Über zusammenstösse zwischen elektronen und den molekülen des quecksilberdampfes und die ionisierungsspannung desselben, *Verh. d. Deutsch. Phys. Ges.* **16**, 457–67.

Gamow, G. (1928) Zur Quantentheorie des Atomkernes, *Z. Phys.* **51**, 204–12.

Gamow, G. (1946) Expanding universe and the origin of elements, *Phys. Rev.* **70**, 572–3.

Gazeau, J. P. (2009) *Coherent States in Quantum Physics* (New York, Wiley).

Gerlach, W. and Stern, O. (1922) Der experimentelle nachweiss der richtungsquantelung im magnetfeld, *Z. Phys.* **9**, 349–52.

Glauber, R. J. (1963) Coherent and incoherent states of the radiation field, *Phys. Rev.* **131**, 2766–88.

Glimme, J. and Jaffe, A. (1987) *Quantum Physics: A Functional Integral Point of View* (Berlin, Springer).

Grabowski, J., Kus, M. and Marmo, G. (2011) Entanglement for multipartite systems of indistinguishable particles, *J. Phys.* **A44**, 175302.

Gradshteyn, I. S. and Ryzhik, I. M. (1965) *Tables of Integrals, Series and Products* (New York, Academic Press).

Grechko, L. G., Sugarov, V. I. and Tomasevich, O. F. (1977) *Problems in Theoretical Physics* (Moscow, MIR).

Green, H. S. (1953) A generalized method of field quantization, *Phys. Rev.* **90**, 270–3.

Haas, A. E. (1910a) *Sitz. Ber. der Wiener Akad* Abt **IIa**, 119–44.

Haas, A. E. (1910b) Über eine neue theoretische methode zur bestimmung des elektrischen elementar quantums und des halbmessers des wasserstoffatoms, *Phys. Z.* **11**, 537–8.

Haas, A. E. (1925) *Introduction to Theoretical Physics* (London, Constable & Company).

Hawking, S. W. (1974) Black hole explosions?, *Nature* **248**, 30–1.

Hawking, S. W. (1975) Particle creation by black holes, *Commun. Math. Phys.* **43**, 199–220.

Heisenberg, W. (1925) Quantum-theoretical re-interpretation of kinematic and mechanical relations, *Z. Phys.* **33**, 879–93.

Herzberg, G. (1944) *Atomic Spectra and Atomic Structure* (New York, Dover).

Holland, P. R. (1993) *The Quantum Theory of Motion* (Cambridge, Cambridge University Press).

Holton, G. (2000) Millikan's struggle with theory, *Europhys. News* **31** No 3, 12–14.

Hörmander, L. (1983) *The Analysis of Linear Partial Differential Operators. I. Distribution Theory and Fourier Analysis* (New York, Springer).

Hughes, A. L. and DuBridge, L. A. (1932) *Photoelectric Phenomena* (New York, McGraw-Hill).

Hunziker, W. (1974) Scattering in classical mechanics, in *Scattering Theory in Mathematical Physics*, eds. J. A. La Vita and J. P. Marchand, 79–96.

Ibort, A., Man'ko, V. I., Marmo, G., Simoni, A. and Ventriglia, F. (2009) An introduction to the tomographic picture of quantum mechanics, *Phys. Scripta* **79**, 065013 (2009).

Ibort, A., Man'ko, V. I., Marmo, G., Simoni, A. and Ventriglia, F. (2010) On the tomographic picture of quantum mechanics, *Phys. Lett.* **A374**, 2614–17.

Ibort, A., Man'ko, V. I., Marmo, G., Simoni, A. and Ventriglia, F. (2011) A pedagogical presentation of a C^*-algebraic approach to quantum tomography, *Phys. Scripta* **84**, 065006 (2011).

Jackson, J. D. (1975) *Classical Electrodynamics* (New York, Wiley).

Jeans, J. H. (1905) On the partition of energy between matter and aether, *Phil. Mag.* **10**, 91–8.

Jeffreys, H. (1923) On certain approximate solutions of linear differential equations of the second order, *Proc. London Math. Soc.* **23**, 428–36.

Jordan, P., von Neumann, J. and Wigner, E. P. (1934) On the algebraic generalization of the quantum mechanical formalism, *Ann. Math.* **34**, 29–64.

Kirchhoff, G. (1860) On the relation between the radiating and absorbing powers of different bodies for light and heat, *Phil. Mag.* (4) **20**, 1–21.

Klauder, J. R. and Skagerstam, B. S. (eds.) (1985) *Coherent States. Applications in Physics and Mathematical Physics* (Singapore, World Scientific).

Kramers, H. A. (1926) Wellenmechanik und halbzahlige Quantisierung, *Z. Phys.* **39**, 828–40.

Landau, L. D. (1930) Diamagnetismus der metalle, *Z. Phys.* **64**, 629–37.

Landau, L. D. and Lifshitz, E. M. (1958) *Course of Theoretical Physics. III: Quantum Mechanics, Non-Relativistic Theory* (Oxford, Pergamon Press).

Lesgourgues, J., Mangano, G., Miele, G. and Pastor, S. (2013) *Neutrino Cosmology* (Cambridge, Cambridge University Press).

Lim, Y. K. (1998) *Problems and Solutions on Quantum Mechanics* (Singapore, World Scientific).

Lippmann, B. A. and Schwinger, J. (1950) Variational principles for scattering processes, I, *Phys. Rev.* **79**, 469–80.

Mandel, L. (1963) Intensity fluctuations of partially polarized light, *Proc. Phys. Soc. (London)* **81**, 1104–14.

Mandel, L., Sudarshan, E. C. G. and Wolf, E. (1964) Theory of photoelectric detection of light fluctuations, *Proc. Phys. Soc. (London)* **84**, 435–44.

Man'ko, M. A., Man'ko, V. I., Marmo, G., Simoni, A. and Ventriglia, F. (2013) Introduction to tomography, classical and quantum, *Nuovo Cimento* **36C**, 163–82.

Marmo, G. and Mukunda, N. (1986) Symmetries and constants of the motion in the Lagrangian formalism on TQ: beyond point transformations, *Nuovo Cimento* **B92**, 1–12.

Marmo, G. and Preziosi, B. (2006) The structure of space-time: relativity groups, *Int. J. Geom. Methods Mod. Phys.* **3**, 591–603.

Marmo, G. and Volkert, G. F. (2010) Geometrical description of quantum mechanics. Transformations and dynamics, *Phys. Scripta* **82**, 038117.

Merli, P. G., Missiroli, G. F. and Pozzi, G. (1974) Electron interferometry with the Elmiskop 101 electron microscope, *J. Phys.* **E7**, 729–33.

Messiah, A. (2014) *Quantum Mechanics* (New York, Dover).

Millikan, R. A. (1916) A direct photoelectric determination of Planck's h, *Phys. Rev.* **7**, 355–88.

Montgomery, D. and Zippin, L. (1940) Topological transformation groups, *Ann. Math.* **41**, 788–91.

Montgomery, D. and Zippin, L. (1955) *Topological Transformation Groups* (New York, Interscience).

Montonen, C. and Olive, D. I. (1977) Magnetic monopoles as gauge particles?, *Phys. Lett.* **B72**, 117–20.

Morandi, G., Ferrario, C., Lo Vecchio, G., Marmo, G. and Rubano, C. (1990) The inverse problem in the calculus of variations and the geometry of the tangent bundle, *Phys. Rep.* **188**, 147–284.

Morette, C. (1951) On the definition and approximation of Feynman's path integral, *Phys. Rev.* **81**, 848–52.

Moyal, J. E. (1949) Quantum mechanics as a statistical theory, *Proc. Camb. Phil. Soc.* **45**, 99–124.

Parisi, G. (2001) Planck's legacy to statistical mechanics, COND-MAT 0101293 (preprint).

Paschen, F. and Back, E. (1912) Normale und anomale Zeemaneffekte, *Ann. Phys.* **39**, 897–932.

Paschen, F. and Back, E. (1913) Normale und anomale Zeemaneffekte, *Nachtrag, Ann. Phys.* **40**, 960–70.

Penrose, R. and Rindler, W. (1984) *Spinors and Space-time. Two-Spinor Calculus and Relativistic Fields* (Cambridge, Cambridge University Press).

Penzias, A. and Wilson, R. W. (1965) A measurement of excess antenna temperature at 4080-Mc/s, *Astrophys. J.* **142**, 419–21.

Perelomov, A. M. (1986) *Generalized Coherent States and Their Applications* (Berlin, Springer-Verlag).

Phipps, T. E. and Taylor, J. B. (1927) The magnetic moment of the hydrogen atom, *Phys. Rev.* **29**, 309–20.

Planck, M. (1900) On the theory of the energy distribution law of the normal spectrum, *Verh. d. Deutsch. Phys. Ges.* **2**, 237–45.

Radon, J. (1917) Über die Bestimmung von Funktionen durch ihre Integralwerte längs gewisser Mannigfaltigkeiten, *Sächs. Akad. Wiss. Leipzig, Math. Nat. Kl.* **69**, 262–77.

Rayleigh, J. W. S. (1900) Remarks upon the law of complete radiation, *Phil. Mag.* **49**, 539–40.

Rayleigh, J. W. S. (1912) On the propagation of waves through a stratified medium, with special reference to the question of reflection, *Proc. R. Soc. London* **A86**, 207–26.

Reed, M. and Simon, B. (1975) *Methods of Modern Mathematical Physics. II: Fourier Analysis and Self-Adjointness* (New York, Academic Press).

Reed, M. and Simon, B. (1979) *Methods of Modern Mathematical Physics. III: Scattering Theory* (New York, Academic Press).

Robertson, H. P. (1930) A general formulation of the uncertainty principle and its classical interpretation, *Phys. Rev.* **35**, 667.

Rollnik, H. (1956) Streumaxima und gebundene Zustande, *Z. Phys.* **145**, 639–53.

Scully, M. O., Englert B. G. and Walther, H. (1991) Quantum optical tests of complementarity, *Nature* **351**, 111–16.

Smithey, D. T., Beck, M., Raymer, M. G. and Faridani, A. (1993) Measurement of the Wigner distribution and the density matrix of a light mode using optical homodyne tomography: application to squeezed states and the vacuum, *Phys. Rev. Lett.* **70**, 1244–7.

Smoot, G. F., Bennett, C.L, Kagut, A. *et al.* (1992) Structure in the COBE differential microwave radiometer first year maps, *Astrophys. J.* **396**, L1–L5.

Sommerfeld, A. and Debye, P. (1913) Theorie des lichtelektrischen effektes vom staud-punkt des wirkumgsquantums, *Ann. der Phys.* **41**, 873–930.

Spergel, D. N., Verde, L., Peiris, H. V., *et al.* (2003) First year Wilkinson microwave anisotropy probe (WMAP) observations: determination of cosmological parameters, *Astrophys. J. Suppl.* **148**, 175–94.

Squires, G. L. (1995) *Problems in Quantum Mechanics* (Cambridge, Cambridge University Press).

Stark, J. (1914) Beobachtungen über den effekt des elektrischen feldes auf spektrallinien, *Ann. Phys.* **43**, 965–82.

Stewart, B. (1858) An account of some experiments on radiant heat, involving an extension of Prévost's theory of exchanges, *Trans. R. Soc. Edin.* **22**, 1–20.

Sudarshan, E. C. G. (1963) Equivalence of semiclassical and quantum mechanical descriptions of statistical light beams, *Phys. Rev. Lett.* **10**, 277–9.

Sudarshan, E. C. G. and Rothman, T. (1991) The two-slit interferometer re-examined, *Am. J. Phys.* **59**, 592–5.

Sudarshan, E. C. G., Chiu, C. B. and Bhamathi, G. (1995) Generalized uncertainty relations and characteristic invariants for the multimode states, *Phys. Rev.* **A52**, 43–54.

Syad, T. A., Antoine, J. P. and Gazeau, J. P. (2000) *Coherent States, Wavelets and Their Generalizations*, Graduate Texts in Contemporary Physics (New York, Springer-Verlag).

Ter Haar, D. (1967) *The Old Quantum Theory* (Glasgow, Pergamon Press).

Thirring, W. (1997) *Classical Mathematical Physics*. Dynamical Systems and Field Theories, 3rd Edition (New York, Springer-Verlag).

Tonomura, A., Endo, J., Matsuda, T., Kawasaki, T. and Ezawa, H. (1989) Demonstration of single-electron buildup of an interference pattern, *Am. J. Phys.* **57**, 117–20.

Uhlenbeck, G. E. and Goudsmit, S. (1926) Spinning electrons and the structure of spectra, *Nature* **117**, 264–5.

Von Neumann, J. (1955) *Mathematical Foundations of Quantum Mechanics* (Princeton, Princeton University Press).

Wentzel, G. (1926) Eine Verallgemeinerung der Quantenbedingungen für die Zwecke der Wellenmechanik, *Z. Phys.* **38**, 518–29.

Weyl, H. (1931) *The Theory of Groups and Quantum Mechanics* (New York, Dover).

Wheeler, J. A. and Zurek, W. H. (1983) *Quantum Theory and Measurement* (Princeton, Princeton University Press).

Wien, W. (1896) Über die energieverteilung im emissionsspektrum eines schwarzen körpers, *Ann. Phys.* **58**, 662–9.

Wiener, N. (1930) Generalized harmonic analysis, *Acta Math.* **55**, 117–258.

Wightman, A. S., ed. (1995). *Wigner Collected Works*, vol. 6, part 4, (Berlin, Springer).

Wightman, A. S. (1996) How it was learned that quantized fields are operator-valued distributions, *Fortschr. Phys.* **44**, 143–78.

Wigner, E. P. (1959) *Group Theory and its Application to the Quantum Mechanics of Atomic Spectra* (New York, Academic Press).

Willmore, T. J. (1960) The definition of Lie derivative, *Proc. Edin. Math. Soc.* **12**, 27–9.

Wintner, A. (1947) The unboundedness of quantum-mechanical matrices, *Phys. Rev.* **71**, 738–9.

Witten, E. (1997) Duality, spacetime and quantum mechanics, *Phys. Today* May, 28–33.

Zeeman, P. (1897a) On the influence of magnetism on the nature of the light emitted by a substance, *Phil. Mag.* **43**, 226–39.

Zeeman, P. (1897b) Doublets and triplets in the spectrum produced by external magnetic forces, *Phil. Mag.* **44**, 55–60, 255–9.

Zhang, W. M., Feng, D. H. and Gilmore, R. (1990) Coherent states: theory and some applications, *Rev. Mod. Phys.* **26**, 867.

Index

Printed in the United States
by Baker & Taylor Publisher Services